THE ABRIDGED
COMPARATIVE PLANT ECOLOGY

TITLES OF RELATED INTEREST

THE ABRIDGED COMPARATIVE PLANT ECOLOGY

J. P. GRIME   J. G. HODGSON   R. HUNT

*The NERC Unit of Comparative Plant Ecology
Department of Animal and Plant Sciences,
University of Sheffield*

CHAPMAN & HALL

London · Glasgow · Weinheim · New York · Tokyo · Melbourne · Madras

Published by Chapman & Hall, 2-6 Boundary Row, London SE1 8HN

Chapman & Hall, 2-6 Boundary Row, London SE1 8HN, UK

Blackie Academic & Professional, Wester Cleddens Road, Bishopbriggs, Glasgow G64 2NZ, UK

Chapman & Hall GmbH, Pappelallee 3, 69469 Weinheim, Germany

Chapman & Hall USA, One Penn Plaza, 41st Floor, New York, NY10119, USA

Chapman & Hall Japan, ITP - Japan, Kyowa Building, 3F, 2-2-1 Hirakawacho, Chiyoda-ku, Tokyo 102, Japan

Chapman & Hall Australia, Thomas Nelson Australia, 102 Dodds Street, South Melbourne, Victoria 3205, Australia

Chapman & Hall India, R. Seshadri, 32 Second Main Road, CIT East, Madras 600 035, India

First edition 1990
Reprinted 1992, 1995

© 1990 J.P. Grime, J.G. Hodgson & R. Hunt

Typset in 9/11pt Times by Columns Design and Production Services Lt, Reading
Printed in Great Britain at the University Press, Cambridge

ISBN 0 412 53250 6

A Catalogue record for this book is available from the British Library
Library of Congress Cataloging-in-Publication Data available

∞ Printed on acid-free text paper, manufactured in accordance with ANSI/NISO Z39.48-1992 and ANSI/NISO Z39.48-1984 (Permanence of Paper).

Preface

This book is a condensation of the much larger work *Comparative plant ecology: a functional approach to common British species* (London: Unwin Hyman, 1988), which provides an alphabetical sequence of standardized comparative accounts of the ecology of common British vascular plants.

As in *CPE*, each account relies heavily upon field surveys, but also contains standardized laboratory data and a distillation of published information. Again, the material is arranged in a standard format, which now includes a drawing of the plant by Gail Furness.

In selecting material for the shortened version we have tried to retain as much as possible of the content of the original work. Approximately one third of the quantitative, factual content of each species account has been preserved, and about two-thirds of the written notes are included. A further economy has been to dispense with all but the most important references to published sources. As a guide to the abridgement, and a pointer to the circumstances where reference to the parent volume may be desirable, a facsimile of one of the full Autecological Accounts from *CPE* has been provided in an Appendix.

The summary tables of *CPE* which describe the essential ecology and biological characteristics of all the common species of the British flora have been retained in full.

We hope that this shortened edition will be useful to individual students, ecologists, botanists, landscape architects and others engaged in the management, conservation or reconstruction of vegetation.

Sheffield, 1989
J. P. Grime
J. G. Hodgson
R. Hunt

Contents

1 Introduction

Aims and methods

The main object of this book is to provide abridged Autecological Accounts of the biology and ecology of common vascular plants of the British flora. The abridged Accounts are drawn from those which appear in the much larger parent volume, *Comparative plant ecology: a functional approach to common British species* by J. P. Grime, J. G. Hodgson & R. Hunt (London: Unwin Hyman, 1988), henceforward referred to as *CPE*.

The structure and content of the full Accounts given in *CPE*, and in particular 'the balance between data presentation and interpretation, were strongly influenced by our concern to meet the specific requirements of academic research, vegetation management and nature conservation. Consequently, both in *CPE* and in the present work, there are special sections which describe in some detail the various ways in which the Accounts may be used in efforts either to understand the ecology of species or to manipulate their distribution and abundance.

The form of both books is a reflection of additional intentions. One is our wish to reaffirm the virtues of a broadly-based comparative approach to the ecological study of the British flora. Another is the commitment of all three authors to the view that the future of ecology as a rigorously predictive science deserves the development of a universal functional classification of organisms, which can be central to an understanding of community and ecosystem processes. This introductory chapter explains how these aims are addressed by the abridged Accounts which follow.

The collection and comparison of standardized information on the species of the British flora strongly parallel the rationale prevailing in the physical sciences. Perhaps the most obvious of these is the role played by the Periodic Table of the Elements in classifying, analysing, and even predicting the structure and properties of chemical elements and compounds.

Initially under the guidance of Professor A. R. Clapham, and funded by the then Nature Conservancy (and latterly by the Natural Environment Research Council), the team at the Unit of Comparative Plant Ecology (UCPE) embarked upon a long-term research programme designed to characterize and compare the field ecology and laboratory characteristics of the more common herbaceous plants of the British flora.

This research consisted of two parallel but independent kinds of programme. The first involved extensive field surveys and observations, and the second a series of laboratory-based screening procedures conducted upon large numbers of species. Both types of programme are outlined later in this chapter; the parts they play in the abridged Accounts and the relevance of the information they contain to the comparative ecology of the subject species are explained in Chapter 2.

The rationale for this whole approach is set out in a series of publications, cited in *CPE*, from which it is evident that the major object was to recognize fundamental determinants of the distribution and ecology of plants by identifying 'design constraints' or 'components of limited potentiality'. These plant attributes, by becoming the foci of conflicting selection pressures, limit the ecological amplitude of geno-types, populations and species. Crucial to this approach is the hypothesis that ecological specializations important to the present or past success of an organism may frequently involve the assumption of genetic characteristics which render the plant unsuited to life in 'other' environments, hence restricting its current ecological range. The research method associated with this approach relies essentially upon large-scale comparisons between groups of plants drawn from contrasted habitats.

The series of comparative experiments conducted at Sheffield has produced a large body of standardized information on various aspects of the biology of native plants of contrasted ecology, much of which appears in summary form in the abridged Accounts. Certain sets of these data reveal recurrent patterns of ecological specialization which are correlated with success in certain types of habitats or niches and failure in others. These patterns allow the abridged Accounts to be examined in relation to a more general field of ecological enquiry which is reviewed in the next section.

Primary ecological strategies

History and semantics

In recent years, evidence from diverse schools of research has begun to point to the existence in animals and plants of primary ecological strategies. These are recurrent types of specialization associated with particular habitat conditions or niches. It is also becoming clear that recognition of these primary strategies can provide a key to understand-ing the structure and dynamics of communities and ecosystems. In the abridged Accounts each species is classified with respect to strategy.

Chapter 2 devotes considerable space to explanation of the inferences which follow from such a classification.

Scientists differ sharply in their attitudes to the use of the term 'strategy' in ecology. Some theorists have used the term freely, while others have taken strong exception to it. With its teleological and anthropomorphic connotations the term is not ideal and it is understandable that some biologists have preferred to use neutral expressions such as 'set of traits' or 'syndrome'.

We feel no special commitment to the term 'strategy', but retain it here as a mark of respect for those who first used it in this context. Their achievement was to recognize that organisms may exhibit sets of traits which are predictably related to their ecology.

Here a *strategy* is defined as 'a grouping of similar or analogous genetic characteristics which recurs widely among species or populations and causes them to exhibit similarities in ecology'. A *primary strategy* is recognized as one which involves the more fundamental activities of the organism (resource capture, growth and reproduction) and recurs widely both in animals and in plants.

The strategy concepts followed in this book involve a triangular array of strategies exhibited in the established (adult) phase together with a quite distinct classification of regenerative (juvenile) stages. The need for this separation has become apparent through the work of several theorists who have recognized the peculiar nature of the selection forces and design constraints which determine the characteristics of the offspring and seen them as distinct from those which shape the characteristics of the adult. In plants this has led to the suggestion that there are distinct *regenerative strategies* which differ in such respects as degree of resource investment, mobility and dormancy, and which confer different but predictable sets of ecological capacities and limitations upon the organism. In many plants, and in some animals, the same genotypes may be capable of regenerating by several quite different mechanisms. This leads to the hypothesis that ecological amplitude is determined not only by genetic variability and by phenotypic plasticity, but also by regenerative flexibility, a function of the number of different regenerative strategies possessed by the species.

The C–S–R model

ESSENTIAL FEATURES

The C–S–R model describing the various types of strategy in the established phase, and to which we refer in each of the abridged Accounts, originates from a classification of the external factors which affect vegetation into two broad categories. The first, which we may describe as *stress*, consists of the phenomena which restrict photosyn-

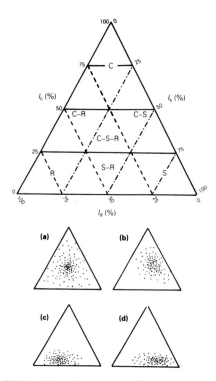

Figure 1.1 A model describing the various equilibria between competition, stress and disturbance in vegetation and the location of primary and secondary strategies. C, competitor; S, stress-tolerator; R, ruderal; C–R, competitive–ruderal; S–R, stress-tolerant ruderal; C–S, stress-tolerant competitor; C–S–R, 'C–S–R strategist'. I_c, relative importance of competition (———); I_s, relative importance of stress (—·—); I_d, relative importance of disturbance (---). The strategic range of four life forms is also shown: (a) herbs, (b) trees and shrubs, (c) bryophytes and (d) lichens.

thetic production, such as shortages of light, water and mineral nutrients, or sub-optimal temperatures. The second, here referred to as *disturbance*, is associated with the partial or total destruction of the plant biomass and arises from the activities of herbivores, pathogens and humans (trampling, mowing and ploughing), and from phenomena such as wind-damage, frosting, droughting, soil erosion and fire.

When the four permutations of high and low stress with high and low disturbance are examined (Table 1.1) it is apparent that only three of these are viable as plant habitats. This is because the effect of continuous and severe stress in highly-disturbed habitats is to prevent a sufficiently rapid recovery or re-establishment of the vegetation.

Table 1.1 The basis for the evolution of three strategies in plants.

Intensity of disturbance	Intensity of stress	
	Low	High
Low	Competitors	Stress-tolerators
High	Ruderals	(No viable strategy)

It is suggested that the three remaining contingencies in Table 1.1 have been associated with the evolution of primary strategies of the established phase which conform to distinct types. These are the *competitors*, exploiting conditions of low stress and low disturbance, the *stress-tolerators*, associated with high stress and low disturbance, and the *ruderals*, characteristic of low stress and high disturbance. The three strategies are, of course, extremes of evolutionary specialization. There are others which exploit the various intermediate conditions which correspond to particular equilibria between stress, disturbance and competition. These can be displayed in a triangular diagram, which can be used also to indicate the strategic range of various life-forms (Fig. 1.1). Evidence relating to the C–S–R model has been reviewed by Grime (in *Plant strategies and vegetation processes*, 1979, Chichester: John Wiley), and current efforts to test the validity of the theory have also been described by Grime (in *Evolutionary plant biology*. 1988, L. D. Gottlieb & S. Jain (eds), 371–93. London: Chapman & Hall).

USES AND LIMITATIONS
Recent studies have revealed scope for using the C–S–R model in relation to the ecological analysis, management and conservation of species and, as the abridged Accounts reveal, considerable progress has been made in relating plant strategies to contemporary changes in species abundance. Strategy concepts have also been applied to larger taxa, including some families of vascular plants. In this research particular emphasis has been placed upon the concept of 'channelling', whereby recent or current evolution appears to be strongly influenced by ancestral patterns of ecological specialization. This principle has been linked to strategy concepts in analysing the mechanisms controlling the ecology of pteridophytes, and in an attempt to explain why families of angiosperms have shown very different responses to recent changes in land-use in Britain. It was concluded that the decline of most pteridophytes, and of many species drawn from relatively primitive families, e.g. Rosaceae and Leguminosae, is related to possession of inflexibly stress-tolerant traits (e.g. low potential relative growth rate and delayed reproduction) and propagule characteristics which are inappropriate (either too small, e.g. Pteridophyta, or too large, e.g. many Rosaceae and Leguminosae) for effective colonization of the productive or disturbed vegetation which is created by modern forms of land use.

Although many inferences follow from the classification of a species within the C–S–R model, and the identification of its regenerative strategies, it would be unreasonable to expect all aspects of the ecology of a species to be predictable from this approach. For this reason each abridged Account also contains additional information and draws attention to gaps in our knowledge and understanding. The uncertain role of many historical factors, the occurrence of genetic variation over the geographical or habitat range of a species, and the complexity of the processes responsible for the 'fine-tuning' of distributions and abundances within individual sites, all set obvious lower limits to the scale at which strategy concepts are able to explain phenomena and generate predictions.

The abridged Autecological Accounts

The full Accounts in *CPE* attempted to carry the 'species biography', pioneered in publications such as *The Biological Flora of the British Isles* (published at intervals in the *Journal of Ecology*, Oxford: Blackwell Scientific Publications) into a further stage of its evolution by enhancing the degree of standardization achieved in the collection, analysis and presentation of laboratory and field data. This aspect is equally important within the abridged Accounts which appear here, as is the role played by strategy theory in providing a common framework within

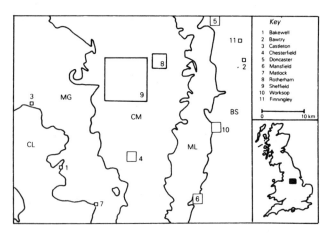

Figure 1.2 The UCPE survey area. The approximate extent of urban development is shown by numbered squares. Geological substrata are: CL, Carboniferous Limestone; MG, Millstone Grit; CM, Coal Measures; ML, Magnesian Limestone; BS, Bunter Sandstone.

which to locate and interpret the more fragmentary classes of information. Strategy concepts have also facilitated comparisons between species, and have introduced priorities into those parts of the abridged Accounts which are concerned with prediction and vegetation management.

Data sources for the abridged Autecological Accounts

Introduction

The abridged Accounts presented in this book were derived directly from the full Accounts in *CPE* and therefore draw on the same sources of information. These include vegetation surveys, additional field observations, laboratory screening experiments and published sources. The first three originate from Sheffield. This section provides background information to each of these principal sources. Chapter 4 provides tables of synopses of the data which they yield, not only for the species dealt with in the abridged Accounts themselves but also for others which are important in the British flora.

Vegetation surveys

In many ways Sheffield is ideally situated for the broad botanical fieldwork which is necessary for the support of generalizing hypotheses. Many local species are at the northern limits of their British and European distributions, and some are at their southern limits. The surrounding area also offers a wide range in geological stratum, altitude, aspect and land use, with a corresponding diversity of plant life.

The UCPE vegetation surveys were conducted within an area of 3000 km² surrounding Sheffield (Fig. 1.2). In all, three surveys were performed. Table 1.2 lists some of their principal features.

Survey I (semi-natural grassland only) has been described by Grime & Lloyd (in *An Ecological Atlas of Grassland Plants*. 1973, London: Edward Arnold). Survey II (all common herbaceous communities) was the chief source of field data for the full Accounts contained in *CPE*, and hence for the abridged Accounts which appear here. Survey III (rarer species and communities) provided ancillary items of ecological information, particularly where other sources were lacking.

In conducting Survey II, quadrats were positioned within each locality so as to provide examples of each of the major plant assemblages within each major habitat. A guide to the habitats recognized in Survey II is given in Figure 1.3. Fieldwork extended over six years, in each of which sampling was restricted to the period April to September inclusive. Because winter annuals and vernal geophytes tend to disappear

Table 1.2 Facts and figures concerning the three UCPE vegetation surveys.

	Survey I	Survey II	Survey III
Objects	(a) An objective description of the ecology of grassland species	(a) A description of the ecology of the most common species of the region	(a) A description of the ecology of species recorded in <20 quadrats in Survey II
	(b) Description and classification of the main types of vegetation of semi-natural grassland	(b) Recognition of the main types of vegetation in each of the major habitats	(b) Records of the less common vegetation
Year commenced	1965	1967	1972
Sampling policy	Random within uniformly selected localities	Subjective within subjectively selected vegetation	Subjective within subjectively selected vegetation
Sampled area*		National Grid references	
NW corner	SE 100020	SE 100020	SE 100050
NE corner	SE 670020	SE 670020	SE 700050
SW corner	SK 100550	SK 100550	SK 100550
SE corner	SK 670550	SK 670550	SK 700550
Number of 1 m^2 quadrats	657	2748	7324
Main habitat types		Quadrats per thousand	
Wetland	—	156.1	237.6
Skeletal	—	130.6	86.3
Grassland	1000	80.8	43.8
Arable	—	44.4	41.2
Spoil	—	172.1	111.7
Wasteland	—	149.6	326.1
Woodland	—	266.4	116.2
Minor habitats	—	—	37.1
Solid geology*		Quadrats per thousand	
Carboniferous limestone	377.5	193.2	162.6
Millstone grit	249.6	245.3	165.2
Coal measures	187.2	283.4	207.7
Magnesian limestone	129.4	129.2	197.2
Bunter sandstone	56.3	141.9	254.4
Keuper marl	—	7.0	12.9

* See also Figure 2.1.

completely during the summer months, vegetation likely to contain these floristic elements was not sampled after the end of May.

Two categories of data from Survey II which provide particularly relevant background information to the abridged Accounts are summarized below.

OUTLINE RESULTS – HABITATS

After a preliminary programme of field sampling, a simple, branched key to major habitat types was devised (Fig. 1.3). This was based almost exclusively upon self-evident physical and physiognomical attributes of

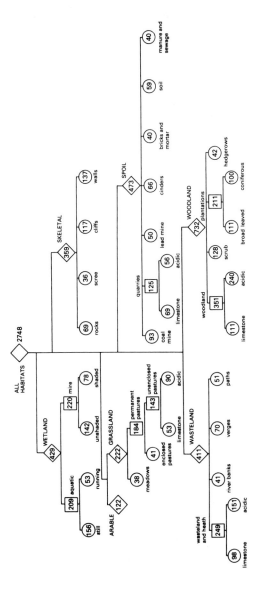

Figure 1.3 Numbers of quadrats in the primary, intermediate and terminal habitats of UCPE Survey II.

the environment and its exploitation, and was constructed in terms designed to be accessible to non-specialist users of the data. The key contained seven 'primary', eight 'intermediate' and 32 'terminal' categories. Knowing the full range of habitats which were likely to be present within the whole survey, it was then possible to carry out the fieldwork in such a way that all of the terminal categories were adequately represented. As far as possible, each type of habitat was sampled throughout the whole region. Samples were taken from the first example of each habitat type encountered within each sector of the survey area.

Figure 1.3 gives the number of quadrats recorded in each of the primary, intermediate and terminal habitats. Most of the terminal categories are common British habitats, though lead-mine spoil is an exception. The hay-meadows of the region are also unusual in that most are grazed after the hay crop has been harvested. Generally, the number of records required reflected the proportion and geographical spread of the habitats within the survey area and the degree of vegetational heterogeneity encountered. A minimum of about 40 records per terminal habitat was considered desirable, though 11 habitats required 100 or more records for satisfactory representation.

OUTLINE RESULTS – FLORISTIC DIVERSITY

In the 2748 quadrats which comprised Survey II (Table 1.2), a total of 629 vascular plant species were recorded within 188 genera. As many as 86 of the species were recorded only once, but the most common, *Poa trivialis*, was present in 605 quadrats. The frequency distribution of numbers of species by numbers of quadrats has a marked leftwards skew, even with geometrically scaled class intervals. This indicates that the greater part of the vegetation of the region consists of a few, relatively common, species. A total of 281 of these are dealt with in detail in the abridged Accounts.

The occurrence of fifteen records or more for any one species in the whole survey guaranteed its inclusion among the Accounts. For less than 15 records a value-judgement was made as to the ecological importance of the species; the minimum number of records for any species included was normally ten, but three exceptions were made to this rule because of exceptional importance or interest: *Phragmites australis*, (9 quadrats); *Carlina vulgaris*, (7); *Reynoutria japonica*, (5). The accounts for these three species rely heavily upon information gained from the literature and from Survey III.

Because most of the species included are very widespread, it was possible to analyse a very large proportion of the region's floristics (as defined by the exhaustively complete Survey III) in terms of the distribution and attributes of no more than 36% of its local species, 51%

of its local genera and 57% of its local families. Nationally, these represent 10, 26 and 46%, respectively, of the native British flora. The coverage of Britain's established alien flora is less complete, with only 3% of species, 5% of genera and 29% of families being included in the abridged Accounts.

Within the 1 m^2 sample, 135 quadrats contained no more than a single species. At the other end of the scale, one quadrat contained 40. Table 1.3 shows how floristic diversity was distributed among the seven primary habitats. Grassland, spoil and wasteland were the most diverse of these, averaging 15% of records with 21 species or more per square metre. Woodlands were least diverse, with 99% of records lying below this figure. Of the terminal habitats, by far the most diverse was limestone pasture, with 51% of records at 21 species or above, followed by the other (always limestone) habitats: quarry heaps (30%) and wasteland and heath (28%). Least diverse were coniferous plantations with 85% of records at 5 species m^{-2} or less, followed by aquatic habitats (average of both variants = 81%) and broad-leaved plantations (74%).

Phenology, demography and regeneration

The vegetation surveys belong to a class of ecological research which is high in realism (being wholly field-based), high in generality (because of the very large number of samples and species included) but low in precision (because of the simple nature of the floristic observations and their lack of continuity, or even repetition, in time). A second kind of fieldwork at UCPE, involving a moderate sacrifice of generality to precision, has concerned itself with more detailed investigations of the phenology, demography and regenerative biology of a more-limited selection of species (see *CPE* for further details). The resulting information, particularly that concerning shoot phenology and seed

Table 1.3 Numbers of quadrats in the seven primary habitats of UCPE Survey II, classified with respect to species diversity.

Habitat	No. of species in the 1 m^2 quadrat							Habitat totals
	<6	6–10	11–15	16–20	21–25	26–30	31+	
Wetland	250	111	43	11	6	5	3	429
Skeletal	188	87	41	21	15	6	1	359
Grassland	61	48	34	32	20	13	14	222
Arable	16	28	39	19	10	9	1	122
Spoil	119	128	109	51	43	17	6	473
Wasteland	121	103	89	54	22	15	7	411
Woodland	411	214	80	22	5	0	0	732
Class totals	1166	719	435	210	121	65	32	2748

persistence in the soil, again plays a prominent part in many of the abridged Accounts.

Screening experiments

The availability to plant ecologists of controlled environment chambers has allowed the germination and growth of large numbers of plant species to be compared under standardized conditions. In such experiments high generality and precision may be maintained, but realism is sacrificed. An additional and important advantage of the approach is that many growth-room investigations can be reproduced or extended wherever there are adequate facilities. Data collected on different species or genotypes and in various laboratories can therefore be compared directly. When comparable data are available for a large number of species and populations drawn from a wide range of habitats,

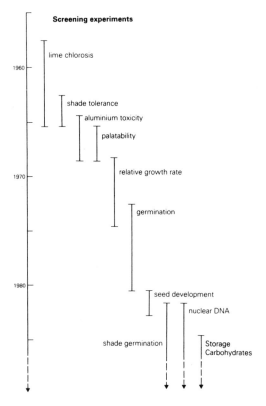

Figure 1.4 A chronology of the UCPE screening experiments.

it is possible to estimate the limits of variation of the plant attribute itself. Hence, it becomes possible to place subsequent individual measurements into context and to begin to assess their ecological significance.

A continuous series of screening experiments has been conducted at UCPE over a 20-year period (Fig. 1.4). All of these sources of data have been drawn upon in the abridged Accounts, if not in the data sections then in the written Synopses.

Although most data have been obtained from only one field population per species, the number of species is large and includes representatives from all major British inland habitats. Uncertainty concerning the extent to which each sample is representative of the species does not, therefore, invalidate the screening, either as an estimate of the overall range of variation or as an attempt to recognize differences between groups of species of contrasted ecology.

The scientific literature

The fourth source of information for the abridged Accounts is the literature already published concerning work done outside Sheffield. Because of the large number of species involved, an exhaustive literature search of the type which is mandatory for *Biological Flora* accounts was not attempted. Also, some reference books in foreign languages were consulted only infrequently because of the difficulty in obtaining a translation. Nevertheless, a formal review of the modern ecological and taxonomic literature was undertaken for each species. Believing that as many as possible of the relevant published sources should be consulted, we drew on all manner of formal and informal reports, papers, theses and books. Though these sources cannot be cited in full here, any relevant information which they provided has been retained within the abridged Accounts. These sources cover an extremely wide range of plant science, with geography, ecology, agronomy, environmental physiology, cytogenetics, plant anatomy, taxonomy, systematics and evolutionary biology all being represented. Complete citations appear, of course, in *CPE*.

In addition to the large number of publications which have yielded points of particular detail, a small number of special sources have been used repeatedly to provide items of standard information. These are identified in Chapter 2, which also explains the contents of the abridged Accounts themselves and provides guidelines for the ecological interpretation of the data which appear in them.

2   Contents of the accounts and how to use them

Introduction

Although the Accounts are of variable length, each follows a standardized layout. This begins with a summary of the major features of the biology and ecology of the species, including information relating to the morphology of the flowers and fruits, derived either from published sources or from our own field observations and laboratory studies. A 'Synopsis' then reviews the distinctive features of the species and its ecology. Sources we have used are cited in full in *CPE*. The Synopsis also includes comments on responses to biotic factors e.g. rabbit grazing, trampling by humans, competition from other plants, and also briefly reviews our knowledge of the impacts of various forms of vegetation management upon the species. Finally, 'Current trends' contain a brief account of changes which may be predicted, or are known to be taking place, in the abundance, distribution or ecology of the species within the British Isles.

This chapter describes in more detail the information contained in each section of an Account. For quicker reference a brief guide appears on the endpapers of the book. As far as possible, the information contained in the Accounts is presented in a form which does not require specialized ecological interpretation. Some technical terms and references to ecological theory are unavoidable, however, and these are explained in the rest of this chapter.

Nomenclature

Explanation

At the beginning of each Account the species is identified by the scientific and common names cited in the *Excursion flora of the British Isles* (3rd edn), 1981, A. R. Clapham, T. G. Tutin & E. F. Warburg, (Cambridge: Cambridge University Press). Any synonyms in common use are added. Other species are sometimes mentioned even though they are not the subject of the Account; these are not provided with authorities to their scientific names, for which reference should be made to Table 4.1. Some closely related taxa could not be separated readily during fieldwork.

Interpretation

Characters utilized in plant taxonomy have not been used extensively in ecological analysis. This is because ecological specialization is not as immediately evident in, or as characteristic of, families of plants as is the case with many of the higher taxa of animals. Lack of specialization is suggested by the wide range of life-forms adopted and diversity of habitats exploited by some families. However, recent research has brought about an increasing recognition that in characters such as association with nitrogen-fixing organisms (in the Leguminosae), seed size and structure, seed germination, plant strategy, and wood structure and its effects on early leafing, the taxonomy coincides with a degree of ecological specialization. It seems likely that as research at the interface of ecology and taxonomy expands, family history will emerge as an important influence on the present characteristics and contemporary ecology of individual species.

Drawings of plants

These have been prepared by Gail Furness from living specimens. They are not to any common scale, nor are they intended to reveal diagnostic structural detail. However, they may serve as useful *aides-mémoire* to the general appearance of the plants and their reproductive structures.

Established strategy

Explanation

As explained in Chapter 1, some of the most important features of the biology and field ecology of a plant can be summarized by reference to its *functional type* or *strategy*. Near the beginning of the Account each species is ascribed a position (or in genetically and ecologically variable species, a range of positions) within the C–S–R model of primary ecological strategies. This theoretical model is an attempt to recognize the main avenues of ecological specialization in the established phase of plant life-histories. In Chapter 1 (under 'The C–S–R model') the nature of the model is examined and its implications for ecology reviewed. Current efforts to test its validity are described in *CPE*. Here, we simply state the main assumptions upon which the theory is based, and describe the criteria used to classify species with respect to strategy.

A plant strategy may be defined as a grouping of similar or analogous genetic characteristics which recurs widely among species or populations, such that they show similarities in ecology. The C–S–R theory argues that there are three primary strategies in plants, because there are three

Table 2.1 Species typical of seven ecological strategies (see Fig. 1.1).

Strategy	Typical species
Competitor	*Chamerion angustifolium*[1-4] *Urtica dioica*[1,5-10] *Phalaris arundinacea*[1,11]
Ruderal	*Capsella bursa-pastoris*[12,13] *Senecio vulgaris*[1,12-14] *Urtica urens*[7,9,15]
Stress-tolerator	*Danthonia decumbens*[1,16,17] *Primula veris*[18] *Sanicula europaea*[1,19]
Competitive-ruderal	*Impatiens glandulifera*[20] *Ranunculus repens*[1,21-23] *Tussilago farfara*[1,2,24,25]
Stress-tolerant ruderal	*Carlina vulgaris*[26-28] *Desmazeria rigida*[1,29] *Linum catharticum*[26,30]
Stress-tolerant competitor	*Dryopteris filix-mas*[31-33] *Mercurialis perennis*[5,34] *Vaccinium myrtillus*[1,35,36]
'C—S—R strategist'	*Holcus lanatus*[1,37,38] *Hypochaeris radicata*[4,39,40] *Rumex acetosa*[1,41]

Publications relevant to this classification are cited in *CPE*:
1, Grime and Hunt 1975; 2, Myerscough and Whitehead 1966, 1967; 3, Myerscough 1980; 4, Van Andel and Vera 1977; 5, Al-Mufti *et al.* 1977; 6, Pigott and Taylor 1964; 7, Boot *et al.* 1986; 8, Wheeler 1981; 9, Greig-Smith 1948a; 10, Bassett *et al.* 1977; 11, Buttery and Lambert 1965; 12, Fryer and Makepeace 1977; 13, Salisbury 1942; 14, Harper and Ogden 1970; 15, Salisbury 1964; 16, Higgs and James 1969; 17, Furness 1980; 18, Tamm 1972; 19, Inghe and Tamme 1985; 20, Al-Mashhadani 1979; 21, Sarukhan 1974; 22, Doust 1981a,b; 23, Sarukhan and Harper 1973; 24, Bakker 1960; 25, Ogden 1974; 26, Verkaar *et al.* 1983; 27, Watt 1981; 28, Verkaar and Schenkeveld 1984b; 29, Clark 1974; 30, During *et al.* 1985; 31, Pogorelova and Rabotnov 1978; 32, Willmott 1985; 33, Page 1982; 34, Hutchings 1978; 35, Ritchie 1956; 36, Pigott 1983; 37, Watt 1978; 38, Beddows 1961; 39, Aarssen 1981; 40, Turkington and Aarssen 1983; 41, Putwain and Harper 1970.

distinct threats to existence: competitive exclusion, chronic stress (usually resulting from limiting effects of mineral nutrient shortages) and repeated severe disturbance (by mechanical damage or climatic events). Each threat occurs under particular types of environmental conditions, and each confers a selective advantage upon a different type of ecological specialization: competitors (C) prevail under the threat of competitive exclusion, stress-tolerators (S) under severe stress, and ruderals (R) in conditions of frequent and severe disturbance.

In order to illustrate the system, typical examples of each strategy are listed in Table 2.1. The full spectrum of habitat conditions and associated plant strategies can be represented by an equilateral triangle (see Fig. 1.2) within which variation in the relative importance of competition, stress and disturbance controls not only the three types of

Table 2.2 Some characteristics of competitive, stress-tolerant and ruderal plants. Characteristics shown in bold type have proved particularly useful in classifying plant strategies.

	Competitive	Stress-tolerant	Ruderal
(i) Morphology			
1. Life-forms	Herbs, shrubs, trees	Lichens, bryophytes, herbs, shrubs and trees	Herbs, bryophytes
2. Morphology of shoot	**High dense canopy of leaves. Extensive lateral spread above and below ground**	Extremely wide range of growth forms	**Small stature, limited lateral spread**
3. Leaf form	Robust, often mesophytic	**Often small or leathery, or needle-like**	Various, often mesoph
4. Canopy	**Rapidly-ascending monolayer**	Often multilayered; if monolayer, not rapidly-ascending	Various
(ii) Life-history			
5. Longevity of established phase	Long or relatively short	**Long to very long**	**Very short**
6. Longevity of roots	Relatively short	**Long**	**Short**
7. Leaf phenology	**Well-defined peaks of leaf production coinciding with periods of maximum potential productivity**	**Evergreens, with various patterns of leaf production**	**Short phase of produ in period of high pote productivity**
8. Phenology of flowering (or sporulation in ferns)	Flowers produced after (or more-rarely before) periods of maximum potential productivity	No general relationship between time of flowering and season	**Flowers produced ea the life-history**
9. Frequency of flowering	Established plants usually flower each year	**Intermittent flowering over a long life-history**	High frequency of flowering
10. Proportion of annual production devoted to seeds	Small	Small	**Large**
11. Perennation	Dormant buds and seeds	**Stress-tolerant leaves and roots**	Dormant seeds
12. Most commonly associated regenerative* strategies	V, S, W, B,	V, W, **B,**	S, W, B,
(iii) Physiology			
13. Mean potential relative growth-rate	**High**	Low	**High**
14. Response to resource depletion	Rapid morphogenetic responses in the form and distribution of leaves and roots	Morphogenetic responses slow and small in magnitude	Rapid curtailment of vegetative growth, diversion of resources flowering
15. Photosynthesis and uptake of mineral nutrients	Strongly seasonal coinciding with long continuous period of vegetative growth	Opportunistic, often uncoupled from vegetative growth	Opportunistic, coinci with vegetative growt

plant specialization described above, but also a range of intermediate strategies (C–R, C–S, S–R and C–S–R), each associated with a less extreme equilibrium between competition, stress and disturbance.

Table 2.2 continued

	Competitive	Stress-tolerant	Ruderal
16. Acclimation of photosynthesis, mineral nutrition and tissue hardiness to seasonal change in temperature, light and moisture supply	Weakly developed	Strongly developed	Weakly developed
17. Storage of photosynthate mineral nutrients	**Most photosynthate and mineral nutrients are rapidly incorporated into vegetative structure, but a proportion is stored and forms the capital for expansion of growth in the following growing season**	Storage systems in leaves, and in stems or roots, or both	**Confined to seeds**
(iv) **Miscellaneous**			
18. Litter	Copious, not usually persistent	Sparse, but often persistent	Sparse, not usually persistent
19. Palatability to unspecialized herbivores	Various	**Low**	Various, often high
20. Nuclear DNA amount	Usually small	Various	Small to very small

* Key to regenerative strategies (see Table 2.3): V, vegetative expansion; S, seasonal regeneration in vegetation gaps; W, numerous small, wind-dispersed seeds or spores; B_s, persistent bank of seeds or spores; B_j, persistent bank of juveniles.

Table 2.2 lists attributes of morphology, life-history and physiology which appear to be associated with the three primary strategies, and identifies those which have proved most useful in classifying the strategies of the species for which Accounts are provided in Chapter 3. It might be considered unlikely that so many variable attributes of plants would conform to these patterns. However, it is the cardinal assertion of the C–S–R model of primary plant strategies that under the distinctive selection pressures associated with the extremes of competition, stress and disturbance the range of adaptive possibilities is extremely constrained and few evolutionary solutions are viable. Further conformity arises from the fact that each of these solutions depends upon an integrated response involving most of the fundamental activities of the plant.

Interpretation

USES OF STRATEGY THEORY

The C–S–R model of primary strategies consists of an attempt to recognize the main avenues of ecological specialization in the established phase of plant life-histories. The sets of attributes associated with the primary strategies (Table 2.2) suggest many opportunities to interpret the distribution and population dynamics of species, and to predict the consequences of changes in their environment or management regime. An identification with respect to strategy often provides the basis for explaining the role of particular species in succession and for predicting both sensitivity to abrupt changes in management or climate and capacity to recover from such disturbance.

COMPETITORS

The list of plant attributes associated with the competitive strategy contains several which are obviously related to the capacity of C-strategists to monopolize resource capture in productive, relatively undisturbed environments. These include a high potential relative growth rate, tall stature and a tendency to form a consolidated growth form by vigorous lateral spread above and below ground. These characteristics are all evident in what we may describe as the 'classical' competitors (e.g. *Chamerion angustifolium*, *Epilobium hirsutum*, *Petasites hybridus*, *Phalaris arundinacea*, *Reynoutria japonica* and *Urtica dioica*).

The high rates of resource capture achieved by these plants are also due to other less obvious but equally important characteristics. These include the formation of substantial underground storage organs which fuel the initial surge in root and shoot growth in spring and allow a very large peak in shoot biomass to be developed in summer (Fig. 2.1). In any attempt to understand the functional characteristics of C-strategists it is vitally important to recognize that these plants have the capacity to withdraw resources from the environment at rapid rates. The effect of this phenomenon is to create a spatially patchy environment in which, despite the general abundance of resources within the habitat at large, zones of severe depletion develop rapidly above and below ground during each growing season in close proximity to functional leaves and roots. It is for this reason that we may suspect that the most important characteristic of the C-strategist is the high degree of morphological plasticity in the development of roots and shoots. This feature, coupled with the normally short life-span of individual leaves and fine roots, brings about a constant readjustment in the spatial distribution of the leaf canopy and in the actively absorbing part of the root system during the growing season. The consequence of this constant deployment of new leaves and roots into unexploited, resource-rich zones of the patchy environments created by competing plants is the phenomenon of 'active-foraging'. This has been analysed experimentally and, in terms of resource capture from productive environments, has been shown to be superior to mechanisms involving less-dynamic root and shoot systems.

Compared with ephemeral plants of frequently-disturbed productive habitats, the 'ruderals', species exhibiting the competitive strategy often show a temporary delay in the onset of seed production. In terms of Darwinian fitness, the advantage of this developmental pattern is clear in the case of habitats normally exploited by these species; in *undisturbed* productive conditions rapid vegetative monopoly is an essential prelude to sustained heavy reproductive output over many years. Equally, however, there can be little doubt that the delayed reproduction of the competitors is one of the major factors restricting their resilience and abundance in *severely disturbed* habitats.

The costs involved in the 'active foraging' for light, mineral nutrients and water, which is characteristic of the competitive strategy, are considerable because of the high rates of reinvestment of captured resources necessary for the construction of new leaves and roots following their rapid senescence. To these costs we must add those associated with the high rates of herbivory experienced by the weakly defended tissues of many competitors. We suspect, therefore, that there

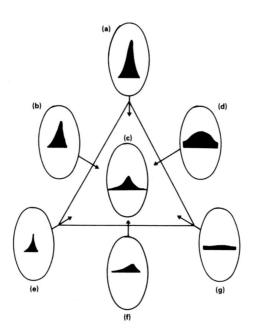

Figure 2.1 A scheme relating the pattern of seasonal change in shoot biomass to strategy type: (a) competitor, (b) competitive ruderal, (c) C–S–R strategist, (d) stress-tolerant competitor, (e) ruderal, (f) stress-tolerant ruderal and (g) stress-tolerator.

are severe penalties attached to 'active foraging' and that these will greatly restrict the success of the competitive strategy in chronically unproductive habitats.

STRESS-TOLERATORS

Under the previous heading we suggested that heavy expenditure of captured resources in new leaves and roots will be of selective advantage only in circumstances where active foraging gains access to large reserves of light energy, water and mineral nutrients. From this argument we predict that competitors will fail in habitats where productivity is low and where resource availability is brief and unpredictable (e.g. where light occurs as sunflecks and/or where mineral nutrients become available as short, rich pulses, as from intermittent decomposition events). In these circumstances, conservation of captured resources is of primary significance and successful plants are likely to be those with the ability to capture *and retain* scarce resources in a continuously hostile physical environment. Conservation of resources is apparent in the low potential relative growth rates and delayed onset of reproduction in many species of unproductive habitats. Also in keeping with these predictions, we find that the leaves of species of unproductive environments tend to be comparatively long-lived, morphologically implastic structures which are strongly defended against herbivory.

Recently, additional significance has been attached to the strong anti-herbivore defences of many stress-tolerators by the suggestion that physical defences which protect the living foliage often remain operational after senescence, retarding the breakdown of litter by decomposing-organisms. This is consistent with the high organic content of the superficial horizons of infertile soils, and perhaps explains also the dense accumulation of litter under trees of comparatively low growth rate (e.g. *Fagus sylvatica* and *Quercus petraea*). As explained in several Accounts, this phenomenon has pronounced effects on the distribution of herbaceous plants on the woodland floor.

Stress-tolerators, despite sharing many common features of life-history and physiology (Table 2.2), are associated with a wide range of life-forms and ecological behaviour. As the Accounts illustrate, there are stress-tolerators characteristic of calcareous soils (e.g. *Koeleria macrantha* and *Primula veris*), of acidic soils (e.g. *Juncus squarrosus* and *Nardus stricta*), of droughted habitats (e.g. *Sedum acre* and *Thymus praecox*), of wetland or damp grassland (e.g. *Carex panicea* and *Succisa pratensis*), of shaded situations (e.g. *Sanicula europaea* and *Viola riviniana*) and of heavy-metal-contaminated spoil (e.g. *Minuartia verna*). Clearly, there are important distinctions to be drawn within the general category 'stress-tolerator', and the Accounts contain an abundance of references to evidence for mechanisms of tolerance which are specific to particular types of stress-tolerators. This diversity prompts the question, 'What is/are the nature of the selection force(s) responsible for the features common to stress-tolerators?'. Two possible answers deserve consideration.

In general terms, one answer proposes that stress-tolerant traits are an inevitable evolutionary response to chronically low productivity, and are relatively independent of the nature of the stresses constraining production. The alternative answer suggests that, despite superficial differences, all the habitats exploited by stress-tolerators share a common underlying stress.

Further research is required to examine these two explanations. However, the Accounts suggest that not only is the balance of evidence currently in favour of the *latter* hypothesis, but that the common underlying stress is likely to be low availability of mineral nutrients, especially phosphorus and nitrogen. The implications of this tentative conclusion are considerable and extend beyond the immediate purposes of this book. However, it is already evident that the notion of limiting mineral nutrient elements as the ultimate selective pressure determining the stress-tolerant strategy is not incompatible with the concept of a recurrent pattern of evolutionary specialization involving all major aspects of the plant. Most plant activities depend upon the level of supply of mineral nutrients and, as proposed in Figure 2.2, it seems likely that conservative mechanisms of mineral nutrient capture and utilization will invariably be associated with constraints upon both carbon assimilation and reproductive activity.

RUDERALS

Two plant characteristics in particular are relevant to an analysis of the population dynamics and ecology of plants exhibiting the ruderal strategy. The first is a potentially high relative growth rate during the seedling phase and the second is an early onset of reproduction, a feature which in Angiosperms often coincides with self-pollination and rapid maturation and release of seeds. In arable weeds, and in ephemeral plants of frequently and severely disturbed habitats such as paths, these characteristics undoubtedly confer resilience and explain the remarkably rapid population fluctuations observed in species such as *Matricaria matricarioides*, *Polygonum aviculare*, *Senecio vulgaris* and *Stellaria media*. As emerges clearly from many of the Accounts, it is also a characteristic of the ruderal strategy that allocation to seed production is sustained (as a proportion of the total biomass) in plants stunted by drought, by mineral nutrient stress or as a consequence of growth at high population densities.

The reproductive imperative evident in the life-histories, physiology and breeding systems of plants classified as ruderals in the Accounts also

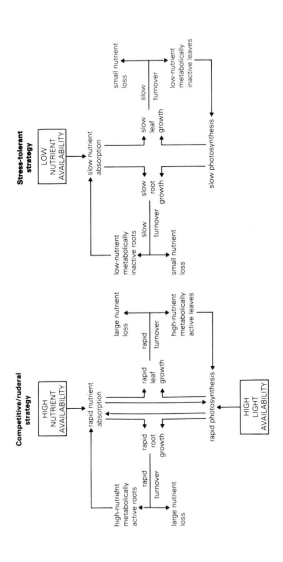

Figure 2.2 Interacting characteristics of plant strategies that appear to confer a selective advantage under conditions of high or low nutrient availability. (From Chapin 1980)

plays an inevitable part in these strategists' failure to exploit relatively undisturbed habitats. The early diversion of captured resources into flowers and seeds is not compatible with the development of the extensive root and shoot systems necessary for dominance and extended occupation of productive stable habitats, nor is it conducive to survival in undisturbed but highly-stressed environments where success is usually associated with very conservative patterns of resource utilization.

INTERMEDIATE STRATEGIES
In addition to the three primary strategies there are four intermediates (C–R, S–R, S-C and C–S–R), corresponding to particular intermediate positions within the triangular model (see Fig. 1.1). Table 2.1 includes examples of the intermediate strategies which, as expected, rely for their classification upon their 'intermediacy' with respect to the criteria used to recognize the primary strategies (Table 2.2). A particularly clear example of intermediacy is provided in Figure 2.1, which summarizes the proposed relationships between plant strategy and seasonal change in shoot biomass.

Gregariousness

Explanation

Here we use survey data to review the range in abundance of the subject species in vegetation samples associated with each of the seven primary habitats (Fig. 1.3). For each habitat we have examined the number of occasions on which the species attains set levels of abundance.

Interpretation

Classifying the occurrences of the subject species in terms of the proportion of quadrat subsections occupied by rooted shoots, has permitted a broad assessment of population structure. Species in which records are concentrated towards the low end of a frequency distribution table include not only the tree seedlings (e.g. of *Acer pseudoplatanus*, *Alnus glutinosa* and *Quercus* spp.), but also various herbaceous species originating from widely scattered propagules and possessing little capacity for vegetative spread (e.g. *Cirsium vulgare*, *Cystopteris fragilis* and *Hypochaeris radicata*).

Species with a large proportion of occurrences in high frequency classes are those capable of developing dense populations. These populations may consist of a large number of discrete individuals originating as seedlings in open habitats (e.g. in *Aira praecox*, *Linum catharticum* and *Matricaria matricarioides*) or they may be the result of vegetative expansion (e.g. in *Agrostis stolonifera*, *Brachypodium pinnatum* and *Mercurialis perennis*).

However, in some large clonal species (e.g. *Pteridium aquilinum*, *Reynoutria japonica* and *Petasites hybridus*) the capacity to dominate herbaceous vegetation is not reflected in a high frequency of rooted shoots. This is because these species produce a relatively low numerical density of individually massive fronds, shoots or leaves.

In addition to the two extremes in gregariousness, an intermediate category can be recognized. This includes a high proportion of species (e.g. *Briza media*, *Campanula rotundifolia* and *Carex caryophyllea*) which are of restricted height and capacity for lateral spread and which normally play a subordinate role in perennial herbaceous vegetation.

Flowers

Explanation

Here we note flower colour, structure of the inflorescence, mode of pollination and breeding system.

In the case of pteridophytes the distribution of the sori is described.

Interpretation

Sexual reproduction is of fundamental importance in creating genetic diversity, both within and between populations, and is exhibited by most species. Flowers, which provide the machinery for this process, are thus of vital significance.

There is, of course, a considerable diversity of floral structure in evidence, even within the British Isles. Although much of this diversity is a legacy of the evolutionary history of the major taxa, a great deal of structural variety can be related to pollination mechanism, even though most British species rely upon generalist pollinators.

The floral character which is most generally regarded as being of highest ecological importance is the breeding system, since this regulates the amount of variation between offspring. In particular, ruderals and stress-tolerant ruderals of the Sheffield flora, and elsewhere, are often in-breeding. This may be of selective importance in short-lived plants of disturbed habitats, either (a) because it reduces the probability that pollination failure will limit the rate or yield of reproduction, or (b) because offspring which are homozygous have, like their parents, a relatively high probability of reproductive success. In contrast, longer-lived competitors and stress-tolerators tend to have a higher incidence of out-breeding.

The effects of flowering upon the growth of the vegetative plant may be considerable. The cost of reproduction by seed (in terms of resources sequestered from the growing vegetative plant) may be much larger than the simple dry-weight yields of seed would imply. In addition, competition for pollinators, or other demands made by the flowering process, may necessitate an allocation of resources into floral development at a time when resources would otherwise by available for vegetative growth. An important role is implicated for flowering in the determination of the ecological distribution of species. However, despite much detailed and innovative recent work on the subject, we have yet to establish the ecological ground-rules governing the form, number and pollination of flowers and the production of seeds.

Regenerative strategies

Explanation

Many previous treatments of regeneration in plants have been particularly concerned with the genetic structure of populations, placing strong emphasis upon the relative importance of sexual and vegetative processes in the regenerative biology of plants. We share this concern, and this is particularly evident in the 'Synopsis' of many Accounts. However, under 'Regenerative strategies' our object is to recognize, and where possible classify, mechanisms of regeneration in *functional* terms by considering the size, number, dispersal, dormancy and degree of independence of the offspring, and the conditions affecting its establishment.

Five major types of regenerative strategy of common occurrence in the British Isles (and the abbreviated symbols used for them in the Accounts) are listed and described in Table 2.3. The absence of reliable demographic information has made it impossible to recognize the strategy involving banks of persistent juveniles with certainty. No references to this strategy have therefore been made in the Accounts. Also, it has been difficult to distinguish between vegetative expansion and seasonal regeneration by vegetative fragmentation in the case of certain species; in such cases of uncertainty we have used the symbol (V).

Interpretation

RELEVANCE
Seedlings and vegetative offspring are comparatively small. They are therefore exposed to hazards which may be much more severe than, and often quite different from, those experienced by established plants. Successful regeneration in many species depends upon exploitation of

Table 2.3 Five regenerative strategies of widespread occurrence in terrestrial vegetation.

Strategy	Functional characteristics	Conditions under which the strategy appears to enjoy a selective advantage
Vegetative expansion (V)	New shoots vegetative in origin and remaining attached to parent plant until well established	Productive or unproductive habitats subject to low intensities of disturbance
Seasonal regeneration (S)	Independent offspring (seeds or vegetative propagules) produced in a single cohort	Habitats subjected to seasonally predictable disturbance by climate or biotic factors
Persistent seed or spore bank (B_s)	Viable but dormant seeds or spores present throughout the year; some persisting more than 12 months	Habitats subjected to temporally unpredictable disturbance
Numerous widely dispersed seeds or spores (W)	Offspring numerous and exceedingly buoyant in air; widely dispersed and often of limited persistence	Habitats subjected to spatially unpredictable disturbance or relatively inaccessible (cliffs, walls, tree trunks, etc.)
Persistent juveniles (B_j)	Offspring derived from an independent propagule but seedling or sporeling capable of long-term persistence in a juvenile state	Unproductive habitats subjected to low intensities of disturbance

locally favourable sites, and plants appear to have evolved a variety of regenerative mechanisms which exploit particular niches. Five major types of regenerative strategy are distinguished in Table 2.3, which also contains a brief description of the conditions under which the particular strategies appear to enjoy a selective advantage. The five strategies will now be considered in turn.

VEGETATIVE EXPANSION

Under this heading it is convenient to assemble many of the regenerative mechanisms which involve the expansion and subsequent fragmentation of the vegetative plant through the formation of persistent rhizomes, stolons or suckers. The most consistent feature of vegetative expansion is the relatively low risk of mortality to the offspring. This is achieved through a prolonged attachment to the parent plant and the mobilization of resources from parent to offspring.

The outstanding feature of the distribution of vegetative expansion in the triangular model is its coincidence with established strategies characteristic of relatively undisturbed habitats (competitors, stress-tolerant competitors and stress-tolerators). This pattern may exist because vegetative expansion involves a period of attachment between parent and offspring, and is therefore not a viable mechanism when vegetation is affected by frequent and severe disturbance.

It seems likely that the role of vegetative expansion may be rather different in competitors and stress-tolerators. In productive, relatively undisturbed habitats, vegetative expansion is often an integral part of a mechanism whereby competitive herbs (e.g. *Phragmites australis*, *Pteridium aquilinum* and *Reynoutria japonica*) rapidly monopolize the environmental resources, and thus suppress the growth and regeneration of neighbouring plants. Under the very different conditions of highly-stressed environments, however, the main advantage of vegetative expansion in species such as *Anemone nemorosa*, *Carex caryophyllea* and *Sanicula europaea* arises from its capacity to sustain the offspring under conditions in which establishment from an independent propagule is a lengthy and hazardous process.

SEASONAL REGENERATION IN VEGETATION GAPS
In a wide range of habitats herbaceous vegetation is subjected to seasonally-predictable damage from phenomena such as temporary drought, flooding, trampling and grazing. Under these conditions the most common regenerative strategy is that in which the areas of bare ground or sparse vegetation cover created every year are recolonized annually during a particularly favourable season. The propagules involved in seasonal regeneration are relatively large, lack long-term dormancy, and may be sexual or asexual in origin. They include cohorts of synchronously germinating seeds (e.g. in *Bromus hordeaceus*, *Heracleum sphondylium* and *Impatiens glandulifera*) and populations of offsets or bulbils (e.g. in *Anthriscus sylvestris* and *Ranunculus ficaria*), or the tips of rapidly-fragmenting rhizomes or stolons (e.g. in *Poa trivialis*, *Ranunculus repens* and *Trifolium repens*).

Seasonal regeneration is particularly associated with competitive ruderals and stress-tolerant ruderals, both of which are restricted to sites at which the intensity of disturbance is sufficient to cause seasonal and temporary gaps in the vegetation. In habitats subjected to forms of disturbance that are more-severe or less-predictable, or both, dependence upon propagules which lack long-term dormancy may be expected to limit regenerative capacity. It is probably for this reason that the incidence of exclusively seasonal forms of regeneration is low among ruderals *sensu stricto*.

The strategy of seasonal regeneration therefore represents a rather unsophisticated mechanism of gap exploitation in which, after propagules have been dispersed locally within the habitat and have germinated simultaneously (or recommenced growth, in the case of fragments of rhizomes or stolons), survival is usually limited to those individuals which occur in gaps. It is evident that this method of regeneration depends upon the creation in the same habitat and in each year of a high density of suitable gaps. Where gaps occur more rarely, or less predictably, seasonal regeneration tends to give way to other types of regenerative strategy.

REGENERATION INVOLVING A PERSISTENT SEED BANK
When flowering plants are compared with respect to the fate of their seeds, two contrasting groups may be recognized. In one, most if not all of the seeds germinate soon after release. In the other, many become incorporated into a bank of dormant seeds which is detectable in the habitat at all times during the year and may represent an accumulation of many years' production. These two groups are, of course, extremes, and between them there are species and populations in which the seed bank, although present throughout the year, shows pronounced seasonal variations in size. Nevertheless, it is convenient to draw an arbitrary distinction between 'transient' and 'persistent' seed banks. A *transient seed bank* may therefore be defined as one in which none of the seed crop remains in the habitat in a viable condition for more than one year, and a *persistent seed bank* as one in which at least some of the component seeds are one year old or more.

Regeneration involving a persistent seed bank commonly occurs in association with each of the established strategies, with the exception of the stress-tolerators, in which the role of a seed bank is often played by a bank of persistent seedlings.

Although some of the plants which develop persistent seed banks have mechanisms which facilitate seed dispersal by animals (e.g. *Danthonia decumbens*, *Rubus fruticosus* and *Vaccinium myrtillus*), most seed banks are located close to their parent plants and may even be situated directly beneath them. It seems therefore likely that the main functional significance of a persistent seed bank is the extent to which it allows regeneration *in situ*. This interpretation is consistent with the fact that seed banks are particularly common in proclimax vegetation, such as grassland, heathland and disturbed marsh, in which the process of vegetational change is cyclical rather than successional. Many forms of proclimax vegetation, especially those in farmland, are subject to alternating patterns of land use, with the result that, at any one location, conditions suitable for the establishment of each potential member of the community are available only intermittently. In this situation the presence in the soil of a persistent seed bank allows the survival of populations during periods in which the management regime is temporarily unfavourable to the established plants.

Persistent seed banks occur in association with a wide range of primary and secondary strategies, and it is clear that in certain respects the role of a seed bank changes according to the established strategy with which it is associated. In ruderals such as the annuals of arable fields and marshland (e.g. *Juncus bufonius*, *Papaver rhoeas*, *Rorippa palustris* and *Veronica persica*) the seed bank may permit rapid recovery from catastrophic mortalities inflicted by cultivation, herbicide treatments or natural phenomena such as flooding. It also allows survival of unfavourable seasons and longer periods of temporary stability during which the ruderals are excluded by perennial species. In competitive ruderals (e.g. *Juncus articulatus* and *Rumex obtusifolius*), in stress-tolerant ruderals (e.g. *Arenaria serpyllifolia* and *Geranium molle*), in C–S–R strategists (e.g. *Anthoxanthum odoratum* and *Holcus lanatus*) and in stress-tolerators (e.g. *Danthonia decumbens* and *Thymus praecox*), the role of the seed bank is similar to that in ruderals, except that it appears to be more concerned with regeneration within gaps in perennial vegetation.

In habitats dominated by competitors and stress-tolerant competitors, the intervals between major disturbances may be very long (perhaps more than 15 years). Competitors and stress-tolerant competitors with persistent seed banks include many of the most familiar herbs of vegetation subject to occasional disturbance, for example by burning or flooding (e.g. *Calluna vulgaris*, *Epilobium hirsutum*, *Juncus effusus* and *Urtica dioica*).

REGENERATION INVOLVING NUMEROUS WIDELY DISPERSED
SEEDS OR SPORES

Regeneration involving numerous widely-dispersed seeds or spores

This mechanism of regeneration occurs in association with the complete range of established strategies, and is conspicuous in two types of situation. The first consists of localized and relatively inaccessible habitats such as crevices in cliffs, walls and tree trunks, where pteridophytes (e.g. *Asplenium trichomanes* and *Cystopteris fragilis*) and wind-dispersed composites (e.g. *Mycelis muralis* and *Hieracium* spp.) are strongly represented. The second arises in circumstances where landscape is subject to spatially unpredictable disturbance. Where disturbance occurs as an exceptional event in an environment of moderate to high productivity, the site often presents a small target for colonization, and usually remains open only for a relatively brief period. Here the most successful invaders include those herbs, shrubs and trees (e.g. *Tussilago farfara*, *Chamerion angustifolium*, *Salix* spp. and *Betula* spp.) which produce locally-saturating densities of mobile seeds.

Because of their prolific seed production and effective dispersal, species in the 'widely dispersed' category are frequently recorded as seedlings or small plants in habitats which are quite unsuitable for the established plant. In only a few instances do species in this grouping form a persistent seed bank (e.g. *Epilobium* spp.).

REGENERATION INVOLVING A BANK OF PERSISTENT JUVENILES

Alternatively described as 'advance reproduction', this mechanism of regeneration appears to be characteristic of species exploiting circumstances where seedling establishment is lengthy and uncertain, and consists essentially of a process of replacement of occasional mortalities in the population of established plants (e.g. *Quercus* spp. and *Sanicula europaea*). Banks of persistent juveniles are especially common among stress-tolerators, many of which tend to produce seeds rather intermittently over the long life-span of the established plant. It may be significant in the maintenance of populations of these plants that the persistent juveniles provide the possibility of recruitment to the breeding population between seed crops. Some persistent seedlings may benefit from assimilate imported by means of mycorrhizal connections from established plants.

FLEXIBILITY IN REGENERATION

In contrast to the strategies associated with the established phases of plants' life-histories, the five regenerative strategies are not primarily determined by inflexible 'design constraints'. They are not, therefore, mutually exclusive. For this reason it is not uncommon for the same genotype to exhibit two or more regenerative strategies simultaneously. This has led to a suggestion that the ecological amplitude of a species may be determined, to some degree, by the *number of regenerative strategies* which it exhibits. Although limited by incomplete data, the Accounts lend strong support to this hypothesis. At one extreme it is possible to recognize species in which a narrowly defined ecological range coincides with a restricted capacity for regeneration (e.g. *Euphrasia officinalis*, *Fagus sylvatica* and *Impatiens glandulifera*). The other extreme is exemplified by species such as *Epilobium hirsutum*, *Holcus lanatus*, *Juncus effusus* and *Poa trivialis*. In these, the ecological versatility evident from a capacity for persistence and rapid population expansion under a variety of forms of vegetation management is in turn supported by a flexibility in mechanisms of regeneration. Polymorphism in seed characteristics, which is particularly associated with plants of disturbed habitats, e.g. *Atriplex*, *Chenopodium*, *Rumex* and many composites, provides a further dimension to regenerative flexibility.

REGENERATIVE FAILURE

Relevance
The success of particular regenerative strategies appears to be strongly dependent upon habitat conditions. We predict that failure of regeneration is a major factor limiting the ecological and geographical range of species.

Failure in plants with a single regenerative strategy
In the British flora there are many species in which regeneration depends

exclusively upon the production of seeds which lack dormancy and which produce a single cohort of seedlings each year. Such reliance upon seasonal regeneration is particularly characteristic of autumn-germinating annual grasses (e.g. *Aira praecox*, *Bromus sterilis* and *Hordeum murinum*) and of a number of spring-germinating trees and shrubs (e.g. *Acer pseudoplatanus*, *Fraxinus excelsior* and *Sambucus nigra*). For both of these groups there is evidence that dependence upon seasonal regeneration is often a potent limiting factor in the field. In *Hordeum murinum*, for example, it has been shown that the altitudinal limit of the species is related to its failure to produce sufficient seed to maintain viable populations. A quite different example of failure is represented by the inability of common trees such as *A. pseudoplatanus* and *F. excelsior* to regenerate in pastures and unfenced woodlands. Here seed production is prolific, but cannot usually compensate for the losses of seedlings inflicted by domestic grazing animals, rodents and deer.

Failure in plants with more than one regenerative strategy

Among the many perennial species which have an ability both to spread vegetatively and to regenerate by seed, it is not uncommon to find populations in which one of these two mechanisms predominates. At the northern limits of their distribution in Britain many species (e.g. *Brachypodium pinnatum* and *Cirsium acaulon*) establish from seed only rarely, but are capable of vigorous clonal expansion. A similar shift away from regeneration by seed often results from management practices: pastures, meadows and frequently mown road verges, lawns and sports fields usually contain populations of grasses (e.g. *Agrostis capillaris* and *Festuca rubra*) in which persistence is relatively independent of seed production. In contrast, situations occur where vegetative expansion is constrained and seed production remains as the only effective means of regeneration. An example of this is provided by the stunted, but flowering, individuals of *Chamerion angustifolium* which are frequently observed in crevices on cliffs and walls.

Seed

Explanation

The word 'seed' is used in the sense of 'germinule', the structure which germinates to give rise to the seedling (or prothallus in the case of ferns). Each Account notes the mean weight of air-dried, viable germinules sampled from natural populations in the British Isles.

Interpretation

The size and shape of the germinule often provides valuable information about function *both before and after germination*. A conflict can be recognized between selection forces favouring the production of numerous offspring and those promoting germinules of large size. In the Accounts the advantages and limitations associated with dependence upon small germinules are evident within the pteridophytes, and in species of *Betula*, *Epilobium* and *Salix*, in which efficient long-range dispersal has been achieved at the expense of the extreme vulnerability of the small offspring to dominance by established plants.

However, it would be a mistake to assume that the functional significance of small germinule size is invariably related to dispersal and to 'escape through space' from the influence of the dominant species of the established vegetation. The production of numerous tiny germinules is also characteristic of species in which the offspring 'escape through time'. Through the agency of rainwater percolation and earthworm activity, the small compact germinules of a wide range of species are incorporated into persistent banks of dormant seeds or spores in the soil. Germination tends to be delayed until disturbances of the vegetation or the soil expose the seeds to unfiltered sunlight or large diurnal fluctuations in temperature, both of which often appear to function as cues for germination. Such 'gap-detection mechanisms' restrict the appearance of seedlings to microsites where conditions may be relatively favourable for establishment.

Persistent seed banks often occur among species of marshland (e.g. *Juncus* spp., *Rorippa palustris* and *Stellaria alsine*), heathland (e.g. *Calluna vulgaris* and *Erica cinerea*), wasteland (e.g. *Hypericum perforatum* and *Origanum vulgare*), disturbed woodland (e.g. *Digitalis purpurea* and *Moehringia trinervia*), grassland (e.g. *Agrostis capillaris* and *Deschampsia cespitosa*) and arable land (e.g. *Capsella bursa-pastoris* and *Chenopodium rubrum*). Tiny seeds are less vulnerable to predation than large ones. It has been suggested that this may have provided an additional selection force, explaining the strong correlation observed between small seed size and seed bank persistence.

The offspring of species producing large germinules appear to be capable of establishing successfully in closed herbaceous vegetation. This is evident from the Accounts for large-seeded tree species (e.g. *Acer pseudoplatanus*, *Fagus sylvatica* and *Quercus* spp.) and from those describing the large-seeded annual herbs (e.g. *Galeopsis tetrahit*, *Galium aparine* and *Impatiens glandulifera*), the obligately or facultatively monocarpic perennials (e.g. *Anthriscus sylvestris*, *Arctium minus* and *Heracleum sphondylium*) and the polycarpic perennials (e.g. *Allium ursinum*, *Myrrhis odorata* and *Vicia cracca*). In these species, however, it is important to recognize that in their natural habitats the benefit of large seed reserves may arise in relation to emergence through plant litter rather than in relation to competition with established plants.

There is a tendency for large seed size to be associated with plants of some droughted habitats. Comparatively large seeds are particularly a feature of those spring-germinating plants which, even in dry habitats, continue to expand their shoot biomass during the late spring and summer. Examples of these include obligate or facultative annual species such as *Medicago lupulina* and *Torilis japonica*, and perennials such as *Centaurea scabiosa*, *Lotus corniculatus*, *Pimpinella saxifraga* and *Sanguisorba minor*, all of which have tap-roots, the initial extension of which is presumably dependent upon mobilization of a large fraction of the considerable seed reserves into the radicle.

British distribution

Explanation

Notes on the British distribution of the species are derived mainly from the *Flora of the British Isles* (cited in full under 'Nomenclature') and from the *Atlas of the British flora* (2nd edn), 1976, F. H. Perring & S. M. Walters (London: Botanical Society of the British Isles), from the *Critical supplement to the Atlas of the British flora*, 1968, F. H. Perring & P. D. Sell (London: Botanical Society of the British Isles), from *Sedges of the British Isles*, 1982, A. C. Jermy, A. O. Chater & R. W. David (London: Botanical Society of the British Isles), and from the *Atlas of ferns of the British Isles*, 1978, A. C. Jermy, H. R. Arnold, L. Farrell & F. H. Perring (London: Botanical Society of the British Isles and The British Pteridological Society). The percentage of British vice-counties in which the subject occurs is noted.

Interpretation

The moist tropics are considered to be the aboriginal home of the angiosperms. Perhaps as a result, the number of angiosperm species per unit area decreases progressively from the tropics to polar regions. Superimposed on this general relationship are the facts that the British flora is of recent origin (from *c.* 1 000 000 BC onwards), and that for the past 7000 years Britain has been geographically isolated from mainland Europe. These features have radically influenced the distribution of species and the composition of the present British flora. However, the influence upon plant distribution which has received greatest attention is climate, studies of which have been regarded as the cornerstone of plant geography.

There is a strong global correlation between climate and life-form. The species of the British flora may also be classified into elements such as 'Arctic–Alpine', 'Atlantic' and 'Continental', which relate to both geographical and climatic distributions. However, in only a few instances has climate actually been shown to regulate plant distribution. Perhaps the best demonstration relates to *Cirsium acaulon* and *Tilia cordata*, where regenerative failure at the edge of their range appears to be due to low summer temperature. An inability to tolerate low winter temperatures can also limit distribution, as in the case of *Ilex aquifolium* and *Hedera helix*, and the same seems true in spring for the maritime subspecies of *Pyrola rotundifolia*.

Several factors account for the failure of climatic studies to explain plant distribution satisfactorily. Measurement of microclimatic variables is beset with difficulties. There are, for example, considerable variations in humidity and temperature associated with microtopography, and the climate experienced at the leaf surface of the plant may be very different from that of the surrounding air. Limits may be imposed by extreme climatic events such as severe droughts or very late frosts which only occur once or twice a century. Moreover, because temperature affects a whole range of metabolic processes, even actual failure due to climate may have many possible mechanisms. These complications make analysis by direct field study extremely uncertain and, often, unrewarding. We conclude that simple comparative experiments defining the climatic tolerances of large numbers of species of contrasted geographical distribution may provide the most reliable way forward.

Habitats

Explanation

Data given here are from the UCPE 'Survey II' (Table 1.2), encompassing all herbaceous vegetation types of the Sheffield region. The survey records may be arranged according to a simple branched key (Fig. 1.3) in which habitats are classified according to physical features of the environment and to forms of land-use. The key contains seven 'primary', eight 'intermediate' and 32 'terminal' categories (indicated in Fig. 1.3 as lozenges, squares and circles, respectively). The values quoted in each case are for the five leading terminal categories of habitat, ranked according to the percentage of samples which contain the subject species.

Interpretation

The survey base for these habitat data consists exclusively of inland habitats sampled in north-central England. The data thus indicate where each species is most commonly found within the Sheffield region, but even on this basis, the data provide useful ecological information.

However, one problem in estimating ecological amplitude from the habitat data arises when considering species regenerating by means of numerous, widely dispersed offspring. The dispersal efficiency of these species is often such as to produce large populations of juveniles in habitats in which there is little prospect of successful establishment. *Betula* spp., *Chamerion angustifolium*, *Dryopteris filix-mas* and *Epilobium hirsutum* provide examples of species for which the inclusion of such transients in the survey records has created a spuriously high impression of ecological amplitude.

The habitat categories used are all extremely broad and some (e.g. river and stream banks) are very heterogeneous in vegetation. It is possible, nevertheless, to use the data as a broad basis upon which to assess the present and future status of a species. In particular, we can differentiate between those (e.g. *Galium sterneri*, *Mercurialis perennis* and *Primula veris*) which have remained confined to habitats that are remnants of the more ancient countryside (unenclosed pastures, woodland, rock outcrops, screes, cliffs) and those (e.g. *Digitalis purpurea*, *Elymus repens* and *Senecio squalidus*) which have successfully colonized recently created or radically transformed habitats (various types of spoil, plantations and arable land). With caution, therefore, the data may be used to detect both threatened and expanding species.

Spp. most similar in habitats

Explanation

Five species are listed which, in terms of their distribution in the terminal categories of the habitat classification, are the ones most similar to the subject-species. The five were identified by means of a similarity analysis conducted with the aid of a multivariate statistical package. The values given are similarity coefficients, expressed as percentages. The primary and intermediate habitat categories were not included in this analysis, since this would involve duplication of data. All of the variables were of the continuous, numerical type and were standardized to zero mean and unit variance.

In some less-common species, particularly those from wetlands (which are poorly represented in the survey area), the similarity coefficients were low and the species list was ecologically heterogeneous. In such cases the results of the similarity analysis would have been potentially misleading, and have therefore been omitted, with a note to that effect.

Interpretation

The 'percentage similarity' returns a high value when there is a good 'match' between the percentage occurrences of species in a large number of the 32 terminal habitat categories and a low value when there is not. Thus, the percentage simply indicates a similarity *in habitats*. It implies neither that species thus identified are necessarily similar in ecology to the subject species, nor that they are necessarily to be found in association with the subject species.

Three general types of result may emerge from the similarity analysis; these should be borne in mind when interpreting the percentage similarity in habitats reported in the Accounts.

First, where the subject species has a relatively specialized and restricted habitat distribution there will be a large number of 'zero matches' in the calculation of percentage similarity. Provided that other species exist with broadly the same limited pattern of positive occurrences, these will emerge as very 'similar' indeed to the subject species. In such cases even the fifth most similar species can exhibit a percentage similarity of almost 90%. For example, the sequence of five similarities to the wetland species *Apium nodiflorum* is *Nasturtium officinale* agg., *Veronica beccabunga*, *Callitriche stagnalis*, *Equisetum palustre* and *Glyceria fluitans*, with percentage similarities ranging from 98.7 down to 89.3. By the same token, in arable habitats the sequence of five species most similar to *Veronica persica* is *Spergula arvensis*, *Fallopia convolvulus*, *Myosotis arvensis*, *Polygonum persicaria* and *Papaver rhoeas*, with percentage similarities from 99.2 down to 88.7. In pteridophyte subjects, too, the specialized and restricted habitats, consisting in this instance of damp and shady situations, give rise to sequences of species of high percentage similarity to the subject.

Second, in this analysis it is difficult for species of genuinely widespread distribution to attain high percentage similarities to their statistical neighbours. Values in the range 60–80% are more common here. For example, the sequence for *Dactylis glomerata* is *Taraxacum* agg., *Poa pratensis*, *Festuca rubra*, *Leucanthemum vulgare* and *Plantago lanceolata*, with percentage similarities ranging from 79.1 down to 67.1. Another example is provided by *Taraxacum* itself, for which the sequence is *Dactylis glomerata*, *Poa pratensis*, *Festuca rubra*, *Plantago lanceolata* and *Leucanthemum vulgare*, with percentage similarities from 79.1 down to 57.2. These two sequences also illustrate the high incidence of networks of mutual similarity among species of widespread distribution.

Third, some species of restricted occurrence have no counterparts capable of returning high percentage similarities in the kind of analysis conducted here. Even their closest statistical neighbours may have percentage similarities which lie only in the range 50–60. For example, the sequence for *Galium cruciata* is *Lathyrus montanus*, *Bromus erectus*, *Trifolium medium*, *Stachys officinalis* and *Brachypodium pinnatum*, with percentage similarities ranging from 56.8 to 47.5. *Vicia sepium* has the sequence *Arctium minus*, *Reynoutria japonica*, *Torilis japonica*, *Vicia cracca* and *Trifolium medium*, with percentage similarities from 52.5 to 41.4. In extreme cases of this kind we have omitted the listing of 'similar species' altogether. This has also been the practice where very few records exist for the subject species.

Full Autecological Account

The page reference here is to *CPE*, where a much fuller autecological account appears. In addition to supplying further information under most of the present categories, the *CPE* accounts include data on many subjects not reviewed here. These are: family, life-form, phenology, height, seedling relative growth rate, nuclear DNA amount, germination, Biological Flora(s), and the distribution of the subject in relation to the axes of C–S–R plant strategy theory and in relation to the environmental variables altitude, slope, aspect, hydrology, soil surface pH and proportion of bare soil.

Synopsis

The aim of each Synopsis is to review the distinctive features of the subject species and its ecology. This involves an interpretation of the standardized field and laboratory data presented in the Account, with an incorporation of the sometimes more fragmentary information which is available from the various published sources. Unpublished field observations from 'Survey III' (Table 1.2) have also been drawn upon.

Inevitably the number of useful data available from the scientific literature varies widely according to species. Agricultural and ecological research workers have dealt exhaustively with commercial forage plants and arable weeds, such as *Lolium perenne*, *Trifolium repens*, *Poa annua* and *Stellaria media*, but at the other extreme there has been great neglect of many common plants of waysides, marginal land and wooded habitats, such as *Leontodon autumnalis*, *Melica uniflora* and *Stellaria holostea*. Hence, for some species our task has been to distil relevant data from an extensive literature, whereas in others we have simply exhausted a woefully incomplete database. In cases where knowledge is exceptionally fragmentary (e.g. *Vicia sepium*) the Synopsis, by its brevity and shallow penetration into ecological analysis, serves mainly as a reminder of our great ignorance of the subject species and its ecology.

However, though the length and content of the Synopses may vary somewhat, we have tried to review what knowledge there is of the species in a standardized and logical sequence. Hence, the initial sections of the Synopses are mainly concerned with the geography and habitat range of the subject, and with its niche within plant communities. This is followed, where possible, by an assessment of the species' tolerance of ecological factors, including those under direct or indirect human control. Information from the UCPE screening programme (Fig. 1.4) is often added at this point. Certain of the plant attributes mentioned here, such as potential relative growth rate and nuclear DNA amount, are not among the standard attributes listed at the beginning of the Account. However, they normally appear linked to a statement or prediction on, for example, the phenology of the species. The reader is referred to *CPE* for fuller information on such attributes *per se*.

Current or past patterns of human utilization of the species in agriculture or otherwise are noted next, and any well-defined responses to vegetation management are described. Any relevant results from the Park Grass Experiment (*PGE*) at Rothamsted are summarized here (from *The park grass plots at Rothamsted*, 1958, by W. E. Brenchley & K. Warington, Harpenden: Rothamsted Experimental Station). The regeneration mechanisms of the species are considered, both in relation to the maintenance of its natural populations and in relation to its potential for colonizing new habitats. Finally, if information is available, we comment on the genetic integrity and taxonomic affinities of the species and cite any evidence of ecotypic differentiation.

In keeping with the aims stated in the Introduction, many of the Synopses include passages in which comparisons have been drawn between the subject species and plants of similar ecology within the British flora. Where such plants occur together, as in the case of the guilds of small winter annuals of droughted, calcareous soils, we have tried to recognize any differences which may permit their coexistence within plant communities (see, for example, the discussion of niche differentiation between *Torilis japonica* and smaller winter annuals).

Current trends

Under this heading any conspicuous changes which are currently taking place in the abundance of the subject species are reported. Based upon UCPE's further and continuing studies of commonness and rarity, we briefly assess the current and future status of the species within the British flora.

3 The abridged accounts

1 *Acer pseudoplatanus*

Sycamore, Great Maple

(As seedlings)

Established strategy Between C and S–C.

Gregariousness Seedlings may occur at high densities.

Flowers Yellowish-green, monoecious, insect-pollinated.

Regenerative strategies W.

Seed 30.6 mg.

British distribution Widespread (100% of VCs).

Commonest habitats Broadleaved plantations 21%, acidic woodland 18%, coniferous plantations/limestone woodland 13%, hedgerows 10%.

Spp. most similar in habitats *Hyacinthoides non-scripta* 91%, *Quercus* sp. (juv.) 81%, *Fagus sylvatica* (juv.) 81%, *Rubus fruticosus* 80%, *Holcus mollis* 80%.

Full Autecological Account *CPE* page 54.

Synopsis *A.p.* is a relatively frost-tolerant, deciduous timber tree with diffuse porous wood, up to 30 m in height, and exceptionally living for 400–600 years. The species is a native of C and S Europe, particularly in mountainous regions. In fact, *A.p.* was introduced into Britain during the 15th or 16th century and has been planted widely from the late 18th century onwards. In 1947 *A.p.* covered *c.* 31 000 ha of woodland (> 4% of the total), making it the fifth most common tree of broad-leaved woodland. *A.p.* produces numerous winged, wind-dispersed fruits and, apart from *Fraxinus excelsior*, is the tree species most commonly recorded as seedlings or saplings within the

Sheffield region. The species is also the most frequent contributor to the shrub or tree canopy, and is infiltrating many older woodlands. *A.p.* is tolerant of coppicing. The largest trees are particularly associated with moist, base-rich, fertile soils, but *A.p.* is widespread on all but the most acidic sites. Damage to trees by grey squirrels may be extensive and causes problems in the management of *A.p.* for timber. Seeds germinate in spring and no persistent seed bank is formed. The seedling is intolerant of competition from herbaceous species and sensitive to grazing; consequently, saplings tend to occur either beneath a tree canopy or in inaccessible rocky sites. A feature which *A.p.* shares with many other trees is the capacity of the saplings to expand leaves before bud-break of the adult trees. As a seedling, *A.p.* is found in sites similar to those exploited by *Fraxinus excelsior*. However, perhaps through mechanisms involving the ring porous wood, bud-break is considerably earlier. Saplings are also moderately shade-tolerant, but exhibit a higher frequency of occurrence on S-facing slopes and tend not to persist under a canopy of *A.p.*. Saplings are capable of more-rapid growth than those of most other British trees and in the Derbyshire Dales establishment of *A.p.* tends to be restricted to highly disturbed sites. The species establishes particularly freely in lightly-shaded sites in disturbed woodland. Leaves expand early in summer to form a rather dense canopy and leaf litter is relatively persistent, though less so than that of *Fagus sylvatica*. Dense accumulations of *A.p.* litter may reduce the floristic diversity of the ground flora of woodlands. The aphid *Drepanosiphum platanoides* may cause considerable reduction in the growth of *A.p.* but very few other insects are associated. However, *A.p.* has been little-studied and it has been suggested that its insect fauna has been underestimated. Populations from different European climatic zones show different phenological patterns. **Current trends** A more economic crop than oak or beech, arguably under-used in modern British forestry. Formerly much planted for forestry and, because of its ease of transplanting, much in demand for amenity use. However, *A.p.* is a weedy species and its rapid and effective regeneration is in contrast to that of *Tilia* spp. in ancient woodland. The relationship between *A.p.* and the species of mixed ancient woodland may be similar to that between *Betula* and *Quercus* (see *Betula* account), and concerted efforts have been made to remove *A.p.* from some ancient woodlands. In this context the indiscriminate planting of *A.p.* for amenity may be viewed as highly irresponsible.

2 *Achillea millefolium*

Yarrow, Milfoil

Established strategy Between C–S–R and C–R.
Gregariousness Intermediate.
Flowers White, pollinated by insects, self-incompatible.
Regenerative strategies V, ?S.
Seed 0.16 mg.
British distribution Widespread (100% of VCs).
Commonest habitats Limestone wasteland 19%, lead mine spoil/limestone pastures 17%, cinder heaps/verges 13%.
Spp. most similar in habitats *Lotus corniculatus* 70%, *Plantago lanceolata* 65%, *Rumex acetosa* 65%, *Cerastium fontanum* 64%, *Festuca rubra* 63%.
Full Autecological Account *CPE* page 56.

Synopsis *A.m.* is a winter-green herb of wide edaphic tolerance found in a range of grassy and open habitats. *A.m.* is equally common and ecologically wide-ranging in N Europe and is included among the six most common species in a survey of Danish roadside habitats. In grazed turf, the leaves are appressed to the ground in basal rosettes which may be aggregated into large clonal patches. In tall grassland, foliage is longer and held erect and, although not a woodland plant, *A.m.* may persist in shade, where it tends not to flower. However, *A.m.* is relatively intolerant of competition from taller and more robust herbs. The root system may descend more than 200 mm. This rooting habit coincides both with a degree of drought tolerance and with vulnerability to waterlogging. A relatively high nuclear DNA amount, and the presence of high concentrations of fructan in the rhizomes during winter, suggests that the species may be capable of growth at low temperatures; this hypothesis needs to be tested. The plants have a pungent odour and bitter taste, and may taint milk. *A.m.* is usually considered to be a grassland weed. In the *PGE* the species is most abundant in plots of low productivity and is suppressed by applications of ammonium salts or by liming. In poorer pasture *A.m.* may be an important food source for sheep. *A.m.* has also been used as a source of food, flavouring and medicine. *A.m.* reproduces vegetatively by means of long rhizomes to form clonal patches which expand at a rate of up to 200 mm year^{-1}. The eventual breakdown of the older rhizomes results in the production of a number of daughter plants. *A.m.* may also become established from pieces of rhizome detached as a result of soil disturbance. In ungrazed communities, seed is often produced in large quantities (1500 per flowering stem), and *A.m.* has potentially a high colonizing ability in ungrazed or lightly grazed habitats. The species is considered an important weed of arable land in New Zealand, but it does not appear to exploit farmland to the same extent in Britain. Seedlings, which are small, tend to establish only in rather open sites and in Britain regeneration by seed may not be important in established populations, but may be critical during initial colonization. Seeds germinate during autumn or the following spring. A transient seed bank has been reported in New Zealand. The flattened achenes are dispersed locally by wind and may survive ingestion by some animals. However, human activities probably provide the most important means of dispersal. The species is polymorphic and in Europe *A.m.* is included in a taxonomically complex grouping of eight species. Heavy-metal-tolerant races have been identified from lead-mine spoil.

Current trends *A.m.* is seldom an important grassland weed in Britain and current agricultural land-use does not favour its increase. Nevertheless, *A.m.* remains a common plant of pastures, waysides and lawns.

3 *Agrostis canina*

Formerly *A. canina* ssp. *canina*

Established strategy C–S–R.
Gregariousness Intermediate.
Flowers Green, hermaphrodite, wind-pollinated, self-pollination initially impeded.

Brown Bent-grass

Regenerative strategies V, B$_s$.
Seed 0.05 mg.
British distribution Widespread (*A.c.* and *A. vinealis* (formerly *A.c.* ssp. *montana*) are in 98% of VCs).

Commonest habitats Unshaded mire
20%, shaded mire 9%, river banks
5%, paths 4%, still aquatic 3%.
Spp. most similar in habitats *Carex
nigra* 80%, *Cirsium palustre* 78%,
Juncus effusus 73%, *Cardamine
pratensis* 72%, *Epilobium palustre*
71%.
Full Autecological Account *CPE* page
58.

Synopsis The most ecologically distinct
of the common *Agrostis* spp., *A.c.* is a
low-growing, stoloniferous wetland
grass virtually restricted to infertile,
acidic, peaty soils. The plant is
essentially a gap-colonizing,
subordinate species rather than a
potential dominant and shoot densities
may fluctuate widely from year to year.
Vegetative spread occurs by means of
long creeping stolons. In the growing
season *A.c.* extends aerial stems into
gaps in the canopy of stands dominated
by more-robust species, such as *Juncus
effusus*. *A.c.* may form floating rafts in
peaty pools from plants rooted on the
bank, and can form diffuse mats over
Sphagnum patches. The species may
increase in response to disturbance
such as ditch-cutting and light
trampling. Locally, *A.c.* forms virtual
monocultures in shaded habitats, but it
seldom flowers or sets seed in heavily-
shaded sites. Fragments of stolons
detached as a result of disturbance
readily root to form new plants. *A.c.*
also regenerates in vegetation gaps by
means of freshly shed seed in autumn
or spring and a persistent seed bank is
formed. The species is frequently an
effective colonist of sites linked by
water and colonization of, for
example, upland reservoirs from
water-borne plant fragments is
suspected. *A.c.* may also colonize
more isolated sites such as gravel pits.
No hybrids have been reliably reported
for Britain.
Current trends Apparently fairly
mobile and potentially favoured by
human activities such as peat-cutting
and sowing as a turf grass.
Nevertheless, *A.c.* is probably
decreasing as the result of habitat
destruction by drainage and
eutrophication.

4 *Agrostis capillaris*

Taxonomy outlined in 'Synopsis'

Established strategy C–S–R.
Gregariousness Patch-forming.
Flowers Brownish, hermaphrodite,
wind-pollinated, outbreeding.
Regenerative strategies V, B_s.
Seed 0.06 mg.
British distribution Widespread (100%
of VCs).

Common Bent-grass

Commonest habitats Limestone
pastures 61%, coal mine waste 50%,
acidic pastures 49%, lead mine spoil
45%, acidic wasteland/limestone
wasteland/meadows 43%.
Spp. most similar in habitats *Rumex
acetosella* 79%, *Luzula campestris*
69%, *Hypochaeris radicata* 68%,

Aira praecox 65%, *Deschampsia flexuosa* 60%.
Full Autecological Account *CPE* page 60.

Synopsis *A.c.* is a tufted, often patch-forming grass with an exceptionally wide ecological range. Grazing land and amenity grassland dominated by *A.c.* cover extensive areas of upland and lowland Britain, but *A.c.* is most characteristic of short turf in upland pastures on moderately acidic, brown earth soils. In this habitat the species is represented by genotypes of moderately high relative growth rate, and is capable of establishing a dense low leaf canopy which under heavy grazing is rapidly renewed during late spring and summer. Defoliation experiments reveal that under fertile conditions the resilience of *A.c.* under frequent cropping is related to the ability to produce very high densities of small tillers, many of which are situated close to the ground surface and escape damage. Although largely restricted to mildly acidic soils, *A.c.* extends on to calcareous N-facing slopes, perhaps for reasons similar to those suggested for *Potentilla erecta*. In the *PGE*, yields of *A.c.* were promoted by ammonium fertilizer and the species was 'very much discouraged by lime', by treatments leading to tall productive turf and by supply of superphosphate alone. Leaves have the low concentrations of Ca typical of many grasses. *A.c.* shows some tolerance of burning. It regenerates by means of rhizomes or stolons to form extensive patches, and both in upland pasture and elsewhere *A.c.* may assume the role of dominant due in large measure to an ability for rapid lateral spread. However, in keeping with a relatively low nuclear DNA amount, leaf growth is delayed compared with many other turf grasses, especially *Festuca rubra* with which the species frequently coexists. Shoots detached as a result of disturbance may give rise to new plants. *A.c.* also regenerates by seeds which germinate in autumn, in spring, or after incorporation into a persistent seed bank. Seed-set, though normally prolific, is irregular at higher altitudes. The persistence and colonizing ability of *A.c.* is in part a product of the diversity of regenerative strategies and of the mobility of seeds which are presumably mainly transported in soil. However, the wide success of the outbreeding *A.c.* is also related to its remarkable ability to develop genetically specialized populations. Populations differ in capacity for lateral spread and seed production. Upland populations and those from sandy areas are mainly rhizomatous, while those from lowland grassland are stoloniferous. Individuals with an inherently slower growth rate have been identified in unproductive habitats such as lead-mine spoil. *A.c.* is both genetically variable and phenotypically plastic in growth form; in consequence it is often difficult to identify taxa by vegetative characteristics. The ecological range of *A.c.* overlaps with all other native British species of *Agrostis* and *A.c.* hybridizes with other species. In particular, the hybrid with *A. stolonifera*, which has been said to be better attuned to certain grazed environments than its parents, may be common. Some field records included

under *A.c.* refer to hybrid taxa or to the recently introduced *A. castellana.*
Current trends *A.c.* has declined, particularly in lowland areas, with the destruction of permanent pasture. The species remains economically important in upland Britain. *A.c.* is a common species and an effective colonist of artificial habitats, particularly on mildly acidic soils. The development of ecotypes specialized towards new artificial habitats is resulting in an increase in the ecological range and genetic diversity of *A.c.*. The disturbance associated with modern land-use is likely to reduce the ecological isolation further and to increase the degree of hybridization between species of *Agrostis.*

5 *Agrostis stolonifera*

Includes some field records for *A. gigantea*

Established strategy C–R.
Gregariousness Potentially patch-forming.
Flowers Green, hermaphrodite, wind-pollinated.
Regenerative strategies V, B_s.
Seed 0.02 mg.
British distribution Widespread, especially in lowland (100% of VCs).
Commonest habitats Arable 66%, bricks and mortar 58%, soil heaps 55%, verges 48%, cinder heaps 46%.
Spp. most similar in habitats *Juncus articulatus* 64%, *Equisetum palustre* 63%, *Veronica beccabunga* 59%, *Stellaria alsine* 58%, *Nasturtium officinale* agg. 58%.
Full Autecological Account *CPE* page 62.

Synopsis *A.s.* is a short, fast-growing, stoloniferous, patch-forming grass which is common in an exceptionally wide range of fertile habitats. It is most abundant where the growth of tall dominants has been restricted by disturbance. *A.s.* exploits aquatic habitats and mire, woodland margins, most types of spoil heaps, moist grassland and arable land. *A.s.* is equally frequent in a range of maritime habitats. The roots and shoots show

Creeping Bent, Fiorin

'active foraging' in space; this explains the marked capacity of *A.s.* to exploit pockets of nutrient enrichment and canopy gaps. *A.s.* is moderately tolerant of ferrous iron toxicity. *A.s.* forms extensive clonal patches by means of long stolons. Detached shoots, which may be transported in soil, readily root to form a new plant, and this mode of regeneration appears to be important in disturbed habitats such as arable fields. The species is more palatable to stock than *A. capillaris*. *A.s.* also regenerates by seed in autumn, in spring or after incorporation into a persistent seed bank. This diversity of regenerative strategies doubtless contributes considerably to the success of *A.s.* as both a colonist of new sites and a survivor in disturbed habitats. The capacity to exploit a wide range of habitats probably also arises from the capacity of *A.s.* to form ecologically specialized populations in a manner similar to that of *A.c.*. Ecotypes tolerant of heavy metals have been detected in N Derbyshire on contaminated soils, and less-winter-green salt-tolerant populations occur on roadsides. It has been suggested that *A.s.* may be subdivided into var. *palustris* (most freshwater mire populations) and var. *stolonifera* (the remainder). No relationships between morphology and chromosome number have been detected. *A.g.* differs in morphology from typical *A.s.* in that (a) shoots are taller and more robust (foliage to 400 mm; flowers to 800 mm) and (b) it produces rhizomes and, consequently, tends to have a somewhat deeper root system. The first of these attributes enhances the capacity of *A.g.* to persist in tall vegetation, while the second may tend to restrict *A.g.* to well-drained soils. *A.g.* reaches its greatest frequency and abundance in cereal crops on sandy soils and exploits lighter, poorer soils than those on which *Elymus repens* is abundant. *A.g.*, unlike *A.s.*, is predominantly a lowland plant, occurring in 31% of British vice-counties, and is most abundant in arable fields and on wasteland on dry soils. *A.g.* produces seeds weighing *c.* 0.09 mg, and has a range of regenerative strategies similar to that of *A.s.*. The temperature range over which seeds germinate (7–34°C) suggests that *A.g.* has a greater potential to germinate in autumn than *A.s.*. Populations from wetlands are less frequent and have been afforded varietal status, and a further ssp. is recorded from Russia. *A.g.*, *A.s.* and *A.c.* are closely related and form hybrids with one another, perhaps quite frequently. *A.s.* also hybridizes with *Polypogon monspeliensis*.

Current trends Despite better prospects for their control by herbicides, both *A.s.* and *A.g.* are apparently increasing as a result of the intensification of cereal growing and the adoption of minimal cultivation. *A.s.* is in addition a common weed of productive pasture systems, a useful turf-grass species and has greatly expanded its habitat range.

6 *Agrostis vinealis*

Brown Bent-grass

Formerly *A. canina* ssp. *montana*

Established strategy C–S–R.
Gregariousness Intermediate.
Flowers Green, hermaphrodite, wind-pollinated, self-pollination initially impeded.
Regenerative strategies V, B$_s$.
Seed 0.06 mg.
British distribution General (98% of VCs).
Commonest habitats Limestone wasteland 22%, limestone pastures 21%, acidic wasteland 8%, acidic pastures 6%, soil heaps 4%.
Spp. most similar in habitats Insufficient data.
Full Autecological Account *CPE* page 64.

Synopsis *A.v.* is a shortly rhizomatous winter-green grass of acidic grassland and open habitats, generally found growing in close association with *A. capillaris*. Apart from the much narrower ecological range of *A.v.*, the biology of the two species is so similar and the overlap in their ecological and geographical distributions so complete that ecological reasons for the separation of the two species are hard to find. *A.v.*, which has lesser capacity for lateral spread than *A.c.*, and an earlier, and possibly less-protracted, flowering period, is considered drought-tolerant. Though obviously responsive to variations in nutrient levels, *A.v.* may have a greater ability to persist on acidic, relatively unproductive soils. Hybrids with *A.c.* and *A. stolonifera* have been reported. Listed as a turf-grass species but seldom seen in amenity grassland. The brevity of these comments is a reflection of our ignorance of the ecology of this elusive species.

Current trends Decreasing as a result of pasture improvement and upland dereliction.

7 *Aira praecox* Early Hair-grass

Established strategy S–R.
Gregariousness Intermediate.
Flowers Silvery-green, hermaphrodite, wind-pollinated.
Regenerative strategies S.
Seed 0.18 mg.
British distribution Widespread (100% of VCs).
Commonest habitats Acidic quarries 9%, coal mine waste/rock outcrops 6%, acidic wasteland/lead mine spoil 3%.
Spp. most similar in habitats *Rumex acetosella* 78%, *Hypochaeris radicata* 76%, *Erica cinerea* 74%, *Calluna vulgaris* 70%, *Carex pilulifera* 65%.
Full Autecological Account *CPE* page 66.

Synopsis *A.p.* is an early-flowering, winter-annual grass of short, open vegetation on thin, rocky or dry, sandy soils where the vigour of perennial species is restricted by the combined effects of summer drought and low soil fertility. Thus, the species is characteristic of habitats such as gritstone outcrops, cinder tips and sand pits, and other habitats with semi-permanent winter-annual communities. *A.p.* is the most common winter annual of calcifuge distribution, and in many acidic upland sites the species is the only winter annual present. In lowland habitats, two winter annuals which occur most commonly with *A.p.* on acidic soils are *Aphanes inexspectata* and *Ornithopus perpusillus*. Both flower and fruit later than *A.p.* and are tap-rooted, extending, for example, into short mown turf. Of the three, *A.p.* appears to adopt the most drought-avoiding strategy; germination occurs in mid-autumn, and the seedlings, which have a short root system, appear to be drought-sensitive. (The ecologically similar *Teesdalia nudicaulis* germinates a little earlier but produces a tap-root.) The persistence of seeds in the soil is not known, but in the absence of any obvious mechanisms of dormancy, we suspect that only transient seed banks occur. *A.p.* is widely distributed, e.g.

beside railways and in sand pits, and we suspect that the awned caryopses of *A.p.* are readily dispersed over long distances, probably through human agency.
Current trends Uncertain. *A.p.* is an effective colonist of artificial sites but has remained restricted to relatively infertile areas.

8 *Alisma plantago-aquatica* — Water Plantain

Established strategy Between C–R and R.
Gregariousness Sparse.
Flowers Pale lilac, hermaphrodite, self-compatible, insect- or wind-pollinated.
Regenerative strategies B_s, V.
Seed 0.27 mg.
British distribution Widespread except in the N (94% of VCs).
Commonest habitats Unshaded mire 12%, still aquatic 6%, shaded mire 3%.
Spp. most similar in habitats Insufficient data.
Full Autecological Account *CPE* page 68.

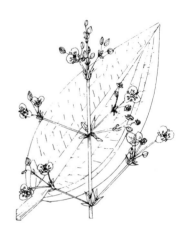

Synopsis *A.p-a.* exploits the boundary between aquatic and mire systems, e.g. ditch and pond margins. The leaves, though moderately large, are all basal, an arrangement which renders the species susceptible to competitive exclusion at sites where taller, productive, clonal, wetland dominants are established. Consequently, *A.p-a.* is promoted by ditch clearance and other disturbance factors. *A.p-a.* is particularly characteristic of sites with a fluctuating water-table. Here, young plants survive inundation as phenotypes in which all the leaves are unusually narrow and submerged. In shallow water young plants with floating leaves are also frequent. Except in the case of some young submerged plants, leaves die back in

autumn. The nuclear DNA amount in *A.p-a.* is relatively high and suggests that leaf growth in the spring may benefit from the expansion of tissues developed during the previous summer (cf. *Ranunculus bulbosus*). *A.p-a.* produces an acrid sap which may confer some defence against herbivory; this was formerly administered medically with dire consequences for the patient. In common with other members of the Alismataceae, a family of wetland species, *A.p-a.* possesses a poorly developed water-conduction system, with vessels confined to the roots. *A.p-a.* regenerates freely by seed, and the average plant produces *c.* 3600 achenes. Freshly-shed seed exhibit hard-coat dormancy. Germination, in autumn or spring, occurs mainly on bare mud, and massive populations have emerged subsequent to summer droughting. Some authors imply the existence of a persistent seed bank. *A.p-a.* is also capable of vegetative propagation through branching of the vegetative axis, but observations suggest that regeneration by seed is more important except perhaps in closed communities. Seeds may be effectively transported by water, can survive ingestion by birds and fish, and may adhere to birds or to clothing. *A.p-a.* is frequently an early colonist of newly-created wetland habitats.

Current trends Because of the extensive destruction of wetlands, *A.p-a.* cannot be considered an increasing species, but the species is not under threat and appears to be favoured by the combined effect of eutrophication and disturbance.

9 *Alliaria petiolata*

Hedge Garlic, Garlic Mustard, Jack-by-the-hedge

Established strategy C–R.
Gregariousness Robust plants occurring at low densities.
Flowers White, hermaphrodite, homogamous, visited by small insects, apparently self-pollinated.
Regenerative strategies S, ?B$_s$.
Seed 2.25 mg.
British distribution Common except in the N and W (91% of VCs).
Commonest habitats River banks 13%, verges 5%, soil heaps 4%, hedgerows 3%, paths/acidic woodland 2%.
Spp. most similar in habitats *Myrrhis odorata* 90%, *Calystegia sepium* 74%, *Impatiens glandulifera* 74%, *Galium aparine* 70%, *Petasites hybridus* 69%.
Full Autecological Account *CPE* page 70.

Synopsis *A.p.* is a tall, short-lived herb characteristic of river banks, woodland and hedgerows on relatively moist, fertile soils where the vigour of more-robust species is restricted by a combination of light to moderate shade and disturbance by factors such as occasional flooding on river banks. Although absent from grazed or trampled sites, *A.p.* responds to cutting by producing lateral flowering shoots and may persist on infrequently mown roadside verges. The leaves are characterized by unusually high concentrations of N, P, Ca, Fe and Na and smell of garlic when crushed. Their use in salads has been suggested. *A.p.* is usually monocarpic, an unusual feature among species of shaded habitats, and occurs in habitats in which tree litter is rapidly broken down. The seed contains a large embryo but a long period of chilling appears necessary to break dormancy.

Seed often remains dormant for 18 months, and sometimes longer. Thus, *A.p.* can be considered to have a moderately persistent seed bank. Seeds germinate in late winter or early spring and plants flower in the following year. Regeneration through the formation of adventitious buds on the roots has also been recorded, but appears to be of rare occurrence. Leaves are formed in both autumn and spring and leaf area is at a minimum in midsummer, when the tree canopy is fully expanded. This phenology is the same as that of *Anthriscus sylvestris*, which exploits a similar range of habitats. However, *A.p.* has a greater dependence upon seed for regeneration, and perhaps for this reason is restricted to more-disturbed sites. The species is also more frequent in sites subject to winter flooding, and extends less frequently into drier and unshaded habitats. Two chromosome numbers are widely recorded, $2n = 36$ and $2n = 42$; these have not been linked to variation in morphology or ecology.

Current trends *A.p.* is found in disturbed, fertile habitats and is more characteristic of secondary than of primary woodland. Furthermore, despite its comparatively large seeds and lack of well-defined mechanism of dispersal, *A.p.* frequently colonizes artificial habitats and is probably increasing.

10   *Allium ursinum*

Ramsons

Established strategy Between C–R and C–S–R.

Gregariousness Patch-forming.

Flowers White, hermaphrodite, protandrous, insect-pollinated or selfed.

Regenerative strategies S.

Seed 4.27 mg.

British distribution Widespread, local in the E (97% of VCs).

Commonest habitats Limestone woodland 21%, river banks 8%, broadleaved plantations/shaded mire 6%, acidic woodland/scrub 5%.

Spp. most similar in habitats *Anemone nemorosa* 88%, *Fraxinus excelsior* (Juv.) 86%, *Festuca gigantea* 84%, *Lamiastrum galeobdolon* 83%, *Sanicula europaea* 82%.

Full Autecological Account *CPE* page 72.

Synopsis *A.u.* is a bulb-forming woodland plant smelling pungently of garlic when crushed. *A.u.* has a specialized phenology and a narrow ecological range, and no ecotypes have been described. *A.u.* is characterized by a very high nuclear DNA amount

(63.0 pg), consistent with the vernal phenology and determinate growth habit. Bulbs and roots analysed in midwinter have exceptionally high concentrations of the reserve carbohydrate fructan. Like other vernal species, *A.u.* is essentially a shade-avoiding species: under continuously dense canopies growth is poor and flowering is often inhibited. *A.u.* is typical of moist, well-drained, base-rich soils and, particularly on alluvial terraces, the foliage contains relatively high concentrations of P. *A.v.* is sensitive both to waterlogging and to drought. Susceptibility to drought and restriction to relatively mild oceanic climates may be related in part to the poor system of water conduction in which vessels are confined to the roots. Further, leaves of *A.u.* persist longer in moist habitats, and during the cool moist summer of 1985 some leaves were still green in a Sheffield woodland during August, nearly two months after normal senescence. In Britain, *A.u.* extends into unshaded habitats in the moist climate of the W. The young shoot of *A.u.* can spear its way through dense litter and *A.u.* is one of the most successful exploiters of beech plantations, where litter accumulation is considerable and shade is dense in summer. Foliage is poisonous to stock if consumed in large quantities. *A.u.* regenerates mainly by seed and annual production may exceed 9000 seeds m^{-2} in dense stands. However, the seeds normally fall within 2.5 m of the parent. Plants usually take three years to flower and persist for *c.* eight years. Roots of seedlings are close to the surface, where supplies of macronutrients are high, while those of plants of reproductive age are concentrated in the less fertile A_2 horizon. *A.u.* is exceptional among species forming monospecific stands in that dominance does not result in differential mortality of the smallest individuals. A further ssp. is recorded from Europe.

Current trends Probably decreasing, since *A.u.* has poor mobility and is not a common colonist of sites distant from existing populations.

11 *Alnus glutinosa*

(Field records as seedlings)

Established strategy S–C.
Gregariousness Sparse.
Flowers Monoecious, wind-pollinated, self-sterile.
Regenerative strategies W.
Seed 1.3 mg.
British distribution Widespread (99% of VCs).
Commonest habitats River banks 8%, walls 6%, shaded mire/ unshaded. mire 4%, acidic woodland 2%.
Spp. most similar in habitats
Epilobium hirsutum 74%, *Cardamine flexuosa* 72%, *Petasites hybridus* 70%, *Phalaris arundinacea* 62%, *Dryopteris filix-mas* 59%.
Full Autecological Account *CPE* page 74.

Synopsis *A.g.* is a deciduous tree with diffuse porous wood, growing up to *c.* 20 m in height and living for a maximum of *c.* 300 years. Alders cover *c.* 11 000 ha of British woodland

Alder

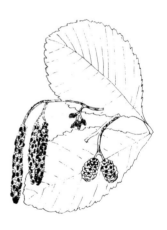

(> 1% of the total). Seedlings are found in the same types of habitat as the adult tree, i.e. on river and stream banks, in soligenous mire or forming more-extensive woodland on river floodplains. Elsewhere the species occurs with other trees in moist plateau

woodland. This close parallel between the distribution of seedlings and that of the canopy tree is related to the conditions required for seedling establishment. Radical elongation is slow in seedlings. As a result, seedlings are very susceptible to droughting as well as to cold periods in early spring. Further, high humidity and high oxygen tension are required for germination, and seedling establishment is dependent upon high light intensity. These critical factors limit the ecological range of the adult tree. Saplings have the potential to survive on substrates, e.g. colliery spoil, where surface layers may become severely desiccated. In general, mature alder woods do not provide conditions suitable for regeneration; even-aged populations of seedlings often appear, but these usually die off, to be replaced by other species. Thus, *A.g.* may be regarded as a seral species. Succession of wetlands involving *A.g.* and *Salix cinerea* tend to occur under more nutrient-rich conditions than those associated with *Betula pubescens*. Patterns of succession are at present poorly understood. Seeds are effectively dispersed either by wind-drift over standing water or in water currents, and consequently linear populations are common along the strand-line. However, it has been suggested that the number of seeds dispersed by wind > 30 m from the parent tree is normally insufficient to ensure establishment. Seedlings are more tolerant of waterlogging than those of *Betula pubescens*, but they are more restricted to continuously moist habitats. Cold treatment reduces the minimum temperature for germination to 7°C and seeds germinate in early spring. No persistent seed bank is formed. The tree roots contain the nitrogen-fixing actinomycete bacterium *Frankia* spp. within root nodules, and the nitrogen content of the tree may double between February and May. The fixation of nitrogen is an oxygen-demanding process, and nodules are restricted to the upper regions of the soil. The roots have the capacity to diffuse oxygen and are at a wide range of depths, including some which are deeply penetrating and tolerant of reducing conditions. As a result *A.g.* can survive in conditions of changing water-table better than most other trees, and produces adventitious roots from the bole in response to flooding. As in many trees, the developmental processes of seed formation are slow. Catkins are formed in July; pollination occurs the following February to March; ovules appear in June; fertilization occurs in July to August; seed ripens at end of September. Unlike *Betula* spp., the female catkins are not enclosed in leaf buds during winter, and this, coupled with the early pollination and extended period before fertilization, may restrict the capacity of *A.g.* to set seed at high altitudes and under severe winter or early spring conditions. The percentage of good seeds may be low, but mast years do occur. Although found over a wide range of pH (3.4–7.2), the species is more frequently observed on non-calcareous strata. This distribution does not appear to hold for continental Europe, where *A. incana* prevails on calcareous soils. Compared with large willows such as *Salix fragilis*, the species appears to be more tolerant of acidic, less-fertile soils and shaded conditions, and is less capable of exploiting fertile, base-rich wetland sites. However, further studies are required to elucidate the factors regulating the distribution of alders and willows. *A.g.* forms a relatively dense canopy and its litter is not usually persistent. The number of insects associated with *A.g.* is rather low for a native deciduous tree, suggesting that the leaves are relatively palatable. No inducible defence system against insect predators has been identified. *A.g.*, which is one of the faster-growing trees, can be used for reclamation of coal-mine spoil, but certain introduced alders which grow even more rapidly are generally

preferred. The hybrid with the introduced *A.i.* is widely planted and may have arisen spontaneously.
Current trends Plantations of *A.g.* occur in certain low-lying areas but the species is no longer widely planted and

is probably decreasing through the destruction of habitats in many lowland districts. Remains relatively common in river- and stream-side habitats.

12 *Alopecurus geniculatus*

Marsh Foxtail

Established strategy C–R.
Gregariousness Intermediate.
Flowers Grey-purple, hermaphrodite, wind-pollinated, self-compatible.
Regenerative strategies V, ?S.
Seed 0.38 mg.
British distribution Widespread (99% of VCs).
Commonest habitats Unshaded mire 10%, still aquatic/coal mine waste/manure and sewage waste/enclosed pastures 3%.
Spp. most similar in habitats Insufficient data.
Full Autecological Account *CPE* page 76.

Synopsis A relatively common species which has not received sufficient attention from plant ecologists. *A.g.* is a low-growing grass, usually associated with fertile sites in which the growth of potential dominants is restricted by inundation by flood water in winter and sometimes also by exposure to summer grazing. *A.g.* closely resembles *Agrostis stolonifera* in growth form, and the two frequently occur together. There is considerable difficulty in distinguishing between seedlings of *A.g.* and *A.s.* under field conditions. However, apart from the fact that *A.s.* is ecologically wide-ranging, the differences between the two species are slight. Within the Sheffield region the only common habitat which *A.g.* regularly exploits to a greater extent than *A.s.* is the margin of reservoirs. *A.g.* is often a dominant species on bare mud exposed as a result of a fall in water level during

summer. We suspect that the higher nuclear DNA amount of *A.g.* (14.2 pg) relative to that of *A.s.* (7.0 pg) is associated with an ability to exploit wetland sites while they are still cool and wet, during spring and early summer. Consistent with this hypothesis is the fact that *A.g.* flowers and sets seed earlier than *A.s.*. Whether the two species differ also in rooting depth and tolerance of anaerobic soil conditions is not known. *A.g.* has also been observed as a dominant of wet mowing meadows, where its abundance may be related to a capacity to set seed early, before the grass is cut. Like *A.s.*, *A.g.* often forms extensive carpets by means of long stolons (up to 450 mm). Shoots detached as a result of disturbance also readily re-root to form new plants. Seeds germinate from April to September. A persistent seed bank might be predicted in view of the disturbed nature of the habitats exploited by *A.g.*. Although one has been recorded, several other studies have failed to confirm such observations. It is also not clear to what extent the shoots and foliage of *A.g.* are capable of overwintering underwater along reservoir margins. Hybrids have been recorded with *A.*

aequalis and *A. pratensis*.

Current trends Uncertain. In view of its wetland distribution the species is likely to decrease as a consequence of habitat destruction. It is not clear to what extent this trend is offset by the capacity of *A.g.* to function as an effective colonist of disturbed artificial habitats.

13   *Alopecurus pratensis*

Meadow Foxtail

Established strategy Between C–S–R and C.

Gregariousness Sparse to intermediate.

Flowers Grey-purple, hermaphrodite, protogynous, wind-pollinated, usually self-incompatible.

Regenerative strategies S, ?V.

Seed 0.71 mg.

British distribution Widespread except in mountainous regions (100% of VCs).

Commonest habitats Meadows 27%, bricks and mortar/enclosed pastures 6%, hedgerows 5%, lead mine spoil 3%.

Spp. most similar in habitats *Phleum pratense* 84%, *Festuca pratensis* 79%, *Rhinanthus minor* 78%, *Cynosurus cristatus* 76%, *Ranunculus acris* 73%.

Full Autecological Account *CPE* page 78.

Synopsis *A.p.* is a loosely-tufted winter-green grass most characteristic of fertile hay meadows on moist soils. *A.p.* is moderately productive and may on occasion outyield *Lolium perenne*. The species responds readily to a plentiful supply of nutrients provided sufficient lime is applied. It requires abundant nitrogen and thrives on the unlimed areas of plots receiving sodium nitrate but requires lime before it can take advantage of nitrogen supplied as ammonium sulphate. *A.p.* grows rapidly in early spring and, except for *Anthoxanthum odoratum*, is

the earliest flowering of the common perennial grasses. Early spring growth and the ability for growth under moderate shade are probably important dimensions of the niche of *A.p.* in grasslands, and may also enable *A.p.* to persist, for example, at woodland margins. *A.p.* has a rather narrow ecological distribution and is very much restricted to moist, well-drained soils, seldom occurring in sites which are waterlogged during summer, and excluded from very dry soils. In a classical competition experiment conducted along an experimental gradient of soil water supply, it was shown that in monoculture *A.p.* attained maximum vigour in moist but not waterlogged soils, but tended to be displaced from such conditions if grown in competition with other species. *A.p.* is moderately tolerant of grazing and cutting, but not of heavy trampling. *A.p.* regenerates by seeds which are shed in summer, often before hay is cut. No buried seed bank has been reported. *A.p.* was formerly sown in seed mixtures to a considerable extent and cultivars were developed, but now *A.p.* has largely been replaced by *Phleum pratense* ssp. *pratense*. In tall vegetation *A.p.* is, however, more persistent than *Phleum pratense*, a capacity which appears to involve some vegetative regeneration. In common with several other aspects of the biology of *A.p.*, this phenomenon has not been adequately investigated. *A.p.* hybridizes with *A. geniculatus* and a further ssp. is recognized in Europe.

Current trends *A.p.* has decreased through a decline both in its agricultural importance and in the acreage of suitable habitats. The amount of seed sown annually is now negligible. Although found to some extent in new artificial environments, the capacity of *A.p.* to exploit recently created environments is likely to be restricted by the absence of both a persistent seed bank and by the lack of an effective method of vegetative spread.

14   *Anagallis arvensis*

Scarlet Pimpernel, Shepherd's Weatherglass

Ssp. *arvensis*

Established strategy Between R and S–R.
Gregariousness Typically scattered.
Flowers Usually red, hermaphrodite, self- or more rarely insect-pollinated.
Regenerative strategies B_s.
Seed 0.40 mg.
British distribution Widespread except in the extreme N (100% of VCs).
Commonest habitats Arable 16%, limestone quarries 5%, paths/rock outcrops 2%.
Spp. most similar in habitats *Spergula arvensis* 91%, *Fallopia convolvulus* 91%, *Veronica persica* 88%, *Myosotis arvensis* 87%, *Sinapis arvensis* 81%.
Full Autecological Account *CPE* page 80.

Synopsis Although restricted to lowland sites in Britain, *A.a.* is relatively cosmopolitan and is classified as one of the world's worst weeds. In warm climates *A.a.* usually behaves as a winter annual, but in cooler regions the species is a summer annual. In Britain *A.a.* is probably a summer annual, and in arable fields, a favoured habitat, it is more frequent in spring-sown crops. The population dynamics of *A.a.* may be complex, since polycarpic perennial individuals have been observed in fallow fields. The possibility that the species may behave locally as a winter annual deserves investigation, particularly since germination may occur in the autumn. *A.a.* is highly toxic to stock. Because

A.a. can germinate and grow at low temperatures, and its seedlings persist under moderate shade, this low-growing species can exploit the phase before the cereal canopy is fully established. *A.a.* is excluded from most broad-leaved crops, perhaps because of shading by the crop and by other weed species. Like *Veronica persica*, *A.a.* typically produces procumbent indeterminate shoot growth, with flowers and fruits formed sequentially in the axils of the leaves. The flowers of *A.a.* tend to open in full sunlight and remain closed on dull or rainy days, earning the species its common name 'Shepherd's Weatherglass'. On average, one plant produces less than 1000 seeds; this appears to be rather few for a common arable weed. Seeds are released close to the soil surface; this probably facilitates their incorporation into a persistent seed bank. Seed is widely dispersed as a result of agricultural management, particularly as a contaminant of crop seeds, and can survive ingestion by birds. Consequently, *A.a.* often becomes established in new open habitats such as quarry spoil, road and rail cuttings and, more rarely, disturbed road verges and other transient habitats. Germination of freshly-shed seeds is said to be prevented by the presence of a water-soluble inhibitor. A range of other environmental controls of germination appear to exist, including a requirement for light, but their exact ecological significance remains unclear. Except when growing at low temperatures, *A.a.* is a long-day plant and the main flowering season, June to August, is more restricted than that of most arable weeds. Exploitation of some sites may be restricted by the strongly seasonal pattern of flowering, which may explain the rarity of *A.a.* in gardens. Factors controlling the distribution of *A.a.* in rocky habitats and sand dunes (the only semi-natural habitat with which the species is commonly associated) are poorly understood. The species is genetically variable, and plants with blue flowers predominate near the Mediterranean. Populations from sand dunes are deep-rooted, fleshy-leaved and usually form compact tufts. The rare ssp. *foemina* (*A.foemina*) with blue flowers, produces fewer, heavier seeds and is a summer annual. Hybrids between the two taxa are reported.

Current trends Often associated with rarer arable weeds, e.g. *Euphorbia exigua*, and probably decreasing as weed-control measures on arable land become more effective.

---

15 *Anemone nemorosa*

Wood Anemone

Established strategy Between S and S–R.
Gregariousness Often patch-forming.
Flowers Typically white, hermaphrodite, self-incompatible, insect-pollinated.
Regenerative strategies V.
Seed 0.99 mg.
British distribution Common except in Ireland (99% of VCs).
Commonest habitats Limestone woodland 40%, limestone pastures 17%, scrub 12%, broadleaved plantations 10%, acidic woodland 8%.
Spp. most similar in habitats *Mercurialis perennis* 94%, *Brachypodium sylvaticum* 92%, *Sanicula europaea* 90%, *Allium ursinum* 88%, *Melica uniflora* 86%.
Full Autecological Account *CPE* page 82.

Synopsis *A.n.* is typically a rhizomatous, patch-forming, vernal herb of less-fertile woodlands. As might be predicted from its high nuclear DNA amount, the phenology of *A.n.* involves shoot expansion during the late winter and early spring and exploitation of the light phase before canopy development by neighbouring trees and shrubs; by mid-July the above-ground portions of the plant have died back. The emergent shoot is crozier-shaped. This affords protection to the flower bud, which was formed in the previous autumn, and often enables the shoot to push its way through tree litter. The same phenomenon occurs in some derelict grasslands, and high population densities have been observed in coarse turf with dense accumulations of *Brachypodium pinnatum* litter. Differences in ecology between *A.n.* and the other common, shallow-rooted vernal species, *Ranunculus ficaria*, are in part attributable to the more erect morphology of *A.n.*: this permits the species to penetrate litter and to rise through closed canopies much more effectively. The species has a shallow root system and is most vigorous under moist conditions where leaves may be produced up to ten days earlier than on adjacent dry soils. Further, in the moist climates of the N and W the ecological range of *A.n.* is greater, and the species extends into grassland and other open habitats. *A.n.* may occur in lightly grazed habitats. The plant contains protoanemonin and is unpalatable to stock, but it can survive occasional defoliation. *A.n.* is also capable of exploiting some grasslands managed by burning and may escape the impact of summer fires. The distribution of *A.n.* shows a considerable overlap with that of *Hyacinthoides non-scripta* and *Mercurialis perennis*. However, *A.n.* is more tolerant of woodland disturbance and responds to coppicing by a marked increase in flowering. In marked contrast to *H.n-s.*, the roots are superficial and *A.n.* frequently exploits soils where the water-table is close to the surface. In common with *H.n-s.*,

but in contrast with *M.p.*, the species extends onto both acidic and calcareous soils. With Rackham, we postulate that *A.n.* is confined to situations where relatively infertile soils coincide with factors (e.g. management regime, hydrology) which restrict the vigour of both *H.n-s.* and *M.p.*. The leaves of *A.n.* are relatively low in P, but often contain rather high concentrations of Ca, Mg, Fe and Al. *A.n.* regenerates mainly by means of rhizome growth to form large but slow-growing clonal patches. It is suspected that in fragmented form genets may survive for hundreds of years and may thus persist for as long as some of the trees and shrubs in their overstorey. Although *A.n.* is self-incompatible, seed-set occurs regularly, but it is higher in woodland than in grassland sites. The achene contains a minute embryo, and has a prolonged requirement for moist conditions before germination will take place. Germination is particularly high after severe winters. No persistent seed bank has been reported. Development of the seedling and young plant is very slow, and individuals may not flower until ten years old. The seed is not dispersed far from the parent plant. A maximum distance of 130 mm has been reported in one population. Thus, *A.n.* has a low colonizing ability and is most typically associated with older areas of woodland or grassland. Several varieties and complements of chromosome number have been reported.

Current trends A poor colonist. Remains relatively common, but is decreasing, certainly in grassland habitats and probably in some woodlands.

16 *Angelica sylvestris*   Wild Angelica

Established strategy Between C and C–R.
Gregariousness Sparse.
Flowers White or pink-tinged, hermaphrodite, protandrous, insect-pollinated.
Regenerative strategies ?S.
Seed 1.15 mg.
British distribution Widespread (100% of VCs).
Commonest habitats River banks 25%, limestone wasteland 18%, shaded mire 12%, limestone woodland 10%, cinder heaps/lead mine spoil/unshaded mire/verges 5%.
Spp. most similar in habitats *Impatiens glandulifera* 81%, *Festuca gigantea* 71%, *Cardamine flexuosa* 65%, *Filipendula ulmaria* 65%, *Solanum dulcamara* 60%.
Full Autecological Account *CPE* page 84.

Synopsis *A.s.* is a robust species of moist, often relatively fertile, sites and has a wide habitat range which includes damp grassland, mire and woodland margins. Although generally considered to be a polycarpic perennial, we suspect that in some shaded habitats *A.s.* is often monocarpic. Observations of the species in winter indicate that a large proportion (sometimes > 50%) of the plants die after flowering. At maturity, individuals produce a small number of massive basal or near-basal leaves up to 600 mm in length and a large flowering stem which may reach 2 m. Despite the large size, the capacity for dominance is low compared with wetland species such as *Epilobium*

hirsutum which have the added advantages of rapid vegetative spread and a taller, more densely packed leaf canopy. *A.s.* achieves rapid early growth through the mobilization of reserves from the stout rootstock, and benefits where management allows the development of a larger summer peak in shoot biomass. Leaf tissue is rich in Ca. In sites subject to frequent mowing or grazing, *A.s.* is usually represented by seedlings or stunted individuals. Sites conducive to large populations of *A.s.* are either periodically disturbed, e.g. river and ditch banks, woodland margins and mire subject to variable winter flooding, or situated in habitats such as coarse grassland where grazing has recently been relaxed. It is not known to what extent the perpetuation of *A.s.* in grassland is due to cutting regimes, but we suspect a dependence similar to that described for another, much rarer, wetland Umbellifer

59

Peucedanum palustre. Regeneration is entirely by seed. Populations often consist of a few mature flowering individuals associated with many young plants, each with one or two small leaves and existing in an apparently suppressed condition beneath the vegetation canopy. Studies are needed to ascertain whether these young plants constitute a bank of persistent seedlings which are dependent upon the subsequent disturbance of the vegetation in order to reach maturity. Seeds, which contain a rather small embryo, have a chilling requirement and germinate in the spring. No persistent seed bank has been reported. Seeds float and long-distance dispersal by this means along waterways may be important in the spread of the species.

Current trends Uncertain, but the future of the species in habitats such as river banks appears secure.

17 *Anthoxanthum odoratum* Sweet Vernal-grass

Established strategy Between S–R and C–S–R.
Gregariousness Intermediate.
Flowers Green, hermaphrodite, protogynous, wind-pollinated, self-incompatible.
Regenerative strategies S, B$_s$.
Seed 0.45 mg.
British distribution Widespread (100% of VCs).
Commonest habitats Limestone pastures 57%, meadows 48%, limestone wasteland 28%, scree 23%, lead mine spoil 19%.
Spp. most similar in habitats *Luzula campestris* 76%, *Ranunculus bulbosus* 74%, *Avenula pubescens* 73%, *Briza media* 71%, *Carex caryophyllea* 69%.
Full Autecological Account *CPE* page 86.

Synopsis *A.o.* is a relatively slow-growing, short, loosely-tufted, winter-green grass typically occurring as scattered individuals in a wide range of grasslands and, to a lesser extent, in open habitats particularly on slightly acidic soils. In the *PGE* the species was said to flourish best on 'well-manured and acid soil' and was 'usually reduced by lime'. From the same source it is also evident that *A.o.* is not encouraged by ammonium salts alone and is suppressed by heavy applications of nitrogenous fertilizer. *A.o.* is thought to be relatively unpalatable but was formerly included in commercial seed mixtures and is preferred by stock to species such as *Festuca ovina*, despite the presence of coumarin. *A.o.* is the earliest-flowering common pasture grass and has distinct spring and autumn peaks of vegetative growth. Perhaps as a result of this phenology and its relatively short life-span. *A.o.* is sensitive to defoliation and to shading in tall derelict grassland. Thus, *A.o.* reaches maximum abundance in damp pastures and meadows of low to moderate fertility. Regeneration is mainly by seed, which is shed during early summer and may be incorporated into a persistent seed bank. Under favourable conditions effective seed production occurs over a wide range of temperatures, and we suspect that seedlings originating from surface-lying or buried seeds play a key role in colonization of gaps where these are caused by factors such as trampling, grazing and urine scorching. Less importantly, *A.o.* has a limited capacity for tillering but individuals retain a distinctly tufted habit. Under laboratory conditions, detached vegetative shoots of *A.o.* readily produce roots and can form new plants, but there is no evidence that this occurs in the field. Ecotypes differing in morphology and mineral nutrition have been reported and evolutionary changes in populations may occur extremely rapidly.

Consistent with classical theory, the genetic variability of *A.o.* and its capacity to exploit a fairly wide range of habitats coincide with a short life-span, rapid population turnover and the tendency to be strongly outbreeding. British plants appear to be tetraploid. Plants with $2n = 10$, $(2x)$, occur mainly in the N part of the European range of *A.o.*.

Current trends Continued abundance is favoured by the capacity of *A.o.* to evolve rapidly in new types of habitat. Opposing this, however, is the association of the species with less fertile habitats. Thus, a maintained genetic diversity, coupled with a decrease in the overall abundance of *A.o.*, is predicted.

18   *Anthriscus sylvestris*

Cow Parsley, Keck

Established strategy C–R.
Gregariousness Intermediate.
Flowers Creamy-white, insect-pollinated.
Regenerative strategies S.
Seed 5.18 mg.
British distribution Widespread (100% of VCs).
Commonest habitats Verges 31%, hedgerows 27%, meadows 24%, river banks 20%, soil heaps 11%.
Spp. most similar in habitats *Galium aparine* 78%, *Stachys sylvatica* 77%, *Heracleum sphondylium* 68%, *Alliaria petiolata* 63%, *Myrrhis odorata* 61%.
Full Autecological Account *CPE* page 88.

Synopsis *A.s.* is a winter-green, tap-rooted competitive-ruderal which exploits a range of moist or shaded fertile habitats. *A.s.* has a cool-season phenology with leaf canopy reaching a minimum during summer. This phenology, coupled with early seed-

set, appears to contribute to the ability of *A.s.* to persist in open woodland, a habitat in which *A.s.* is an important constituent in parts of N Europe. *A.s.* occurs in hay meadows and is able to coexist on roadsides and in hedgerows with tall grasses such as *Arrhenatherum elatius* and *Elymus repens*, which have later (summer) peaks in biomass. However, *A.s.* is sensitive to regular cutting, and in roadsides and hedgerows – the major habitats of the species – is often restricted to the less-intensively managed areas close to the hedge, where it may form a conspicuous white zone of flowering shoots in early spring. *A.s.*, which is eaten by cattle and rabbits, is also intolerant of heavy grazing and trampling, and does not occur on heavily droughted or waterlogged soils, although it is characteristic of productive flooded meadows in reclaimed land in Holland. Regeneration is rather specialized. *A.s.* is a 'biennial' which perennates by buds in the axils of the basal leaves. These produce new plants when the flowering stem dies, and thus represent a means both of perennation and of vegetative reproduction. The tall flowering stems dry out during the summer and seed is lost over a prolonged period. Seeds contain a poorly differentiated embryo when shed, have a chilling requirement and germinate synchronously in spring. Under productive conditions the relative growth rate of the seedling is slow, but this has been shown to result from the early allocation of photosynthate to the swollen root-stock which forms the capital for shoot expansion during the second growing season. The leaves are relatively rich in P and Ca. *A.s.* is a 'follower of humans' and river banks are perhaps the only semi-natural habitat in which the species occurs. It has been speculated that the common northern form, which has been called var. *angustisecta*, may be indigenous, while the southern plant (var. *latisecta*) may have immigrated later as a result of human activities.

Current trends The specialized phenology and regenerative biology, including the absence of a persistent seed bank, renders *A.s.* rather vulnerable to seasonally unpredictable disturbance. Nevertheless, *A.s.* is common and likely to increase due to its capacity to exploit artificial linear sites such as roadsides, which provide corridors for population expansion.

19   *Aphanes arvensis*   Parsley Piert

Established strategy Between R and S–R.
Gregariousness Sparse.
Flowers Green, hermaphrodite, apomictic.
Regenerative strategies S, B_s.
Seed 0.18 mg.
British distribution Widespread (100% of VCs).
Commonest habitats Arable 14%, rock outcrops 9%, limestone quarries 8%, limestone pastures/paths 4%.
Spp. most similar in habitats *Anagallis arvensis* 80%, *Veronica arvensis* 70%, *Spergula arvensis* 64%, *Veronica persica* 63%, *Fallopia convolvulus* 63%.
Full Autecological Account *CPE* page 90.

Synopsis *A.a.* is a polyploid, annual herb associated with a wide range of disturbed habitats. These include relatively infertile, shallow calcareous and dry, sandy soils on which the cover of perennials is restricted by summer drought. Immediate germination is prevented by an after-ripening requirement, and germination is delayed until autumn. The overwintering plant flowers apomictically in spring and early summer. Like *Trifolium dubium*, *A.a.* appears to occupy microsites with deeper soil, and is one of the last species to succumb to drought. Some plants may even persist until autumn during wet summers. With *T.d.*, *A.a.* extends into more-fertile, managed habitats, e.g. lawns, where the vigour of perennials is restricted by frequent close mowing. *A.a.* also exploits sites where the cover of perennials is restricted by annual cultivation, particularly for winter cereals. In this habitat *A.a.* appears to occupy a low, shaded stratum during summer, and it seems probable that growth occurs mainly in the autumn and spring before the crop is fully established, and again in late summer after the crop has been harvested. In further contrast with many other small winter annuals, *A.a.* often colonizes heavy, water-retaining clays. The capacity to exploit lawns and arable fields, and habitats subject to summer drought, may stem from the fact that unlike (for example) *Myosotis ramosissima*, *A.a.* is not obligately winter-annual. A secondary peak in germination often occurs during spring on arable land. Even so, *A.a.* is probably discouraged by the cultivation of spring-sown crops. *A.a.* is entirely dependent upon seed for regeneration, and produces seeds which are large compared with those of most winter-annuals but smaller than average among arable weeds. Buried seeds of *A.a.* show seasonal changes in germination requirements and the species forms a large and persistent seed bank. It is not known whether the ability of *A.a.* to exploit three very different types of habitat in Britain is under genotypic or phenotypic control. The related sexual, diploid species, *A. inexspectata*, occurs on dry sandy soils. Except in arable fields, where the distributions of the two overlap, *A.i.* largely replaces *A.a.* on the sandy soils on the Bunter sandstone of NE England. *A.i.* occurs within the pH range 4.0–7.5 and is more frequent towards the middle to upper end of this range.

Current trends Uncertain, but apparently favoured by reduced cultivation and direct drilling.

20 *Apium nodiflorum*

Fool's Watercress

Established strategy C–R.
Gregariousness Intermediate, sometimes stand-forming.
Flowers Greenish-white, hermaphrodite, insect-pollinated.
Regenerative strategies V, ?S.
Seed 0.53 mg.
British distribution Common except in Scotland (91% of VCs).
Commonest habitats Running aquatic 19%, shaded mire 9%, still aquatic 8%, unshaded mire 5%, river banks 3%.

Spp. most similar in habitats
  Nasturtium officinale agg. 99%,
  Veronica beccabunga 93%,
  Callitriche stagnalis 92%, *Equisetum palustre* 91%, *Glyceria fluitans* 89%.
Full Autecological Account *CPE* page 92.

Synopsis *A.n.* is a moderately robust, prostrate to ascending stand-forming species which straddles the boundary between aquatic and mire habitats, and may behave as an emergent or, more rarely, submerged aquatic. The species occurs in fertile sites where the growth of potential dominants is restricted by disturbances such as ditch clearance and erosion by water currents. *A.n.* is frequent in ditches, at the margins of ponds, lakes, canals and slow-moving rivers and streams, on both calcareous and non-calcareous soils. Roots, mainly located at the basal internodes, provide good anchorage, and the species is also characteristic of shallow, potentially fast-flowing streams on chalk and limestone. Many of these streams often dry up, partly or completely, during summer, providing conditions which *A.n.* appears better able to exploit than *Nasturtium officinale*. The species is infrequent in grazed sites. Typically *A.n.* develops an erect growth form, but shoots in contact with the ground readily produce adventitious roots. Seedlings may become established either in mud or under water from early spring onwards, but seldom survive except in very open areas and appear to be vulnerable to competition from established plants. Consequently, effective regeneration is mainly by vegetative means and detached vegetative shoots, which in water develop new roots within two days under laboratory conditions, are probably widely dispersed to new sites in water currents. The commonest semi-aquatics of similar stature and

ecological distribution to *A.n.* are *N. officinale* agg. and *Veronica beccabunga*. *A.n.* appears more commonly associated with fluctuating water levels in still-water habitats than *V.b.*. Perhaps associated with this latter difference, *A.n.* shows much greater die-back of shoots in autumn than *V.b.*. Differences from *N.o.* are described in the account for this species. *A.n.* is phenotypically plastic, and phenotypes resembling *A. repens* may occur in short turf. *A.n.* hybridizes with *A. inundatum*.
Current trends Uncertain. Tolerant of eutrophic conditions, but many habitats formerly occupied by the species have been destroyed by drainage and other factors related to land-use.

21   *Arabidopsis thaliana*

Thale Cress

Established strategy S–R.
Gregariousness Sparse.
Flowers White, hermaphrodite, homogamous, sometimes gynomonoecious, insect- or self-fertilized.
Regenerative strategies S, B$_s$.
Seed 0.02 mg.
British distribution Common except in the N and W (93% of VCs).

Commonest habitats Rock outcrops 9%, limestone quarries 3%, limestone pastures/paths/soil heaps/acidic wasteland 2%.
Spp. most similar in habitats *Saxifraga tridactylites* 88%, *Trifolium dubium* 82%, *Myosotis ramosissima* 79%, *Sedum acre* 78%, *Geranium molle* 78%.
Full Autecological Account *CPE* page 94.

Synopsis *A.t.*, which is normally a winter annual, is most characteristic of shallow rocky or sandy soils subject to summer drought. Both the association with S-facing slopes and an extremely low nuclear DNA amount suggest that vegetative development, which occurs in the cool parts of the year, depends upon the capacity for opportunistic growth in brief warmer periods. Flowering and seed-set occurs later than in some other winter annuals, such as *Cardamine hirsuta*, *Erophila verna*, *Myosotis ramosissima*, *Saxifraga tridactylites* and *Veronica arvensis*. Accordingly, *A.t.* persists only in less-droughted areas, often on deeper soil, and in open microsites within perennial grassland. Populations in grasslands fluctuate from year to year in inverse relation to the cover of perennials. In years following a severe drought, large populations, often containing individuals of unusually large size, may be observed. Elsewhere *A.t.* occurs sporadically as a weed of gardens and park flower-beds. The species also occurs in tree nurseries and on railway ballast, two habitats in which colonization by perennials is often restricted by herbicide treatment. Seed is released in early summer and is prevented from germinating in summer by dormancy, which is removed by exposure to warm temperatures for several weeks. A persistent seed bank is also formed, and this appears to act as a buffer against local extinction. According to season and site, individuals vary considerably in size and seed production; much of this variation appears to be phenotypic but equally there is much genetic variation, particularly between populations. Thus, plants from a limestone daleside, a habitat where the period suitable for growth varies little from year to year, have been shown to require

vernalization before flowering whereas a park population showed no such requirement. Plants of *A.t.* occasionally seen in flower during late summer on cultivated ground in Britain may belong to this latter type of population, arising from seed germinating from a buried seed bank following spring or summer disturbances of the soil.

Summer-annual races of *A.t.* have been recorded, and it is suspected that in some tree nurseries *A.t.* may have been introduced from continental Europe along with container plants. However, in some populations (perhaps even a majority) seed dormancy is induced during winter, and as a result germination only occurs in autumn. The small size and short life-cycle of some genotypes, coupled with the low chromosome number ($2n = 10$), have made *A.t.* a useful subject for genetic studies, and a range of genotypes and mutants circulates widely between laboratories.

Current trends *A.t.* is likely to increase further, both in abundance and in the range of habitats exploited.

Established strategy Between S and
  S–R.
Gregariousness Sparse.
Flowers White, hermaphrodite,
  homogamous, normally
  self-pollinated.
Regenerative strategies S, ?B_s.
Seed 0.09 mg.
British distribution Scattered (86% of
  VCs).
Commonest habitats Scree 14%, rock
  outcrops 11%, bricks and mortar
  6%, limestone quarries 5%,
  cliffs/limestone pastures/limestone
  wasteland 4%.
Spp. most similar in habitats *Sedum
  acre* 76%, *Scabiosa columbaria* 70%,
  Senecio jacobaea 65%, *Leontodon
  hispidus* 64%, *Hieracium pilosella*
  63%.
Full Autecological Account *CPE* page
  96.

Synopsis *A.h.* is a winter-green herb of
infertile, calcareous soils. *A.h.* forms a
low-growing rosette and is confined to
relatively open microsites. The species
lacks the capacity for lateral vegetative
spread and regeneration of this short-
lived perennial is entirely by seed,
which germinates mainly in autumn.
As in many other plants of disturbed
habitats, large quantities of seed are
produced and a robust plant may
release in excess of 2000 seeds. *A.h.*
may increase in abundance following
burning and the development of a
persistent seed bank is suspected.
However, flowers and seeds are borne
on a tall stem, up to *c.* 600 mm. Thus,
flowering individuals of *A.h.* are
vulnerable to the effects of grazing and
in pasture the species is particularly
associated with those outcrops and
scree margins which are less accessible
to stock. The low stature of the leaf
canopy renders *A.h.* liable to
dominance by taller species, and

probably accounts in part for the
restriction to rocky sites and grassland
on shallow soils. *A.h.* attains peak
above-ground biomass in summer and
we suspect that in drier sites growth is
sustained during this period through
the capacity of the tap-root to exploit
subsoil water in rock crevices.
Unusually among species of infertile
calcareous habitats, *A.h.* lacks
mycorrhizas. Locally *A.h.* extends into
lightly shaded habitats, where
flowering and seed-set are reduced.
A.h. is mainly self-pollinated and
populations are often topographically
isolated. As a result, *A.h.* is genetically
polymorphic. Populations from Irish
sand dunes were formerly separated as
A. brownii. On a broader European
scale, *A.h.* is part of a closely related
group in which five species have been
recognized.
Current trends Likely to decrease. The
capacity of *A.h.* to exploit artificial
habitats is probably restricted more by
narrow ecological niche than by
powers of seed production and
dispersal.

23 *Arctium minus*   **Burdock**

Includes sspp. *minus, nemorosum* and
pubescens

Established strategy C–R.
Gregariousness Sparse.
Flowers Florets purple, tubular,
hermaphrodite, protandrous, self- or
insect-pollinated.
Regenerative strategies S.
Seed 11.47 mg.
British distribution Widespread, most
common in England (95% of VCs).
Commonest habitats Soil heaps 7%,
river banks/verges 3%, cinder
heaps/scree/limestone
wasteland/limestone woodland 2%.
Spp. most similar in habitats
Reynoutria japonica 62%, *Vicia
sepium* 53%, *Lamium album* 52%,
Calystegia sepium 51%, *Lamium
purpureum* 50%.
Full Autecological Account *CPE* page
98.

Synopsis *A.m.* is a tall, monocarpic,
perennial herb associated with a wide
range of fertile, disturbed habitats.
When close to maturity after two, or
usually more, years, *A.m.* forms an
exceptionally large and robust plant.
Plants observed in Michigan did not
flower until the five-leaved stage, not
attained for at least four years. The
leaves of the mature plant are
reminiscent of, and only slightly
smaller than, those of rhubarb *(Rheum
rhaponticum)* and cast a dense shade
on neighbouring plants. The openings
in the vegetation created by adult
plants may be important as sites for
colonization by their own seedlings, as
described for *Dipsacus fullonum* and
Senecio jacobaea. However, *A.m.* is
perhaps most characteristically an
opportunist colonizer of fertile
disturbed sites within or adjacent to
woodland. *A.m.* is not, however, a
true woodland species and, under
moderate shade, plants remain in a
suppressed state consisting of only a
few, relatively small, basal leaves.

Flowering occurs only in light shade or
in the open. *A.m.* extends almost to its
British altitudinal limit (380 m) within
the Sheffield region. Thus, its
distribution is probably limited to some
extent by climatic factors.
Regeneration is entirely by seed which
appears to be shortly persistent in the
soil. Like many woodland species and
unlike most common Compositae,
which are wind-dispersed, the hooked
fruits, with enclosed seeds, are
dispersed by animals. Three ssp. are
recognized but these are inter-fertile,
both with each other and with *A. lappa*
and are often difficult to distinguish.
Consequently, differences in ecology
between the ssp. remain obscure. In
Britain *A.l.* is largely restricted to
disturbed, unshaded alluvial sites in S
England, where it is represented by
large, persistent populations with less-
ephemeral seed banks. The burrs of
A.l. open at maturity and seeds fall
near the parents. The seeds of *A.l.* are
also less heavily predated.
Current trends Likely to become more
common in woodland areas as a result
of clear felling and other disturbances
related to modern forestry; perhaps
also increasing on roadsides and in
wasteland generally.

24   *Arenaria serpyllifolia*   Thyme-leaved Sandwort

Data include *A. leptoclados*

Established strategy S–R.
Gregariousness Intermediate.
Flowers White, hermaphrodite,
   homogamous, occasionally insect-
   otherwise self-pollinated.
Regenerative strategies S, B$_s$.
Seed 0.06 mg.
British distribution Widespread (100%
   of VCs).
Commonest habitats Rock outcrops
   28%, limestone quarries 14%, lead
   mine spoil 11%, cinder heaps 8%,
   paths/soil heaps 4%.
Spp. most similar in habitats *Saxifraga
   tridactylites* 88%, *Sedum acre* 86%,
   Myosotis ramosissima 85%,
   Arabidopsis thaliana 78%, *Medicago
   lupulina* 78%.
Full Autecological Account *CPE* page
   100.

Synopsis *A.s.* is typically a small
winter-annual herb of droughted or
disturbed habitats. Most
characteristically the species exploits
rock outcrops on dry sandy soils but,
compared with co-occurring winter
annuals such as *Saxifraga tridactylites*,
A.s. is relatively late-flowering (May or
June to August) and tends to exploit
less-droughted microsites. The species
also requires relatively high levels of
soil moisture for germination. *A.s.*
does not persist in shaded sites, is
absent from mown grassland and in
pasture is restricted to rocky sites
inaccessible to grazing stock. The
seedlings are grazed by sheep and
snails but are apparently little predated
by rabbits. The species extends into
some relatively fertile but highly
disturbed habitats, e.g. arable fields.
Flowering is induced by vernalization
and long days and it is not certain
whether the late-flowering plants of
A.s. frequently observed in the field
originate by spring germination.
Freshly-shed seed has an after-ripening
requirement and germination is

inhibited by canopy shade, providing a
possible mechanism of 'gap detection'
and imposed dormancy. These
characteristics facilitate the
incorporation of seeds into a persistent
seed bank. Populations fluctuate
widely from year to year and the
presence of the persistent seed bank
probably enables populations in more
'stable' sites, such as rock outcrops, to
be maintained following years of poor
seed-set. Although as few as 100 seeds
may be borne on small plants on ant
hills, seeds are frequently produced in
much larger numbers. A mean number
per plant of > 3000 has been suggested
and, although lacking obvious dispersal
mechanisms, *A.s.*, which is tetraploid,
appears highly mobile. Thus, in
addition to exploiting sites which
remain open for many years, the
species frequently occupies transient
areas of exposed soil in grassland or
wasteland. On anthills, a favoured
habitat on chalk in S England, seedling
mortality is very high and has a variety
of causes. In particular, seedlings
formed late in the autumn are very
vulnerable to winter fatalities and
drought. A large-seeded form, var.
macrocarpa, is found on the Atlantic
coast. The diploid *A. leptoclados*, with

$2n = 20$, $2x$ and a more-lowland distribution, also occurs in Britain. *A.l.* is much less common than *A.s.*, and in our survey area is only recorded with certainty from lowland sites on the Bunter sandstone.

25   *Arrhenatherum elatius*

Includes ssp. *bulbosum*

Established strategy Ssp. *b*. C; ssp. *e.* C–S–R.
Gregariousness Intermediate.
Flowers Green, wind-pollinated; self-incompatible.
Regenerative strategies S, (V).
Seed 2.39 mg.
British distribution Widespread (100% of VCs).
Commonest habitats Scree 98%, limestone quarries 57%, hedgerows 39%, verges 41%, limestone wasteland 38%.
Spp. most similar in habitats *Geranium robertianum* 75%, *Senecio jacobaea* 68%, *Festuca rubra* 65%, *Carex flacca* 59%, *Teucrium scorodonia* 58%.
Full Autecological Account *CPE* page 102.

Synopsis *A.e.* is a tall, tetraploid, tussock-forming grass found in a wide range of unshaded or lightly shaded habitats. It has extensive roots and exploits habitats with a deep water-table better than grasses such as *Alopecurus pratensis*. However, both ssp. appear to have a high moisture demand and the constancy in limestone screes may be in part a result of continuously moist soil maintained by the 'mulch effect' of talus. In meadows the species thrives best when 'receiving heavy complete manures whether as organic or artificial fertilisers' (*PGE*). The leaves of *A.e.* have unusually high P levels and their Ca status is greater than that of the foliage of most common grasses. Ecologically, the two subspecies are poorly distinguished.

Current trends Has decreased on arable land, but is a common plant, particularly beside railways and in quarries. On balance it is probably increasing.

Oat-grass

Ssp. *b.*, with bulbous stem bases during winter, grows predominantly on fertile soils and was formerly a common arable weed. Ssp. *e.* occurs on limestone scree and on other less fertile sites and shows a greater tendency to winter greenness. Plants from fertile wasteland and uncut road verges (mainly ssp.*b.*) are tall; they may achieve co-dominance with broad-leaved herbs. On less-fertile soils (e.g. scree) *A.e.* (mainly ssp. *e.*) also may have a robust morphology and may replace shorter, slow-growing grasses, such as *Festuca ovina*, when grazing ceases. Both ssp. are particularly susceptible to grazing since (a) stems are upright and nodal, bearing several leaves, (b) only a small number of basal axillary buds are available for regeneration and (c) regeneration by seed is almost entirely from summer and autumn germination, no seed bank is formed and the seed is large and perhaps poorly dispersed. The seed of *A.e.* is moderately large and under shade the first leaf is capable of considerable extension. This may facilitate establishment on deep block scree. *A.e.* is shortly rhizomatous, with a limited capacity for lateral vegetative spread, and rhizome connections may break, resulting in closely associated daughter plants. In ssp. *b.* regeneration by means of detached stem bases can occur during autumn or winter disturbance, e.g. in ploughed fields. This ssp., 'Onion Couch', used to be a frequent agricultural weed. In some lightly grazed pastures semi-prostrate phenotypes or genotypes that are more resilient under

defoliation have been recorded. Plants perhaps referrable to *A.e.* with $2n = 14$, $2x$ are recorded from SW Europe.

Current trends *A.e.* exploits a wide range of derelict, fertile environments. It has, for example, increased in response to a relaxation of management of road verges and railway banks, and to withdrawal of sheep grazing on some limestone dales. Thus, given current trends, *A.e.* is likely to become even more common. With the greater use of minimum tillage and direct drilling, an increase in ssp. *b.* on arable land has also been predicted.

26 *Artemisia absinthium*   **Wormwood**

Established strategy C–R.
Gregariousness Individuals usually
    scattered.
Flowers Florets yellow, central
    hermaphrodite, outer female, wind-
    pollinated.
Regenerative strategies S, ?B$_s$.
Seed 0.10 mg.
British distribution Largely in England
    (73% of VCs).
Commonest habitats Bricks and mortar
    11%, rock outcrops/verges 3%,
    cinder heaps/coal mine
    waste/limestone quarries/soil heaps
    2%.
Spp. most similar in habitats *Senecio
    squalidus* 78%, *Sonchus oleraceus*
    76%, *Senecio viscosus* 68%, *Atriplex
    prostrata* 59%, *Lamium purpureum*
    55%.
Full Autecological Account *CPE* page
    104.

Synopsis *A.a.* is a tall, bushy, aromatic herb formerly much cultivated for cuisine and medicine. *A.a.* is probably as abundant within the Sheffield region as anywhere in Britain; it is largely restricted to areas of urban dereliction.

A.a. does not exploit semi-natural habitats and is probably not native to Britain. Colonizes relatively fertile soils and reproduces entirely by seed in vegetation gaps. Populations tend to be common in recently disturbed areas, and in situations where opportunities for vegetation consolidation are restricted by the terrain. Demolition sites, where pieces of brick or concrete rubble reduce access to the soil and pavement edges, are frequently exploited. Elsewhere individuals of *A.a.* may occur in isolation in, for example, tall grassland at sites of previous disturbance. Little else is known of the ecological foundation of the distribution of *A.a.*, but its occurrence in demolition sites and in areas close to the sea suggests some nutritional specialization. *A.a.* is known to extract large quantities of N from the soil, possibly to the detriment of potential competitors. The association of *A.a.* with large urban areas and maritime regions may also indicate a sensitivity to harsh winter climates. *A.a.* has a predominantly S distribution in Britain and plants often show signs of winter damage. The significance of the winter growth form,

typically an aggregation of leaves at the top of a woody stem, is obscure, but it is a form shared with the maritime *Brassica oleracea*. *A.a.* is a weed of overgrown pasture in Canada. This is not the case in Britain, although *A.a.* has been observed to persist on derelict rubbish tips reclaimed for horse-grazing. *A.a.*, which may taint milk, is rejected by cattle in Canada and can withstand infrequent cutting.

Current trends *A.a.* appears to have increased as a result of urban dereliction and industrial spoilage. Whether this increase will be sustained is uncertain, but it seems improbable that *A.a.* will become an important grassland weed in Britain.

27 *Artemisia vulgaris*   Mugwort

Established strategy Between C and C–R.
Gregariousness Sparse to intermediate.
Flowers Yellow or purplish, tubular, wind-pollinated.
Regenerative strategies S, B$_s$, V.
Seed 0.12 mg.
British distribution Widespread except in Scotland and Ireland (82% of VCs).
Commonest habitats Soil heaps 19%, cinder heaps 16%, bricks and mortar/coal mine waste 13%, arable 9%.
Spp. most similar in habitats *Tripleurospermum inodorum* 79%, *Atriplex prostrata* 75%, *Senecio squalidus* 75%, *Senecio viscosus* 73%, *Rumex obtusifolius* 72%.
Full Autecological Account *CPE* page 106.

Synopsis *A.v.* is an erect, tufted, aromatic herb which occurs most frequently in disturbed, relatively fertile, urban sites. The species produces a lower canopy than that of

species such as *Cirsium arvense* and *Urtica dioica*, and does not form extensive stands. Thus, *A.v.* is particularly associated with sites where consolidation by more-robust species is prevented by the rocky substrata, e.g. cinders, brick rubble and paving stones. In such sites the deep root system of *A.v.* may allow access to mineral soil beneath the spoil. Colonies may also persist for many years on roadsides, particularly on steep slopes or sandy soils, which are liable to some soil movement. However, in many habitats, e.g. soil heaps, *A.v.* may be an early colonist which is later replaced by taller species with greater potential for dominance. *A.v.* is absent from shaded, grazed and waterlogged sites, and is potentially toxic to livestock. The species has a lowland distribution in Britain; this is determined in part by patterns of land-use, but a restriction by climatic factors is also likely. The colonizing potential of *A.v.* is related to its high level of seed production, which may exceed 9000 per flowering stem. Achenes are not plumed but may be wind-dispersed on a local scale. Seeds also appear to be widely distributed through human activities and many field records from bared ground refer to isolated seedlings which may not reach maturity. In particular, *A.v.* appears to exploit road and railway systems as corridors of dispersal. Survival of disturbance in sites where *A.v.* is established is facilitated by the presence of a persistent seed bank and rhizomes and patches of *A.v.* may expand at a rate of 300 mm year^{-1} by vegetative means. Seed germination, which occurs mainly in spring, is partially inhibited in the presence of a leaf canopy, and a light requirement can be induced in some seeds by storage in the soil. There is much apparently genotypic variation within the species, e.g. in leaf dissection and flower colour. The extent to which this variation is ecologically important is uncertain.

Current trends Probably still increasing in urban habitats and in other disturbed sites in lowland Britain.

---

28 *Arum maculatum*   **Lords-and-Ladies, Cuckoo-pint**

Established strategy S–R.
Gregariousness Sparse.
Flowers Unisexual flowers at base of spadix, female below, male above, all in yellow-green spathe, insect pollinated.
Regenerative strategies V.
Seed 31.93 mg.
British distribution Widespread except in N Scotland (86% of VCs).
Commonest habitats Limestone woodland 18%, scrub 11%, hedgerows 8%, broadleaved plantations/acidic woodland 3%.
Spp. most similar in habitats *Mercurialis perennis* 94%, *Brachypodium sylvaticum* 90%, *Geum urbanum* 89%, *Sanicula europaea* 89%, *Lamiastrum galeobdolon* 88%.
Full Autecological Account *CPE* page 108.

Synopsis *A.m.* is a vernal, monocotyledonous herb of relatively fertile soils in woodland and hedgerows, and shows poor growth in full sunlight. *A.m.* reaches the N edge of its distribution in Britain and is primarily a lowland species in N England. The young shoot is said to be tolerant of low temperatures, but inflorescences are sometimes damaged by frost. The corm becomes buried to depths of 200–300 mm by means of contractile roots. *A.m.* is largely confined to moist but well-drained soils. The buried corm and deep root system are disadvantaged by anaerobic soils, and the poorly differentiated conductive tissue may preclude establishment in droughted sites. Foliage of *A.m.* is toxic and is generally avoided by stock. *A.m.* is slow to establish from seed and is heavily dependent upon its mycorrhizal associate. Foliage is not produced until the second or third year's growth; seed is rarely produced before the seventh year. In view of this extended and vulnerable juvenile phase, and the fact that one plant often produces < 30 seeds, it is surprising that *A.m.* is frequent in habitats subject to intermittent disturbance, such as plantations and hedgerows. It seems likely that the reduced competition from other species following disturbance promotes *A.m.*, as it does many orchids which are similarly dependent upon mycorrhizas and which exploit bare areas of derelict quarries and gravel pits. Regeneration by vegetative means occurs through proliferation of daughter tubers. However, regeneration by seed dispersed by birds is equally important. The floral structure and mechanisms promoting cross-pollination by insects, described by Lamarck and others, involve thermogenic respiration and release of an attractive odour. Flowers are as elaborate and spectacular as any within the British Flora. Considerable morphological variation exists, but the extent of ecotypic differentiation is uncertain. *A.m.* hybridizes with *A. italicum*.

Current trends Uncertain. Vulnerable to sustained and frequent disturbance, but able to exploit broad-leaved plantations if subject to normal forestry management.

29 *Asplenium ruta-muraria*   Wall-rue

Established strategy Between S–R and S.

Gregariousness Sparse to intermediate.

Sporangia In sori on underside of fronds.

Regenerative strategies W.

Spore 46 × 33 μm.

British distribution Widespread (100% of VCs).

Commonest habitats Walls 6%, cliffs 5%, rock outcrops 3%, limestone quarries 2%.

Spp. most similar in habitats
Cystopteris fragilis 97%, *Asplenium trichomanes* 90%, *Mycelis muralis* 86%, *Athyrium filix-femina* 86%, *Dryopteris filix-mas* 80%.

Full Autecological Account *CPE* page 110.

Synopsis *A.r-m.* is a small, winter-green fern with thick leathery leaves. The species is characteristic of crevices, usually in unshaded calcareous cliffs and mortared walls. The species is the only common fern in which the distribution shows a S-facing bias. *A.r-m.* is also the fern which most consistently survives to maturity on walls in the lowland parts of N England. There can be little doubt that *A.r-m.* is the most drought-tolerant of the common ferns. However, in the context of the vascular flora of Britain, this represents only a modest degree of tolerance. In common with other widespread ferns of skeletal habitats, the species is more frequent in the moist W part of the British Isles. Further, the frond canopy, or even the whole plant, may be periodically destroyed during summer drought. This vulnerability may be partly compensated for by the facts that the species produces large quantities of spores from an early age and that most of its fronds are fertile. It is not known whether a persistent spore bank is formed. As in *Asplenium trichomanes* and *Cystopteris fragilis*, spores are slow to germinate (see *C.f.*). It is not known whether the prothallus itself is exceptionally drought-resistant, but the spores of *A.r-m.* have a higher temperature requirement for germination than those of *A.t.*, and *A.r-m.* tends to be found in shallower crevices than *A.t.*. *A.r-m.* has a strictly limited capacity for lateral vegetative spread. The species is phenotypically plastic, with leaves produced early in the season being considerably less divided than those produced later. The species hybridizes, very rarely, with *A.t.* and with one other species. Diploid populations of this predominantly tetraploid species have been reported from Italy.

Current trends Although a very effective colonist by virtue of its wind-dispersed spores, *A.r-m.* appears to have declined considerably during the last 50 years in industrial areas, possibly due to atmospheric pollution. However, there are signs, in Sheffield at least, of rehabilitation, and *A.r-m.* is occurring with increasing frequency on stone and even on relatively modern brick walls in moister residential parts of the city. With respect to *A.r-m.* and other ferns, two possible causal factors may be suggested; first, the level of the atmospheric toxin sulphur dioxide has declined, and, second, the concentration of oxides of nitrogen has increased. The latter, by increasing the nitrogen status of the substratum, may at least in theory accelerate the process of colonization. However, the species is probably still declining in dry lowland areas, due to the loss of old mortared stone walls, a favoured habitat.

30 *Asplenium trichomanes*

Ssp. *quadrivalens*

Established strategy S.
Gregariousness Sparse to intermediate.
Sporangia In sori on underside of fronds.
Regenerative strategies W.
Spore 37 × 28 μm, brown.
British distribution Widespread especially in the N and W (99% of VCs).

Maidenhair Spleenwort

Commonest habitats Cliffs 6%, limestone quarries 5%, rock outcrops/walls 3%, scrub 1%.
Spp. most similar in habitats *Asplenium ruta-muraria* 90%, *Cystopteris fragilis* 89%, *Athyrium filix-femina* 87%, *Mycelis muralis* 86%, *Elymus caninus* 78%.
Full Autecological Account *CPE* page 112.

Synopsis *A.t.* is a small winter-green fern which is most characteristically associated with crevices in calcareous cliffs and mortared walls. In form and distribution the species resembles *A. ruta-muraria*, with which it frequently occurs. However, *A.t.* is more frequent on N-facing slopes, regularly occurs in shaded sites and is infrequent within the lowland parts of the region. The species appears to require moister habitats and exhibits greater shade-tolerance than *A.r-m.*. In N Derbyshire, *A.t.* is frequent on walls of carboniferous limestone but, unlike *A.r-m.*, the species is unusually scarce on adjacent walls of mortared gritstone. It remains uncertain whether this distribution pattern is a reflection of a more calcicolous habit on the part of *A.t.* or is related to other factors such as the potential of limestone to counteract the acidity associated with sulphur dioxide pollution. *A.t.* has only a restricted capacity for lateral spread (by means of short rhizomes) and regeneration is mainly from spores. A single frond of *A.t.* may produce *c.* 750 000 spores, and the species is an effective long-distance colonist. As with *A.r-m.*, spores are slow to germinate. Studies are required to measure the persistence of spores in natural environments. It is unclear to what extent the distribution of the sporophyte generation of *A.t.* (and other British ferns) is restricted by the ecological requirements of the

prothallus. On less-calcareous rocks in montane regions the diploid ssp. *trichomanes* ($2n = 72$) largely replaces the more widespread tetraploid, ssp. *quadrivalens*; rarely, the two form a sterile hybrid. Also, very rarely the species hybridizes with *A.r-m.* and with *Phyllitis scolopendrium*.

Current trends In lowland areas, now largely restricted to old walls and natural rock faces. Remains an effective colonist in upland areas, and here the species is not under threat. Like *A.r-m.*, the species may have decreased in industrial areas as a result of the effects of atmospheric pollution. Currently, as with *A.r-m.*, *A.t.* appears to be increasing in some parts of residential Sheffield.

31 *Athyrium filix-femina*

Lady-fern

Established strategy Between S–C and C.
Gregariousness Can achieve dominance at low frequency.
Sporangia In sori on underside of frond.
Regenerative strategies W, V.
Spore 41 × 25 μm, brown.

British distribution Widespread (99% of VCs).
Commonest habitats Cliffs 9%, walls 6%, shaded mire 4%, river banks/rock outcrops/acidic woodland 3%.
Spp. most similar in habitats *Cystopteris fragilis* 89%, *Mycelis*

muralis 88%, *Asplenium trichomanes* 87%, *Asplenium ruta-muraria* 86%, *Elymus caninus* 83%.
Full Autecological Account *CPE* page 114.

Synopsis *A.f-f.* is a potential dominant of moist woodland, and shares many of the attributes of *Dryopteris dilatata*. However, the species commonly occurs on river and stream banks, and on woodland floodplains, and is thus more frequently associated with fertile, regularly disturbed habitats than *D.d.*. Indeed, *A.f-f.* has a shoot phenology typical of the 'competitive strategy'. The fronds expand mainly during a single flush in early summer, attain maximum size in July, and die back with the onset of autumn frosts. Again, unlike *D.d.*, old plants of *A.f-f.* may consist of a branched rhizome and a clump of numerous crowns; this morphology, coupled with the tendency to accumulate a dense layer of litter, confers upon *A.f-f.* a strong capacity for vegetation dominance. Vigorous stands of *A.f-f.* contain few associated species. The possibility that rhizomes detached by disturbance may be transported in water to form new colonies elsewhere deserves investigation. The species is not recorded from highly calcareous soils, but occurs at its most luxuriant between pH 5.0 and 7.0, the upper end of its pH range. Unlike *D.d.* and *D. filix-mas*, the fronds are not cited as toxic. In rocky sites many plants die before reaching reproductive maturity, often apparently because crevices colonized by *A.f-f.* provide insufficient soil and moisture to sustain an adult plant. A single frond of *A.f-f.* may produce 2 million spores. It is not known whether *A.f-f.* forms a persistent spore bank, but the spores are widely dispersed and the species is a common colonist of ditch banks and wet walls. Much of the considerable morphological variation observed between populations is maintained in cultivation and the presence of heavy-metal-tolerant populations is suspected.
Current trends Appears to be decreasing in many lowland areas as a consequence of habitat destruction.

32   *Atriplex patula*

Iron-root, Common Orache

Established strategy R.
Gregariousness Sparse to intermediate.
Flowers Green, monoecious, facultatively autogamous, slightly protogynous, mostly wind-pollinated.
Regenerative strategies S, B$_s$.
Seed 1.33 mg.

British distribution Widespread except in N Scotland (97% of VCs).
Commonest habitats Bricks and mortar/manure and sewage waste 31%, arable 30%, soil heaps 14%, lead mine spoil/river banks 5%.
Spp. most similar in habitats *Chenopodium album* 94%, *Senecio*

vulgaris 92%, *Stellaria media* 82%, *Capsella bursa-pastoris* 81%, *Polygonum aviculare* 78%.
Full Autecological Account *CPE* page 116.

Synopsis *A.p.* is an erect or prostrate summer-annual with a growth form intermediate between those of *Polygonum aviculare* and *Chenopodium album*. The species colonizes a range of artificial, moist, unshaded, relatively fertile sites following periods of severe disturbance. The three main habitats with which the species is associated are demolition sites, manure heaps and arable land, particularly in broad-leaved crops or fallow areas. *A.p.* is usually much branched from near the base, and control measures involving cutting may encourage the species at the expense of taller, more-dominant weeds. The species is largely absent from trampled and from grazed sites. *A.p.* shows a greater restriction to fertile mineral soils and is less gregarious than *A. prostrata*. Like other ephemerals, plant size and reproductive output are very variable. Plants with less than 10 and nearly 1000 flowers have been observed at a single site. Regeneration is entirely by seed, which germinates between late winter and early summer, and a very persistent seed bank may be formed. Most common weeds have two peaks of germination, in autumn and in spring, and the restricted germination period of *A.p.* must limit its regenerative flexibility compared with

these species, even though seed is polymorphic, both with respect to size and to germination requirements. Brown seeds are produced, which are typically present in small quantities, and will germinate in the laboratory within two weeks, whereas black seeds require scarification and alternating temperatures. The disadvantages of this strongly seasonal pattern of germination are perhaps offset by the widespread occurrence of *A.p.* as an impurity in crop seed; in this respect *Fallopia convolvulus* and *Polygonum* spp. show similar characters. Seeds are probably dispersed through human activities and they can survive ingestion by a range of mammals and birds and may be dispersed through these agencies. The hybrid with *A. littoralis* is recorded from Britain.
Current trends Remains an effective colonist with persistent populations in artificial habitats.

33   *Atriplex prostrata*

Formerly *A. hastata*

Established strategy R.
Gregariousness Intermediate.
Flowers Green, monoecious, facultatively autogamous, wind-pollinated but also visited by insects.

Hastate Orache

Regenerative strategies S_s, B_s.
Seed 0.86 mg.
British distribution Widespread near coast. Only local inland, except in the S and E.
Commonest habitats Bricks and mortar

16%, coal mine waste 13%, soil heaps 9%, river banks 5%, cinder heaps 4%.

Spp. most similar in habitats *Artemisia vulgaris* 75%, *Cirsium arvense* 75%, *Rumex crispus* 70%, *Senecio viscosus* 69%, *Senecio squalidus* 68%.

Full Autecological Account *CPE* page 118.

Synopsis *A.p.* is an erect or prostrate summer annual which exploits open, moist, unshaded sites, either (a) as an early colonist following disturbance and before the establishment of potentially-dominant perennial species, e.g. on demolition sites and coal-mine spoil, or (b) as a persistent component of habitats where the growth of perennials is restricted by regular disturbance, e.g. on the sea shore. Although perhaps originally a maritime species exploiting the upper reaches of salt marshes and sand, silt and shingle beaches, *A.p.* is now widespread inland. The origin of *A.p.* as a plant of maritime soils high in available sodium presumably explains the exceptional ability of the species to exploit certain nutritionally-extreme, inland habitats; *A.p.* is particularly common on demolition sites and coal-mine spoil. Demolition sites frequently exhibit very high pH (> 8.0) and coal-mine spoil is often characterized by high sodium and chloride concentrations. *A.p.* also exploits pulverized fly-ash, another substrate with a very high pH. Not all of the sites colonized by *A.p.* are fertile, and the species can effectively utilize the low levels of nitrogen available in fly-ash. Although growth is better where nitrogen is supplied in the form of nitrate, *A.p.* is also tolerant of the ammonium ion. This may be important in enabling the species to exploit manure heaps and shorelines where detritus accumulates at the high-tide mark. When grown in solution culture *A.p.* tends to absorb a relatively low concentration of chloride ion.

However, the presence of increased salt does lead to increased succulence. *A.p.* is not exclusively confined to extreme habitats of the type described above. The factors regulating the distribution of *A.p.* in more 'normal' ruderal habitats require investigation. *A.p.* is a C_3 species and, of British species of *Atriplex*, only *A. laciniata* exhibits the Kranztypus leaf venation of C_4. *A.p.* is largely absent from trampled and from grazed sites. The species regenerates entirely by seed, which germinates in spring. Mature seeds of *A.p.* are shed from within green tissues and, consequently, all require subsequent exposure of the imbibed seed to unfiltered sunlight before germination can occur. The seeds are polymorphic but, unlike those of *A. patula*, the large brown and the (more frequent) small black seeds germinate at the same rate. A buried seed bank is suspected. Seeds can survive ingestion by cattle and are presumably spread by mammals and birds as well as by human activities. *A.p.* is polymorphic, often with several genotypes occurring within one population, and is highly plastic. A genetically-fixed, prostrate form is restricted to certain exposed maritime sites. Three other British species are recognized within the *prostrata* group. *A.p.* forms hybrids with two of these, and with one other species also.

Current trends An apparently increasing colonist of artificial habitats such as spoil heaps and derelict land. Perhaps increasing also in maritime areas and on verges beside roads salted in winter. Disturbance of maritime areas may be expected to increase the extent of hybridization with other species.

34 *Avenula pratensis*   **Meadow Oat**

Established strategy Between S and
  S–C.
Gregariousness Intermediate.
Flowers Green, hermaphrodite, wind-
  pollinated.
Regenerative strategies S.
Seed 2.08 mg.
British distribution Widespread (54%
  of VCs).
Commonest habitats Limestone
  pastures 46%, scree 26%, limestone
  wasteland 13%, lead mine spoil 7%,
  rock outcrops 6%.
Spp. most similar in habitats
  Helianthemum nummularium 94%,
  Koeleria macrantha 93%,
  Sanguisorba minor 93%, *Carex
  caryophyllea* 92%, *Polygala vulgaris*
  89%.
Full Autecological Account *CPE* page
  120.

Synopsis *A.p.* is a potentially robust but slow-growing, tufted, winter-green grass restricted to infertile, calcareous grassland and rocky habitats. *A.p.* shows some ecological affinities with calcareous races of *Festuca ovina*, with which it is often associated, and shares several attributes with *F.o.* and *Koeleria macrantha*, including (a) high leaf extension rate in early spring, which in the case of *A.p.* is predictable from an exceptionally high value for nuclear DNA amount, (b) a short, and early, flowering period, (c) a dependence upon regeneration by seed, which germinate in autumn, soon after dispersal, (d) a wide temperature range for seed germination and (e) the absence of a persistent seed bank. However, *A.p.* is potentially taller and

more robust than either of the other two species, and has a greater capacity to dominate derelict limestone grassland where, as in many other situations inaccessible to grazing animals, *A.p.* can develop large winter-green tussocks with long spreading leaves. Compared with many other grasses, including *F.o.*, *A.p.* suffers less damage during fires, probably because meristems near the centre of tussocks are insulated by green leaf tissue. In grazed sites *A.p.* is reduced in stature and remains a relatively inconspicuous component. We suspect that its leaves are more palatable to sheep than those of *F.o.*. Foliar concentrations of N and Ca are relatively low. *A.p.* appears to have low mobility and is restricted to grasslands of some antiquity and is absent from Ireland. It is one of only four common calcicolous species which are strictly confined to limestone strata in S and C England, the other three being *Helianthemum nummularium*, *Scabiosa columbaria* and *Viola hirta*. The full extent of ecotypic differentiation within *A.p.* remains to be investigated although montane populations with larger spikelets may be taxonomically distinct.

Current trends Already close to extinction in many lowland areas and likely to decrease further as a result of habitat destruction.

35 *Avenula pubescens*

Hairy Oat-grass

Established strategy Between S and C–S–R.
Gregariousness Intermediate.
Flowers Green, hermaphrodite, wind-pollinated, possibly self-incompatible.
Regenerative strategies S, (V).
Seed 1.92 mg.
British distribution Widespread (95% of VCs).
Commonest habitats Limestone wasteland 19%, limestone pastures 17%, scree 6%, lead mine spoil/meadows/enclosed pastures/rock outcrops 3%.
Spp. most similar in habitats *Pimpinella saxifraga* 88%, *Lathyrus montanus* 88%, *Stachys officinalis* 83%, *Briza media* 82%, *Galium verum* 82%.
Full Autecological Account *CPE* page 122.

Synopsis *A.p.* is a loosely-tufted winter-green grass mainly restricted to moist but well-drained, relatively unproductive, calcareous grassland. Considered worthless for agriculture. The calcicolous habit of the species is evident in the *PGE*, where it is 'very intolerant of ammonium salts' and 'considerably increased by lime even to the extent of becoming one of the three chief grass species'. In chalk downland *A.p.* was found to be relatively shallow-rooted. *A.p.* produces tillers with long ascending leaves not unlike those found in some forms of *Poa pratensis*. This appears to allow the persistence of *A.p.* in relatively tall grassland. However, *A.p.*, which has little capacity for lateral vegetative spread, is usually a minor component of such vegetation. With similar frequency *A.p.* also occurs as a minor component in grazed and burned sites. Thus, *A.p.* is a morphologically 'intermediate' species with no obvious specialization for any one regime of grassland management and in the *PGE* is described as 'a rather insignificant member of various mixed associations'. Thus, *A.p.* may be regarded as occupying a subordinate niche within moist, less-productive grassland analogous to that of *Trisetum flavescens* in drier sites. Like those of other 'intermediate' species, populations may be expected, under fluctuating conditions of climate, soil fertility and management, to show a greater degree of homeostasis than those of potential dominants. In this context it is interesting to note that *A.p.* reaches maximum local abundance in upland parts of the Sheffield region, growing on roadsides which are infrequently cut and which in many instances are subject to light sheep-grazing during summer. *A.p.* forms a transient seed bank with germination in late summer and early autumn. Within the Sheffield region *A.p.* has spread extensively from calcareous geological strata, but only along roadsides where, because of the use of limestone in road-making, edaphic conditions are closely similar to those in calcareous grassland. An additional subspecies is recorded from Europe.

Current trends *A.p.* is largely restricted to older grassland although it has colonized road verges extensively in the past. *A.p.* shows no sign of spreading into new artificial habitats, and seems destined to decline further.

36 *Bellis perennis*   Daisy

Established strategy Between R and C–S–R.
Gregariousness Intermediate.
Flowers Florets insect-pollinated, self-compatible.
Regenerative strategies V.
Seed 0.09 mg.
British distribution Widespread (100% of VCs).
Commonest habitats Meadows 43%, rock outcrops 14%, verges 13%, enclosed pastures/limestone pastures/paths 10%.
Spp. most similar in habitats *Cynosurus cristatus* 89%, *Prunella vulgaris* 85%, *Phleum pratense* ssp. *bertollonii* 85%, *Trifolium repens* 78%, *Cerastium fontanum* 73%.
Full Autecological Account CPE page 124.

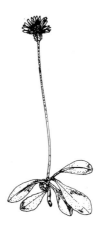

Synopsis *B.p.* is a winter-green, low-growing rosette-forming herb producing clonal patches in short turf on fertile soils, and is strongly associated with trampled habitats. *B.p.* is essentially a 'cool season' plant capable of flowering throughout most of the year, particularly in lawns. During winter the roots of *B.p.* contain large amounts of the storage carbohydrate fructan. Summer growth is often checked by drought. Despite the S-facing bias in field distribution,

B.p. is most characteristic of moist soils, particularly where the effects of trampling and other forms of disturbance create local patches of bare soil. *B.p.* is most familiar as a weed of lawns and sports fields, but is also capable of persisting at reduced population densities in meadows, provided that there is a post-harvest grazing period. In meadows, before the hay is cut, *B.p.* produces upright leaves. However, when the field is grazed the leaves of *B.p.*, which are palatable to stock, form an appressed rosette. *B.p.* is phenotypically plastic in many morphological features, including the number of florets within each capitulum, but the genetic differences between British populations appear to be insufficient to allow the recognition of distinct ecotypes. *B.p.* regenerates vegetatively by means of stolons to form patches. Breakdown of stolons results in the isolation of daughter plants from the parent. Regeneration by seed is generally less important, but is more common in hay meadows than in lawns. Seed germinates over an extended period during summer and autumn, a reflection of the long flowering season and the capacity for germination over a wide range of temperatures. Indeed, flowers are present in most months of the year; during the spring and summer they develop extremely rapidly and consequently seeds can escape predation even in heavily grazed and frequently mown grasslands. Buried viable seeds have been detected, but this report should be treated with caution as many other workers have failed to find them and, according to some, the seed of *B.p.* is only briefly persistent. Seeds appear to be dispersed in mud on human feet and on animals and vehicles.

Current trends The abundance of *B.p.* is maintained at a high level through human activities and, as a result of modern land use, *B.p.* may be expected to increase still further.

37 *Betula*

Birch, Silver Birch

Field data refer to seedlings/saplings; spp. include *pubescens pendula* and their hybrids

Established strategy Between C and S–C.

Gregariousness Seedlings may occur at high densities.

Flowers Green, monoecious, wind-pollinated.

Regenerative strategies W.

Seed 0.12 mg.

British distribution Widespread, but more common in the N and W (95% of VCs).

Commonest habitats Scrub 15%, limestone quarries 11%, cinder heaps/rock outcrops/acidic woodland 8%.

Spp. most similar in habitats *Milium effusum* 73%, *Hyacinthoides non-scripta* 67%, *Ulmus glabra* (juv.) 66%, *Quercus* sp. (juv.) 64%, *Crataegus monogyna* (juv.) 63%.

Full Autecological Account *CPE* page 126.

Synopsis *B.* spp. are fast-growing deciduous trees which may reach 25 m in height and have a life-span of *c.* 60–70 years. Wood is diffuse-porous and bud-break comparatively early. In a forestry survey, *B.* spp. were found to cover 157 000 ha of woodland (> 21% of the total), making them the second most common broad-leaved tree of British woodland and the most common in Scotland. *B.* spp. were exceedingly common in Britain in the pre-Boreal period. The extremely high abundance of *B.* spp., particularly *B.pub.*, in Scotland today probably relates in part to a tolerance of cold climate. *B.* spp. may commence flowering within its first five to ten years. Small, winged, wind-dispersed fruits are produced in large quantities (up to 43 000 m^{-2} of ground), although a good seed year is generally followed by a number of years with lower production. Seed viability is particularly high in good seed years. Seeds have a chilling requirement and may then germinate in darkness. However, germination may be inhibited by placement under a leaf canopy. These findings are consistent with the frequent observation that young birch seedlings tend to be restricted to areas of bare soil, especially where there are canopy gaps in grassland or woodland. Seedlings are small and very susceptible to drought, although less so than those of *Alnus glutinosa*. In sharp contrast with oak, the seedling root:shoot ratio never exceeds unity. It has been suggested that root competition from heather plays a major role in the prevention of *B.* spp. from establishing in Callunetum. Establishment is often also unsuccessful in the vicinity of mature birch trees. In soils of low nutrient status, seedling establishment is much enhanced by mycorrhizal infection of the roots. Many fatalities occur during the first winter and subsequently, due to the effects of shade. Establishment from seed is poor in heavily-grazed sites. A few viable seeds are often found in soils but, with one exception, there are no reports of a persistent seed bank. Seedlings and young saplings have been observed in most of the major habitats of the Sheffield region, often in sites quite remote from established trees and in areas unsuitable for subsequent establishment of trees. *B.pen.* is often found at high pH, and may be an early colonist of limestone quarries. Nevertheless, birch is most conspicuous as a colonist of open ground, burned areas and forest

clearance on acidic soils. The soil under *B.* spp. may be converted from mor to mull humus following the penetration of the iron pan by the root system and the uptake of calcium and other nutrients from lower horizons of the soil profile. Breakdown of litter is relatively rapid. *B.pen.* survives swift-moving fires and is tolerant of coppicing. *B.pub.* is relatively palatable to the snail *Helix aspersa*, and supports a diverse insect fauna. An inducible defence against foliar predation has been described in which decreased palatability is correlated with increased levels of phenolics. Often, birch scrub does not give rise to oak woodland, as might have been predicted, but appears in many places to be able to perpetuate itself. Indeed, it has been suggested that the high reproductive capacity of *B.* spp. offsets their short life-span, and that oak-to-birch succession frequently occurs. The two segregates *B.pen.* and *B.pub.* frequently occur together. *B.pub.* is more characteristic of moist soils, but the distribution of the two in woodlands may be determined by chance factors of colonization. This is probably due to the high level of disturbance associated with British woodlands; fertile hybrids between the two species are frequent. A greater association of *B.pub.* with ancient woodland has been noted. *B.pub.* shows some evidence of less efficient seed mobility, is shorter-lived, and is more tolerant of cold, of acidic and of nutrient-deficient conditions. The relative growth rate of *B.pen.* exceeds that of *B.pub.* over a wide range of light intensities. In Scotland *B.pub.* hybridizes also with *B.nana*.

Current trends The capacity for effective long-distance dispersal and regeneration by seed enabled birch to spread rapidly at the end of the last glaciation. This aggressive species is showing a similar capacity for expansion in the present disturbed landscapes, both in areas of former heath or grassland and (as a weed) in forestry. Some now consider birch a major threat to loss of open heathland in the Brecklands of East Anglia. Two suggestions have been made: (a) that birch should not normally be planted for conservation (or amenity) purposes, nor should tree planting be undertaken in ancient woodland, since, whatever is planted, birches normally appear, and (b) allowing for the fact that birch is likely to increase further, attempts should be made to find an economic use for it. (The species was probably formerly of considerable economic importance in N England and Scotland.) Both suggestions seem eminently reasonable.

38 *Brachypodium pinnatum*

Heath False-brome, Tor Grass

Established strategy S–C.
Gregariousness Stand-forming.
Flowers Green, hermaphrodite, wind-pollinated.
Regenerative strategies V, ?S.
Seed 2.85 mg.
British distribution Mainly in the S and E (41% of VCs).
Commonest habitats Limestone wasteland 48%, limestone quarries 12%, rock outcrops 5%, limestone pastures 4%, hedgerows/scrub 3%.
Spp. most similar in habitats *Bromus erectus* 95%, *Centaurea nigra* 88%, *Trifolium medium* 86%, *Hypericum perforatum* 81%, *Valeriana officinalis* 80%.
Full Autecological Account *CPE* page 128.

Synopsis *B.p.* is an aggressively rhizomatous, potentially dominant, stand-forming grass of infertile lowland pasture and wasteland, usually on calcareous soil. As might be predicted from its low nuclear DNA amount, *B.p.* has a summer peak in growth. This is in marked contrast to *Bromus erectus*, with which the species often grows. *B.p.* has a deep root system and tends to attain maximum vigour on terraces with an accumulation of mineral subsoil. In N and upland Britain *B.p.* is rare and is often confined to S-facing slopes. This geographical restriction appears to arise from problems in initial establishment. In common with certain other grasses which form large clonal patches (e.g. *Elymus repens* and *Glyceria maxima*), *B.p.* often produces little viable seed. A seed set of 20% is considered a good average in lowland sites although, somewhat paradoxically, generally high amounts of good seed were observed in the cool summer of 1985. In upland sites seed is not normally produced at all; sometimes even the inflorescences are not formed. Accordingly, at higher altitudes new colonies of *B.p.* are often the result of unwitting human introduction; however, the species is frost-resistant and, once established, may be locally invasive even in N-facing grasslands. Seeds germinate in the spring and do not form a persistent seed bank. Regeneration by means of transported rhizome fragments requires investigation, but may have been important during the colonization of quarries. *B.p.* is resistant to trampling, and only the young leaves are eaten by sheep and cattle. In many unmanaged areas *B.p.* has become more prominent following myxomatosis and the decline in rabbit populations. Increases in area and biomass of *B.p.* have brought about a decrease in the floristic diversity of the associated vegetation, since the relatively dense canopy of *B.p.* and the often thick layer of persistent leaf litter which accumulates beneath it inhibit the survival of many of the smaller herbaceous species of calcareous grassland. *B.p.*, with rhizomes down to 20 mm below the soil surface, is tolerant of burning; this frequently allows *B.p.* to displace the more fire-sensitive *Bromus erectus*. Although a

relatively strict calcicole in native British sites, *B.p.* also occurs on more-acidic serpentine soils in France and, as an adventive, is not restricted to calcareous soils. Another subspecies, ssp. *rupestre*, occurs in Europe.

Current trends *B.p.* is likely to decrease in frequency of occurrence in lowland areas as a result of the further destruction of semi-natural calcareous grassland. Nevertheless, the species has increased in abundance within many of its extant sites due to a relaxation of grazing pressure. In uplands *B.p.* has spread along railway banks and roadsides, sometimes on non-calcareous strata, and has also expanded following quarrying activity. Thus, the range of *B.p.* appears to be slowly increasing. The species is apparently also expanding in Europe.

39 *Brachypodium sylvaticum*

Slender False-brome

Established strategy Between S and S–C.
Gregariousness Intermediate.
Flowers Green, hermaphrodite, wind-pollinated, probably self-compatible.
Regenerative strategies S, (V).
Seed 0.62 mg.
British distribution Widespread, particularly in S and C England (99% of VCs).
Commonest habitats Limestone woodland 37%, scrub 11%, scree 9%, hedgerows/limestone pastures 8%.
Spp. most similar in habitats *Mercurialis perennis* 97%, *Anemone nemorosa* 92%, *Sanicula europaea* 91%, *Arum maculatum* 90%, *Melica uniflora* 89%.
Full Autecological Account *CPE* page 130.

Synopsis *B.s.* is typically a tufted grass of woodland habitats. However, in the Sheffield region the species shows a

significant S-facing bias in woodlands, and is generally absent from deep shade. Further, tussocks of *B.s.* in woodland may produce as few as 30 seeds or may even fail to flower. In keeping with a nuclear DNA amount which is exceptionally low among native grasses, leaf expansion in *B.s.* is somewhat delayed, and shoot biomass reaches its maximum in summer. *B.s.* is a common constituent of coppiced woodland. Particularly in upland areas, *B.s.* is also frequent in grassland and scree habitats, and here one vigorous plant may produce as many as 2700 seeds. Plants in open sites show signs of photo-bleaching and, as a result, their leaves are distinctly more yellow-green than those of adjacent grasses. It is not clear whether this yellowing is associated with any diminution of growth rate. In grassland sites *B.s.* is regarded as a relic of former woodland. The species is found under a wide range of edaphic conditions, and is particularly characteristic of less fertile, calcareous soils. *B.s.* loses water more rapidly than *B. pinnatum* during both the stomatal and the cuticular phases of transpiration, and its frequency on scree slopes in the Derbyshire dales may be, in part, a result of the continuously moist soil maintained by the 'mulch effect' of talus. *B.s.* may be eaten by sheep and rabbits, and is absent from heavily grazed grassland. However, the species is frequently found in sites subject to burning. *B.s.* probably regenerates mainly by seed, but its limited capacity for lateral vegetative spread may be of significance in shaded sites, where seed-set is frequently poor. No persistent seed bank has been reported. Despite low seed mobility, *B.s.* is also frequently found along the edges of woodland rides in modern forests. The species is morphologically uniform in Britain. Elsewhere the species is more variable, and a further ssp. is recorded from Europe.

Current trends Persistence in woodland does not appear to be severely threatened by modern forestry practice. However, *B.s.* is not a regular colonist of artificial habitats, and a continuing decrease in its abundance is predicted.

40 *Briza media*

Quaking Grass, Doddering Dillies

Established strategy S.
Gregariousness Intermediate.
Flowers Purplish, hermaphrodite, homogamous, wind-pollinated, strongly self-incompatible.
Regenerative strategies S, V.
Seed 0.23 mg.
British distribution Widespread except in N Scotland (95% of VCs).
Commonest habitats Limestone pastures 29%, limestone wasteland 13%, limestone quarries/scree 9%, verges 8%.
Spp. most similar in habitats *Carex caryophyllea* 94%, *Pimpinella saxifraga* 86%, *Polygala vulgaris* 85%, *Veronica chamaedrys* 84%, *Avenula pubescens* 82%.
Full Autecological Account *CPE* page 132.

Synopsis *B.m.* is a slow-growing, winter-green grass largely restricted to unfertilized, often species-rich grassland, particularly on calcareous soils. Agricultural experience suggests that *B.m.* is generally an indicator of poverty or exhaustion of soil and disappears when conditions are improved. It is usually increased by liming. In turf microcosms, roots of this species became heavily colonized by VA mycorrhizas with increases of up to fourfold in seedling yield. *B.m.* produces little foliage, and typically in short turf the leaves are held close to the ground surface. In this respect *B.m.* resembles *Cynosurus cristatus*. The species is susceptible to submergence by litter and to shading from taller species, and is a poor competitor, failing to persist in derelict grassland. The low growth habit may afford some protection from mammalian grazing. Experiments with snails suggest that the leaves possess no major chemical deterrent to herbivory. In keeping with a fairly high nuclear DNA amount, *B.m.* has a vernal phenology and rapid leaf extension occurs during early spring. This early growth contrasts with the delayed phenologies of many of the species, e.g. *Carex flacca*, with which *B.m.* frequently occurs. *B.m.* is significantly more abundant in N-facing grassland, is characteristic of moist but unproductive habitats and often extends into soligenous mire. Seeds germinate soon after release in late summer and autumn, and the species does not develop a persistent seed bank. In marked contrast with many other grasses of calcareous habitats, the seeds and young seedlings are relatively small and the latter suffer heavy mortalities during the initial phase of establishment. It is not clear whether regeneration by seed or by vegetative means is the more important. Cytological variation occurs with 4x cytotypes having a predominantly E and 2x plants W European distributions, a pattern which recurs in Britain. Although mainly restricted to calcareous soils and to short turf, *B.m.* is wide-ranging with respect to soil type and may occur on heavy non-calcareous clays and on sandy soils. Ecotypes differing in their mineral nutrition have been identified, and it is suspected that much of the edaphic tolerance of this outbreeding species results from the formation of ecotypes. Another subspecies is reported from SE Europe.

Current trends *B.m.* is largely restricted to older, unproductive, managed grasslands and a further contraction in abundance of the species seems likely, particularly in lowlands. In the past *B.m.* has colonized artificial habitats such as roadsides and old quarries, but the species does not appear to be effectively colonizing recently-created environments. With the continuing fragmentation of suitable habitats the evolution of in-breeding populations of *B.m.* is predicted.

---

41   *Bromus erectus*

Upright Brome

Established strategy Between C–S–R and S–C.
Gregariousness Tussock- or stand-forming.
Flowers Green, hermaphrodite, wind-pollinated, self-incompatible.
Regenerative strategies S, V.

Seed 4.23 mg.
British distribution Common in the SE, local elsewhere (50% of VCs).
Commonest habitats Limestone wasteland 23%, limestone quarries 6%, limestone pastures 4%, hedgerows 3%, paths/rock outcrops 2%.

Brachypodium pinnatum 95%,
Trifolium medium 94%, *Centaurea
nigra* 83%, *Hypericum perforatum*
83%, *Lathyrus montanus* 79%.
Full Autecological Account *CPE* page
134.

Synopsis *B.e.* is a tufted winter-green
grass which, like *Brachypodium
pinnatum*, is largely restricted to semi-
natural, infertile, lowland grassland on
dry calcareous soils. *B.e.* is not
rhizomatous and consequently is a less
effective dominant than *B.p.*. Despite
the relatively shallow root system, *B.e.*
tends to be restricted to drier, shallow
soils which *B.p.* is less able to exploit.
As might be predicted from its high
nuclear DNA amount, *B.e.* shows
rapid extension growth in early spring.
The capacity to exploit dry habitats
during a cool, moist period of the year
(*B.e.* also grows to some extent during
winter) constitutes a critical
phenological difference between *B.e.*
and *B.p.*. In studies of European
populations *B.e.* has been found to be
favoured by a moist spring, but the
balance in mixed communities is
shifted in favour of *B.p.* if the weather
is dry. More rarely, *B.e.* is a local
dominant on moist N-facing slopes; its
status in these sites requires
investigation. Because of its
morphology, *B.e.* is less protected
from, and less tolerant of, burning
than *B.p.* and, although unpalatable,
tends to be replaced by *Festuca ovina*
under heavy grazing. *B.e.* regenerates
effectively by seed, and appears to
form large colonies only in habitats in
which open areas are intermittently
available for seedling establishment.
Thus, although locally *B.e.* and *B.p.*
appear equally well-dispersed in
derelict lowland calcareous grasslands,
B.e. is less common as a vegetation
dominant, probably because of its
higher level of dependence upon
regeneration by seed and its greater
sensitivity to burning. A small number
of native, upland populations of *B.e.*
occur on the carboniferous limestone.
Here they set good seed, suggesting
that restriction of *B.e.* at higher
altitudes may arise from problems of
seedling establishment rather than seed
production. In Europe, *B.e.* is more
ecologically wide ranging and four ssp.
are recognized.

Current trends Like most other species
of lowland, calcareous semi-natural
grassland, *B.e.* is decreasing in
abundance. Relaxation of management
in sites where it grows with
Brachypodium pinnatum has generally
led to a shift in favour of the latter.
However, the edaphic range of *B.e.*
appears to be increasing, at least
temporarily, as a result of its
establishment on roadsides and on
railway banks on non-calcareous
strata.

Established strategy R.
Gregariousness Intermediate.
Flowers Green, hermaphrodite, wind-pollinated, facultatively autogamous.
Regenerative strategies S.
Seed 2.9 mg.
British distribution Widespread except in the N (100% of VCs).
Commonest habitats Meadows 45%, rock outcrops 5%, bricks and mortar/lead mine spoil/manure and sewage waste 3%.
Spp. most similar in habitats *Rhinanthus minor* 93%, *Phleum pratense* 78%, *Festuca pratensis* 74%, *Alopecurus pratensis* 73%, *Trifolium pratense* 72%.
Full Autecological Account *CPE* page 136.

Synopsis *B.h.* is a tufted winter-annual grass of hay meadows, roadsides, waste places and rock outcrops, particularly on moderately fertile soils. *B.h.* is very much a 'follower of humans' and is often dispersed as an impurity in grass seed. In the *PGE*, *B.h.* was increased by applications of sodium nitrate and minerals but declined in plots where the soils were untreated or received ammonium salts. The species 'shows a specially close connection with certain Leguminosae, notably *Lathyrus pratensis*'. *B.h.* is dependent upon seed for regeneration, and no persistent seed bank has been reported. *B.h.* is particularly susceptible to heavy grazing, frequent mowing and trampling, and seldom persists in pastures or lawns. Seeds ripen in midsummer at or slightly before the time at which hay meadows are cropped. The seeds, which have no dormancy, germinate equally rapidly in darkness or in light, and under constant or fluctuating temperatures. The caryopses are large, and germination occurs underneath the grass canopy during summer and autumn. *B.h.* is phenotypically highly plastic; robust plants produce many flowering culms, but in unfavourable sites an individual may bear only a solitary spikelet with just six florets. *B.h.* is also genotypically variable, but much of its variation appears to stem from hybridization and introgression with *B. lepidus*, producing the hybrid, *B. × pseudothominii*. In upland areas of the Sheffield region the smaller-seeded *B. × p.* shows little persistence in older hay meadows, and is found at higher altitudes only as a rarity on rock outcrops. In lowland areas, however, *B. × p.* is well established, particularly in dry, open sites, and *B.h.* var. *leiostachys* with hairless spikelets is also relatively frequent. Hybridization may be less frequent in much of Europe, where self-fertilization is less common. Rare ecotypes, with a very restricted geographical distribution in the British Isles, have been given subspecific rank. Ssp. *thominii* includes dwarf sand-dune populations, ssp. *ferronii* occurs on cliff tops in S England, and ssp. *molliformis* is a casual from S Europe.
Current trends Uncertain. In lowland areas *B. × pseudothominii* may be increasing relative to *B.h.*, and the ecological differences between the two taxa are becoming increasingly obscure.

Established strategy C–S–R.
Gregariousness Sparse to intermediate.
Flowers Green, hermaphrodite, wind-pollinated, self-compatible.
Regenerative strategies S.
Seed 7.37 mg.
British distribution Widespread except in the far N (97% of VCs).
Commonest habitats Limestone woodland 10%, hedgerows 8%, cliffs/acidic woodland 6%, river banks 5%.
Spp. most similar in habitats *Hedera helix* 88%, *Geum urbanum* 83%, *Melica uniflora* 74%, *Silene dioica* 74%, *Elymus caninus* 72%.
Full Autecological Account *CPE* page 138.

Synopsis *B.r.* is a tufted, winter-green grass of base-rich lowland woodlands and has a Sub-Atlantic distribution within Europe. *B.r.* shares many attributes with *Festuca gigantea*. The species are remarkably similar in their field distributions, and frequently occur together. The main differences between the two are that *B.r.* tends to occur in drier habitats than *F.g.* and is less abundant in deep shade. Consistent with these observations, *B.r.* is recorded more frequently from cliff ledges and less commonly from mire than *F.g.*, and its distribution shows a greater bias towards S-facing woodland. These differences in distribution coincide with a higher DNA amount in *B.r.*, allowing differences in shoot phenology between the two species to be predicted. The regenerative strategies of *B.r.* are also similar to those of *F.g.*.

B.r. shows a similar dependence on seedlings, which appear in spring and rarely establish in sites with a large accumulation of tree litter. The seed of *B.r.* is larger than that of *F.g.*, but the awn is shorter both proportionally and in absolute terms than that of *F.g.*. Nevertheless, the seed is effectively dispersed by attachment to fur or clothing, and *B.r.*, like *F.g.*, is a frequent colonist of secondary woodland in Lincolnshire. In data collected from the Sheffield region, *F.g.* is recorded in 70% more quadrats than *B.r.*, and the possibility that the seed of *B.r.* is less mobile requires investigation. Both *B.r.* and *F.g.* are in urgent need of thorough ecological study.
Current trends Status uncertain but appears favoured by modern methods of management of broad-leaved woodland.

Established strategy Between R and
C–R.
Gregariousness Typically sparse but
can form dense patches.
Flowers Green, hermaphrodite, mainly
selfed but can be cross-fertilized by
wind.
Regenerative strategies S.
Seed 8.4 mg.
British distribution Widespread except
in Scotland and Ireland (92% of
VCs).
Commonest habitats Hedgerows 12%,
bricks and mortar 8%, cinder heaps
4%, cliffs/meadows/river banks/rock
outcrops 3%.
Spp. most similar in habitats *Sambucus
nigra* (juv.) 57%, *Convolvulus
arvensis* 55%, *Anthriscus sylvestris*
53%, *Tamus communis* 53%,
Stachys sylvatica 52%.
Full Autecological Account *CPE* page
140.

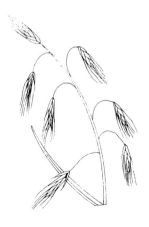

Synopsis *B.s.* is a winter- or more
rarely summer-annual grass which is
similar in many ecological respects to
Galium aparine, though it is unable to
persist in tall herb communities and
tends to germinate earlier, before the
onset of winter. *B.s.* is found in fertile
habitats where, as a result of summer
shade in the case of hedgerows or
summer drought in the case of rock
outcrops, the vigorous growth of
polycarpic perennial herbs is
prevented. The species is also
becoming increasingly important as a
weed of winter-sown cereals. Growth
of *B.s.* is restricted to the period
between autumn and early summer,
and *B.s.* flowers before the onset in
summer of the period, with greatest
risk of disturbance in these and other
related habitats. Flowering is
controlled by a weak vernalization
requirement, and even seedlings which
are produced in early spring may
flower. Panicle extension takes place
only in long days. The seeds are
exceptionally large and the number
produced (often *c.* 200 per plant) is
lower than in most ruderal species.
Germination of seeds, which are shed
from June to August, is frequently
delayed until autumn, apparently
through the absence of adequate
moisture, or sufficiently close contact
between the seed and the soil.
Seedlings can establish successfully
through up to 130 mm of soil.
However, at 150 mm depth, although
seeds still germinate, the seedlings fail
to survive. Germination is known to be
inhibited by white and by red light, but
the ecological significance of this
behaviour remains obscure. No
persistent seed bank is formed. In
arable fields, seed is dispersed within
the crop, but in other habitats the
caryopses, which have a long barbed
awn, adhere to the coats of animals
and to clothing. *B.s.* is a contaminant
of pasture-grass seed, and seeds may
survive ingestion by stock. The species
has an Atlantic distribution and is
largely restricted to lowland sites. It is
sometimes considered to have been
introduced.

Current trends Encouraged by disturbance and by high fertility; apparently increasing. In particular, *B.s.* is becoming a serious weed in fields where winter cereals are grown repeatedly using minimum cultivation techniques.

45 *Callitriche stagnalis*   **Common Water-starwort**

Established strategy Between R and C–R.

Gregariousness Potentially patch-forming.

Flowers Green, monoecious, protandrous, perhaps pollinated by wind, water, or selfed.

Regenerative strategies V, ?B_s.

Seed 0.09 mg.

British distribution Widespread (100% of VCs).

Commonest habitats Running aquatic 21%, still aquatic 20%, unshaded mire 11%, shaded mire 7%, soil heaps 2%.

Spp. most similar in habitats *Glyceria fluitans* 94%, *Juncus bulbosus* 93%, *Apium nodiflorum* 92%, *Sparganium erectum* 92%, *Equisetum palustre* 91%.

Full Autecological Account *CPE* page 142.

Synopsis *C.s.* is a patch-forming herb associated with relatively fertile wetland habitats. In shallow ditches and muddy cart-tracks *C.s.* occurs on mud with species such as *Juncus bufonius* and in soligenous mire with *Montia fontana*. The species also occurs in aquatic habitats (deeper ditches and ponds) typically in the shallows but sometimes in water up to 1 m deep. In mire and in sites normally flooded only in winter *C.s.* is a small, prostrate plant with short leaves and internodes, while in aquatic situations the species produces long leafy stems, often terminating in a floating leaf rosette. In both types of site *C.s.* often displays considerable resilience, populations recovering from major impacts of disturbance whether mechanical or resulting from large fluctuations in water-table. Thus, in aquatic systems *C.s.* remains abundant in large drainage ditches which are regularly cleared of foliage. Like many other aquatic macrophytes, *C.s.* is heterophyllous. The first leaves formed are narrowly elliptical and are often short-lived. Subsequent leaves are ovate and the terminal ones, in the leaf rosette, may be almost circular. The form of these mature leaves is not greatly altered by immersion or emergence. Even by the standard of aquatic plants, the vascular tissue of *C.s.* is very much reduced and terrestrial phenotypes are persistent only in very wet habitats. The apparent absence of epidermal chloroplasts in *Callitriche* is unusual amongst aquatic macrophytes. *C.s.* is wide-ranging with respect to soil type, and a degree of salt tolerance is evident. The species is most frequent on non-calcareous soils and may achieve abundance with *Juncus bulbosus* and *Potamogeton polygonifolius* in peaty drainage ditches. Regeneration is mainly by vegetative means. Stems readily root on contact with mud, and clonal patches are formed in both terrestrial and aquatic systems. Detached shoots readily regenerate to form new plants. Seed production does not occur in submerged plants, and appears to be greater in plants growing on mud than in aquatic specimens (with floating leaves). We suspect that further research will confirm the existence of a persistent seed bank, and on the basis of laboratory responses it seems likely that germination will usually occur in spring. Flowers are perhaps typically wind-pollinated, but a remarkable form of self-pollination, where the pollen tube grows from the anther of the male flower through the vegetative plant to a female flower, has been described for several species of *Callitriche*. *C.s.* is replaced in relatively base-poor (often peaty) aquatic habitats particularly in upland areas by *C. hamulata* and in lowland water enriched by agricultural run-off by *C. platycarpa*. Both species are European endemics and are more consistently aquatic than *C.s.*

Current trends Apparently a highly mobile species of disturbed fertile wetlands. Not under threat and perhaps even increasing.

46 *Calluna vulgaris*

Ling, Heather

Established strategy S–C.
Gregariousness Large, stand-forming individuals.
Flowers Purple, hermaphrodite, weakly protandrous, insect- or wind-pollinated.
Regenerative strategies B_s.
Seed 0.03 mg.
British distribution Widespread (100% of VCs).
Commonest habitats Acidic quarries 47%, acidic pastures/acidic wasteland 22%, limestone pastures 21%, rock outcrops 19%.
Spp. most similar in habitats *Erica cinerea* 87%, *Deschampsia flexuosa* 84%, *Carex pilulifera* 76%, *Vaccinium myrtillus* 74%, *Aira praecox* 70%.
Full Autecological Account *CPE* page 144.
Synopsis *C.v.* is a relatively short-lived (normally < 30 years), evergreen shrub found in a wide range of habitats on acidic, usually well-drained soils. It is by far the most important species of heathland in Britain. *C.v.* is a valuable food plant for hill sheep and grouse, particularly during the first 15–20 years

of its life, when its canopy is increasing. Leaves are low in P and Ca, but often contain relatively high amounts of Mn and Al. Later, the major shoot branches tend to spread apart, leaving a gap in the canopy above the centre of the plant, and 'leggy' older plants are little-grazed. Controlled burning, preferably in spring, is used to rejuvenate older plants which subsequently resprout, particularly if < 15 years old. However, C.v. is killed by fierce fire and is sensitive to overgrazing. C.v. is sometimes regarded as a weed of forestry, since its presence restricts the growth and establishment of some coniferous species. C.v. shows physiological adaptations to an oceanic type of climate. C.v. is potentially taller than most heathland species, has a dense canopy during much of its life and forms persistent litter. In consequence, sites with vigorous, mature C.v. are often virtually devoid of other herbaceous species. C.v. also has some more ruderal characteristics: the life-span is relatively short for a woody species, there is prodigious seed output and under favourable conditions flowering may occur in the second year. C.v. produces numerous minute seeds (up to 1 million m^{-2}). These vary in morphology and germination characteristics, and may be incorporated into a buried seed bank in which they are capable of long-term survival, eventually germinating at any time from spring until autumn. Seedling establishment appears to depend upon the creation of bare ground. Seeds may be wind-dispersed. The growth form shows a degree of plasticity (e.g. dwarf, prostrate forms occur on exposed sites). Some populations exhibit pronounced genotypic variation in morphology. On acidic soils contaminated by lead and other heavy metals C.v. may be virtually the only colonist. Survival here depends heavily upon the capacity of the mycorrhizal associate to bind heavy metals rather than the tolerance of C.v. itself. The soil around the roots is frequently more acidic than the surrounding soil.

Current trends Many lowland heaths in Britain have now been destroyed. In some upland areas C.v. has decreased through the conversion of moorland to commercial forests, and locally through overgrazing and other practices. Thus, although still an abundant upland species, C.v. is decreasing in its original habitats. However, the species appears to be extending its range into acidic, artificial habitats such as roadsides and a variety of types of spoil.

47 *Caltha palustris*   Kingcup, Marsh Marigold

Established strategy Between S and C–S–R.

Gregariousness Individuals usually isolated.

Flowers Yellow, hermaphrodite, homogamous, insect-pollinated, self-incompatible.

Regenerative strategies V, ?S.

Seed 0.99 mg.

British distribution Widespread (99% of VCs).

Commonest habitats Shaded mire 12%, unshaded mire/river banks 3%, still aquatic 2%.

Spp. most similar in habitats *Mentha aquatica* 96%, *Filipendula ulmaria* 93%, *Solanum dulcamara* 92%, *Chrysosplenium oppositifolium* 89%,

Cardamine amara 85%.
Full Autecological Account *CPE* page 146.

Synopsis *C.p.* is a wide-ranging, wetland herb with a variable but restricted capacity for vegetative spread. Individual plants appear to be capable of surviving for at least 50 years. The species is particularly characteristic of topogenous mire on moderately fertile soils in more-open, tall-herb communities in sites where the vigour of potential dominants is restricted by summer shade and winter flooding. *C.p.* also occurs on river banks close to the water level, a habitat which is frequently too disturbed to allow consolidated growth of stand-forming dominants. With increasing altitude *C.p.* becomes particularly frequent in unshaded situations; the species extends into montane habitats and, in Europe, into Arctic regions. Consistent with a high DNA value, shoot extension commences in late winter and plants produce leaves and flowers in early spring before the expansion of deciduous tree canopies. As in other early flowering species of shaded wetlands, such as *Cardamine amara* and *Chrysosplenium oppositifolium*,

the leaves of *C.p.* persist throughout the summer; they are characterized by high concentrations of N and Na, and low levels of P and Fe. *C.p.* occurs in grazed sites and contains protoanemonin, which is toxic to stock. Shoots detached during disturbance are capable of regeneration. Although typically plants occur in discrete clumps, forms with prostrate stems rooting at the nodes are also observed, particularly in upland areas (var. *radicans*). Viable seed is produced regularly and has a well-defined chilling requirement; germination occurs in spring. It is uncertain whether seeds persist in the soil, but it is known that they may float for up to four weeks, and widespread dispersal by water is possible.
Although *C.p.* is variable in the field in Europe and N America, there are no major morphological discontinuities between British populations, and many differences between populations disappear on cultivation. *C.p.* is represented by a polyploid complex with, in addition, aneuploidy, but chromosome races are not morphologically distinct. Plants with $2n = 56$ are by far the commonest in Britain.
Current trends Decreasing with the destruction of wetlands.

48 *Calystegia sepium*

Ssp. *sepium*

Established strategy Between C and C–R.
Gregariousness Dense canopy formed at low density.
Flowers Usually white, hermaphrodite, homogamous, insect-pollinated.
Regenerative strategies (V), B$_s$.
Seed 33.9 mg.
British distribution Common except in the N.
Commonest habitats River banks 8%, soil heaps 6%, bricks and mortar/hedgerows/manure and

Bellbine, Larger Bindweed

sewage waste/verges 3%.
Spp. most similar in habitats *Alliaria petiolata* 74%, *Reynoutria japonica* 69%, *Equisetum arvense* 69%, *Myrrhis odorata* 68%, *Impatiens glandulifera* 61%.
Full Autecological Account *CPE* page 148.

Synopsis *C.s.* is a robust, rhizomatous, twining herb of moist, fertile, disturbed habitats. In the less disturbed and more semi-natural of its habitats, e.g. shaded mire and reed beds, *C.s.* is often an inconspicuous, non-flowering component of the vegetation. However, on river banks, roadsides and other, more-frequently disturbed, productive sites the shoots of very large plants may occur as extensive curtains supported by tall herbs, fences or hedges, sometimes ascending to > 2 m. In recently abandoned or neglected gardens the species has been observed to develop near-monocultures in a short time. Field observations indicate that moist conditions and a degree of disturbance are necessary to sustain populations of the species; where populations of *C.s.* remain undisturbed there is usually a decline in vigour and an expansion of dominance by other clonal herbs or shrubs. *C.s.* is a lowland species and its range is presumably restricted by climatic factors. *C.s.* is also vulnerable to regular grazing or cutting. The species shows aggressive vegetative spread by means of a network of rhizomes which extend to a depth of *c.* 0.3 m in the soil. In disturbed sites vigorous regeneration occurs through both rhizome and root fragments, both of which readily form new plants. The species frequently colonizes roadsides and spoil tips by means of detached plant fragments in dumped soil. Regeneration by seed appears to be of lesser importance. *C.s.* is self-incompatible, and in small colonies, which are often single clones, seed set is frequently poor. In larger colonies dependence on vegetative

regeneration may be less extreme. The large seeds germinate in autumn or in spring and may retain a high level of viability even when buried for as long as 39 years. Seed of riverside populations may be transported by water. Ssp. *s.* has been given specific rank and subdivided into four ssp., one of which is a coastal ecotype found in Britain. In the Sheffield region, pink-flowered genotypes are restricted to mire. Ssp. *pulchra* (*C.p.*) and ssp. *silvatica* (*C.s.*) are aliens occurring in waste places, mainly in urban areas. They both hybridize locally with the native taxon and with each other, and form extensive and at least partially fertile clones. Ssp. *sylvatica* is the most robust taxon and is particularly characteristic of the most fertile sites.
Current trends Uncertain. A common colonist of artificial, disturbed habitats; may be increasing locally.

---

49 *Campanula rotundifolia*

Harebell, Bluebell (Scotland)

Established strategy S.
Gregariousness Intermediate.
Flowers Blue, rarely white, hermaphrodite, protandrous, insect-pollinated, self-incompatible.
Regenerative strategies V, B$_s$.
Seed 0.07 mg.

British distribution Widespread (91% of VCs).
Commonest habitats Limestone pastures 53%, lead mine spoil 43%, scree 39%, limestone wasteland 34%, rock outcrops 16%.
Spp. most similar in habitats *Thymus*

praecox 87%, *Galium sterneri* 86%, *Sanguisorba minor* 86%, *Linum catharticum* 85%, *Koeleria macrantha* 83%.
Full Autecological Account *CPE* page 150.

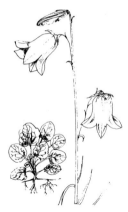

Synopsis *C.r.* is a diminutive, slow-growing, rhizomatous herb found usually as a minor component of vegetation in a wide range of unproductive, grassy and rocky habitats. *C.r.* usually occupies the lower levels of the canopy, and we suspect that it is shade-tolerant and occupies a niche in infertile habitats not dissimilar to that of *Poa trivialis* in moist, fertile ones. *C.r.* is heterophyllous with leaf form dependent largely on illumination. Ovate, shade leaves are found on non-flowering shoots, while more-linear sun leaves are produced on the disproportionately robust flowering stems. During the winter, roots of *C.r.* contain high concentrations of the reserve carbohydrate fructan. In turf microcosms, roots of this species become colonized by VA mycorrhizas with increases of up to sixfold in seedling yield. Regeneration amongst closed communities is facilitated by the production of creeping stems which ramify locally at the soil surface and within the bryophyte layer. Seed germination is stimulated by the creation of vegetation gaps. A persistent seed bank may be formed. The season of germination varies with latitude and aspect. At one site in N Derbyshire, germination on a N-facing slope was delayed until spring whereas seedlings appeared in autumn and spring on an adjacent slope of S-facing aspect. In some habitats, seeds of *C.r.* are said to be dispersed by ants. Most British populations are 4*x* except in the W and N, where they are apparently replaced by 6*x* plants. Plants with higher ploidy levels have larger seeds, but ecological differences associated with the two ploidy levels have not been detected. *C.r.* is an outbreeding species which is extremely variable within Europe and is associated with a fairly wide range of edaphic and climatic conditions in Britain. Heavy-metal-tolerant races have been recognized within the Sheffield region.
Current trends *C.r.* is a plant of infertile habitats. Therefore, despite its widespread distribution and apparently high mobility, we suspect that *C.r.* is likely to continue to decrease unless current trends in land use are reversed.

50   *Capsella bursa-pastoris*

Shepherd's Purse

Established strategy R.
Gregariousness Individuals typically scattered.
Flowers White, hermaphrodite, homogamous, mainly self-pollinating.

Regenerative strategies B$_s$.
Seed 0.11 mg.
British distribution Widespread (100% of VCs).
Commonest habitats Arable 30%, bricks and mortar 18%, paths/soil

heaps 14%, manure and sewage waste 11%.

Spp. most similar in habitats *Matricaria matricarioides* 96%, *Polygonum aviculare* 96%, *Poa annua* 88%, *Stellaria media* 82%, *Chenopodium album* 82%.

Full Autecological Account *CPE* page 152.

Synopsis *C.b-p.* is an annual herb which exploits disturbed, fertile, artificial habitats. In arable fields it is most frequent in broad-leaved crops. The status of *C.b-p.* as one of the 'world's worst weeds' and as an ardent 'follower of man' may be related to the short life-span, coupled with the capacity to germinate over much of the year. Seed of this facultatively long-day plant may be produced within six weeks, and up to three generations may be formed each year. *C.b-p.* thrives on soils of high mineral status and an average plant may produce *c.* 3000 seeds. However, seed number is variable, with robust plants producing *c.* 17 000 or more and stunted plants from droughted or nutrient-stressed habitats forming less than 50 seeds. Seeds are shed slowly from each siliqua and have a variety of mechanisms for long-distance dispersal. They are small and flattened, and may be dispersed locally by wind. The seed testa is mucilaginous, and transport in mud on footwear and agricultural machinery appears to be the major form of seed dispersal. Seeds may survive ingestion by stock, birds and earthworms. An exceptionally persistent seed bank is formed in the soil. Seeds require chilling before germination is possible.

Because freshly shed seed is dormant, *C.b-p.* is limited in its capacity rapidly to exploit impermanent areas of bare soil. *C.b-p.* is more characteristic of continuously disturbed sites, e.g. gardens and arable fields. In these habitats, moreover, the semi-rosette growth form enables the species to survive regimes of trampling, grazing and mowing to a much greater extent than erect ruderals such as *Senecio vulgaris*. *C.b-p.* is very variable and largely in-breeding; many morphologically distinctive local populations were formerly separated. Although in less-disturbed sites particular genotypes tend to predominate, in highly disturbed sites where selection pressures may be weaker populations are genetically variable. *C.b-p.* hybridizes with the alien *C. rubella*.

Current trends Very common in a range of artificial habitats, and a further increase in its abundance and ecological range appears likely.

---

51 *Cardamine amara*   **Large Bitter-cress**

Established strategy C–R.
Gregariousness Intermediate or patch-forming.

Flowers White, hermaphrodite, homogamous, at least partially self-incompatible, insect-pollinated.

Regenerative strategies V, ?B$_s$.
Seed 0.26 mg.
British distribution Not in the W (56% of VCs).
Commonest habitats Shaded mire 7%, unshaded mire/river banks 5%, running aquatic 2%, still aquatic 1%.
Spp. most similar in habitats *Solanum dulcamara* 91%, *Myosotis scorpioides* 89%, *Phalaris arundinacea* 88%, *Carex acutiformis* 88%, *Galium palustre* 87%.
Full Autecological Account *CPE* page 154.

Synopsis *C.a.* occurs typically in soligenous mire or 'semi-aquatic' habitats, and is characteristic of moderately fertile soils where the vigour of the more robust wetland species is restricted by summer shade and often also by winter flood damage. The species exhibits rapid shoot growth in spring, and flowering is completed before the tree canopy has fully formed. In upland areas *C.a.* frequently occurs in the open. Unlike other native members of its genus, *C.a.* is not truly winter green, but leaves are formed during winter. *C.a.* forms extensive patches by means of stolons. Breakdown of older stolons by mechanical damage during winter flooding results in the isolation of daughter plants. In the laboratory, shoot pieces readily re-root to form new plants, and this process may enable *C.a.* to colonize whole stream or river systems. The species is variously reported as self-incompatible and partly autogamous, and it has been reported that stream-side populations, which may consist of a single clone, often produce virtually no seed. Little is known about regeneration by seed, which is recorded rarely, probably from a bank of persistent seeds. The laboratory characteristics of the seed suggest that germination occurs in spring and, like a number of other wetland species, fluctuating temperatures are required. It is not clear whether the rarity of *C.a.* in W Britain and Ireland is related to ineffective dispersal of the species or to other ecological factors. Two other members of the *C. amara* group are recognized in Europe.

Current trends Probably decreasing as a result of habitat destruction.

52   *Cardamine flexuosa*   Wood Bitter-cress

Established strategy Between S–R and R.
Gregariousness Sparse to intermediate.
Flowers White, hermaphrodite, inbreeding.
Regenerative strategies S, B$_s$.
Seed 0.13 mg.

British distribution Widespread (99% of VCs).
Commonest habitats Shaded mire 29%, river banks 22%, walls 9%, running aquatic/unshaded mire 6%.
Spp. most similar in habitats *Chrysosplenium oppositifolium* 92%,

Filipendula ulmaria 86%, *Solanum dulcamara* 85%, *Caltha palustris* 84%, *Cardamine amara* 81%.
Full Autecological Account *CPE* page 156.

Synopsis *C.f.* is a small, rosette-forming herb of moist, usually shaded habitats. The species is particularly characteristic of those fertile sites in which the vigour of the more robust colonizing perennials is restricted by seasonal shade or some form of physical disturbance. Thus, *C.f.* occurs in mire, stream banks and river margins which may be subject to scouring and flooding in winter. The species is found in gardens, and also colonizes crevices in brickwork and cliffs. The explosive discharge of the narrowly-winged seeds may be important as a means of locating these relatively inaccessible microsites. As illustrated by the triangular ordination and the list of species 'most similar in habitats', despite the relatively small stature and rather ephemeral life-history, the species occurs most consistently in vegetation dominated by competitive ruderals. Several features of *C.f.* appear to determine this behaviour. They include (a) the ability of seedlings and overwintering shoots to commence growth and produce seeds early in the season before consolidation of the leaf canopy, (b) flexibility in life-history which allows swift fruiting in temporary situations and continued reproduction in more secure locations, (c) tolerance of moderate intensities of

shade and (d) flexibility in seed germination biology. Seeds germinate over a wide range of temperatures, and are often prominent in the buried seed bank, in locations where *C.f.* is only a minor component of the established vegetation. Shoot fragments detached as a result of disturbance are capable of rooting to form new plants; the significance of this mode of regeneration remains uncertain. In Japan, ssp. are recognized which differ in their longevity, capacity of the stem segments to root and distance to which seed is explosively dispersed. *C.f.* hybridizes with *C. hirsuta* and with *C. pratensis*, and may itself be an allotetraploid resulting from hybridization between *C.h.* and *C. impatiens*.
Current trends An effective colonist of disturbed fertile habitats, and probably increasing.

53 *Cardamine hirsuta*

Hairy Bitter-cress

Established strategy S–R.
Gregariousness Sparse.
Flowers White, hermaphrodite, homogamous, rarely insect-, automatically self-pollinated.

Regenerative strategies S, B$_s$.
Seed 0.09 mg.
British distribution Widespread (99% of VCs).
Commonest habitats Limestone

quarries 11%, cinder heaps/lead mine spoil/rock outcrops 5%, limestone pastures/soil heaps/verges/limestone wasteland 2%.

Spp. most similar in habitats *Crepis capillaris* 82%, *Erigeron acer* 77%, *Desmazeria rigida* 75%, *Solidago virgaurea* 71%, *Arenaria serpyllifolia* 68%.

Full Autecological Account *CPE* page 158.

Synopsis *C.h.* is most characteristically a winter annual on well-drained sandy or calcareous soils subject to summer drought. After *Erophila verna*, *C.h.* is the earliest flowering of the common winter annuals, and occupies the shallowest soils and other particularly drought-prone sites. Regeneration is entirely by seed, which is explosively discharged up to *c.* 1 m from the plant in still conditions. Plant size and seed production vary considerably with soil fertility and prevailing climate; an average of *c.* 100 seeds per plant have been quoted for a sand-dune population, but > 50 000 seeds were found on a nearby garden plant. Seeds are often shed as early as May. However, germination is delayed until autumn as a result of an after-ripening requirement and, typically of many winter annuals, *C.h.* forms a persistent seed bank. Fragments of leaves with pieces of stem attached are capable of rooting; however this is seldom important as a means of regeneration. Like *Arabidopsis thaliana*, the species has expanded its range to become an abundant weed of tree nurseries. Here the factor restricting the incursion of perennials and favouring this ephemeral is not drought but the herbicide applied during early summer. *C.h.* also occurs as a winter annual on shaded rock outcrops and is also found in moist, often shaded, gardens and in other waste places; here two generations may occur within a single year, one in spring and another in autumn. Thus, *C.h.* is associated with a wide range of edaphic conditions and produces populations which differ in phenology. Whether these differences have a genetic basis, as in the case of *A.t.*, requires investigation. Hybrids have been recorded with *C. flexuosa*. **Current trends** An effective colonist of artificial habitats, this species appears to be increasing.

54   *Cardamine pratensis* — Cuckoo Flower, Lady's Smock

Established strategy Between C–S–R and R.

Gregariousness Sparse to intermediate.

Flowers Lilac, hermaphrodite, protogynous, self-incompatible, insect-pollinated.

Regenerative strategies (V), ?B_s.

Seed 0.60 mg.

British distribution Widespread (100% of VCs).

Commonest habitats Unshaded mire 20%, shaded mire 11%, enclosed pastures 5%, limestone pastures 4%, meadows/river banks 3%.

Spp. most similar in habitats *Juncus effusus* 90%, *Galium palustre* 89%, *Epilobium palustre* 88%, *Myosotis scorpioides* 87%, *Stellaria alsine* 81%.

Full Autecological Account *CPE* page 160.

Synopsis *C.p.* is an early-flowering, winter-green, rosette-forming herb characteristic of wet grassland and mire on moderately fertile soils. The species, like several others characteristic of soligenous mire (e.g. *Carex panicea*), is also frequent on N-facing limestone grasslands in upland areas. The basal leaves generally occupy the lower levels of the leaf canopy and *C.p.* may persist in open woodland. Thus, we suspect that *C.p.* is to some extent shade-tolerant. However, the species is less common in tall vegetation and is most frequent where the vigour of potentially dominant species is restricted by grazing. *C.p.* exploits vegetation gaps in a manner similar to *Ranunculus repens* except that new plants arise from leaflets in contact with the soil rather than from stolons. The resulting plantlets may be as much as 200 mm from the centre of the parent plant, and reproduction by this means may constitute the main method of regeneration in moist sites. Leaves and shoot pieces detached as a result of disturbance are capable of producing new plants and in the absence of competition *C.p.* may also extend through the development of branched rhizomes at the rate of 40–50 mm year^{-1}. Seed-set is frequently low in grazed sites, and is probably most important in drier sites where vegetative regeneration is less easily effected. Seeds are very variable in size and are explosively discharged up to 2.5 m from the parent plant. Germination occurs mainly in spring, and a persistent seed bank is suspected. Despite being an outbreeding species, much of the locally-occurring morphological variation within the species is attributable to phenotypic plasticity. Six species are included within the *C. pratensis* group in Europe: *C. pratensis* $2n = 16, 28–34, 38–44, 48$ and *C. palustris* $2n = 56, 64, 72, 76, 80, 84$, c. 96 are recognized for Britain. They cannot be recognized by morphological characters in Britain and have not been distinguished, either in British floras or here. While populations in Europe often have a range of chromosome numbers, those from the British Isles are more constant. The $8x$ karyotype tends to be the commonest within N Britain, with the $4x$ plant appearing to be largely restricted to S England. Hybridizes with *C. flexuosa*.

Current trends Probably not as common as formerly in lowland pastures, but likely to remain as a frequent component of marshland and damp grassland in upland Britain.

Established strategy S–C.
Gregariousness Often stand-forming.
Flowers Red or purple-brown,
  monoecious, wind-pollinated.
Regenerative strategies V.
Seed 1.11 mg.
British distribution Widespread except
  in N Scotland (83% of VCs).
Commonest habitats River banks 8%,
  unshaded mire 5%, shaded mire
  4%, still aquatic 3%, running
  aquatic 2%.
Spp. most similar in habitats
  Insufficient data.
Full Autecological Account *CPE* page
  162.

Synopsis A species in urgent need of
further study. *C.a.* is a tall (to *c.*
1.5 m), wetland sedge forming
extensive stands where the vigour of
other potential dominants is reduced
by intermittent grazing, by light shade
from a tree canopy or by low soil
fertility. *C.a.* also frequently grows at
the water's edge, particularly beside
large ponds, lakes and rivers, and the
possibility that it may persist in these
sites because the growth of other
dominants is suppressed by the
disturbance associated with wave
action or water currents requires
investigation. *C.a.* is not normally
found in water systems subject to large
fluctuations in water level and is absent
from reservoir margins. Although
extending from mire into aquatic
systems, in no site is *C.a.* exclusively
restricted to submerged soil (see notes
on *C. riparia* below). *C.a.* is primarily
a lowland species, with a maximum
altitude in the British Isles of only
370 m. *C.a.*, which is moderately
tolerant of ferrous iron toxicity, is
found in a wide range of edaphic
conditions including calcareous mire
but is absent from acidic peat. In
lowland Britain *C.a.* is primarily found
in water-margin communities and fen-
meadows, whereas in the uplands it is
almost exclusively a species of
soligenous mire. It is assumed that this
difference relates to the availability of
suitable habitats at different altitudes,
rather than to any major ecological
discontinuity in the behaviour of *C.a.*.
Species of *Carex* often contain
cyanogenic glycosides and other toxic
principles. The role of these
constituents in defence against
herbivory remains uncertain for *C.a.*
and for other British species of *Carex*.
C.a. shows effective vegetative
regeneration, forming extensive
patches by means of rhizomatous
growth. Rhizome fragments detached
during periods of disturbance may be
transported to other sites and form
new colonies. This may be the most
efficient method of long-distance
dispersal, since seed-set is frequently
poor. We suspect that germination
takes place in spring. The nature of the
seed bank remains uncertain. The
dispersule floats on water and may also
be dispersed by waterfowl, but there is
no evidence that *C.a.* is an early
colonist of artificial habitats. A
frequent associate of *C.a.* in waterside
habitats is *C. riparia*. The distribution
of the two overlap but *C.r.*, which

commences growth and flowers slightly earlier, consistently extends further into the water than *C.a.*.
Current trends Probably decreasing.

56   *Carex caryophyllea*   Spring Sedge

Established strategy S.
Gregariousness Intermediate.
Flowers Red-brown, monoecious, wind-pollinated.
Regenerative strategies V.
Seed 1.19 mg.
British distribution Widespread (96% of VCs).
Commonest habitats Limestone pastures 42%, limestone wasteland 14%, scree 12%, limestone quarries 6%, lead mine spoil 3%.
Spp. most similar in habitats *Briza media* 94%, *Avenula pratensis* 92%, *Koeleria macrantha* 89%, *Polygala vulgaris* 88%, *Sanguisorba minor* 87%.
Full Autecological Account *CPE* page 164.

Synopsis *C.c.* is an inconspicuous, but locally abundant, winter-green creeping sedge of unproductive low turf, and usually only reaches high frequencies in situations where dominance by taller and more-robust species is prevented by mineral nutrient stress and grazing. In pastures the canopy of *C.c.* may become submerged between grazing episodes, and the species shows considerable shade-tolerance. However, in derelict calcareous pastures where the leaves of *C.c.*, which do not exceed 200 mm, become heavily shaded, the species tends slowly to disappear. In common with most sedges, *C.c.* is non-mycorrhizal; this dependence upon the root system for the uptake of mineral nutrients is unusual among species from unproductive calcareous grasslands. Though largely restricted to calcareous soils, *C.c.* shows wide edaphic tolerance. In the Sheffield region the species is restricted to calcareous sites within the lowland area; additional fieldwork has revealed that in the uplands *C.c.* extends onto moderately acidic podsolized soils, of pH as low as 4.0. This accords with the British distribution, where *C.c.* is frequent on calcareous soils in the S and E but extends onto more acidic soils in mountain areas. This pattern is reminiscent of that of *Helianthemum nummularium* and *Thymus praecox*. With *Viola hirta*, *C.c.* is the earliest-flowering common species of calcareous grassland. Nevertheless, like other sedges *C.c.* has small cells and a low DNA amount and leaf growth is delayed until late spring. Seed production is often extremely poor and fruits, which are ripe by May to July, are frequently retained on the plant until August to September, leading to losses during grazing. Seedlings are rarely seen in the field, and time of seed germination is uncertain, although the known behaviour of similar grassland sedges predicts spring germination. The information needed to characterize the seed bank is lacking. However, it is clear that regeneration is mainly by vegetative means from clonal patches which may fragment to form daughter colonies. Although *C.c.* has colonized lime-enriched roadsides across acidic moorland, the species appears to have a low potential for colonizing new sites and is characteristic of old grasslands.

There is no information on the range of ecotypes formed by *C.c.*.

Current trends *C.c.* shows low seed production and poor mobility and is vulnerable both to increased fertility and to the relaxation of grazing pressure. Thus, *C.c.* is likely to decrease still further, particularly in lowland areas.

57 *Carex flacca*   Carnation-grass, Glaucous Sedge

Established strategy S.
Gregariousness Intermediate.
Flowers Dark brown, purplish-brown or black, monoecious, wind-pollinated.
Regenerative strategies V, B_s.
Seed 0.37 mg.
British distribution Widespread (100% of VCs).
Commonest habitats Scree 37%, limestone pastures 33%, limestone wasteland 12%, limestone quarries 11%, verges 6%.
Spp. most similar in habitats *Helianthemum nummularium* 90%, *Polygala vulgaris* 87%, *Viola riviniana* 86%, *Avenula pratensis* 85%, *Potentilla sterilis* 82%.
Full Autecological Account *CPE* page 166.

Synopsis *C.f.* is the commonest British sedge of nutrient-deficient calcareous soils and is particularly frequent in short turf. The species is widespread but generally less abundant on non-calcareous strata. Studies of this presumably long-lived, stress-tolerant species in derelict limestone grassland revealed little seasonal change in shoot biomass. This is due to leaves surviving for one year and being replaced annually in late spring. Thus the leaves appear to have a longer life-span than many of the species with which *C.f.* is commonly associated, and we predict that *C.f.* is among the less palatable

species of calcareous grassland. *C.f.* exhibits a N-facing bias in grassland. Infrequent occurrence in shaded habitats may be related more to susceptibility to fungal attack than to intolerance of low light *per se*, since experimental plants grow successfully in sand-culture at 1.6% daylight while failing at this light intensity under the more humid conditions associated with solution culture. Once established on a site *C.f.* forms large clonal patches. A persistent seed bank has been reported but regeneration by seed appears to be a rare event. *C.f.* extends into a wide range of calcareous habitats including soligenous mire (often with *C. panicea*). Survival in wetland habitats may be related to the capacity of the roots to diffuse oxygen from shoot to root to soil. The extent to which ecotypes have been formed in this morphologically variable and ecologically wide-ranging species appears not to have been investigated. However, relative growth rates of comparable ramets cloned from a single genet of *C.f.* have been shown to vary by more than twofold.

Current trends *C.f.* is a species of infertile, often calcareous soils and is declining, particularly in lowland areas, as a result of the replacement of unproductive pasture by a range of disturbed and fertile habitats.

58   *Carex nigra*   Common Sedge

Established strategy Between S–C and S.

Gregariousness Intermediate to stand-forming.

Flowers Black, monoecious, wind-pollinated, almost completely self-incompatible.

Regenerative strategies V, B_s.

Seed 0.81 mg.

British distribution Widespread (100% of VCs).

Commonest habitats Unshaded mire 10%, shaded mire/acidic pastures 7%, limestone wasteland 5%, paths 4%.

Spp. most similar in habitats Insufficient data.

Full Autecological Account *CPE* page 168.

Synopsis An ecologically wide-ranging species in need of further investigation. *C.n.* is a rhizomatous, stress-tolerant sedge most typical of soligenous mire in sites where the growth of potential dominants is suppressed by low fertility (and often also by grazing pressure). *C.n.* also occurs beside upland streams, in moist grassland and flood hollows, but is virtually absent from mire which is adjacent to open water. Although typically a minor component of vegetation, *C.n.*, which is taller than *C. panicea* and shows winter die-back, may exert local dominance particularly on acidic peaty soils. *C.n.* also extends into calcareous mire. Regeneration is mainly by vegetative means, and large clonal patches may be formed as a result of rhizome growth. A persistent seed bank has been reported, but regeneration by seed is probably a rare event. Further, *C.n.* rarely colonizes recent artificial habitats. *C.n.* hybridizes with four other members of the *Carex nigra* group. Some of the variation within *C.n.* in Britain may be due to subsequent introgression; this, in turn, may be responsible in part for the wide habitat range and the lack of cohesive ecological identity in the species. In topogenous mire subject to large fluctuations in water-table, *C.n.* occasionally forms large tussocks, perhaps for reasons similar to those advanced for *Molinea caerulea*. In dune slacks and around sheep shelters the leaves are short and broad, while those in nutrient-poor mire are long and narrow.

59   *Carex panicea*

Carnation-grass, Carnation Sedge

Established strategy S.
Gregariousness Intermediate, patch-forming in mire.
Flowers Purplish or brown, monoecious, wind-pollinated.
Regenerative strategies V, ?S.
Seed 1.88 mg.
British distribution Widespread (100% of VCs).
Commonest habitats Limestone pastures 10%, limestone wastelanu 7%, unshaded mire 3%.
Spp. most similar in habitats Insufficient data.
Full Autecological Account *CPE* page 170.

Synopsis *C.p.* is a winter-green, low-growing, rhizomatous sedge most characteristic of moist or waterlogged sites where the growth of potential dominants is suppressed by low soil fertility and by grazing or cutting. The species, which is virtually absent from shaded habitats, is most typical of soligenous mire, both on moderately base-poor soils with *C. demissa* and *C. echinata* and in highly calcareous sites, often with the smaller-seeded *C. flacca*. *C.p.* also occurs in wet pasture and dune slacks, and within the Sheffield region is particularly characteristic of calcareous grassland on moist N-facing slopes. A similar capacity to exploit both soligenous mire and limestone grassland in Northern Britain is exhibited by a number of grassland species, e.g. *Anthoxanthum odoratum*, *Briza media*, *C.f.* and *Festuca rubra*, and by some predominantly wetland species (*Anagallis tenella*, *C. hostiana*, *C.*

pulicaris, *Parnassia palustris*, *Pinguicula vulgaris* and *Valeriana dioica*). At small springs within soligenous mire, *C.p.* is often replaced close to the water's edge by species more exclusively restricted to wetland, such as *C.d.* and *C.e.* in base-poor sites, or by *Eleocharis quinqueflora* in base-rich areas. *C.p.* frequently forms diffuse but extensive clonal patches, usually in species-rich vegetation. Regeneration by means of the rather large seeds, which germinate in the spring, is probably infrequent. The utricle surrounding the nut gives the dispersule prolonged buoyancy in water, and where the species occurs close to water this may be an important method of colonizing new sites. However, in lowland areas at least, *C.p.* appears a poor colonist and is indicative of older vegetation types.
Current trends Decreasing as a result of habitat destruction. Already becoming rare in some lowland areas.

Established strategy S.
Gregariousness Sparse to intermediate.
Flowers Brown, mainly
   wind-pollinated.
Regenerative strategies B_s, V.
Seed 1.17 mg.
British distribution Widespread (98%
   of VCs).
Commonest habitats Acidic pastures
   9%, acidic quarries 8%, paths/acidic
   wasteland 2%, cliffs/scrub 1%.
Spp. most similar in habitats *Erica
   cinerea* 93%, *Vaccinium vitis-idaea*
   87%, *Vaccinium myrtillus* 82%,
   Empetrum nigrum 76%,
   Deschampsia flexuosa 76%.
Full Autecological Account *CPE* page
   172.

Synopsis *C.p.* is a tufted winter-green sedge most typically found in unproductive grassland and heathland on leached, moderately acidic soils. The species has also been observed, albeit rarely, on calcareous clay soils on limestone, and its distribution extends into dry habitats to a greater extent than that of any other common sedge. The tussock growth form of *C.p.* affords the young shoots of the mature plant some protection against burning and the shoot apices are protected by leaf bases. Although moderately resistant to trampling and able to persist and set seed in light shade, *C.p.* is more susceptible to close grazing than more-productive species such as *Agrostis capillaris*, and is often replaced in little-managed sites by more-robust species such as *Nardus stricta*. However, *C.p.* produces a persistent seed bank. and would be expected to be most abundant following extensive disturbance. This suggestion is consistent with the behaviour of *C.p.* described in *Calluna* heath burned about every 15 years. Here, abundant seed is produced only during the early phase of the *Calluna* regeneration cycle, at which time the seed bank of *C.p.* rapidly increases in size with stand age. The common occurrence of *C.p.* in old gritstone quarries is also strongly indicative of the importance of past disturbance events, in allowing the species to establish as an opportunist colonist of bare areas. Some lateral vegetative expansion of established plants occurs and patches may occasionally exceed 0.5 m in diameter. However, *C.p.* retains its densely tufted habit and there is little evidence of extensive clonal development. Thus, effective regeneration is by seed. Both the mechanisms breaking seed dormancy and the seasonality of natural germination require further study. Seeds temporarily possess an elaiosome and are dispersed by ants up to *c.* 1.4 m away from the immediate canopy of the parent plant, to sites which may be more conducive to seedling establishment. Heavy seed predation by mice and beetles has been noted. Geographically isolated plants are sometimes found by paths and occasional long-distance dispersal by humans or vehicles is suspected. Another ssp. of *C.p.* is endemic to the Azores.

Current trends Although *C.p.* can colonize some artificial habitats, few suitable habitats for *C.p.* are being created, and the species is probably decreasing as a result of the destruction of heaths and acidic grassland.

61   *Carlina vulgaris*   Carline Thistle

Established strategy S–R.
Gregariousness Sparse.
Flowers Florets yellow-brown, tubular, homogamous, insect-pollinated or selfed.
Regenerative strategies Limited dispersal by wind.
Seed 1.53 mg.
British distribution Lowland only (77% of VCs).
Commonest habitats Lead mine spoil 5%, limestone pastures 4%, limestone wasteland 3%, rock outcrops 2%.
Spp. most similar in habitats Insufficient data.
Full Autecological Account *CPE* page 174.

Synopsis *C.v.* is a monocarpic, perennial thistle of unproductive calcareous grassland, open rocky habitats and stabilized sand dunes. Generally only a small proportion of individuals flower in any one season, and rosettes usually persist in the vegetative state for several years, some needing six years to flower. The low-growing rosettes appear particularly vulnerable to overshading, and tend to be restricted to open microsites. All parts of the shoot, including the bracts protecting the inflorescence, bear stiff spines which are effective deterrents to mammalian herbivores. *C.v.* occurs in dry habitats, and we suspect that its relatively deep tap-root may allow *C.v.* to utilize subsoil moisture during periods of drought. Plants growing in N England are typically small and with few capitula. Those from, for example, S England are often taller, and under very favourable conditions *C.v.* may produce 1–30 capitula which may each contain up to 300 achenes. Regeneration is entirely by seeds, which are shed late in the year and germinate mainly during spring. Studies in Holland suggest that low numbers of seeds persist in the seed bank. Population densities in some sites may be regulated primarily by severe seed predation. The size of the achene is large relative to that of the easily detachable pappus, and *C.v.* has a relatively short dispersal distance. Two other subspecies are recognized from Europe, including one which exploits damp grassland.
Current trends. The creation of bare, infertile areas through quarrying and lead mining has probably led to an increase in *C.v.*. However, *C.v.* rarely shows evidence of colonization over long distances, and appears to have a

narrow ecological range. *C.v.* is likely to decrease as many of the unproductive calcareous pasture sites in which it occurs continue to become overgrown or are destroyed.

62 *Centaurea nigra*

Lesser Knapweed, Hardheads

Includes sspp. *nigra* and *nemoralis*

Established strategy Variable, mostly between C–S–R and S.
Gregariousness Intermediate.
Flowers Florets purple, tubular, hermaphrodite, insect-pollinated, self-incompatible.
Regenerative strategies S, V.
Seed 2.55 mg.
British distribution Widespread (99% of VCs).
Commonest habitats Limestone wasteland 45%, limestone quarries 19%, meadows/rock outcrops 11%, limestone pastures 8%.
Spp. most similar in habitats *Brachypodium pinnatum* 88%, *Bromus erectus* 83%, *Hypericum perforatum* 82%, *Trifolium medium* 76%, *Pimpinella saxifraga* 70%.
Full Autecological Account *CPE* page 176.

Synopsis *C.n.* is a genetically-variable, short- or longer-lived, tufted herb found in a wide range of grasslands on moderately fertile or infertile soils. In the *PGE*, the species occurs 'in the very mixed associations of plots receiving no manure or incomplete fertilisers' and 'is seldom found on well-manured soils'. In turf microcosms, roots of this species became heavily colonized by VA mycorrhizas with increases of up to eightfold in seedling yield. Frequency, vigour and flowering are much greater in grassland experiencing occasional burning or infrequent mowing than in pastures. However, *C.n.* can persist in grazed turf, where its leaves may be eaten by sheep, or more rarely by cattle. The tough wiry stems are normally avoided by stock. *C.n.* shows a summer peak in shoot production and in more-droughted sites such as shallow calcareous soils on S-facing slopes; we suspect that this phenology is sustained by access to subsoil moisture. The species may be partially winter green at low altitudes but not in upland areas. Regeneration is primarily by seeds, and a robust plant may produce > 1000 each year. Much seed is retained in the fruit head after ripening and is shed at irregular times during autumn and winter. Seeds are heavily predated by insects in the capitula and by small rodents on the ground. Germination probably takes place mainly in the spring but some autumn germination may occur. No substantial persistent seed bank is formed but a few ungerminated seeds may survive for several years. Vegetative spread seldom appears important. However, new side-shoots may be produced up to 25 mm from the main stem and these may become detached from the parent plant within one to two years, occasionally giving

110

rise to a 'fairy ring' type of spread. *C.n.* comprises a complex group in need of taxonomic revision and ecological study. Early work suggested that segregates with different ecological and geographical ranges could be identified. Ssp. *nigra* is considered the prevalent northern form, found particularly on heavy moist soils, while ssp. *nemoralis* has a southern distribution and is associated with lighter, often calcareous soils. These segregates intergrade within the Sheffield area and elsewhere and their validity has been questioned. *C.n.* has not been subdivided in this account but a large number of spp. and sspp. have been separated elsewhere. Doubtless the considerable genetic variation within the aggregate taxa contributes to the ecological range of *C.n.* and to the conspicuously wide variation in flowering times observed between populations. In S England populations with radiate heads (*C. debeauxii* ssp. *thuillieri*) have distributions similar to that of *C. scabiosa*.

Current trends More common in older pastures, and probably decreasing in farmland. However, this outbreeding and genetically variable species has perhaps increased its ecological amplitude by colonizing a range of artificial grassy or forb-rich communities.

63 *Centaurea scabiosa* Greater Knapweed

Established strategy Between S and C–S–R.

Gregariousness Occurs as isolated individuals.

Flowers Florets purple, insect-pollinated; self-incompatible.

Regenerative strategies S.

Seed 7.46 mg.

British distribution Lowland, particularly in the SE (70% of VCs).

Commonest habitats Limestone wasteland 15%, cliffs 6%, cinder heaps 4%, limestone quarries/rock outcrops 2%.

Spp. most similar in habitats *Daucus carota* 70%, *Centaurea nigra* 66%, *Brachypodium pinnatum* 61%, *Asplenium trichomanes* 60%, *Hypericum perforatum* 60%.

Full Autecological Account *CPE* page 178.

Synopsis *C.s.* is an apparently long-lived, tufted, winter-green herb, largely restricted to dry and, to a lesser extent, open grassland habitats and rock outcrops, all nearly always on calcareous soils. *C.s.* is more common on S-facing slopes and, despite occupying habitats subject to drought, *C.s.* develops its maximum leaf canopy during midsummer. It seems likely that the deep tap-root allows evasion of drought by exploitation of moisture in deep crevices. *C.s.* is uncommon on the upland carboniferous limestone of Derbyshire, despite an abundance of apparently suitable habitats. We suspect that the rarity of *C.s.*, which has a southerly distribution, may be determined at least in part by climatic factors. *C.s.* is infrequent in heavily grazed or trampled areas, but appears to be tolerant of burning.

Regeneration is primarily by seed, which is produced in abundance. In Derbyshire populations, seed production may be affected by heavy infestation of the flowering stems by aphids, and the seeds are heavily predated by small rodents. The seed is large and the temperatures required for germination are high compared with those of the autumn-germinating grasses with which *C.s.* is often associated. This suggests that the strategy of establishment in *C.s.* involves spring germination and rapid penetration of the radicle into the subsoil before summer desiccation of the surface soil. Observations suggest that *C.s.* is a better competitor for insect pollinators than *C. nigra. C.s.* may be subdivided into a large number of varieties. Recently, variation with respect to growth form and leaf characters has been shown to have both a genetic and an ecological basis, although *C.s.* also shows some phenotypic plasticity. Plants from the upland carboniferous limestone are smaller and less robust than those from the magnesian limestone of N England and tend to be restricted to more skeletal soils.

Current trends *C.s.* has some capacity to exploit new disturbed habitats as an impurity in grass seed. Despite this, a decrease in the abundance of *C.s.* is predicted as a consequence of habitat destruction. As populations decrease in size, the incidence of self-compatibility may be expected to increase.

64 *Centaurium erythraea*

Common Centaury

Established strategy S–R.
Gregariousness Sparse.
Flowers Pink, hermaphrodite, usually self-pollinated.
Regenerative strategies S, B$_s$.
Seed 0.01 mg.
British distribution Widespread (93% of VCs).
Commonest habitats Limestone wasteland 7%, limestone quarries 6%, rock outcrops 5%, lead mine spoil 3%, paths/verges 2%.
Spp. most similar in habitats *Hypericum perforatum* 82%, *Inula conyza* 74%, *Centaurea nigra* 65%, *Origanum vulgare* 61%, *Cardamine hirsuta* 60%.
Full Autecological Account *CPE* page 180.

Synopsis *C.e.* is a rosette-forming winter-annual, colonizing a range of open unproductive habitats, particularly on calcareous soils.

Overwinters as a flattened rosette which contains exceptionally high concentrations of glucose and fructose. As might be predicted from its late-summer flowering period, *C.e.* is absent from frequently droughted soils, and the low-growing rosettes are susceptible to the effects of shade from surrounding species. A virtual absence of the species from pasture is presumably a reflection of the vulnerability of its erect flowering stems to grazing by stock. *C.e.* is mainly found in lowland sites and on S-facing slopes, and the distribution in upland parts of Britain may be limited by climatic factors. *C.e.* regenerates exclusively by means of minute seeds and a single plant of average size may produce *c.* 40 000. A persistent seed bank has been recorded.

Establishment requires the existence of bare areas in which the seedling experiences minimal competition. Disturbed unproductive sites associated with calcareous quarry spoil or sand and gravel pits, or even the clay overburden of coal-mine heaps, may support large colonies of *C.e.* during the earlier stages of colonization. Similarly, less-transient habitats such as outcrops, sand-dune grassland and other sites subject to a degree of soil erosion or disturbance, may have sizeable populations of *C.e.*. In laboratory experiments with turf microcosms, seedling establishment on infertile calcareous soil fails completely in the absence of VA mycorrhizas, whereas infection leads to the development of robust plants. *C.e.* is phenotypically plastic with respect to size and flower number. It also shows much genotypic variation in Europe, where six ssp. and numerous varieties have been recognized. British material has been subdivided into five varieties. Two refer to sea-cliff populations, and only one, var. *erythraea*, is typical of inland situations. Some dwarf populations from sea cliffs have been separated as var. *capitatum*. The ecology of *C.e.* is further obscured near the sea by the presence of other species of *Centaurium*, and hybrids have been reported. The ecology and life-history of *C.e.* is also very similar to that of *Blackstonia perfoliata*, another annual member of the Gentianaceae; mixed populations of *C.e.* and *B.p.* are frequently observed in lowland calcareous habitats.

Current trends A relatively mobile species which is ecologically more wide-ranging than formerly and which may become more common.

65   *Cerastium fontanum*

Common Mouse-ear Chickweed

Ssp. *glabrescens*, formerly known as ssp. *triviale*

Established strategy Between R and C–S–R.
Gregariousness Sparse to intermediate.
Flowers White, hermaphrodite, protandrous, selfed or insect-pollinated.
Regenerative strategies B_s, (V).
Seed 0.16 mg.
British distribution Widespread (100% of VCs).
Commonest habitats Meadows 85%, lead mine spoil 47%, cinder heaps 34%, limestone pastures 31%, rock outcrops 28%.
Spp. most similar in habitats *Rumex acetosa* 75%, *Trifolium repens* 74%,

Plantago lanceolata 74%, *Bellis perennis* 73%, *Ranunculus acris* 73%.

Full Autecological Account *CPE* page 182.

Synopsis *C.f.* is a low-growing, winter-green herb typically of moderately fertile but disturbed habitats. In many respects its ecological characteristics are similar to those of *Sagina procumbens*, although the latter is a much smaller plant. *C.f.* is a perennial which may form patches in excess of 0.5 m² in lawns. These may persist for periods in excess of six years. However, in some circumstances *C.f.* is capable of flowering within nine weeks of germination and can function as an annual in highly disturbed habitats, e.g. arable fields. Field observations suggest that, particularly in dry years, the species behaves as a winter annual in habitats such as rock outcrops. Invariably *C.f.* is a subordinate component of plant communities and is dependent upon the absence or debilitation of potential dominants for survival; accordingly *C.f.* is most frequent in grazed or mown grassland, and in the *PGE* the species is prominent only in unproductive plots. *C.f.* is widespread on forms of wasteland liable to occasional mechanical disturbance and shoots can grow through the heaped soil of anthills. *C.f.* is consumed by cattle and appears relatively sensitive to trampling. In the absence of competition *C.f.* persists in shaded conditions and the capacity to exploit hay-meadows, which are grazed after cutting, suggests also a modest degree of shade tolerance in the early part of the growing season. Much of the success of *C.f.* relates to its regenerative biology. An average plant may produce 6500 seeds in a year, although plants on infertile soils produce many fewer. *C.f.* forms a persistent seed bank, in which some buried seeds may survive for 40 years. As in the case of *Agrostis capillaris*, freshly-dispersed seeds often show no evidence of a light-requirement for germination and it seems likely that this characteristic is acquired during the phase in which the seed lies on the soil surface beneath a leaf canopy before burial. Germination occurs from early spring to late autumn. In addition, *C.f.* also has the capacity for lateral spread since its decumbent vegetative shoots root freely. Detached shoot apices are also capable of regeneration to form a new plant. The small seeds appear to be very mobile, and *C.f.* is a listed impurity of agricultural seed of *Phleum pratense* and *Trifolium repens*. Although the distribution is centred on moist, fertile, disturbed sites, *C.f.* is ecologically wide-ranging. The distribution overlaps those of winter-annual *Cerastium* spp. in areas subject to summer drought, that of the more robust species *C. arvense* L. in dry, less fertile, grassland and that of the Arctic *Cerastium* spp. in montane areas. Hybrids with species in the latter two groups have been recorded. The species is very variable in Europe, where it is subdivided into five ssp.

Current trends An effective and probably increasing colonist of artificial habitats. Likely to be encouraged on arable land by reduced cultivation and direct drilling.

66 *Chaenorhinum minus*   Small Toadflax

Established strategy Between R and S–R.
Gregariousness Sparse to intermediate.
Flowers Purple, hermaphrodite, usually selfed.
Regenerative strategies S.
Seed 0.07 mg.
British distribution Common only in the S (58% of VCs).
Commonest habitats Cinder heaps 14%.
Spp. most similar in habitats *Senecio viscosus* 66%, *Linaria vulgaris* 65%, *Sagina procumbens* 64%, *Daucus carota* 60%, *Epilobium ciliatum* 55%.
Full Autecological Account *CPE* page 184.

Synopsis *C.m.* is a small, summer-annual herb, which most typically grows on ballast, particularly cinders, beside railway lines in sites where the lateral vegetative spread of perennials is restricted by the rocky nature of the substrate. In habitats on actively-used track the growth of perennials may be further constrained by the application of herbicides. *C.m.* can persist in such sprayed sites because it is capable of setting seed relatively early in summer, but in untreated areas, where plants may persist until autumn, seed-set and population density are generally greater. *C.m.* seeds abundantly, an average plant produces > 2000. The seeds are minute and ribbed, and are shed from pores in a many-seeded capsule. Along railway tracks, the seeds are dispersed most effectively as a result of air currents caused by passing trains. However, the mobility of the seeds is generally low, and the dispersal corridors provided along this linear habitat appear critical for the spread of the species. The plants remain erect for a while after their death, and thus the period of dispersal may be prolonged. Seeds may also be dispersed by the plant behaving as a 'tumble-weed'. *C.m.* is also found more occasionally in a range of intermittently disturbed situations, but usually these populations are impermanent, a characteristic which almost certainly arises from the transient nature of the seed bank. The lack of persistent seeds in the soil, coupled with restriction of the germination period, may have contributed to the present rarity of *C.m.* on arable land, a habitat in which it was formerly widespread. In farmland *C.m.* is now restricted to scattered occurrences in cereal fields on calcareous soils. The species is restricted to lowland habitats in the British Isles. It is not known whether arable populations differ genetically from those on railways, but further investigation seems justified in a species capable of exploiting two such contrasting habitats. Two additional subspecies are recorded from Europe.
Current trends Formerly a frequent arable weed but becoming increasingly rare in this habitat. Now an expanding species on cinders beside railways, both in Britain and in North America.

115

Rose-bay Willow-herb, Fireweed

Also known as *Epilobium angustifolium*

Established strategy C.
Gregariousness Intermediate to stand-forming.
Flowers Purple, hermaphrodite, protandrous, self-compatible, insect-pollinated.
Regenerative strategies W, V.
Seed 0.05 mg.
British distribution Scarce only in NW Scotland and Ireland (78% of VCs).
Commonest habitats Cinder heaps 64%, bricks and mortar 53%, coal mine waste 39%, limestone quarries 37%, soil heaps 28%.
Spp. most similar in habitats *Senecio squalidus* 70%, *Senecio viscosus* 69%, *Linaria vulgaris* 61%, *Artemisia vulgaris* 60%, *Tussilago farfara* 60%.
Full Autecological Account *CPE* page 186.

Synopsis *C.a.* is a tall, potentially dominant colonizing herb rarely found in long-established plant communities and most characteristically associated with derelict land. A peak of shoot biomass occurs in midsummer. This involves a rapidly-ascending monolayer of short-lived leaves and results in a dense canopy. *C.a.* can form local monocultures with considerable accumulation of stem litter. The litter is not colonized by bryophytes, and new robust shoots of *C.a.* may penetrate it annually for many years. The habitat range of *C.a.* is restricted by the palatability of the shoots to stock and *C.a.* is also susceptible to shoot loss resulting from cutting or even frequent burning. *C.a.* is found under a wide range of edaphic conditions, from sandy soils to clay and from low to high pH. However, it is usually absent from waterlogged sites and, other than on demolition sites (of pH > 7.5), is most characteristic of slightly acidic soils. In the *PGE* the

species temporarily appeared in plots receiving ammonium salts and the 'inhibiting effect of lime on establishment' was noted. On soils of higher pH we suspect that *C.a.* is often replaced as an early colonist by *Tussilago farfara*. The foliage contains high concentrations of N and P. The current widespread abundance may be attributed to the numerous widely-dispersed seeds which frequently approach 'saturating densities' in lowland habitats and provide access not only to relatively inaccessible regeneration niches on cliffs and walls, but also to open sites recently created by local fires or earth-moving operations. The seeds are relatively short-lived and do not persist in the soil. The small seedling is highly susceptible to dominance by established vegetation. Seedling establishment is usually restricted to circumstances where, after the creation of bare soil, the habitat remains relatively undisturbed. During the first year the seedling develops into a flattened rosette, and in the second and later years of secondary succession on disturbed fertile soils *C.a.* rosettes rapidly consolidate colonies by the development of horizontal roots generating tall leafy stems. At woodland margins and in clearings *C.a.* produces phenotypes with an efficient display of thin leaves and fewer flowers. *C.a.* was formerly a rare native species and was grown in gardens for ornament (where it may still be seen in Derbyshire). Several early local records are described as garden escapees. A North American origin for the later invasive spread in this century has been suggested, but this has not been verified.

Current trends *C.a.* has increased as a result of such factors as mechanized disturbance and industrial and urban dereliction. Further increases in abundance appear likely.

68 *Chenopodium album* — Fat Hen

Established strategy Between R and C–R.

Gregariousness Individuals robust and scattered.

Flowers Green, hermaphrodite or female, wind- or self-pollinated.

Regenerative strategies B_s.

Seed Dimorphic; brown or black *c.* 1.5 mg.

British distribution Widespread but less frequent in the N and W (100% of VCs).

Commonest habitats Arable 22%, manure and sewage waste 21%, soil heaps 14%, bricks and mortar 11%, cinder heaps/paths 2%.

Spp. most similar in habitats *Atriplex patula* 94%, *Stellaria media* 88%, *Senecio vulgaris* 88%, *Capsella bursa-pastoris* 82%, *Polygonum persicaria* 79%.

Full Autecological Account *CPE* page 188.

Synopsis *C.a.* is an often robust summer annual found in a wide range of disturbed, relatively fertile habitats. It is ranked as the world's worst weed of potatoes and sugar beet. *C.a.* shares with many other troublesome weeds, e.g. *Sinapis arvensis*, (a) the ability to form a buried and highly persistent seed bank, (b) a tendency for some seed to be retained on the plant at harvest time, leading to contamination of crop seed, and (c) the tendency for dispersal by birds and mammals. *C.a.* is characteristic of nitrogen-rich habitats, a feature even more marked in *C. rubrum*. *C.a.* is palatable to livestock, though toxic in large quantities. Both the foliage and the seeds have been eaten by humans and the shoot has exceptionally high protein concentrations. *C.a.* is not, however, found generally in grazed habitats, as it does not readily recover from damage to the shoot. Regeneration is entirely by seed, and *C.a.* takes several months to set seed, which prevents its successful establishment in the most frequently disturbed sites. Although some seed germination occurs in autumn, only seedlings originating in spring survive to flower. A number of genetic variants of *C.a.* have been identified, including populations resistant to atrazine. In common with other ruderals *C.a.* also shows considerable phenotypic plasticity; this attribute is particularly marked with respect to seed characters. Under long days many small, black, dormant seeds are produced whereas under short days most seeds are brown and non-dormant. The ecological significance of this particular seed polymorphism remains unclear, but the proportion of each type of seed is known to vary from year to year within a single population, and the differences in germination requirements between the different types are as great as those occurring between some species.

Current trends Now primarily a species of broad-leaved crops, although formerly widespread in barley and other cereals. The range of *C.a.* in agricultural habitats thus appears to be declining, but the species remains common on urban wasteland. On balance *C.a.*, though still a very common and troublesome weed, may be decreasing in abundance.

69 *Chenopodium rubrum*   **Red Goosefoot**

Established strategy Between R and C–R.
Gregariousness Intermediate.
Flowers Hermaphrodite or more rarely predominantly male, usually protogynous, mainly wind- or self-pollinated.
Regenerative strategies B_s.
Seed 0.09 mg.
British distribution Mainly in the S and E (63% of VCs).
Commonest habitats Manure and sewage waste 13%, bricks and mortar 8%, coal mine waste 6%, unshaded mire 3%, soil heaps/walls 2%.
Spp. most similar in habitats *Atriplex patula* 75%, *Chenopodium album* 69%, *Senecio vulgaris* 67%, *Rorippa palustris* 63%, *Atriplex prostrata* 57%.
Full Autecological Account *CPE* page 190.

Synopsis Although widely distributed in fertile disturbed habitats, *C.r.* has a more restricted ecology than *C. album*. *C.r.* is most characteristic of nitrogen-rich environments, and in the Sheffield region reaches its maximum abundance on farmyard manure and sewage residues. Occurrences with *Lycopersicum esculentum* (tomato) on

river shingle may provide a useful indicator of pollution of waterways; both species are often observed where raw sewage is discharged into the water. *C.r.* is also common by eutrophic lakes and ponds, and may develop high even-aged populations in shallow water bodies on mud exposed during the summer months. Occurrences at the upper reaches of salt marshes imply that at least certain populations of *C.r.* exhibit salt tolerance. Like *C. album*, *C.r.* is absent from continuously and severely disturbed sites. *C.r.* rarely occurs in grazed habitats. *C.r.* shows extreme phenotypic plasticity in reproductive effort. Stunted plants may reach only 30 mm in height and produce c. 20 seeds, while robust plants may be nearly 30 times as tall and produce 10 000 times as many seeds. The seeds have a hard coat and may pass unchanged through the digestive tract of grazing animals and presumably also resist digestion by seed-eating birds. Seeds are persistent in the soil. Germination is strongly promoted by fluctuating temperatures; this effect is most pronounced in darkness and probably functions as a 'depth-sensing' and 'gap-detecting' mechanism in buried seeds. *C.r.* is often used as an experimental subject in physiological investigations of flowering. Floral induction can be effected at the cotyledon stage, and races differ in the range of day-length under which flowering is initiated. Most other aspects of the biology have not been studied intensively, although *C.r.* clearly shares many of the distinctive features of *C. album*, a species with which it shows marked ecological similarities.

Current trends Likely to increase in lowland regions, particularly in urban and probably in some agricultural areas. *C.r.* is also a regular colonist of artificial habitats in upland regions, and an increase in its altitudinal maximum is predicted.

70 *Chrysosplenium oppositifolium*   Golden Saxifrage

Established strategy C–S–R.
Gregariousness Intermediate, potentially carpet-forming.
Flowers Yellow, hermaphrodite, protogynous, insect- or self-pollinated.
Regenerative strategies (V), B_s.
Seed 0.04 mg.
British distribution Widespread (98% of VCs).
Commonest habitats Shaded mire 21%, river banks 8%, limestone woodland 5%, running aquatic 4%, cliffs/unshaded mire 3%.
Spp. most similar in habitats *Cardamine flexuosa* 92%, *Filipendula ulmaria* 90%, *Caltha palustris* 89%, *Solanum dulcamara* 83%, *Mentha aquatica* 83%.
Full Autecological Account *CPE* page 192.

Synopsis *C.o.* is a low-growing, mat-forming, winter-green herb most typically associated with soligenous mire and river banks. The species is a European endemic and part of the sub-Atlantic element of the British flora.

Typically *C.o.* forms an understorey beneath taller vegetation, and in lowland habitats *C.o.* is largely restricted to sites shaded by trees. On river floodplains in upland areas *C.o.* is most frequent beneath tall-herb communities, whereas at higher altitudes the species forms continuous carpets in the open on wet rocks and along mountain stream sides associated with species such as *Epilobium*

obscurum and *Montia fontana*. Foliage of *C.o.* is characterized by low concentrations of N and P and elevated levels of Na and Fe. *C.o* maintains an extensive leaf canopy throughout the year and is the earliest flowering species of shaded mire. These two attributes are consistent with the ability of the low-growing *C.o.* to coexist with tall herbs such as *Epilobium hirsutum* and *Petasites hybridus*, which do not expand their foliage fully until early summer. The species has a wide edaphic range, occurring both on limestone cliffs and in base-poor flushes on gritstone. *C.o.* forms large clonal patches by means of creeping stoloniferous growth. In addition, shoots detached as a result of disturbance readily re-establish, and this probably constitutes an important means of colonizing river systems. The species also produces numerous small seeds, which may be incorporated into a buried seed bank. However, regeneration by seed appears to be infrequent, and the species shows a poor colonizing ability in new artificial habitats. *C.o.* often coexists with *C. alternifolium* which is rhizomatous and tends to occur on slightly better drained soils.

Current trends Probably decreasing, at least in lowland areas.

---

71 *Circaea lutetiana* — Enchanter's Nightshade

Established strategy C–R.
Gregariousness Potentially patch-forming.
Flowers White, hermaphrodite, usually insect-pollinated but self-compatible (Raven 1963); > 50 in a raceme.
Regenerative strategies (V), S.
Seed 0.88 mg.
British distribution Infrequent only in N Scotland (92% of VCs).
Commonest habitats Shaded mire 9%, limestone woodland 6%, acidic woodland 5%, scrub 4%, broadleaved plantations 2%.
Spp. most similar in habitats *Veronica montana* 91%, *Silene dioica* 78%, *Ranunculus ficaria* 78%, *Festuca gigantea* 77%, *Lamiastrum galeobdolon* 74%.
Full Autecological Account *CPE* page 194.

Synopsis *C.l.* is a medium-sized stand-forming herb of fertile woodland in sites where the vigour of potential dominants is limited by shade. The species produces a network of short-lived shoots and stolons, and has been classified as a 'pseudo-annual'. The rapid growth and long stolons (to 300 mm) enable the species to persist as a weed of shrubberies and shaded gardens. More importantly, these attributes may also enable *C.l.* to achieve rapid local dominance during the formation of secondary woodland, and may facilitate the active foraging for optimal sites within a changing or disturbed environment. Unlike most woodland species, *C.l.* is not evergreen and shows no pronounced vernal growth; it achieves its peak in above-ground biomass during summer. Thus, *C.l.* is similar to two other species which exploit disturbed woodland: *Chamerion angustifolium* and *Stachys sylvatica*. All three exhibit a similar phenology, coupled with a well-developed capacity for vegetative spread. Their occurrence within secondary woodland is likely to be related to a high colonizing ability in disturbed sites. We suspect that their persistence during more-stable late phases of woodland development depends in part upon the poor colonizing ability of many of the woodland herbs with vernal or winter-green phenologies which are complementary to those of trees. *C.l.* does, however, differ from *C.a.* and

S.s. in two important respects: (a) it is virtually restricted to shaded habitats and is thus more truly a woodland species and (b) the fruits are clothed in hooked bristles and are, like a number of other woodland plants, dispersed by attachment to animals (and to clothing). *C.l.* does not form a persistent seed bank. A hybrid, *C.l.* × *intermedia*, is formed with *C. alpina* L. Vegetatively, this plant is perhaps even more vigorous than *C.l.* and is widespread in N Britain.

Current trends A woodland species capable of exploiting disturbed habitats and of surviving in deep shade. An expansion in plantations and urban woodland is forecast.

72   *Cirsium arvense*

Creeping Thistle, Field Thistle

Established strategy C.
Gregariousness Potentially stand-forming.
Flowers Florets tubular, pale purple or whitish, incompletely dioecious (see 'Synopsis'), insect-pollinated and partially self-fertile.
Regenerative strategies V, W, B$_s$.
Seed 1.17 mg.
British distribution Widespread (100% of VCs).
Commonest habitats Coal mine waste 31%, meadows/soil heaps 24%, bricks and mortar 23%, river banks 20%.
Spp. most similar in habitats *Holcus lanatus* 81%, *Atriplex prostrata* 75%, *Artemisia vulgaris* 71%, *Tussilago farfara* 63%, *Cirsium vulgare* 61%.
Full Autecological Account *CPE* page 196.

Synopsis *C.a.* is a tall, locally dominant, broad-leaved herb. It is designated a noxious weed under the 1959 Weeds Act and is also one of the world's worst weeds. In grassland the spiny leaves are avoided by grazing

animals, and the species regenerates aggressively by means of an extensive system of branched, lateral horizontal roots. Each of these may give rise to shoots, resulting in the formation of large clonal patches which may expand at up to 6 m year^{-1}. Roots degenerate after about two years to produce isolated daughter colonies. The success of *C.a.* on arable land, and in other disturbed environments, arises from its capacity to regenerate from root fragments. *C.a.* is also difficult to eradicate since its root system may extend several metres in depth. *C.a.* is generally believed to be dioecious, but some 'male' clones in Britain are hermaphrodite and some are subhermaphrodite. *C.a.* is also self-fertile, although seed set is relatively low. Where the species is dioecious, seed set usually requires male and female plants to be within a few hundred metres of each other. Since the pappus is readily detached, seed dispersal is often poor. Nevertheless, colonization of new sites by seedlings is frequently seen in the field. Scarification improves the germination of some thistles, suggesting that the diversity in reported germination behaviour may relate to variation in the condition of the seed coat when the test was carried out (cf. many legumes which germinate well if tested before hardening of the seed coat). *C.a.* may form a bank of persistent seeds in the soil. Both the root and young shoot were formerly eaten by humans. The species hybridizes with *C. palustre* and one further species in Britain.

Current trends *C.a.* successfully exploits agricultural land and urban wasteground. Despite its status as a noxious weed and the fact that it may be controlled by combinations of herbicides, ploughing and crop rotation, *C.a.* remains common and may even be increasing in grassland and wasteland.

73   *Cirsium palustre*   Marsh Thistle

Established strategy C–S–R.
Gregariousness Individuals usually scattered.
Flowers Florets normally purple, hermaphrodite or functionally male, moderately self-compatible, insect-pollinated.
Regenerative strategies W, B$_s$.
Seed 2.0 mg.
British distribution Widespread (100% of VCs).
Commonest habitats Limestone pastures 19%, unshaded mire 10%, shaded mire/limestone wasteland 7%, lead mine spoil/river banks 5%.
Spp. most similar in habitats *Agrostis canina* 78%, *Cardamine pratensis* 77%, *Galium palustre* 72%, *Potentilla erecta* 71%, *Juncus effusus* 68%.
Full Autecological Account *CPE* page 198.

Synopsis *C.p.* is a spiny, winter-green, monocarpic, perennial thistle of mire and moist grassland, particularly on mildly acidic soils of moderate fertility.

C.p. is infrequent in permanently waterlogged topogenous mire, and is more characteristic of soligenous mire or sites where the surface soil is moist in summer and flooded in winter. In upland regions *C.p.* extends into damp daleside grassland, and we suspect that *C.p.* does not persist in a strongly anaerobic rooting medium. The low-growing rosettes of *C.p.*, which contrast with the tall flowering stem, are very dependent upon short turf or open vegetation for establishment, and may develop for several years before flowering. Seed production is often high (*c.* 700 per plant). *C.p.* has some shade-tolerant attributes and may persist without flowering in lightly shaded woodland. *C.p.* is little grazed by stock, and is particularly frequent in grazed habitats and on the banks of rivers, ditches, etc. The capacity of *C.p.* to exploit these regularly disturbed sites is enhanced by the formation of a persistent seed bank. Buried seeds are particularly important during the 'closed' phases of coppice cycles. In coppice, seed burial and persistence appear to involve canopy-induced inhibition of germination in surface-lying seeds and thermal reversion of phytochrome in buried

seeds. It has been observed that self-pollinated plants produce fewer, larger achenes than cross-pollinated individuals. The seedlings from larger achenes have a lower risk of mortality, and it is suggested that this may facilitate the establishment of colonies from single isolated individuals. White-flowered plants occur amongst normal purple-flowered individuals in sea-cliff and mountain populations. This white-flowered form is preferentially pollinated, and, in theory, may enjoy a selective advantage under conditions of low pollination. *C.p.* forms hybrids with certain other species of *Cirsium*. **Current trends** Uncertain. *C.p.* is favoured by the increased disturbance associated with modern land-use, but not by the effects of increased drainage on wetland communities.

74 *Cirsium vulgare* Spear Thistle

Established strategy C–R.
Gregariousness Sparse.
Flowers Florets tubular, purple, hermaphrodite, aggregated into heads, insect-pollinated.
Regenerative strategies W, ?B$_s$.
Seed 2.64 mg.
British distribution Widespread (100% of VCs).
Commonest habitats Limestone pastures 14%, bricks and mortar/cinder heaps/verges 13%, soil heaps 12%.
Spp. most similar in habitats *Senecio viscosus* 62%, *Cirsium arvense* 61%, *Tripleurospermum inodorum* 60%, *Senecio squalidus* 60%, *Senecio vulgaris* 60%.
Full Autecological Account *CPE* page 200.

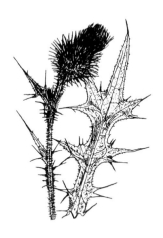

Synopsis *C.v.* is a typically robust, monocarpic perennial herb which, because of its spiny stems and leaves, is unpalatable to stock. It can colonize

the bare areas created by overgrazing in pasture and, for this reason, *C.v.* is listed under the 1959 Weeds Act as a noxious weed which landowners are obliged to control. The species is among the earlier colonists of a range of other fertile disturbed habitats, e.g.

spoil heaps. The species has a deep tap-root and is intolerant of waterlogging. *C.v.* does not prosper in shaded habitats and is less persistent in droughted sites, e.g. rock outcrops, than *Carduus nutans*. However, young plants can withstand a degree of damage and can persist in lawns. Immature plants consist of a rosette, up to 1.5 m or more in diameter, of spiny leaves and can have a stout tap-root, and flowering is delayed until the plant attains a size threshold. The dry matter allocated to seeds is *c.* 10% of the total plant dry weight, irrespective of plant size. Where productivity is lower or defoliation is experienced, flowering may be delayed for several years. A single plant may produce up to 8000 large seeds on a tall, erect flowering shoot. Because of the large pappus, seeds are very buoyant in air. However, it has been found that only 10% are dispersed outside the parental population, and seed predation is between 72 and 93%. *C.v.* seeds show little innate dormancy, may germinate in autumn or spring and are highly effective colonists of transient vegetation gaps in a disturbed landscape. One source suggested the formation of a persistent seed bank, but this was not confirmed by other studies. In fertile sites the species may even behave as a winter annual, completing its life-cycle in ten months. As with *C. palustre*, it has been suggested that self-pollinated plants produce a smaller number of larger achenes than cross-pollinated individuals. The species is very variable, but the taxon *C.v.* has not yet been satisfactorily subdivided. *C.v.* forms hybrids in Britain with two other species of *Cirsium*.

Current trends A common colonist of disturbed artificial habitats. Perhaps still increasing overall, despite the control measures applied to populations on agricultural land.

75 *Conopodium majus*

Pignut, Earthnut

Established strategy S–R.
Gregariousness Intermediate.
Flowers White, insect-pollinated, mostly hermaphrodite.
Regenerative strategies S.
Seed 2.26 mg.
British distribution Widespread (100% of VCs).
Commonest habitats Limestone pastures 17%, meadows 16%, limestone woodland 11%, limestone wasteland 8%, enclosed pastures/lead mine spoil/scrub 3%.
Spp. most similar in habitats *Anemone nemorosa* 67%, *Ranunculus bulbosus* 64%, *Viola hirta* 62%, *Brachypodium sylvaticum* 60%, *Mercurialis perennis* 59%.
Full Autecological Account *CPE* page 202.

Synopsis *C.m.*, a European endemic with a sub-Atlantic distribution, is a vernal herb with a long-lived, edible, underground tuber. The species occurs in woodland and grassland habitats. In both, *C.m.* is associated with relatively infertile, deep, mildly acidic, mineral soils. Leaves are high in N and Ca but low in P. Shoots emerge from the soil in early spring; flowering and fruiting are essentially over by midsummer; by July only the tuber remains alive. This phenology closely parallels that of *Hyacinthoides non-scripta*, and in woodlands shoots die soon after full tree-canopy expansion. The nuclear DNA amount is lower than would be expected for a vernal species, and may reflect a tendency for the shoot to grow underground throughout most of the winter and be dependent upon warm intervals and microsites for spring growth. However, the species is not exceptionally thermophilous and seedling growth rate under 20°C day and 10°C night is little different from that under a 20/15°C regime. While the vernal phenology represents a shade-avoidance mechanism in woodland, its role in grassland systems is less clear. As suggested for *Oxalis acetosella*, *C.m.* may be a woodland relic in these sites. The tendency for grassland occurrences to be restricted to N-facing upland slopes may indicate a requirement for continuously moist conditions. *C.m.* is seldom found in burned sites and the basal leaves, which usually senesce by May or early June, are vulnerable to spring fires. In common with *Lathyrus montanus*, *C.m.* is susceptible to heavy grazing during early spring. The species relies upon regeneration by seed and does not form a persistent seed bank. Only *c.* 7% of biomass was allocated to ripe fruits in one study. Clonal growth is rare. Like many members of the Umbelliferae, the embryo is relatively small and chilling is required to break dormancy. Seeds germinate in spring and the seedlings observed in the Sheffield region possess a single cotyledon. This unusual feature for a dicot herb is also shared by *Ranunculus ficaria*. *C.m.* is seldom found as a colonist of new habitats, and is usually indicative of sites with habitat continuity. Whether this restriction is exercised through failure of seedling establishment or lack of seed mobility is unknown. Studies carried out on the Iberian Peninsula suggest the existence of two sspp.

Current trends Apparently decreasing, and likely to decline still further.

76   *Convolvulus arvensis*

Bindweed, Cornbine

Established strategy C–R.
Gregariousness Forms extensive canopy at low frequencies.
Flowers Pink or white, hermaphrodite, insect-pollinated.
Regenerative strategies (V), B$_s$.
Seed 10.1 mg.
British distribution Widespread, but rarer in Scotland (98% of VCs).
Commonest habitats Arable 13%, hedgerows 10%, limestone wasteland 8%, soil heaps 4%, river banks 3%.
Spp. most similar in habitats *Lapsana communis* 71%, *Elymus repens*

67%, *Veronica persica* 60%,
Spergula arvensis 60%, *Fallopia
convolvulus* 59%.

Full Autecological Account *CPE* page
204.

Synopsis *C.a.* is a rhizomatous,
prostrate or erect, twining herb of
disturbed, fertile habitats. Like
Calystegia sepium, which it resembles
in habit, *C.a.* exploits both productive
open sites, e.g. arable land, and less-
disturbed terrain such as unmanaged
roadsides and railway or field banks,
where it may twine around the stems
of the taller grasses such as
Arrhenatherum elatius. The species is
most conspicuous, and probably most
abundant, in recently disturbed sites,
and is particularly troublesome as an
arable weed. Crop yield may be
severely reduced and, where the shoots
envelop the stems of the crop plant,
difficulties in harvesting may arise. The
species, formerly widespread in all
crops, is now mainly associated with
cereals and is most common in fields
which are winter-sown. However, it
has been suggested that, because
above-ground shoot elongation is
delayed until late spring, *C.a.* is an
ineffective competitor with those crops
which grow in early spring. Many of
the roots are in the top 600 mm of soil,
but some may penetrate to 9 m and the
species is notoriously difficult to
eradicate. The leaves and stems are
difficult to wet with herbicide sprays,
and control by herbicides is often
ineffective in sites where shoots are
afforded protection by rocky ground.
As a result *C.a.* is also especially
frequent on railway ballast. *C.a.* is a
lowland species with a bias towards S-
facing slopes, and is seldom recorded
above 200 m. Overwintering roots may
suffer damage if the ground freezes
and both distribution and seed-set
appear to be limited by climatic
factors. As we might expect from the
deep root system, *C.a.* is absent from
wetlands and is strongly persistent on
droughted cliffs and outcrops. *C.a.* is
absent from regularly grazed or mown
situations, and is also excluded from
heavily shaded habitats. In common
with *C.s.*, the species exhibits
aggressive vegetative spread in
gardens; this is due in large measure to
the ability to regenerate from a rapidly
extending network of horizontal roots,
which also regenerate effectively if
fragmented by, for example, digging.
C.a. produces seeds which are long-
persistent in the soil and have the
potential to germinate during the
warmer part of the year, particularly in
late spring. Flowers persist for only
one day, and in Britain regeneration
by seed is probably infrequent since
seed is set only in good summers.
However, seedling establishment is
very rapid and the importance of seed
in colonizing new sites is often
underestimated. Rhizome fragments
and seeds are probably transported to
new sites in soil, and seeds are also
known to be dispersed by birds.
Populations vary in leaf shape and
flower colour, and within Europe
ecotypes have been recognized. Some
plants growing beside railway tracks
may be referable to var. *linearifolius*,
which is native only to S Europe.
Populations of *C.a.* resistant to 2,4-D
have been reported.

Current trends Probably decreasing on
arable land as a result of weed control
but remaining common in other types
of disturbed habitats.

---

77   *Crataegus monogyna*

(Field records refer to seedlings)

Established strategy S–C.
Gregariousness Sparse.

Hawthorn

Flowers Usually white, hermaphrodite,
self-incompatible, insect-pollinated.

Regenerative strategies S.
Seed 8.1 × 5.1 mm.
British distribution Widespread (100%
  of VCs).
Commonest habitats Limestone
  pastures 21%, meadows 19%, scrub
  18%, limestone woodland 17%, rock
  outcrops/limestone wasteland 14%.
Spp. most similar in habitats
  Mercurialis perennis 68%, *Arum
  maculatum* 65%, *Betula* spp. (juv.)
  63%, *Fraxinus excelsior* (juv.) 65%,
  Anemone nemorosa 62%.
Full Autecological Account *CPE* page
  206.

Synopsis *C.m.* is a deciduous shrub or small tree with diffuse, porous wood and, predictably from this feature and from a high nuclear DNA amount, bud-break is relatively early in spring. One specimen in the Derbyshire Dales was found to be 165 years old. The species occurs in a wide range of habitats as a seedling or young sapling. As a taller plant contributing to the tree or under-shrub canopy, *C.m.* is less widespread and is mainly found on soils of pH > 4.5, showing a peak within the pH range 6.5–7.5. In Britain, *C.m.* is the most common colonizing shrub of grazed and ungrazed grassland on base-rich soils. It has even been suggested that hawthorn is the commonest woody species in Britain. *C.m.* is by far the most abundant hedgerow species. Apart from *Betula* spp., which exploit more acidic soils, *C.m.* is the most frequently recorded canopy species in scrub. However, *C.m.* is less frequent as a constituent of woodland, and the species is only the fourth most common canopy species of limestone woodlands of the Sheffield region. It is scarcely recorded in the remaining woodland categories. Classical studies indicate that invasion by *C.m.* following a relaxation in grazing pressure represents a first stage in the development of secondary woodland. The distribution of *C.m.* within the Derbyshire Dales consistently suggests that *C.m.* is a seral species of developing woodland. Typically, in wooded habitats *C.m.* is restricted to the understorey of secondary woodland and does not appear to regenerate effectively under shaded conditions. However, this is rather an oversimplification, since some hawthorn thickets are long-persistent and the species shows some capacity to regenerate in gaps in long-established woodland. Sites dominated by *C.m.* seldom accumulate persistent litter. The species is thorny and may even be toxic to stock. The foliage is also relatively unpalatable to the snail *Helix aspersa* and, like other unpalatable species, *C.m.* supports a large insect fauna. Recent experimental evidence suggests that *C.m.* possesses a mechanism of inducible defence against foliar predation by insects. Consequently, *C.m.* is better able to establish from seed in grazed areas than many other woody species. However, heavy predation of foliage has been observed and may contribute to the transitory nature of the dominance exerted by the species. *C.m.* is self-incompatible. Immature fruits may be predated and shaded bushes tend not to flower. Although these factors reduce seed-set, the species displays an impressive capacity for dispersal and regeneration by seed. The fleshy fruits are consumed both by resident and migrating birds, and by mammals, and seedlings are common in most types of herbaceous vegetation. Seeds are enclosed in a lignified endocarp and show embryo dormancy. Germination requires exposure to warm moist conditions to break hard-coat dormancy, followed by chilling. As a result germination in the field does not occur until 18 months after seeds are shed. No persistent seed bank has been reported. The variability of the species in Europe is such that six ssp. are recognized. In parts of S England *C.m.* shows extensive introgression with *C.*

laevigata. The latter is typical of woodland conditions and heavy soils. *C.m.* also hybridizes with *Mespilus germanica.*
Current trends Aerial photographs indicate substantial increases in this species during recent years in the semi-natural grasslands of the Derbyshire Dales. *C.m.* has shown a similar increase in many other grassland areas when the severity of grazing has been relaxed. The species is also increasingly prominent in the landscape as a result of activities such as planting of *C.m.* on motorway verges and in new hedgerows.

78 *Crepis capillaris*

Smooth Hawk's-beard

Established strategy Between R and S–R.
Gregariousness Sparse to intermediate.
Flowers Florets yellow, ligulate, hermaphrodite, self-incompatible, insect-pollinated.
Regenerative strategies W, B$_s$.
Seed 0.21 mg.
British distribution Widespread (100% of VCs).
Commonest habitats Limestone quarries 31%, rock outcrops 11%, cinder heaps/coal mine waste 8%, lead mine spoil 5%.
Spp. most similar in habitats *Desmazeria rigida* 87%, *Erigeron acer* 83%, *Cardamine hirsuta* 82%, *Leucanthemum vulgare* 72%, *Hieracium* spp. 69%.
Full Autecological Account *CPE* page 208.

Synopsis In view of its widespread occurrence, *C.c.* deserves more study than it has so far received. The species is most typically a relatively robust winter-annual herb, often exploiting open, moderately productive habitats where the cover of perennials is reduced by drought or mechanical disturbance. However, some

populations appear to contain, and may even consist entirely of, polycarpic perennial individuals. Damaged stems may also reflower in autumn. The species is associated with two main types of habitat: (a) semi-permanent habitats such as droughted rock outcrops, sandy wasteland, and steep road and railway cuttings or embankments subject to soil creep, and (b) disturbed, often transient habitats such as soil heaps and localized areas of bare ground in pasture and on road verges. *C.c.*

appears to avoid heavily droughted sites and does not normally occur in association with the early-flowering winter annuals, such as *Cardamine hirsuta* and *Saxifraga tridactylites*. Seed is set later than that of most other winter annuals, and it is capable of rapid and immediate germination. The species is frequently the only ruderal present in relatively closed vegetation dominated by perennials, and on sandy soils it may even persist in lawns. Plants without a main central stem (var. *capillaris*) are distinguishable from those with one (var. *agrestis*), though the ecologies of these varieties have not been compared. In theory, var. *capillaris* would be at a selective advantage in the short open vegetation of habitats similar to those exploited by *Geranium molle*, *Medicago lupulina* and *Trifolium dubium*, while the taller

and more erect var. *agrestis*, which also has larger involucres, would show a stronger affinity with annuals of taller vegetation such as *Torilis japonica* and *Vicia sativa* ssp. *nigra*. The species displays a strong colonizing ability, which is related to the production of numerous wind-dispersed seeds with a limited capacity for persistence at the soil surface. *C.c.* has fewer chromosomes than any other common vascular plant in the British flora, a specialization which, according to Stebbins, facilitates rapid evolution. Perhaps, therefore, we should expect considerable ecotypic differentiation within this species.

Current trends Often a rapid colonist of dry, relatively unproductive, disturbed habitats, and probably still increasing.

79 *Cynosurus cristatus*   Crested Dog's-tail

Established strategy C–S–R.
Gregariousness Intermediate.
Flowers Green, hermaphrodite, homogamous, wind-pollinated, self-incompatible.
Regenerative strategies S.
Seed 0.70 mg.
British distribution Widespread (100% of VCs).
Commonest habitats Meadows 48%, verges 19%, paths 16%, enclosed pastures 11%, limestone pastures 10%.
Spp. most similar in habitats *Bellis perennis* 89%, *Phleum pratense* 87%, *Trifolium repens* 85%, *Lolium perenne* 79%, *Alopecurus pratensis* 76%.
Full Autecological Account *CPE* page 210.

Synopsis *C.c.* is a small, short-lived, tufted grass of relatively fertile grasslands, but is now little used in agriculture. Morphologically, *C.c.* is

rather uniform and of narrow ecological range, although ecotypes may occur which are influenced in their distribution by land management. The species is susceptible both to drought and to waterlogging, and exhibits low tolerance of shade. *C.c.* shows a marked restriction to heavily grazed short turf. Although the foliage is very palatable, the wiry flowering stem is rejected by stock and the conspicuous dead flowering stems often overwinter. Flowering may occur in the first growing season and seed production may reach 1100 year^{-1} in a mature plant. Provided that the plant is not subject to heavy spring grazing, seed production may be maintained at a high level each year. However, as in a number of other sown pasture grasses (including *Dactylis glomerata*, *Lolium perenne* and *Phleum pratense*), there is no persistent seed bank. With its rather low canopy, *C.c.* performs better in pastures than in permanent hay fields. The abundance of *C.c.* in productive pastures appears to be limited by the dominant effects of more-robust species.

Current trends Formerly sown for pasture on soils of relatively low fertility, but now probably decreasing as a result of reductions in the acreage of permanent pasture. Still grown in small quantities, chiefly in lawn-grass mixtures, but since flower stems are only partially resistant to mowing, the capacity for *C.c.* to persist in lawns is probably limited. Sowings for amenity are unlikely to offset the declining abundance of *C.c.* in pastures.

80   *Cystopteris fragilis*

Brittle Bladder-fern

Established strategy Probably between S and S–R.
Gregariousness Usually sparse.
Sporangia In sori on underside of frond.
Regenerative strategies W.
Spore $44 \times 29 \mu$m, black, spiny.
British distribution Frequent only in NW Britain (82% of VCs).
Commonest habitats Cliffs 6%.
Spp. most similar in habitats
  Asplenium ruta-muraria 97%,
  Asplenium trichomanes 89%,
  Athyrium filix-femina 89%, *Elymus caninus* 85%, *Mycelis muralis* 82%.
Full Autecological Account *CPE* page 212.

Synopsis A species which has not been studied intensively. *C.f.* is a small fern of moist, skeletal habitats. Growth is more rapid than that of most ferns. The fronds are unusually thin and the species can survive at exceptionally low light intensities in, for example, cave entrances. *C.f.* is also frequent on unshaded cliffs and walls. In N England the species is almost totally restricted to upland areas, and nationally *C.f.* is distributed mainly in the N and W. It is not known whether this restriction is related to the susceptibility of established plants to summer drought or to problems of establishment from spores in drier climates. Each fertile frond produces an estimated 1 500 000 wind-dispersed spores, which are the most important means of regeneration. Even under ideal conditions, germination rate of the spores is lower than that recorded for *Athyrium filix-femina*, *Dryopteris* *dilatata* or *D. filix-mas*. It is possible that this feature, which is shared by *Asplenium ruta-muraria* and *A. trichomanes*, may be a specialization reducing the risk of premature spore germination following summer showers in otherwise droughted skeletal habitats. It is also possible that the spiny spore is hydrophobic, and thus less easily hydrated under intermittently dry conditions. It is not known whether a persistent spore bank is formed in the soil. *C.f.* produces a short branching rhizome but has little capacity for lateral, vegetative spread. *C.f.* has brittle stipes, hence its English name 'Brittle Bladder-fern'. The

species is capable of regeneration from detached fronds. There is much genetically determined morphological variation within the species, and tetraploid and hexaploid chromosome races have also been identified. These form a sterile hybrid. It is unclear to what extent the distribution of *C.f.* is determined by the (unknown) ecology of the prothallial stage and it is not known whether plants from different habitats, and of different cytotypes, are ecotypically distinct.

Current trends Uncertain, but apparently decreasing on walls through destruction or alteration of this habitat.

81   *Dactylis glomerata*   Cock's-foot

Established strategy Mostly between C–S–R and C.
Gregariousness Intermediate.
Flowers Green, hermaphrodite, homogamous, wind-pollinated, usually cross-fertilized.
Regenerative strategies S.
Seed 0.51 mg.
British distribution Widespread except in parts of Scotland (100% of VCs).
Commonest habitats Meadows 85%, verges/limestone wasteland 58%, limestone quarries 53%, soil heaps 43%.
Spp. most similar in habitats
  Taraxacum agg. 79%, *Poa pratensis* 78%, *Festuca rubra* 69%, *Leucanthemum vulgare* 68%, *Plantago lanceolata* 67%.
Full Autecological Account *CPE* page 214.

Synopsis *D.g.* is a potentially robust, winter-green grass forming clumps on a wide range of soil types and under management regimes extending from intensive to lax. *D.g.* is palatable to stock and tolerant of summer grazing. However, *D.g.* is most successful in grasslands where defoliation takes

place infrequently and a tall canopy is established. In derelict grasslands, e.g. neglected meadows and roadsides, *D.g.* may develop relatively massive tussocks which exert local dominance upon smaller herbs. In unproductive, base-rich pastures and particularly on S-facing slopes *D.g.* is represented by small-tussocked phenotypes. In turf microcosms, roots of this species became heavily colonized by VA mycorrhizas with increases of up to 80% in seedling yield. *D.g.* exhibits some tolerance of drought and also survives light burning. Despite its widespread occurrence on paths, the species is usually considered to be susceptible to trampling. In the *PGE*, *D.g.* is described as widespread and plentiful across the range of treatments 'except those inducing very acid or starved conditions'. Foliage is characterized by high Na and low Ca and Fe. The tissues of *D.g.* are rich in fructan. In common with *Lolium perenne* and several other sown species, *D.g.* has only a restricted capacity for lateral tussocking and regenerates almost exclusively by seed in the spring, primarily in vegetation gaps. Seed may persist for 2–3 years near the soil surface but *D.g* does not form a substantial persistent seed bank. This feature appears to reduce the resilience of *D.g.* in heavily grazed and trampled pastures where bare ground is quickly colonized by species such as *Poa annua*. However, seed-set is often high and the stiff cilia on the dispersule may enhance mobility. The capacity to exploit a wide range of habitats appears to stem both from phenotypic plasticity and ecotypic differentiation. Particularly in fertile sites, the distribution of *D.g.* has been profoundly influenced by the use of its seed in agriculture since the early 19th century. Over 250 t are now sown annually.

Current trends Uncertain, owing to the difficulty of distinguishing between native and introduced populations and in predicting the fate of a grass which regenerates solely by seed but which has little capacity for seed persistence. The use of *D.g.* in agriculture appears to be declining, and the abundance of *D.g.* in farmland is likely to decrease.

---

82 *Danthonia decumbens*   **Heath Grass**

Established strategy S.
Gregariousness Sparse to intermediate.
Flowers Green to purple, hermaphrodite, usually cleistogamous.
Regenerative strategies B_s.
Seed 0.87 mg.
British distribution Widespread (100% of VCs).
Commonest habitats Limestone pastures 19%, limestone wasteland 8%, paths 4%, scree 3%, enclosed pastures/verges/acidic wasteland 2%.
Spp. most similar in habitats *Succisa pratensis* 88%, *Carex caryophyllea* 84%, *Stachys officinalis* 83%, *Primula veris* 82%, *Briza media* 81%.
Full Autecological Account *CPE* page 216.

Synopsis *D.d.* is a slow-growing, loosely-tufted, winter-green grass restricted to unproductive grassland and heathland, and present on both mildly acidic and calcareous soils. *D.d.* is vulnerable to dominance by more-productive species and is totally excluded from fertile grassland. It has been suggested that the palatability of *D.d.* is greater than that of either *Agrostis capillaris* or *Festuca ovina*. In pastures, damage by grazing may have severe effects as a result of the slow rate of replacement of the relatively long-lived leaves. Where pasture productivity is high, *D.d.* is rapidly displaced by species such as *A.c.* which have a more dynamic leaf canopy. *D.d.* is more abundant in N-facing grassland, a pattern consistent with its frequent association with moist soils and local excursions into soligenous mire. *D.d.* shows only modest resistance to trampling and its distribution appears little affected by the presence or absence of burning. Compared with those of other common grasses forming persistent seed banks, the 'seeds' of *D.d.* are unusually large. Buried seed populations of *D.d.* are significantly clumped. This may be related to the role of ants in dispersal of the seed, which has an elaiosome. It seems possible that burial and persistence of seeds in the soil may be related to their late maturation, delayed release and high temperature

requirement for germination, all of which militate against immediate germination. *D.d.* exhibits a limited capacity for clonal growth. Little is known of the genetic and physiological factors underlying the bimodal distribution with respect to soil pH. It appears that populations from calcareous soils may be ssp. *decipiens*, which has a more slender habit, while those from mildly acidic soils belong to ssp. *decumbens*, but the ecological distribution of these taxa requires further study. Flowers are often cleistogamous. and it is tempting to conclude that the rather uniform morphology and restricted ecological distribution of the species may be related to in-breeding. The measurements given for seed size and germination requirements refer to populations associated with acidic soils.

Current trends *D.d.* is confined to unproductive grassland. The species is indicative of old or semi-natural grassland and has decreased, particularly in lowland areas as a result of intensive methods of farming.

83   *Daucus carota*   Wild Carrot

Ssp. *carota*

Established strategy Between S–R and C–S–R.
Gregariousness Sparse.
Flowers White, hermaphrodite, protandrous, insect-pollinated, subsequently self-fertile.
Regenerative strategies ?S, B_s.
Seed From terminal umbel.
British distribution Aggregate widespread, but mainly S and coastal (99% of VCs).
Commonest habitats Limestone wasteland 7%, cinder heaps 5%, rock outcrops 2%, cliffs 1%.
Spp. most similar in habitats *Centaurea scabiosa* 70%, *Hypericum perforatum* 63%, *Chaenorhinum minus* 60%, *Trifolium medium* 59%,

Brachypodium pinnatum 57%.
Full Autecological Account *CPE* page 218.

Synopsis *D.c.* is a monocarpic herb of dry, relatively unproductive, grassland and open habitats, often taking 3–4 years to reach maturity. *D.c.* has sometimes been considered a 'moisture-loving' plant but, presumably because of its stout tap-root, can exploit drier environments. In Britain, *D.c.* is characteristic of calcareous soils but it is also widely distributed as an adventive on other well-drained soil types. In Europe the calcicolous tendency is less pronounced. *D.c.* is vulnerable to ploughing and cutting, particularly in late summer, and grows poorly with delayed flowering in shaded habitats. *D.c.* taints milk and is not grazed preferentially by cattle. Nevertheless, it is eaten to some extent by cows, horses and sheep, and the flowering shoot is particularly vulnerable to grazing. *D.c.* regenerates entirely by seed and until maturity consists of a compact rosette of basal leaves easily overgrown by more-robust vegetation dominants. Thus, *D.c.* tends to be persistent only in short turf or open sites in ungrazed or under-utilized pasture. There is a close correlation between rosette size and probability of flowering. Seed production in large plants has been estimated at between 1000 and 40 000, but is often considerably less. Numbers as low as *c.* 80 are reported, with low rates of seedling mortality. Seeds vary in size and germination characteristics even when derived from the same plant. In chalk grassland, detached germinable seeds are detectable on or near the soil surface throughout the year, although their numbers fall in summer. Seedlings tend to survive better on more-productive microsites. Seed is dispersed long distances, largely by humans although the spiky mericarps are carried locally by wind and in animal fur and may survive ingestion. The longevity of plants of *D.c.* is under both environmental and genetic control, and annual, biennial and longer-lived populations have been identified. *D.c.* is polymorphic and 11 ssp. have been described for Europe. Another taxon, ssp. *gummifer* is found on sea cliffs and sand dunes on Atlantic coasts. Ssp. *carota* is capable of interbreeding with ssp. *g.* and the cultivated carrot (ssp. *sativus*). Whether there is any appreciable gene flow from cultivated carrot crops is doubtful, however, since crops and wild populations are seldom in close contact.

Current trends Probably decreasing as a consequence of the declining acreage of infertile habitats. *D.c.* may have increased in genetic diversity as a result of the introduction of alien genotypes along railways and in grass seed.

84 *Deschampsia cespitosa*

Excluding the montane ssp. *alpina*

Established strategy Between C–S–R and S–C.
Gregariousness Normally intermediate.
Flowers Silvery-green, hermaphrodite, protandrous, wind-pollinated.
Regenerative strategies S, B_s, V.
Seed 0.31 mg.
British distribution Widespread (99% of VCs).

Tufted Hair-grass

Commonest habitats Limestone woodland 31%, limestone pastures 25%, river banks 15%, bricks and mortar/shaded mire 11%.
Spp. most similar in habitats *Lamiastrum galeobdolon* 78%, *Festuca gigantea* 75%, *Allium ursinum* 75%, *Anemone nemorosa* 74%, *Ranunculus ficaria* 68%.

Full Autecological Account *CPE* page
220.

Synopsis *D.c* is a relatively persistent
tufted grass often forming a dense
tussock with persistent leaf litter.
These characteristics often allow local
dominance of communities by the
species. The principal habitats are
grassland and woodland. In both, *D.c.*
is most frequent on poorly drained,
slightly acidic soils. The coarse leaves
have a high silica content, and mature
leaves are usually avoided by
herbivores. As a result, *D.c.* may
thrive as a weed of lowland cattle
pasture. However, young foliage may
be eaten by horses and rabbits, and in
upland areas, where leaves tend to
have less silica, the species is grazed
freely by cattle, sheep and deer, and is
sometimes considered a useful pasture
species. Foliar levels of N, P and Ca
are low. Survival in woodland, where
flowering and seed production is often
poor, may be related to (a) the
considerable longevity of established
plants (> 30 years), (b) the capacity to
produce phenotypes with long narrow
leaves, reducing the degree of self-
shading usually characteristic of
tussock grasses and (c) persistence as
seeds in the soil through the densely-
shaded phase of coppice cycles. Under
favourable conditions *D.c.* may flower
in its second year, producing up to
500 000 small seeds per plant and
forming a persistent buried seed bank.
Regeneration by seed appears to occur
only in areas of open ground mainly
during the period February to June.
D.c. has only a restricted capacity for
lateral vegetative spread, although as a
result of breakdown of the rhizomes at
the centre of the tussock, *D.c.* may
occasionally fragment into a number of

daughter plants. Edaphic and climatic
ecotypes have been reported and
heavy-metal-tolerant populations have
been identified. There are variations in
ploidy and plants with $2n = 52$ (4x) are
more common than the diploids, which
are mainly restricted to semi-natural
woodland. This taxonomically complex
species has been subdivided into seven
ssp., many of which hybridize freely.
Current trends *D.c.* may have
decreased in grasslands as a result of
improved drainage and management,
and the conversion of pasture to arable
in lowland areas. However, in
woodlands and many other artificial
habitats a high level of mechanized
disturbance appears beneficial to *D.c.*,
and tetraploids of the species may be
increasing. The future status of
diploids, which have a restricted
habitat range, remains uncertain. *D.c.*
is frequently a colonist of many
artificial habitats and is common on
derelict land. Changes in the status of
the species are therefore more likely to
be expressed as fluctuations in relative
abundance in particular habitats rather
than more generally.

Established strategy Between S and
S–C.
Gregariousness Patch-forming.
Flowers Silvery green, hermaphrodite,
wind-pollinated, self-incompatible.
Regenerative strategies V, S.
Seed 0.43 mg.
British distribution Widespread (98%
of VCs).
Commonest habitats Acidic quarries
99%, acidic pastures 97%, acidic
wasteland 66%, limestone
pastures/scrub 36%.
Spp. most similar in habitats
Vaccinium myrtillus 89%, *Calluna
vulgaris* 84%, *Vaccinium vitis-idaea*
81%, *Galium saxatile* 80%, *Erica
cinerea* 78%.
Full Autecological Account *CPE* page
222.

Synopsis *D.f.* is a slow-growing,
evergreen, clump- or carpet-forming
grass tolerant of low pH, low mineral
nutrient supply and high external
concentrations of aluminium and
manganese. The leaves contain low
amounts of N, P, Mg and Ca and *D.f.*
responds to low levels of nitrogen by
more-efficient utilization and
redistribution of dry matter in the form
of finer roots and root hairs. The
species is typical of cold climates and
podzolic soils and produces humus
which is persistent and inhibitory to
plant growth. *D.f.* is the most
successful calcifuge grass in Britain,
and combines a very wide geographical
distribution with the capacity to exploit
a diversity of acidic dryland habitats.
The species is particularly
shade-tolerant, and seedlings are
capable of persistence in total darkness
for three months. However, vegetative
vigour and flowering are both inhibited
by dense shade, and *D.f.* is also rather
vulnerable to submergence beneath
deciduous tree litter. In unshaded
habitats in Britain differences in
performance may be associated with
aspect. On N-facing slopes, vegetative

expansion may result in continuous
monocultures, while on adjacent S-
facing slopes tussocks are often stunted
but produce abundant inflorescences.
D.f. is eaten by sheep and rabbits, but
in moorland habitats new shoots of
Calluna vulgaris are preferred and
grazing here can lead to an increase in
D.f.. However, under conditions of
intense grazing, mowing or trampling,
or if the soil is dry, *Festuca ovina* tends
to be more prevalent. The species is
capable of forming large clonal patches
as a result of rhizome growth. Freshly
shed seed germinates in autumn. The
seedlings so formed may be long-lived,
often persisting in a stunted form until
conditions become more favourable for
growth, but no persistent seed bank is
accumulated. In consequence, *D.f.* is
not an effective colonist after severe
fires. However, the species often
persists vegetatively in habitats subject
to regular fires of low intensity. The
species is occasionally planted as a
shade-tolerant amenity grass.
Current trends Although *D.f.*
colonizes a number of artificial
habitats, e.g. railway banks and coal-
mine spoil, the species is decreasing in
lowland areas, due to the destruction
of heathland and acidic grassland. *D.f.*
is likely to remain abundant in upland
areas.

Established strategy S–R.
Gregariousness Intermediate.
Flowers Green, hermaphrodite, wind-pollinated, generally self-fertilizing.
Regenerative strategies S.
Seed 0.19 mg.
British distribution Mainly in the S and E (75% of VCs).
Commonest habitats Limestone quarries 9%, rock outcrops 5%, cinder heaps/coal mine waste 2%.
Spp. most similar in habitats *Erigeron acer* 88%, *Crepis capillaris* 87%, *Cardamine hirsuta* 75%, *Saxifraga tridactylites* 73%, *Inula conyza* 67%.
Full Autecological Account *CPE* page 224.

Synopsis *D.r.* is a small, mainly winter-annual grass of relatively infertile droughted soils, either on calcareous strata or near the sea. Often the species produces a single inflorescence containing less than ten caryopses per plant. In view of this exceptionally low level of seed production it is not surprising that *D.r.* tends to be restricted to continuously open sites, e.g. rock outcrops and sand dunes. As might be expected for a relatively late-flowering winter annual, floral development shows some requirement for vernalization and long days. However, *D.r.* shows regenerative flexibility, in that (a) the species is not an obligate winter annual, although the spring-germinating plants which occur in certain populations produce even fewer seeds, (b) the after-ripening requirement is short and seedlings, many of which do not survive, may be found after any wet period between spring and autumn, and (c) additional later-flowering, secondary inflorescences may be formed. These characteristics suggest that the habitats exploitable by *D.r.* are not necessarily heavily droughted every year and, in the most common habitat for the species, limestone quarries, *D.r.* often occurs in the absence of other winter annuals. The species, which often grows to some extent during winter, is susceptible to low winter and spring temperatures, and has a predominantly lowland distribution in Britain which lies at the N limit of its European range. The majority of plants overwinter as small individuals and are particularly vulnerable to frost-heave. Because of the small and localized nature of many colonies, the species is also potentially vulnerable to the effects of heavy grazing. Like other stress-tolerant ruderals, *D.r.* responds to added nutrients. Thus, on ground manured by rabbits, seed set per plant may be twice as great as on unfertilized sites and cultivation on fertile soil has been shown to cause nearly a 30-fold increase in the number of inflorescences. More-robust populations, formerly treated as ssp. *majus*, occur in the S and W of the British Isles but appear to deserve only varietal status. Other studies have failed to find evidence of local specialization between populations. Another ssp. has been identified from maritime sand in the Mediterranean region. *D. marina* replaces *D.r.* in areas closest to the sea and a sterile hybrid between the two has once been recorded.

Current trends *D.r.* has effectively colonized many calcareous quarries in lowland England; despite the low seed production, examples occur where the species appears to have been dispersed to sites at least 10 km from established populations. On this basis, a further expansion of the species is anticipated.

87 *Digitalis purpurea*   Foxglove

Established strategy Between C–R and C–S–R.
Gregariousness Sparse.
Flowers Usually purple, hermaphrodite, protandrous, insect-pollinated but automatically selfed if not.
Seed 0.07 mg.
British distribution Widespread (99% of VCs).
Commonest habitats Broadleaved plantations 7%, hedgerows/river banks 5%, soil heaps/acidic woodland 4%.
Spp. most similar in habitats *Acer pseudoplatanus* (juv.) 76%, *Fagus sylvatica* (juv.) 76%, *Hyacinthoides non-scripta* 69%, *Quercus* spp. (juv.) 69%, *Holcus mollis* 68%.
Full Autecological Account *CPE* page 226.

Synopsis *D.p.* is a winter-green, rosette-forming, polycarpic (or more rarely monocarpic) perennial with an Atlantic distribution. It exploits lightly-shaded (often S-facing) sites usually on disturbed acidic soils of moderate fertility. Particularly within the upland part of the region, *D.p.* is characteristic of perennial plant communities on river banks, rocky habitats and steep hedgebanks. Here flooding and soil creep at intervals give rise to areas of bared soil suitable for regeneration by seed, and dictate that the growth of potential dominants is to some extent restricted. *D.p.* also occurs in recent plantations, burned areas in woods and around the exposed roots of windthrown trees. Expansions in *D.p.* occur after coppicing. Here populations are short-lived, becoming replaced in a few years by taller, more-dominant herbaceous species. In comparison with most calcifuge herbs, *D.p.* is relatively fast-growing, although under unfavourable conditions development may be slow. On fertile soils the species may form a large rosette within one year and flower in its second season. Dense populations on acidic soils may be indicative of locally enhanced soil fertility. Foliage of *D.p.* is characterized by high concentrations of P, Na, Fe and Mn. Both fresh and after drying and storage, foliage of the species is toxic to stock and humans. The poisonous principle (digitalin) is used in the treatment of heart complaints, and *D.p.* is grown as a crop. Despite its toxicity, *D.p.* is infrequent in grazed habitats. The species may occasionally produce daughter rosettes, but regenerates mainly by seed. An average plant may produce > 70 000 seeds. Freshly-shed seed has a light requirement, and germination is inhibited by a leaf canopy. As a result large quantities of seed are incorporated into a persistent seed bank, with the result that *D.p.* may remain in sites subject to intermittent disturbance, and often develops high population densities following disturbance. In an outdoor pot experiment, germination was observed in all months except December. Seed may be dispersed by wind and other agencies, and the species is a frequent colonist of a range of artificial habitats. *D.p.* is extremely variable in several of its morphological characters, and dwarf alpine ecotypes occur. A further two ssp. are recognized in Europe.

Current trends Remains common in uplands, but perhaps decreasing in the lowland region since the increased level of habitat disturbance (a feature favouring *D.p.*) is probably outweighed by the loss of habitat to agriculture and coniferous woodland.

88 *Dryopteris dilatata* Broad Buckler-fern

Established strategy Between S–C and C–S–R.

Gregariousness High dominance at low population density.

Sporangia In sori on underside of fronds.

Regenerative strategies W, V, ?B$_s$.

Spore $50 \times 33\ \mu m$, black.

British distribution Widespread (100% of VCs).

Commonest habitats Coniferous plantations 26%, cliffs 18%, acidic woodland 15%, broadleaved plantations 14%, walls 8%.

Spp. most similar in habitats *Luzula pilosa* 89%, *Pteridium aquilinum* 71%, *Rubus fruticosus* 66%, *Acer pseudoplatanus* (Juv.) 64%, *Hyacinthoides non-scripta* 63%.

Full Autecological Account *CPE* page 228.

Synopsis Long-lived and highly shade-tolerant, this fern forms a tall crown of fronds from an erect or ascending rhizome, and often functions as a woodland herb layer dominant on moist, acidic, moderately fertile, peaty soils. *D.d.* extends more frequently into unshaded habitats at higher altitudes. The fronds, particularly when young, contain thiamase, which induces thiamine deficiency. This provides an effective defence against a majority of insects, and the fronds are potentially toxic to stock. Occasionally plants produce a number of small crowns from a slender, long-creeping rhizome, but it is more typical for vegetative spread to be strongly confined. An estimated 13.5 million minute wind-borne spores are produced by each frond, and these are released at 'saturating densities' into the landscape. Large numbers of viable spores, which have a light requirement for germination, occur throughout the top 120 mm of the soil profile in many woodland sites, and the presence of a shortly-persistent spore bank is suspected. Interpretation of the ecological range of *D.d.* is complicated by the fact that the distribution of the long-lived sporophyte is determined to a considerable extent by the habitat requirements of the ephemeral gametophytic prothallus. Since the prothallus has little effective control of water-loss and requires moisture for the fertilization of the gametes, establishment of sporelings is largely confined to damp microsites. In the event of soil movement or litter decomposition, prothalli and sporelings are often killed, and they are also susceptible to competition from bryophytes and higher plants. Young sporelings tend to be observed mainly on moist, bare soil on the woodland floor, in rock crevices and on fallen trees or rotting stumps, particularly in heavy shade. The specialized nature of the microsites required for the establishment of the sporeling may account for the tendency of *D.d.* to occur as widely-spaced individuals rather than as a stand-forming species. Moreover, as a result of the ecological dissimilarity between the gametophyte and sporophyte generations, *D.d.* often persists as a young sporeling in sites which are unsuitable for the adult sporophyte. In a study in Derbyshire woodlands it was found that populations of *D.d.* exhibited a much greater level of recruitment and a higher rate of mortality than *D. filix-mas*. This 'weedy' behaviour is most marked in the cooler and moister conditions of N Britain. Consistent with its wide habitat range, including sites as distinct as montane rock ledges and lowland woodland, the species is morphologically very variable and is believed to be an auto-tetraploid derived as a result of hybridization between *D. expansa*, and a montane species *D. intermedia*, or related species. *D.d.* hybridizes with four other related species in Britain.

Current trends A rapid colonist of new plantations and rocky habitats; apparently increasing.

---

89 *Dryopteris filix-mas*   **Male Fern**

Established strategy S–C.
Gregariousness Potentially stand-forming.
Sporangia In sori on underside of fronds.
Regenerative strategies W, V.
Spore 44 × 30 μm, black.
British distribution Widespread (100% of VCs).
Commonest habitats Walls 14%, limestone woodland 6%, cliffs/coniferous plantations 5%, scrub 4%.
Spp. most similar in habitats *Asplenium ruta-muraria* 80%, *Cystopteris fragilis* 78%, *Epilobium montanum* 78%, *Elymus caninus* 74%, *Mycelis muralis* 71%.
Full Autecological Account *CPE* page 230.

Synopsis In common with *D. dilatata*, with which it frequently occurs, *D.f-m.* has the potential for functioning as a woodland-floor dominant. However, the species is less restricted to soils of low pH than *D.d.*, and is frequent on calcareous rocks and in limestone woodland. The fronds of *D.f-m.* are less persistent in winter, and expand slightly later in spring. A further ecological difference from *D.d.* may be recognized from the fact that *D.f-m.* occurs primarily on drier, well-drained soils and is completely absent from wetlands. In addition, *D.f-m.* is more common than *D.d.* in lightly or even unshaded habitats. *D.f-m.* is better able than *D.d.* to colonize mortared walls which are the major habitat for the species in some urban areas. *D.f-m.* is grown in rockeries, and some populations in urban areas are likely to be naturalized genotypes of cultivated stock. On free-standing walls, plants often do not grow large enough to sporulate, either because the crevice in which *D.f-m.* is rooted is too small to allow a large plant to establish or because development is interrupted by pointing. However, the species produces spores moderately frequently in other rocky habitats such as screes, cliffs and outcrops. Studies in Russian woodland suggest that plants of *D.f-m.* may live for at least 30–40 years and do not sporulate until 6–7 years old. Establishment in Derbyshire woodland is comparatively uncommon but, once established, *D.f-m.* is long-lived. This contrasts with the more ruderal population dynamics of *D.d.*. The species declines in abundance at high altitudes. In upland woodlands on acidic strata *D.f-m.* is largely restricted to nutrient-enriched areas of higher pH close to the streams in the valley bottoms. In these woodlands *D.f-m.* is replaced on more acidic soils by *D. affinis*, a species with a more oceanic distribution. In unshaded acidic sites at high altitudes replacement of *D.f-m.* by *D.d.* is almost complete. It is not known whether the ecological differences between *D.f-m.* and *D.d.* arise from characteristics of the sporophyte or relate to differences in requirements for prothallial establishment. It may be significant that, whereas *D.a.* is apomictic, *D.f-m.* is sexual; more research is required to examine the effect of this difference during the establishment of new colonies. There is often considerable variation, presumably genetic in origin, between populations. *D.f-m.* hybridizes frequently with *D.a.* and rarely with *D.d.* and two other species.

Current trends Considered to have been decreasing since Roman times through the destruction of woodland habitats, and to be becoming still rarer for the same reason. Page suggests that much of the genetic variation still present in the species is attributable to the former existence of more extensive populations; this characteristic is likely to be lost as *D.f-m.* becomes increasingly restricted to plantations and rocky habitats.

90   *Eleocharis palustris*

Ssp. *vulgaris*; field records may include ssp. *palustris*

Established strategy C–S–R.
Gregariousness Often patch-forming.
Flowers Brownish, hermaphrodite, protogynous, wind-pollinated, self-compatible.
Regenerative strategies V, ?B_s.
Seed 0.96 mg.

Common Spike-rush

British distribution Aggregate species widespread (100% of VCs).
Commonest habitats Unshaded mire 12%, still aquatic 8%, river banks 3%.
Spp. most similar in habitats Insufficient data.
Full Autecological Account *CPE* page 232.

Synopsis In common with all other common angiosperms of the British flora, in which the stem is the only major photosynthetic organ, *E.p.* is confined to wetland. *E.p.* is much shorter than many other water-margin plants, and tends to be restricted to relatively open situations where the growth of more-robust species is restricted by factors such as disturbance by water currents and a degree of soil infertility. Its tolerance of these factors is particularly marked at the margin of Scottish lochs, where *E.p.* is characteristic of shallow water on exposed sandy shores. According to habitat and phenotype, the stems vary between 1 and 8 mm in diameter, and their height is also very variable. *E.p.* does not extend far into drier habitats where the growth form tends to become tufted and less vigorous. Thus, the species is generally absent from small ponds with a large fluctuation between summer and winter water-table. *E.p.* may persist in a non-flowering condition in lightly grazed habitats. Vegetative growth by means of rhizomes gives rise to extensive stands. These may break into daughter plants either through the breakdown of old rhizomes during the second year or through natural disturbance caused by, for example, flood damage. Detached plantlets can re-root to form new colonies. Seedlings are rarely seen, and regeneration from seed appears to play little part in the maintenance of populations, but is presumably involved in the initial colonization of sites. The nut of *E.p.* has barbed bristles which appear to facilitate

dispersal by animals. *E.p.* is eaten by water-fowl, and seed discharged in faeces appears to be responsible for long-distance dispersal. No seed bank has been detected during the limited studies so far undertaken. Field records may include the diploid ssp. *palustris* with smaller seeds and a restricted British distribution. This ssp. is self-incompatible. Ssp. *vulgaris* is a polyploid which may have arisen through hybridization between ancestors similar to *E. palustris* ssp. *palustris* and *E. uniglumis*. The two ssp. of *E.p.* form hybrids, and ssp. *v.* also hybridizes with *E.u.*.

Current trends Not as regular a colonist of new artificial wetland sites as *Juncus* spp. and *Typha latifolia*. Because of its small stature, *E.p.* is vulnerable to dominance by larger, fast-growing clonal species in habitats subject to eutrophication. Probably decreasing.

91 *Elodea canadensis*   Canadian Pondweed

Established strategy C–R.
Gregariousness Patch-forming.
Flowers Greenish-purple, dioecious, water-pollinated.
Regenerative strategies V.
Seed None.

British distribution Absent only from the extreme N (82% of VCs).
Commonest habitats Still aquatic 14%, running aquatic 6%, unshaded mire 1%.

Full Autecological Account *CPE* page
234.

Synopsis A native of N America, *E.c.*
is now naturalized and widespread in
the British Isles, where it was first
recorded in the early 19th century.
E.c. is a submerged aquatic of still or
slow-moving water, and the plant is
often rooted in silt. In some habitats
E.c. is the dominant, or even the only,
hydrophyte present; in others it forms
a lower layer beneath the canopy of
larger aquatic macrophytes such as
Potamogeton natans and *Sparganium
erectum*. However, the species is
usually suppressed if the water has a
floating carpet of *Lemna minor* or if
there is a high level of water
turbulence. The pattern of colonization
is described thus: (a) build-up of the
initial colony, often to the exclusion of
other aquatics (perhaps over a period
of 3–4 years); (b) attainment of an
equilibrium state (next 3–10 years); (c)
decline (next 7–15 years); (d) the
presence of a small residual
population, or complete
disappearance, of *E.c.*, with possibly a
return later. *E.c.* became a serious
weed following its naturalization,
choking waterways before declining to
its present less-abundant (but still
common) level. It has been suggested
that the decline may be associated with
some form of nutrient depletion, but
the exact cause is unknown and
vulnerability of the genetically uniform
populations to some form of pathogen
attack seems worthy of consideration.
The plant consists of an extensive
system of prostrate stems and erect
leafy shoots. The leaves are only two
cells in thickness and do not bear
stomata. *E.c.* can utilize both dissolved
carbon dioxide and the bicarbonate ion

for photosynthesis. The plant lacks
vessels. The roots are adventitious and
unbranched, and usually contribute
less than 3% of the dry weight of the
plant. Nevertheless, inorganic
nutrients are absorbed through the
roots and are translocated to the shoot.
The roots are characterized by large
polyploid root hairs. The species does
not exhibit aquatic acid metabolism to
any appreciable extent (see account for
Ranunculus flammula). In a Canadian
study, shoots were found to lack
dormancy in winter and although the
light-compensation point was low, a
thick covering of ice in winter resulted
in some shoot death. Re-establishment
in iced-up areas occurred by means of
rapid growth in summer, but the
probability of survival, and the
longevity, of such populations was low.
The mechanism of the apparent
restriction of *E.c.* to lowland sites in
Britain may have a similar climatic
component. Regeneration is entirely
vegetative. Patches increase in size as a
result of shoot extension *in situ* and
shoot pieces, carried by water or
attached to wild-fowl, may form new
colonies. Shoots may continue growing
even when not rooted. Plants
overwinter as dormant buds which
separate as the old stems decay. These
dormant buds, which normally sink,
may be transported to new sites. *E.c.*

is shy-flowering, and no populations in very deep or in flowing water have been observed to flower. The production of flowers has little ecological significance in any event, since the species is dioecious and the male plant has been reported only once in Britain. *E.c.* has a wide latitudinal range both as a native species and as an established alien. The species extends from peaty waters to calcareous sites and, judging by its associated species, the plant is found in both mesotrophic and eutrophic waters. *E. nuttallii*, another N American species, first recorded in 1966, is at an earlier stage in colonizing Britain, and is still spreading rapidly within the Sheffield region. As in *E.c.*, populations of *E.n.* appear to consist exclusively of female plants. *E.n.*

differs from *E.c.* in a number of ways. First, it is usually more robust, and shoots often extend to the water surface. Second, *E.n.* is more phenotypically plastic; growth forms in flowing water differ from those in still water, and the species also adopts an unusual growth form in deep lakes. Third, *E.n.* flourishes in nutrient-rich situations and is not recorded from peaty waters. The two species often occur together, with *E.n.* usually assuming the dominant role. Both locally and nationally, *E.n.* is replacing *E.c.* in many sites. The continued spread of *E.n.* is likely to be encouraged by eutrophication.
Current trends Probably decreasing as a result of the destruction of aquatic habitats and the rapid spread of *E. nuttallii* (see above).

92 *Elymus caninus*   Bearded Couch-grass

Established strategy Between C and C–S–R.
Gregariousness Intermediate.
Flowers Green, hermaphrodite, wind-pollinated, self-compatible.
Regenerative strategies S.
Seed 4.04 mg.
British distribution Widespread, uncommon in N Scotland and Ireland (81% of VCs).
Commonest habitats Cliffs/limestone woodland 4%, walls 3%, verges/acidic woodland 2%.
Spp. most similar in habitats *Hedera helix* 63%, *Poa trivialis* 56%, *Silene dioica* 38%, *Bromus ramosus* 31%, *Mercurialis perennis* 31%, *Taraxacum* agg. 31%.
Full Autecological Account *CPE* page 236.

Synopsis *E.c.* is a tufted grass most typical of base-rich woodland margins in sites where the vigour of potential dominants is reduced not only by light-to-moderate shade but also by

infertility (as in some woodland sites) or by disturbance (as on river banks and roadsides), or more usually by a combination of the two. The tendency to occur on S-facing slopes is a reflection of an association with more-open woodland sites. The annual increase in shoot biomass continues into the shaded phase, resulting from expansion of the tree canopy. The species produces new shoots in autumn and is winter green. *E.c.* is characteristic of freely-drained soils as the frequent occurrences on rock ledges and the absence from mires testify. *E.c.* is non-rhizomatous and largely dependent upon seed for regeneration. The species is virtually absent from sites with large quantities of tree litter, and the possibility that regeneration by seed is poor in the presence of tree litter requires investigation. Some seed germinates directly in autumn and the remainder in the following spring. No persistent seed bank has been reported, although a small number of seeds germinated after two years of burial in a pot experiment. It is not known whether the caryopsis floats, but the species extends along many kilometres of tree-lined river bank within the upland portion of the Sheffield region. This is the habitat that *E.c.* occupies most consistently here, and in which the species reaches its maximum abundance in N England. A number of variants have been described, and distinctive populations recorded from montane habitats in Scotland were formerly separated as *Agropyron donianum*.

Current trends *E.c.* does not appear to be a very effective colonist of new artificial habitats, including secondary woodland, and is probably decreasing.

93 *Elymus repens*   Couch-grass, Scutch, Twitch

Established strategy Between S–C and C–R.

Gregariousness Intermediate to stand-forming.

Flowers Green, hermaphrodite, largely self-sterile, wind-pollinated.

Regenerative strategies (V), (B_s).

Seed 2.02 mg.

British distribution Widespread, particularly in lowland (100% of VCs).

Commonest habitats Arable 57%, hedgerows 55%, soil heaps 53%, verges 43%, manure and sewage waste 38%.

Spp. most similar in habitats *Stellaria media* 68%, *Convolvulus arvensis* 67%, *Lapsana communis* 65%, *Chenopodium album* 61%, *Urtica dioica* 60%.

Full Autecological Account *CPE* page 238.

Synopsis *E.r.* is a rhizomatous grass associated with a wide range of fertile, disturbed habitats. *E.r.*, classified as one of the world's worst weeds, is perhaps the most serious perennial

weed of the cooler regions of the N temperate zone. *E.r.* is recorded from 32 crops in more than 40 countries, and in Britain is most important as a weed of cereals. The success of *E.r.* on arable land arises particularly from the following characteristics. (a) An extensive system of rhizomes which in undisturbed sites give rise to large stands with a dense tall leaf canopy and which, after disturbance by ploughing, cause extensive regeneration from rhizome fragments. In some places rhizomes may reach 400 mm in depth and are thus difficult to eradicate. Rhizome fragments may also be transplanted to new sites as a result of human activities. (b) *E.r.*, which has a high nuclear DNA amount and contains abundant fructans in winter, is a 'cool season' grass; tillering and photosynthesis are most active during spring and autumn. This phenology enables *E.r.* to compete with the young cereal crop and to exploit the site after the crop has been harvested. Further, *E.r.* is a 'luxury consumer' of major nutrients although, in common with most grasses, the species has a low Ca content in the leaves. (c) *E.r.* is unaffected by many types of herbicides used to control dicotyledonous weeds. Its control is hindered by growing cereals continually on one site rather than in rotation. (d) Its capacity to exploit many field banks and road verges enables *E.r.* to re-invade arable land following effective weed control. Seed production is often poor, and regeneration by seed generally has a subsidiary role to rhizome production. *E.r.* may form a short-lived seed bank. Like other relatively palatable grasses with tall nodal stems, *E.r.* is usually absent from heavily grazed habitats and although relatively cold-tolerant, the young autumn shoots are sometimes killed under severe winter conditions. Claims that *E.r.* is allelopathic require further investigation. *E.r.* is out-breeding and genetic variation in several important ecological attributes has been described. Five ssp. are recognized from Europe. Known to hybridize with two other species of *Elymus* and one of *Hordeum*.

Current trends Despite its poor ability to regenerate from seed, *E.r.* is, thanks to human agency, abundant and probably still increasing. Genotypes exploiting arable land and those tolerant of salt-spray on roadsides appear particularly favoured at present.

---

94 *Empetrum nigrum*

Ssp. *nigrum*

Crowberry

Established strategy S–C.
Gregariousness Sometimes patch-forming.
Flowers Purple, usually dioecious and wind-pollinated.
Regenerative strategies V, S, ?B$_s$.
Seed 0.75 mg.
British distribution W and N Britain, absent from the SE (73% of VCs).
Commonest habitats Acidic pastures 8%, rock outcrops 5%, acidic wasteland 4%, limestone pastures/acidic quarries 2%.
Spp. most similar in habitats *Vaccinium myrtillus* 85%, *Nardus stricta* 82%, *Galium saxatile* 80%, *Juncus squarrosus* 79%, *Carex pilulifera* 76%.
Full Autecological Account *CPE* page 240.

Synopsis *E.n.* is a low-growing, long-lived (to 140 years), evergreen shrub restricted to areas with a cool, moist climate. In N England, which lies at the S edge of its distribution, the species is confined to upland sites and shows a bias towards N-facing slopes. The lower altitudinal limit decreases in N Britain. *E.n.* is relatively tolerant of snow cover and severe winter conditions. Shoot growth commences in early spring and involves the emergence of flowers which were fully formed in the preceding autumn. In the nutrient-deficient habitats of which *E.n.* is characteristic, the evergreen habit of *E.n.* may be important in the retention and efficient utilization of limiting elements such as N and P. The species is perhaps most typical of acidic peaty soils, but is also recorded on highly calcareous soils in N Britain and W Ireland. Though frequent in moist habitats, *E.n.* is intolerant of severe waterlogging, other than in soligenous mire. *E.n.* is tolerant of controlled burning but is destroyed by severe fires. The foliage is eaten by grouse but not by sheep and seed is dispersed by a variety of birds and animals and fruit is eaten by humans. Regeneration is mainly by means of vegetative growth. Prostrate stems may root on contact with the ground, and as a result *E.n.* may form large clonal patches which may increase at a rate of up to 100 mm year^{-1}. Male and female plants are often found in association and fruiting is of frequent occurrence. Flowering is usually confined to open habitats. Most seeds germinate in the first spring following their release. The remainder germinate in small numbers over several years. However, establishment from seed appears to be infrequent. Work in Alaska suggests that *E.n. sensu lato* may form a small bank of persistent buried seeds. This combination of regenerative strategies, involving extensive clonal growth and the production of only moderate numbers of relatively large seeds, is not unlike that of *Vaccinium myrtillus* and *V. vitis-idaea*, either of which *E.n.* may replace in moister, more northerly or montane sites. However, there is a marked contrast with the biology of *Calluna vulgaris* and *Erica cinerea*, which tend to form relatively short-lived plants and are dependent upon the production of numerous minute seeds and large persistent seed banks. Ssp. *hermaphroditum* replaces ssp. *nigrum* in some mountainous regions of N Britain, parts of the Alps and the extreme N of Europe.

Current trends *E.n.* is now extinct in many lowland parts of Britain through the destruction of habitats, but perhaps also through an increase in winter temperatures since the 19th century. In upland areas *E.n.* appears more common than formerly. This may be due to overgrazing by sheep, and to other forms of land management which have resulted in increased bare ground for seedling establishment and have reduced the vigour of some potential competitors.

---

95 *Epilobium ciliatum*

Formerly *E. adenocaulon*

Established strategy C–R.
Gregariousness Sparse.
Flowers Pink, hermaphrodite, automatically self-pollinated.
Regenerative strategies W, (V), B$_s$.
Seed 0.06 mg.
British distribution Commonest in the SE (> 20% of VCs).

American Willow-herb

Commonest habitats Cinder heaps 17%, soil heaps 11%, river banks 10%, bricks and mortar 8%, shaded mire 7%.
Spp. most similar in habitats *Artemisia vulgaris* 65%, *Rumex obtusifolius* 62%, *Senecio viscosus* 61%, *Rumex crispus* 56%, *Chaenorhinum minus* 55%.

Full Autecological Account *CPE* page 242.

Synopsis A native of N America, *E.c.* is a tall, fast-growing polycarpic perennial herb which colonizes a range of disturbed, often relatively fertile sites. In some of its habitats, such as gardens, stony river margins and railway ballast beside the permanent way, the spread of potential dominants is limited by regular disturbance in the form of weeding, flooding and the application of herbicides respectively. In others, e.g. demolition sites, coal-mine spoil and wasteland, *E.c.* is an early colonist, but in time is replaced by more-dominant species. *E.c.* was first recorded in Britain in 1891, and since 1932 has spread rapidly. First observed in the Sheffield region in 1959, *E.c.* has since spread to such an extent that it is now the most abundant willow-herb in urban areas. The success of *E.c.* appears to be related to the considerable habitat range. Despite being an in-breeding species, *E.c.* is able to colonize a wide variety of environments and appears better able than most native species to exploit artificial habitats such as cinders, demolition sites, gravel pits and disturbed ground beside forestry rides. Effective colonization appears to relate to the exceptional regenerative versatility of *E.c.*. Although normally a polycarpic perennial, seedlings develop rapidly and under long day conditions may set seed within ten weeks; thus the species has the potential to function as an annual. Seed production is prolific; a large plant may produce over 10 000 plumed and wind-dispersed seeds.
Germination occurs over a wide range of temperatures, and establishment from seed may be initiated over much of the year. Seeds accumulate in large numbers on or near the soil surface, and a seed bank is often still present in the soil after *E.c.* has been displaced by later successional species. Although lacking the capacity for extensive lateral vegetative spread, the species is able to regenerate vegetatively to a limited extent by the leafy stolons, which are produced at the base of the stem in autumn; these may become detached as a result of disturbance and give rise to new plants. However, in undisturbed sites the leafy stolons appear to function primarily as organs of perennation. The stolons may grow to some extent during winter, suggesting that in upland regions the distribution of *E.c.* may be restricted by climate as well as by the availability of suitable habitats. Like *Senecio squalidus*, which also produces plumed seeds, *E.c.* has spread widely along the network of railways. *E.c.* forms hybrids with a wide range of other *Epilobium* species and *E.c.* × *E. montanum* appears to be the commonest hybrid in N England.
Current trends An alien which appears to be still increasing, particularly in urban and industrial areas. Now the most abundant and ecologically wide-ranging species of *Epilobium* in lowland Britain.

Established strategy C.
Gregariousness Stand-forming even at intermediate frequencies.
Flowers Purple, hermaphrodite, often protandrous, insect-pollinated, generally out-breeding.
Regenerative strategies V, W, B_s.
Seed 0.05 mg.
British distribution Widespread except for much of Scotland (92% of VCs).
Commonest habitats River banks 22%, unshaded mire 16%, shaded mire 15%, walls 9%, still aquatic 8%.
Spp. most similar in habitats *Phalaris arundinacea* 82%, *Solanum dulcamara* 81%, *Carex acutiformis* 80%, *Galium palustre* 78%, *Myosotis scorpioides* 77%.
Full Autecological Account *CPE* page 244.

Synopsis *E.h.* is a tall, stand-forming herb particularly associated with river banks and productive mire. Although characteristic of sites with high net production, *E.h.* is sometimes capable of dominance in sites of slightly lower fertility, but it is sensitive to grazing and mowing, both of which lead to rapid displacement by shorter species. The germination rate and total germination percentage of seeds is increased by ethene, and in response to flooding, shallow adventitious roots are formed at the expense of the deep primary root. Both responses may be interpreted as specializations associated with waterlogged habitats, especially since neither is shown by the dryland species *Chamerion angustifolium*. Soils colonized by *E.h.* are often characterized by low levels of extractable iron and manganese, and *E.h.* is, within the context of wetland species, very susceptible to ferrous iron toxicity. This sensitivity may explain the greater frequency of *E.h.* on river banks (flooded mainly during winter) than in mire (which is permanently waterlogged). The plant produces leafy stolons during the later part of winter, and leaves damaged by winter conditions may be replaced in early spring. Whether this behaviour plays a part in the tendency for *E.h.* to be restricted to lowland areas (maximum British altitude 370 m) requires investigation. In this context it may be significant that seed set was very low in the cool wet summer of 1985. The species has a formidable array of regenerative strategies, including the production of numerous wind-dispersed seeds which have the capacity to germinate over a wide range of temperatures. Some seeds germinate in autumn soon after dispersal, but in many others germination may be delayed until the following spring. *E.h.* also accumulates considerable reserves of buried seeds, and forms large clones by means of rhizomes, which may extend at rates of 0.5 m year^{-1} and die in the year following their initiation. Regeneration may also occur from detached rhizome fragments. These attributes, coupled with the high potential growth rate and capacity to form a dense, rapidly ascending, leaf canopy during summer, confer upon *E.h.* the ability for

effective colonization of fertile habitats, high rates of resource capture and dominance of perennial communities in productive habitats. Hybrids with other species of *Epilobium* are infrequent.
Current trends Increasing. An effective colonist of artificial habitats, highly responsive to eutrophication.

97   *Epilobium montanum*　　　Broad-leaved Willow-herb

Established strategy C–S–R.
Gregariousness Sparse.
Flowers Pink, hermaphrodite, commonly self-pollinated.
Regenerative strategies W, ?B_s.
Seed 0.13 mg.
British distribution Widespread (100% of VCs).
Commonest habitats Limestone quarries 22%, bricks and mortar 21%, walls 19%, cinder heaps/limestone woodland 10%.
Spp. most similar in habitats *Dryopteris filix-mas* 78%, *Asplenium ruta-muraria* 75%, *Cystopteris fragilis* 70%, *Mycelis muralis* 68%, *Asplenium trichomanes* 61%.
Full Autecological Account *CPE* page 246.

Synopsis *E.m.* is an erect, relatively fast-growing polycarpic perennial herb which colonizes a range of disturbed and skeletal sites. In habitat range and life-history *E.m.* is very similar to *E. ciliatum*. However, for a number of reasons, *E.m.* is much less abundant. First, its biology and ecology are less ruderal in character. Thus, the species takes *c.* 10 days longer to set seed; plants are smaller and few-flowered, and therefore produce smaller numbers of seeds; these germinate over a narrower range of temperatures. Second, *E.m.* is most characteristic of lightly shaded conditions and only attains abundance as a colonist of moist habitats. Thus, *E.m.* has a narrower habitat range than *E.c.*. However, *E.m.*, whose stolons bear less leaf area and grow less during winter than those of *E.c.*, extends extensively into upland regions. Also, judging by its habitat range, *E.m.* exhibits greater shade-tolerance. The production of numerous wind-dispersed propagules and the association with moist, sometimes shaded conditions are shared with many ferns and with *Mycelis muralis*, all of which exploit walls and other steep, rocky habitats (see 'Spp. most similar in habitats'). A degree of ecotypic differentiation has been reported in this predominantly in-breeding species. Garden populations tend to allocate more reserves to seeds, whereas woodland populations produce more stolons. As in the case of *E.c.*, *E.m.* forms many hybrids.
Current trends Uncertain, although as an effective colonist of artificial habitats, *E.m.* is likely to remain common.

Established strategy C–S–R.
Gregariousness Sparse.
Flowers Pink, hermaphrodite,
  self-pollinated.
Regenerative strategies W, B$_s$.
Seed 0.05 mg.
British distribution Widespread (99%
  of VCs).
Commonest habitats Unshaded mire
  6%, running aquatic/shaded
  mire/walls 2%.
Spp. most similar in habitats
  Epilobium palustre 84%, *Stellaria
  alsine* 82%, *Juncus articulatus* 78%,
  Equisetum palustre 78%,
  Ranunculus flammula 78%.
Full Autecological Account *CPE* page
  248.

Synopsis *E.o.* is an erect herb which
produces numerous wind-dispersed
seeds, and is an effective colonist of a
wide range of moist habitats. The
species is most common in two
relatively fertile types of environment:
(a) mire, river banks and wet rocks,
often in situations where the vigour of
potential dominants is restricted by
winter-flooding and sometimes by
shade; and (b) disturbed ground,
including garden plots. In wetlands the
species is absent from highly
calcareous soils and is most frequent in
soligenous mire. *E.o.* frequently
exploits sites which are intermediate in
soil fertility between productive
habitats capable of supporting *E.
hirsutum* and the impoverished sites of
E. palustre. In highly disturbed
wetland habitats, including those with
wide fluctuations in water-table, e.g.
river shingle, *E.o.* is invariably
replaced by *E. ciliatum*. In gardens and
waste places the species is restricted to
moist, often shaded sites which are
infrequently disturbed. In such sites
E.o. is less frequent than either *E.c.* or

E. montanum. As in the case of other
species of *Epilobium*, the tall stems of
E.o. are vulnerable to grazing. A
robust plant of *E.o.* produces many
thousands of small, wind-dispersed
seeds. These do not float and often
form a large bank of persistent seeds in
the soil. The species is stoloniferous,
and additional plants may be formed
by this means in closed vegetation.
However, despite this capacity for
vegetative spread, *E.o.* is sparsely
distributed in vegetation. Detached
shoot segments are also capable of
regeneration, and plant fragments
washed away during flooding may
prove to be of importance in colonizing
new sites downstream. The distribution
of *E.o.* overlaps those of several other
species of *Epilobium* and those of
some of the many hybrids recorded
within the British Isles.
Current trends Uncertain. Perhaps
increasing as a result of disturbance in
artificial habitats, but decreasing as a
wetland plant due both to drainage and
to eutrophication.

Established strategy Between C–S–R
  and S.
Gregariousness Sparse.
Flowers Pink, hermaphrodite, usually
  selfed.
Regenerative strategies W, (V), ?B_s.
Seed 0.04 mg.
British distribution Widespread (100%
  of VCs).
Commonest habitats Unshaded mire
  9%, shaded mire 2%.
Spp. most similar in habitats
  Insufficient data.
Full Autecological Account *CPE* page
  250.

Synopsis *E.p.* is an erect herb growing
at low densities in a range of wetland
habitats, where the growth of potential
dominants is restricted by an
intermediate level of soil fertility and
sometimes also by grazing. Sites with
E.p. are further characterized by low
levels of extractable macronutrients
and low values of net above-ground
production. In many upland areas the
species is most characteristic of
soligenous mire, but is also found in
topogenous mire, flood hollows and
ditch banks. Although recorded from
calcareous sites in some parts of
Britain, and indifferent in its response
to calcium level when grown
experimentally, *E.p.* is virtually
restricted to mildly acidic soils in the
Sheffield region. On calcareous strata
the species is largely confined to acidic
drift or to leached soils, a pattern
similar to that shown by *Ranunculus
flammula*. On highly calcareous soils
E.p. appears to be replaced by *E.
parviflorum*. At lowland sites the plant
has a tall leafy stem (to 600 mm) and,
when growing in pasture, is often
found in areas of *Juncus effusus* where
it may receive some protection from
grazing. In upland areas plants are
smaller (often < 150 mm) and few-
flowered, and tend to occur in lightly

grazed habitats. In the more base-rich
and productive non-calcareous mire
sites (characterized, for example, by
Cirsium palustre and *Rumex acetosa*)
the species is apparently replaced by
E. obscurum. Though less susceptible
than *E. hirsutum*, *E.p.* is relatively
intolerant of ferrous iron toxicity and is
characteristic of soils with moderate
levels of extractable iron. Effective
regeneration is probably mainly by
means of stolons. Swollen terminal
buds are formed at the end of thread-
like stolons, and these give rise to new
plants, often over 100 mm away from
the parent plant. Detached segments
of stem are also capable of
regeneration. *E.p.* produces small,
plumed seeds and annually an average
plant may release *c.* 14 000, although
the output from the smaller plants
frequently observed in upland
populations is likely to fall short of
1000. No seed bank studies appear to
have been carried out, but *E.p.* would
be exceptional for the genus if it did
not maintain some dormant seeds in
the surface soil horizons. The seeds do
not float but wind-dispersal in *E.p.*
appears to be effective, and the species
has colonized gravel pits and colliery
subsidence areas remote from some

established populations. Although *E.p.* is known to form hybrids with most other British members of the genus, the species is often ecologically isolated from other species.
Current trends Although an effective colonist of new sites, *E.p.* is largely restricted to less-fertile soils and, at least in lowland areas, is likely to decline as a consequence of drainage, eutrophication and habitat destruction.

100 *Epilobium parviflorum* — Lesser Hairy Willow-herb

Established strategy Between R and C–S–R.
Gregariousness Sparse to intermediate.
Flowers Rose, hermaphrodite, often self-pollinated.
Regenerative strategies W, B_s, ?(V).
Seed 0.11 mg.
British distribution Widespread, but less abundant in the N (99% of VCs).
Commonest habitats Limestone quarries 9%, running aquatic 8%, unshaded mire 6%, shaded mire 4%, walls 1%.
Spp. most similar in habitats *Veronica beccabunga* 91%, *Nasturtium officinale* agg. 88%, *Equisetum palustre* 86%, *Apium nodiflorum* 85%, *Juncus articulatus* 81%.
Full Autecological Account *CPE* page 252.

Synopsis A species which requires further study. *E.p.* colonizes a wide range of relatively moist habitats, and is persistent in circumstances where the vigour of potential dominants is restricted by intermediate levels of soil fertility and occasional disturbance. The species occurs primarily in two types of habitat: (a) derelict land and (b) soligenous mire and seasonally exposed river beds on calcareous strata. On derelict land the species is associated with a wide range of soil conditions and is unusual in its capacity to grow on finely-powdered magnesian limestone quarry spoil of pH 8.0. In wetlands, the species' commonest habitat, *E.p.* is rare on non-calcareous soils and, consequently, there is little overlap with either *E. obscurum* or *E. palustre*. Moreover, in regions where *E. palustre* occurs frequently on calcareous soils, *E.p.* exploits the more productive areas. In topogenous mire, *E.p.* tends to occupy drier upper reaches of the wetland habitat. *E.p.* is relatively tall, and the shoots are vulnerable to trampling and heavy grazing. However, *E.p.* is not infrequent in grazed mire, perhaps because trampling results in the formation of open microsites suitable for colonization by seedlings. In common with *E. hirsutum*, *E.p.* is a lowland species which is close to its altitudinal limit over much of N Britain. *E.p.* produces numerous wind-dispersed seeds which germinate directly in vegetation gaps or after incorporation into a persistent seed bank. The species also forms short overwintering stolons, but these are produced close to the base of the flowering stem and it is uncertain whether these operate primarily as a means of regeneration or merely allow perennation. *E.p.* hybridizes with many other species, but none of the crosses achieves widespread abundance.

Current trends Overall and as a colonist of artificial habitats, *E.p.* is probably increasing. However, the species may be decreasing in wetlands as a result of habitat destruction.

101 *Equisetum arvense*   Common Horsetail

Established strategy C–R.
Gregariousness Intermediate, but can form patches.
Sporangia Brown, in a terminal cone.
Regenerative strategies V, W.
Spore $34 \times 33\ \mu$m, green.
British distribution Widespread (100% of VCs).
Commonest habitats River banks 18%, lead mine heaps 15%, coal mine heaps 7%, cinders/shaded mire 4%.
Spp. most similar in habitats Insufficient data.
Full Autecological Account *CPE* page 254.

Synopsis *E.a.* resembles *Tussilago farfara* in both ecology and life-history. Each is a colonist of moist disturbed areas, and both are capable of persisting long after perennial vegetation has been re-established. Morphological similarities include an extensive, potentially fast-growing and often deep rhizome system and the production of large photosynthetic organs (in the case of *E.a.*, branched stems) which persist throughout the summer. As with *T.f.*, the rhizome system exploits a wide range of soil depths, enabling *E.a.* to persist in sites prone to soil slippage. However, the most remarkable parallel lies in regeneration. *E.a.* produces cones on short-lived achlorophyllous stems in early spring before the green shoots have developed (similarly *T.f.* flowers before the leaves appear). Like the seeds of *T.f.*, the numerous spores soon lose viability. Cones are usually only produced in abundance in open habitats, and are seldom observed in well-vegetated sites. *E.a.*, which is toxic to stock, is a weed of over 25 broad-leaved, cereal and pasture crops, and is regarded as one of the world's worst weeds. However, this designation seems unjustified within much of Britain. Distinctive features which relate generally to the genus *Equisetum* are described under *E. palustre*.

Current trends Perhaps increasing in disturbed artificial habitats, particularly beside railways, roads and ditches, and to a lesser extent in gardens and on arable land.

Established strategy C–R.
Gregariousness Stand-forming.
Sporangia Brown, in terminal cone.
Regenerative strategies V, W.
Spore Green.
British distribution Widespread (99% of VCs).
Commonest habitats Still aquatic 13%, unshaded mire 6%.
Spp. most similar in habitats Insufficient data.
Full Autecological Account *CPE* page 256.

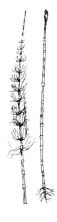

Synopsis (See also that of *E. palustre*.) *E.f.* is a rhizomatous, often stand-forming species which most characteristically behaves as an emergent aquatic in water depths of *c.* 1.5 m at the margins of lakes or ponds and in ditches. In such sites the growth of other potential dominants is usually restricted by soil infertility and sometimes by shade. In N England the species is typically associated with mildly acidic soils, though it may occur locally on highly calcareous soils. The species is most frequent where there are comparatively minor fluctuations in water table during summer. *E.f.* shows some vulnerability to the effects of wave action, but may occur on relatively exposed shores where it replaces *Eleocharis palustris* as the water-table during summer. *E.f.* shows abundant as an aquatic, *E.f.* may persist in mire during pond siltation, and may increase again when a tree canopy is formed. *E.f.* also occurs as a usually minor constituent in soligenous mire and wet grazed meadows. The species lacks tubers, but the rhizome system is extensive and up to 80% of biomass may be below ground. The length of shoot projecting above water is proportional to the depth of the water. The aerial stems have a central hollow which may reach 90% of the diameter of the shoot. Unusually, this central cavity extends into the rhizome and may facilitate oxygen diffusion to the roots, apparently allowing *E.f.* to persist in highly anaerobic soils. Cones are frequently produced, but establishment of new plants from spores is probably very rare. *E.f.* can form large clonal patches by means of rhizome growth, and detached rhizomes or pieces of shoot may regenerate either *in situ* or after being transported by water to a new site. The species is unusually plastic. Typically the stems are unbranched, but in sheltered or shaded sites plants may have whorls of lateral branches. Hybrids are formed with *E. arvense* and probably also with *E.p.*.
Current trends Apparently decreasing in lowland areas as a result of habitat destruction and eutrophication.

Established strategy Between C–R and C–S–R.
Gregariousness Intermediate, sometimes stand-forming.
Sporangia Brown, in a terminal cone.
Regenerative strategies V, W.
Spore Green.
British distribution Widespread (99% of VCs).
Commonest habitats Running aquatic 14%, unshaded mire 12%, still aquatic 6%, shaded mire/soil heaps 2%.
Spp. most similar in habitats *Juncus articulatus* 95%, *Glyceria fluitans* 92%, *Callitriche stagnalis* 91%, *Apium nodiflorum* 91%, *Nasturtium officinale* agg. 89%.
Full Autecological Account *CPE* page 258.

Synopsis The genus *Equisetum* belongs to an ancient section of the Pteridophyta, and has no closely related extant relatives. Over 30% of the world taxa are found in the British Isles. Plants lack secondary thickening, with the leaves reduced to scales. The stem, which is often highly siliceous and unpalatable to stock, is the main photosynthetic organ. The green spores are short-lived, and development of the green, delicate gametophyte is contingent upon dispersal to moist, bared ground. The gametophytes of the three commonest species, *E. arvense*, *E. fluviatile* and *E. palustre*, are seldom recorded, but may form large populations on bare mud beside lakes and reservoirs. They are very sensitive to competition, and apparently also to allelopathic substances produced by the sporophyte, but mature rapidly and are typically dioecious. In the three common species, *E.a.*, *E.f.* and *E.p.*, the vulnerability of the gametophyte is offset by vigorous vegetative regeneration. This is unusual in the Pteridophytes; here, as in *Pteridium*, it is associated with the presence of vessels in the vascular tissue. *E.p.* exploits a variety of moderately fertile wetland habitats. The rhizomes are often deep, enabling *E.p.* to persist on drained land and to locate roots in soil horizons deeper than those exploited by most of its associates. The species has a rather sparse canopy and is not normally a dominant of plant communities. Instead, it tends to occur either as a colonist of disturbed habitats, such as ditch sides and recently excavated gravel pits, or as a subordinate component of communities in which the growth of potential dominants is restricted by grazing or cutting. Apart from being unpalatable, *E.p.* is also toxic to stock and 2 g day^{-1} in hay may reduce the milk yield of cattle. The toxic principle involved appears to be an alkaloid rather than the enzyme thiaminase, which is very important in other species. Both *E.a.* and *E.p.* are relatively frequent on metalliferous spoil. However, reports that these species accumulate gold in their tissues are grossly exaggerated, resulting from a failure to take account of the presence of arsenic, which affects some analytical procedures and may lead to overestimation of gold content. *E.p.* spreads by means of stout rhizomes to form extensive clonal patches, and may also regenerate from plant fragments. It is suspected that plant parts are occasionally transported by water to form new colonies. In low-growing vegetation in soligenous mire, particularly on less-fertile calcareous soils, *E.p.* may be replaced by *E.a.*. The species, which is named as one of the world's worst weeds, varies markedly in habit and branch form according to habitat, and hybridizes with three other species of *Equisetum*. **Current trends** Uncertain. Capable of long-distance dispersal, and both invasive and persistent once established, but not usually amongst the first colonists of new wetland habitats, presumably because of problems in establishment from spores.

104   *Erica cinerea*   **Bell-heather**

Established strategy S.
Gregariousness Intermediate.
Flowers Purple, hermaphrodite, insect-pollinated or selfed.
Regenerative strategies B$_s$.
Seed 0.04 mg.
British distribution Widespread (97% of VCs).
Commonest habitats Acidic quarries 9%, acidic pastures 5%, cliffs/rock outcrops/acidic wasteland 2%.
Spp. most similar in habitats *Carex pilulifera* 93%, *Calluna vulgaris* 87%, *Vaccinium vitis-idaea* 82%, *Deschampsia flexuosa* 78%, *Aira praecox* 74%.
Full Autecological Account *CPE* page 260.

Synopsis *E.c.* is a small evergreen shrub which is found primarily on well-drained acidic mineral soils, with *Calluna vulgaris*. Unlike *C.v.*, *E.c.* has a narrow ecological range and tends not to vary greatly in morphology. The species is usually shorter in stature than *C.v*, and has a lower altitudinal limit which decreases with increasing latitude. *E.c.* is associated with oceanic climates, has a capacity greater than that of *C.v.* to exploit dry habitats, and shows a significant bias towards S-facing slopes in the Sheffield region. *E.c.* is also more shade-tolerant. Elsewhere in heathland *E.c.* is less successful, tending to survive under the canopy of *C.v.* or as scattered individuals. *E.c.* often expands following burning and in these circumstances may achieve temporary dominance. The stem bases appear to re-sprout more readily than those of *C.v.*. Germination is stimulated by heat treatment (a short period at 100°C may overcome the requirement for vernalization) and is frequently superior to that of *C.v.*. If sites are grazed, *E.c.* is less successful despite its low palatability to sheep and a majority of our survey records are from ungrazed sites. *E.c.* shows some winter growth and is more frost-sensitive than *C.v.*. Like *C.v.*, the species is unusual among woody plants of infertile sites in being relatively short-lived (perhaps *c.* 20 years) and in its potential to flower in the second or third year. More research is needed to examine the extent to which the life histories and reproductive effort of *E.c.* and *C.v.* are dependent upon the pulses of mineral nutrient release associated with heathland burning. The species produces numerous minute seeds which germinate in vegetation gaps during autumn and intermittently during the following growing season. Seeds are also incorporated into a persistent seed bank. The plant forms compact tufts, but vegetative spread is not sufficiently extensive to be regarded as an important means of regeneration. Populations may exhibit some attunement to the climate of their site of origin and some are physiologically adapted to alkaline heath soils.

Current trends *E.c.* has been drastically reduced in the lowlands by destruction of heathland and *E.c.* is now uncommon and largely restricted to roadsides, railway banks and golf courses. Trends within the uplands are uncertain.

---

105 *Erigeron acer* — Blue Fleabane

Established strategy S–R.
Gregariousness Sparse to intermediate.
Flowers Outer florets purple, ligulate, female; disc florets yellow, outer female, inner hermaphrodite; insect-pollinated, self-incompatible.
Regenerative strategies W.
Seed 0.11 mg.
British distribution Mainly in the S and E (59% of VCs).
Commonest habitats Limestone quarries 15%, rock outcrops 6%, soil heaps 2%.
Spp. most similar in habitats *Desmazeria rigida* 88%, *Crepis capillaris* 83%, *Inula conyza* 81%, *Cardamine hirsuta* 77%, *Origanum vulgare* 72%.
Full Autecological Account *CPE* page 262.

Synopsis A species which requires further study in field and laboratory. Populations appear to include monocarpic and polycarpic individuals. *E.a.* is a short-lived herb which is found on sand dunes and on shallow calcareous soils where the cover of polycarpic perennials is restricted by soil infertility and summer drought. Some sites are also liable to soil creep or to sand blow, but only seldom are other annual species well represented in the vegetation. The species is restricted, presumably by climatic factors, to lowland sites and is very scarce, for example, on the upland carboniferous limestone of Derbyshire, despite being frequent at the same latitude on the lowland magnesian limestone of S Yorkshire. In favourable conditions, individuals may become large and survive to flower in a second season, but the capacity for vegetative spread is slight. Perhaps because of the relatively erect form of the flowering shoot and the dependence upon seed for regeneration, *E.a.* is generally absent from grazed sites. An average plant with 6–8 capitula may produce over 1000 small achenes, and robust specimens may produce five times this number. Achenes have a large plume, and individuals of *E.a.* have been observed both on roadsides and beside railways 15 km from the nearest known populations. The consisten occurrence of *E.a.* in lowland limestone quarries further illustra the high mobility of the species. However, in a majority of sites with large populations *E.a.* is long-established. No seed bank studies appear to have been undertaken. Four ssp. are described from other parts of Europe, *E.a.* hybridizes with *E. canadensis* (*Conyza canadensis*).

Current trends Uncertain. An effective colonist, but narrowly restricted to dry, unproductive, usually calcareous habitats.

---

106 *Eriophorum angustifolium* Common Cotton-grass, Bog Cotton

Established strategy S.
Gregariousness Intermediate to stand-forming.
Flowers Brownish-green, hermaphrodite, protogynous, wind-pollinated.
Regenerative strategies V, W.
Seed 0.44 mg.
British distribution Widespread and locally abundant (100% of VCs).
Commonest habitats Unshaded mire 13%, still aquatic/limestone pastures 4%, rock outcrops 3%.
Spp. most similar in habitats Insufficient data.
Full Autecological Account *CPE* page 264.

Synopsis *E.a.* is most typical of infertile ombrogenous mire, although its habitat range includes soligenous mire and shallow water. *E.a.* forms large, often diffuse, clonal patches by means of rhizomes which in exceptional cases may extend by up to 1 m year^{-1}. As a result of this rhizome growth, *E.a.* is able to colonize eroding peat in overgrazed habitats, and in this situation may have a niche not unlike that of *Carex arenaria* in sand dunes. Breakdown of older rhizomes after 3–4 years leads to the isolation of daughter plants. Detached shoots readily re-root and may form new plants on eroding peat or in areas subject to peat cutting. Although *E.a.* produces large amounts of wind-dispersed fruit, establishment by seed in spring is very rare, and there is no evidence of a persistent seed bank. *E.a.* is tolerant of grazing by sheep and of burning, but rhizome growth is poor in dry habitats. *E.a.* is one of the very few species which exploit both highly acidic and calcareous sites, and there is evidence that clones from plants on acidic soils show normal growth on calcareous soils. Leaves have very low Ca content.

Current trends In Britain as a whole the destruction of wetlands has led to a marked decline of *E.a.* in lowland areas, and in some regions the species is close to extinction. However, *E.a.*

remains abundant in many upland areas, and is particularly abundant on the millstone grit within the Sheffield region. This abundance, which is greater than is typical for British ombrogenous mires, may be due to an ability to colonize the extensive areas of eroded peat arising from overgrazing and burning.

---

107 *Eriophorum vaginatum* — Cotton-grass, Hare's tail

Established strategy Between S–C and S.
Gregariousness Stand-forming.
Flowers Black, hermaphrodite, wind-pollinated.
Regenerative strategies W, ?B$_s$.
Seed 1.02 mg.
British distribution Widespread except in the S and E (87% of VCs).
Commonest habitats Unshaded mire/limestone pastures 4%, still aquatic/acidic pastures 2%.
Spp. most similar in habitats Insufficient data.
Full Autecological Account *CPE* page 266.

Synopsis *E.v.* is a long-lived (> 100 years) tussock-forming species developing extensive stands on wet, acidic, peaty soils of low potential productivity. The species is particularly characteristic of Arctic tundra. In Britain, *E.v.* has a northerly and an upland bias in its distribution. *E.v.* is particularly prominent in sites which are waterlogged in spring but drier in summer. The species is tolerant of light grazing by sheep or cattle, and survives superficial burning. In Alaska, growth is severely limited by nutrient availability, particularly nitrogen, and marked seasonal changes in the levels of nutrients and carbohydrates have been described. In Alaskan populations (ssp. *spissum*) leaf production has been found to be

coupled with the remobilization of mineral nutrients from senescing leaves. This minimizes the dependence of *E.v.* upon the current supply of soil nutrients, and may be important in facilitating dominance in the chronically unproductive habitats in which the species grows. The leaves are particularly low in Ca. Early flowering is facilitated by development of the inflorescence during the previous year's growth, and seeds germinate in spring. A bank of persistent seed has been recorded in Alaska. Under favourable conditions *E.v.* may produce over 1000 wind-dispersed seeds m^{-2} and is an effective colonist of areas of peat-cutting and peat-erosion both in Alaska and in the S Pennines. Disturbances such as fire, grazing, human activity and frost action result in the formation of microsites suitable for establishment by seed. Seedling establishment in tundra is particularly associated with microsites covered with liverworts, since bared areas are subject to a high level of soil instability and in other sites there are dominant impacts by vascular plants and mosses. There is no evidence of budding of tussocks. Detached plantlets can be grown in the field, and may afford a means of regeneration in the eroding peat that is common in many higher moorland areas. However, at least one authority contends that the only effective mechanism of regeneration is by seed.

Current trends In the lowland areas of Britain *E.v.* is decreasing as a result of habitat destruction. However, the macro-fossil record in peat suggests that *E.v.* has flourished during past episodes of artificial disturbance. *E.v.* may expand at the expense of *Calluna vulgaris* in response to increased grazing, and may be colonizing in areas of eroding peat in upland Britain.

108 *Euphrasia officinalis*   **Eyebright**

Taxonomically complex (see 'Synopsis')

Established strategy S–R.
Gregariousness Sparse to intermediate.
Flowers White, hermaphrodite, moderately self-fertile, insect- or self-pollinated.
Regenerative strategies S.
Seed 0.13 mg.
British distribution Widespread, particularly in the N and W.
Commonest habitats Lead mine spoil 35%, limestone pastures 16%, scree 6%, limestone quarries 5%, paths 4%.
Spp. most similar in habitats *Minuartia verna* 93%, *Thymus praecox* 75%, *Carlina vulgaris* 71%, *Galium sterneri* 67%, *Rumex acetosa* 66%.
Full Autecological Account *CPE* page 268.

Synopsis *E.o.* is a root hemiparasite found in a wide range of less-fertile habitats. *E.o.* is represented in Britain by 19 microspecies, seven of which are endemic to Britain and a further nine endemic to Europe. Highest concentrations of species occur in the N and W. All have a narrow habitat range coupled with a well-defined geographical distribution. Thus, each is potentially distinct ecologically, except where hybridization occurs when taxa of the same ploidy level are found in close proximity. The commonest taxa in N England (*E. confusa* and *E. nemorosa*) are particularly associated with dry habitats on calcareous strata. However, other taxa are widespread on heathland and at other moderately acidic sites, particularly in the N and W. Specialization also occurs with respect to land management (tall microspecies in meadows, short in sheep pastures), with respect to pollination mechanism (large-flowered forms insect-pollinated, small often selfed) and with respect to habitat (maritime, montane and wetland taxa are recognized). Although capable of autotrophic existence, *E.o.* is essentially hemiparasitic and may be attached experimentally to the roots of a wide range of hosts. *E.o.* probably also has a wide host range in the field, perhaps even parasitizing more than one species simultaneously. No single potential host is present at all sites with *E.o.*, but the commonest hosts may be members of the Gramineae and Leguminosae. The early germination of the seed and the complementary early spring growth of many grasses also point to the importance of the Gramineae as host plants. *E.o.*, which has a poorly-developed root system, receives carbohydrates (and presumably also water and mineral nutrients) from its host; it grows considerably better attached than unattached. The possibility that the relatively high levels of mannitol and galactitol in *E.o.* play an important osmotic role in the maintenance of hemiparasitism deserves investigation. However, the origin and ecological advantages of this species' hemiparasitic life-history are unknown, and it is not clear to what extent host plants are debilitated by its presence. Like most ruderals, *E.o.* is phenotypically plastic and its size and reproductive effort vary considerably according to habitat and host. Large vigorous plants form on a good host; under less-favourable conditions a small stunted plant is produced with, on occasion, just a single capsule. *E.o.* is considered unpalatable to grazing animals. Regeneration is entirely by seed, and seedling mortalities are high. There is apparently no persistent seed bank.

Current trends Like other species of less-fertile habitats, *E.o.* is probably decreasing.

109 *Fagus sylvatica*

Beech

(Field records as seedlings)

Established strategy S–C.

Gregariousness Seedlings sometimes in small clusters.

Flowers Green, monoecious, protogynous, wind-pollinated; male flowers numerous on tassel-like heads, female flowers usually in pairs, wind-pollinated.

Regenerative strategies S.

Seed 22.5 mg.

British distribution Native in the SE, planted elsewhere.

Commonest habitats Broadleaved plantations 16%, coniferous plantations 5%, shaded mire/acidic woodland 2%, scrub/limestone woodland 1%.

Spp. most similar in habitats
  Hyacinthoides non-scripta 81%, *Acer pseudoplatanus* (juv.) 81%, *Digitalis purpurea* 76%, *Quercus* spp. (juv.) 71%, *Luzula pilosa* 58%.

Full Autecological Account *CPE* page 270.

Synopsis *F.s.* is a tall, deciduous, dominant forest timber tree which extends to 30 m in height and exploits fertile conditions more effectively than oak. *F.s.* has a life-span which rarely exceeds 300 years. As a native tree the species is restricted, apparently by climatic factors, to SE England. Indigenous populations occur on a wide range of soils particularly over acidic gravels, sands and sandstones and only rarely over calcareous strata. *F.s.* occurs over an exceptionally wide range of soil pH (< 3.5 to > 7.5), but maximum height growth tends to occur at *c.* pH 5.0. *F.s.* is relatively shallow-rooted and vulnerable to strong winds. The species is also unable to exploit dry sites with deep subsoil water. The species is widely planted for commercial forestry and for ornament, and occupies *c.* 74 000 ha of woodland ($> 10\%$ of total), making it the third most common tree of British broad-leaved woodlands. The species is not resilient to coppice management. Planted *F.s.* can regenerate outside the present geographical range of the species. Despite the role of birds and squirrels as vectors, beech nuts are poorly dispersed in native populations. In the Sheffield region the distribution of seedlings is closely correlated with the presence of *F.s.* in the woodland canopy. Annual seed production is often low. Mast years, in which heavy crops of seed are produced, tend to follow warm summers and are dependent also upon the absence of spring frost, to which the young shoots are sensitive. Heavy seed crops tend to occur once every 5–10 years, are more frequent in S England than the N, and may be associated with a narrow annual growth ring. Seeds may be killed by excessive water loss, but in this respect *F.s.* is less vulnerable than oak. Moist conditions are required for establishment, but nuts may loose viability under waterlogged conditions. Subsequent establishment of seedlings, which germinate in spring, usually takes place in the shelter of bushes or trees. Seedlings are highly shade-tolerant. However, under a full canopy of *F.s.*, saplings only persist for a few years. Under more-favourable

conditions they can produce seed when only 28 years old, although 60 years may be a more typical age for first fruiting. Seed production may continue for more than 200 years. No persistent seed bank is formed, and nuts are toxic to stock. The insect fauna of *F.s.* is of intermediate diversity compared with that associated with other broad-leaved species. Seed predation is high but the synchronization of leaf and fruit fall may afford the seed some protection from predators. Variation in annual amount of seed set may also minimize any build-up of potential predators. A sufficient number of seedlings are normally produced in mast years to ensure regeneration of the woodland, but seedlings and saplings are subsequently exposed to a variety of environmental and biotic hazards. Monocultures of *F.s.* are particularly characteristic of modern forestry, and are characterized by heavy shade and accumulation of the exceptionally persistent leaf litter, with the result that the ground flora is often sparse and species-poor. *F.s.* is more productive than oak on fertile soils, and the concept of oak woodland as the climax vegetation of much of the British Isles appears ill-founded; indeed the notion of a single predictable stable forest climax is becoming increasingly suspect.

Current trends Many beech woods were originally planted, but there is some evidence that this potentially dominant species is increasing. Native beech woods may be less under threat than many other woodland types, although their composition may be altering through the cessation of traditional methods of woodland management. Some of the seed used in cultivation in this country originated in Romania. Thus, the genetic diversity of *F.s.* in Britain has presumably been increased, although the fitness of these 'foreign' genotypes in Britain remains uncertain.

110 *Fallopia convolvulus*   **Black Bindweed**

Formerly known as *Bilderdykia convolvulus*

Established strategy R.
Gregariousness Individuals usually scattered.
Flowers Greenish-white, hermaphrodite, insect- or self-pollinated, sometimes cleistogamous.
Regenerative strategies B$_s$.
Seed 1.28 mg.
British distribution Widespread but less common in uplands and in Ireland (99% of VCs).
Commonest habitats Arable 41%, soil heaps 4%, bricks and mortar/manure and sewage waste 3%, cinder heaps/verges 2%.
Spp. most similar in habitats *Spergula arvensis* 98%, *Veronica persica* 97%, *Polygonum persicaria* 94%, *Myosotis arvensis* 93%, *Anagallis arvensis* 91%.
Full Autecological Account *CPE* page 272.

Synopsis *F.c.* is a summer-annual species which is considered only doubtfully native and is in need of further study. Unlike other annual arable weeds, the species produces weak stems which scramble over or twine around adjacent vegetation. This growth form is likely to minimize the usage of carbohydrates for structural tissue, thus allowing *F.c.* to grow exceedingly rapidly and to allocate resources first to flowers and then to the unusually large seeds at an early stage of development. *F.c.* occurs frequently, but at low population densities on arable land, mainly in cereal crops, and is equally frequent in spring- and autumn-sown cereals. *F.c.* is also found to a lesser extent in other unshaded, disturbed habitats on moist, relatively fertile soils. Seedlings are large and have the potential for a high relative growth rate. In favourable situations, early-germinating seedlings

may form large, much-branched individuals producing several thousand seeds. Reductions of up to 25% in crop yield have been reported where *F.c.* is the dominant weed. Under water- or mineral nutrient-stress, vegetative growth is curtailed, and small plants bearing flowers and seed are frequently observed. Seeds, which are incorporated into a buried seed bank, survive particularly well at great depths in the soil. The seeds have a chilling requirement, with germination occurring mainly in spring. Seed banks of *F.c.* tend to increase if cereals are grown for several years in the same field. We suspect that, in common with several other British arable weeds in the Polygonaceae, dormancy is re-imposed annually in response to rising summer temperatures. Although surviving ingestion by birds, seeds are probably dispersed mainly by human activity. The species is a widespread contaminant of wheat, other cereal grains and a variety of broad-leaved crops, and its frequency in arable fields is perhaps due more to the difficulty in cleaning crop seeds than to any other aspects of the biology of *F.c.*.
Current trends A common impurity of

crop seeds, but producing fewer, larger seeds than many arable weeds, and consequently potentially vulnerable to improved methods of cultivation and seed cleaning.

111 *Festuca gigantea*

Tall Brome, Giant Fescue

Established strategy C–S–R.
Gregariousness Individuals usually scattered.
Flowers Green, hermaphrodite, wind-pollinated, self-compatible.
Regenerative strategies S.
Seed 3.12 mg.
British distribution Widespread except in the extreme N of Scotland (94% of VCs).
Commonest habitats River banks 22%, limestone woodland 21%, shaded mire 16%, scrub/acidic woodland 10%.
Spp. most similar in habitats *Ranunculus ficaria* 89%, *Silene dioica* 86%, *Allium ursinum* 84%, *Veronica montana* 81%, *Circaea lutetiana* 77%.
Full Autecological Account *CPE* page 274.

Synopsis *F.g.* is a tufted grass of moist, base-rich sites in lowland woodland, often growing in association with *Bromus ramosus*. The species occurs in moderate shade and is particularly frequent beside streams and other disturbed fertile sites at woodland margins and in clearings. *F.g.* is not a potential dominant and is usually a minor component of vegetation. *F.g.* is winter green, with an annual increase in shoot biomass beginning in early spring and continuing into the shaded phase. Germination occurs in spring and no persistent seed bank has been reported. *F.g.* is largely restricted to sites with little accumulation of tree litter, perhaps because the species is non-rhizomatous and is largely dependent upon seed for regeneration. The species is a widespread colonist of secondary woodland in Lincolnshire and appears to be relatively mobile. The dispersule has much longer awns than those of other British species of *Festuca* and fruits adhere tenaciously to fur and clothing. *F.g.* often occurs close to streams and rivers, and the possibility that seeds may be water-dispersed requires investigation. The species is rather isolated ecologically from other species of *Festuca*, but hybrids with *F. arundinacea*, *F. pratensis* and *Lolium perenne* have been recorded within the British Isles.
Current trends Status uncertain. Certainly not under threat, and apparently favoured by modern methods of managing broad-leaved woodland.

Field records may include *F. tenuifolia*

Established strategy S.
Gregariousness Patch-forming.
Flowers Green, hermaphrodite, wind-pollinated, strongly self-incompatible.
Regenerative strategies S, V.
Seed 0.35 mg.
British distribution Widespread (100% of VCs).
Commonest habitats Limestone pastures 74%, acidic pastures 67%, scree 53%, lead mineheaps 49%, rock ourcrops 38%.
Spp. most similar in habitats
   Campanula rotundifolia 70%,
   Thymus praecox 68%, *Koeleria
   macrantha* 65%, *Hieracium pilosella*
   65%, *Sanguisorba minor* 64%.
Full Autecological Account *CPE* page 276.

Synopsis *F.o.* is a slow-growing, tufted to mat-forming, winter-green grass found in a wide range of unproductive grasslands and rocky habitats. In particular, *F.o.* is the most consistent component species of infertile pasture in Britain, and is unique amongst our grassland species in its capacity to achieve high frequency and abundance in both highly acidic and strongly calcareous sites. However, the species most similar in habitat range (see above) are calcicolous. In sites of greater fertility, *F.o.* declines in importance relative to faster-growing and broader-leaved grasses, e.g. *Agrostis capillaris* and *F. rubra*. The failure of *F.o.* in competition with *A.c.* has been demonstrated experimentally. *F.o.* is less palatable than some of the pasture grasses with which it occurs. Nevertheless, the species is an important food plant in hill pastures extensively grazed by sheep, comparing very favourably with, for example, *Nardus stricta*. *F.o.* can withstand moderate trampling and is often found along the edges of paths and sheep walks, but it is sensitive to

burning. *F.o.* is cold-tolerant but tends to be replaced in moist montane habitats by the proliferous *F. vivipara*. Calcium status of the leaves is low. Dereliction of pastures leads to the suppression of *F.o.* by taller species, but on shallow, nutrient-deficient soils *F.o.* may persist indefinitely. Despite the relatively shallow root system of *F.o.*, many of the habitats it exploits are subject to severe drought. Features of *F.o.* which appear to contribute to its success in dry habitats include xerophylly and an early shoot phenology and seed-set. On shallow soils, tussocks are frequently killed by drought, and in very dry years populations may be severely reduced. However, re-establishment occurs from the autumn cohorts of seedlings. Seeds lack dormancy and tend to germinate synchronously in late summer, although at northern stations a small proportion may remain ungerminated until spring. In moister habitats, where longevity of plants is likely to be greater, the capacity for slow vegetative expansion may also be important and, exceptionally, clones *c.* 10 m in extent and apparently hundreds of years old may develop. However, in *F.o.* there is no evidence for the well-defined capacity for rapid vegetative proliferation exhibited by *F.r.* ssp. *rubra*. *F.o.* has colonized a number of artificial habitats, but does not appear to be very mobile. The capacity of *F.o.* to exploit a wide range of habitats appears to be related to its outbreeding habit, which promotes the emergence of distinct ecotypes. Thus, for example, acidic and calcareous races have been shown to differ in their calcium nutrition and heavy-metal-tolerant races have also been identified. However, more work is required to characterize the edaphic ecotypes of *F.o.*. The genetic and ecological implications of the existence of long-lived individuals in some habitats and relatively short-lived ones in others requires investigation. Similarly, the ecological differences between *F.o.* and the closely related *F.*

tenuifolia, which appears to be relatively rare in the Sheffield region, deserve further study. *F.t.* is largely, or possibly exclusively, restricted to acidic soils, and has a smaller germinule. Tetraploids are moderately self-fertile. *F.o.* hybridizes, though rarely, with *F.t.*, with *F. rubra*, and with one other species.

Current trends *F.o.* (and *F. tenuifolia*) are sown in small quantities (< 50 t year^{-1}), primarily for turf of amenity grassland. However, this does not compensate for the decrease in abundance of *F.o.* due to the destruction of older grassland systems, and *F.o.* is becoming uncommon in some lowland areas. As its lowland sites become smaller and more isolated *F.o.* might be expected to evolve a greater capacity for in-breeding.

---

113 *Festuca pratensis*

Meadow Fescue

Established strategy C–S–R.
Gregariousness Intermediate or scattered.
Flowers Green, hermaphrodite, wind-pollinated, rarely selfed.
Regenerative strategies S, V.
Seed 1.53 mg.
British distribution Widespread (93% of VCs).
Commonest habitats Meadows 24%, verges 9%, enclosed pastures 3%, running aquatic/limestone wasteland 2%.
Spp. most similar in habitats *Rhinanthus minor* 82%, *Phleum pratense* 80%, *Alopecurus pratensis* 79%, *Cynosurus cristatus* 76%, *Bromus hordeaceus* 74%.
Full Autecological Account *CPE* page 278.

Synopsis *F.p.* is a tufted, often relatively short-lived grassland species which is palatable to livestock. The species is of agricultural importance, particularly in meadows, and over

100 t are sown annually. *F.p.* is in many respects similar to *Lolium perenne* and frequently forms the sterile hybrid × *Festulolium loliaceum*. Unlike *L.p.*, however, *F.p.* is most frequent on moist to waterlogged soils and, in keeping with a lower nuclear DNA amount, the shoot phenology is

delayed in comparison with that of *L.p.*. Also, *F.p.* is less productive than *L.p.* and, as an apparently native species, frequently occurs on soils of intermediate fertility, often in relatively species-rich vegetation. Characteristically *F.p.* occurs as a subordinate component of turf, and is usually represented by scattered individuals. *F.p.*, which does not persist either in tall grassland or in intensively grazed pasture, has been most frequently recorded from hay meadows. Like *L.p.*, *F.p.* appears largely dependent upon seed for regeneration and does not form a persistent seed bank. As in the case of the annual grass, *Bromus hordeaceus*, the capacity to persist in hay meadows probably results from early shedding of the seed, a feature facilitating colonization of the open ground

available immediately after the hay has been cropped. The species hybridizes with *F. gigantea* and *F. arundinaceae*, as well as with *L.p.*. The ecology of the ssp. *apennina*, with $2n = 28, 4x$, from continental Europe makes an interesting contrast with that of the more widespread and British ssp. *pratensis*, which is diploid. Ssp. *a.* occurs at higher elevations in the mountains than ssp. *p.*, where it survives the severe winters by dying back in autumn to become winter-dormant. Seeds of ssp. *a.* require cold-treatment for germination.

Current trends Decreasing as a native species through the destruction of alluvial grasslands and other suitable habitats. Also becoming agronomically less important, and expected to decrease further in popularity.

114 *Festuca rubra*

Red Fescue, Creeping Fescue

Ssp. *rubra*. Field records may include *F. nigrescens*

Established strategy C–S–R, some populations tending to C, S–C or S.
Gregariousness Intermediate to stand-forming.
Flowers Green, hermaphrodite, wind-pollinated, mainly self-incompatible.
Regenerative strategies V, S.
Seed 0.79 mg.
British distribution Widespread (100% of VCs).
Commonest habitats Verges 71%, scree/limestone wasteland 67%, limestone pastures 65%, limestone quarry heaps 63%.
Spp. most similar in habitats *Senecio jacobaea* 76%, *Plantago lanceolata* 75%, *Trisetum flavescens* 75%, *Poa pratensis* 72%, *Dactylis glomerata* 69%.
Full Autecological Account *CPE* page 280.

Synopsis Ssp. *r.* is a rather slow-growing, winter-green grass forming

large clones by means of extensive rhizome growth in a wide range of grasslands and other habitats, particularly on base-rich soils. *F.r.* occurs in 'every plot, limed and unlimed' in the *PGE*, where the species' abundance is increased at low levels of fertilizer input and decreased

by heavy nitrogenous dressings. Comparatively rapid rates of leaf growth are achieved in early spring, suggesting that temporal niche differentiation plays a critical role in coexistence with *Agrostis capillaris*, which has a much later pattern of shoot development in the agriculturally important 'Red Fescue–Brown Bent' pastures of intermediate productivity in Britain. Ssp. *r.* can also occupy a somewhat different niche in more-productive tall-grass communities, forming a non-flowering understorey. Levels of N and Ca in the leaves are usually low. In turf microcosms, roots of *F.r.* became heavily colonized by VA mycorrhizas but infection brought about no significant increase in seedling yield. Although readily grazed by sheep, ssp. *r.* is considered less palatable than pasture grasses such as *Lolium perenne*. This may contribute to the ability of ssp. *r.* to persist as scattered individuals in many productive pastures. The distribution of ssp. *r.* in limestone grassland strongly overlaps with that of *F.ovina*, essentially a species of infertile soils and readily grazed by sheep. Unlike some other ssp., *r.* is tolerant of cutting but does not persist under close (5 mm) mowing. Ssp. *r.* is an important stabilizer of limestone screes onto which it encroaches vegetatively. Rhizome growth also plays an important role in regeneration, and clones perhaps 1000 years old have been recorded. Rhizomes break down rapidly, leaving nutritionally-independent daughter plants. Seeds are released in summer and germinate mainly in autumn. Ssp. *r.* only forms a transient seed bank. The *F.r.* group is extremely complex, and a number of morphologically distinct taxa of *F.r. sensu stricto*, with specialized ecological distributions, have been recognized. These include ssp. *arenaria* (sand dunes, shingle and wasteland by the sea), ssp. *arctica* (mountains above 600 m), ssp. *litoralis* (salt marsh) and ssp. *pruinosa* (sea cliffs, salt marshes and stony sea shores). This taxonomic treatment may prove unsatisfactory. In most ecological studies it is not clear which segregates have been studied. The maritime ssp. are presumably salt-tolerant. Others, found in sown grassland, are more robust than native populations, and may contribute to the extensive monospecific stands on motorway verges in which there is a dense build-up of persistent litter. However, evolution within the species is potentially rapid; lead-tolerant ecotypes have been detected on motorway verges. It seems probable that the exceptionally wide range of habitats exploited by *F.r.* may result from ecotypic differentiation. The problems concerning ecological distribution and taxonomy are compounded by the fact that cultivars of two maritime ssp., as well as *F. nigrescens*, are now recommended for use in turfgrass. *F.r.* hybridizes frequently with *Lolium perenne* and crosses with three species of *Vulpia*.
Current trends Though still common, *F.r.* has decreased in abundance due to the destruction of older grassland systems. However, this effective colonist of artificial habitats is probably found in a more diverse range of environments than was formerly the case. Seeds of *F.r.* agg. are sown at over 1500 t year^{-1}. It is predicted that even larger quantities will be sown in the future, since *F.r.* is still underused in amenity grassland. *F.r.* may also have more agronomical potential than is at present appreciated.

Established strategy Between C and
  S–C.
Gregariousness Forms dense stands.
Flowers Cream, hermaphrodite, insect-
  pollinated.
Regenerative strategies V, B_s.
Seed 0.99 mg.
British distribution Widespread (100%
  of VCs).
Commonest habitats Shaded mire 18%,
  river banks 10%,
  hedgerows/unshaded mire/limestone
  woodland 3%.
Spp. most similar in habitats *Caltha
  palustris* 93%, *Solanum dulcamara*
  91%, *Chrysosplenium oppositifolium*
  90%, *Cardamine flexuosa* 86%,
  Mentha aquatica 85%.
Full Autecological Account *CPE* page
  282.

Synopsis *F.u.* is a tall herb, particularly
characteristic of moderately fertile sites
at the transition between wetland and
dryland. Habitats include river banks
and floodplain terraces. The absence
from sites which remain waterlogged
throughout the year may be related to
the sensitivity of *F.u.* to ferrous iron
toxicity. The aerenchyma of the root is
poorly developed compared with that
of many wetland species and, for a
wetland species, *F.u.* shows an
unusually high level of drought
resistance in the field. Grazing and
frequent mowing eliminate the species.
The plant shows rapid extension
growth in the late spring and reaches a
peak in above-ground biomass in July.
Unlike many stand-forming species
(e.g. *Epilobium hirsutum* and *Urtica
dioica*) which have a high relative
growth rate (RGR) and the capacity to
form a rapidly ascending canopy of
short-lived foliage, the RGR of *F.u.* is
low, and most of the leaves survive
until the end of the growing season.
Leaf persistence in shade is particularly
marked in the case of the foliage of
non-flowering shoots, which occur as a
basal rosette, often submerged beneath
a dense canopy. The above-ground
biomass in stands of *F.u.* tends to be
less than that in other commonly
occurring tall-herb communities, and
F.u. tends to occur in sites where the
vigour of these more productive
species is reduced by mineral-nutrient
stress or shade. Thus, *F.u.* is frequent
in shaded mire and widespread in open
woodland on wet clayey soils.
Although this is not apparent from the
triangular ordination, we also suspect
that some of the sites exploited by *F.u.*
are less rich in nutrients than those
occupied by, for example, *E.h.*.
Foliage of *F.u.* contains high
concentrations of N, P and Mg and a
relatively low level of Fe. *F.u.*
regenerates mainly by means of
creeping rhizomes to form extensive
patches. In open habitats an
abundance of seed may be produced,
but in shaded sites flowering and seed-
set are frequently reduced.
Germination occurs in spring, and
seeds may be incorporated into a large
persistent seed bank. Seeds are
capable of floating for several weeks;
this may enable *F.u.* to colonize water
systems. It is suspected that vegetative
portions detached by disturbance are

also capable of regeneration. Leaves are plastic with respect to hairiness, hairs being developed under dry atmospheric conditions. In Europe, the existence of numerous ecotypes has been demonstrated and *F.u.* has been formally subdivided into three ssp.

Current trends Not generally a rapid colonist of artificial habitats and, like most species associated with wetlands, *F.u.* is probably decreasing.

116   *Fragaria vesca* — Wild Strawberry

Established strategy C–S–R.
Gregariousness Intermediate.
Flowers White, hermaphrodite, insect-pollinated or selfed.
Regenerative strategies V, B_s.
Seed 0.31 mg.
British distribution Widespread (100% of VCs).
Commonest habitats Scree 6%, lead mine spoil/limestone quarries/limestone wasteland 5%, bricks and mortar/rock outcrops/limestone woodland 3%.
Spp. most similar in habitats
   Hypericum hirsutum 63%, *Solidago virgaurea* 58%, *Carlina vulgaris* 55%, *Teucrium scorodonia* 55%, *Galium sterneri* 53%.
Full Autecological Account *CPE* page 284.

Synopsis *F.v.* is a low-growing, winter-green herb which exploits sites where the vigour of potential dominants is suppressed by moderately low fertility and where there is often a rocky substrate or moderate shade from a tree canopy, or both. The species is particularly characteristic of open turf in situations where, despite the shallow soil, the incidence of droughting is slight. In this respect *F.v.* resembles several other broad-leaved, rosette plants such as *Potentilla sterilis*, *Prunella vulgaris*, *Primula veris* and *Ajuga reptans*. Thus, the habitat range of *F.v.* includes sites such as stabilized scree, quarry spoil and ballast beside the permanent way, and woodland margins, scrub and hedgerows. In upland areas the species is more frequent in unshaded sites, and throughout the habitat range flowers and fruits are more prolific in unshaded habitats. In N Britain this edaphically wide-ranging species is particularly frequent on limestone strata. Regeneration is mainly vegetative and *F.v.* produces long stolons. The runners also allow ramification over rocky substrata and experiments are required to examine the extent to which they enable *F.v.* to exploit sites of uneven soil depth or with a patchy light environment. The turnover of rosettes is relatively slow. Plantlets detached as a result of disturbance may also play a role in the spread of the species. Vigorous clonal spread is frequently observed, and regeneration by seed is probably mainly of importance during the colonization of new sites. Strawberries are dispersed by birds and seed is incorporated into a persistent seed bank. There is a high level of ecotypic differentiation in N American populations, and a large number of varieties and forms, including cultivars, have been described. However, the most commonly cultivated strawberry is *F.* × *ananassa* (*F. chiloensis*) × *F. virginiana*; $2n = 56$, $6x$, which is more robust in all its parts and, when naturalized, occurs on deeper moister soils than *F.v.*, usually in unshaded sites.

Current trends Has spread in the past beside railways and hedgerows but is not locally a rapid colonist of new artificial habitats and is perhaps decreasing.

117 *Fraxinus excelsior* Ash

(Field records as seedlings)

Established strategy C.
Gregariousness Seedlings often clumped in shaded habitats.
Flowers Purplish, partially dioecious, wind-pollinated.
Regenerative strategies S.
Seed 29.3 mg.
British distribution Widespread except in the far N (99% of VCs).
Commonest habitats Limestone woodland 37%, meadows 19%, scrub 16%, shaded mire 13%, hedgerows/broadleaved plantations 12%.
Spp. most similar in habitats *Mercurialis perennis* 87%, *Allium ursinum* 86%, *Anemone nemorosa* 85%, *Brachypodium sylvaticum* 84%, *Lamiastrum galeobdolon* 83%.
Full Autecological Account *CPE* page 286.

Synopsis *F.e.* is a timber tree or shrub with ring-porous wood. New shoots are vulnerable to frost and bud-break is delayed until late spring. The species occupies 56 000 ha and is the fourth most common broad-leaved tree in British woodlands. In optimal sites, which are sheltered, moist, well-drained and fertile, the species at maturity may extend to 35 m but lives for only *c*. 180 years. However, coppiced specimens may survive for over 300 years, and in less-favourable habitats *F.e.* may persist as a shrub. The species regenerates freely through the production, after 30–40 years, of numerous wind-dispersed fruits (up to 100 000 every second year). Although fruits are less mobile than those of *Betula* spp., seedlings are widely distributed. However, many of the sites in which seedlings germinate are unsuitable for establishment. Seedlings are sensitive to shade and summer-green woodland herbs, e.g. *Mercurialis perennis*, may provide an effective barrier against establishment of saplings. In turf *F.e.* becomes established only in vegetation gaps, and effective increase in height is dependent upon high light intensities. However, once the sapling has penetrated the herb layer the species may be long-persistent (to at least 28 years in shaded conditions). Seedlings are vulnerable to grazing, but establishment may take place within the protection of bushes of *Crataegus monogyna* or on screes if inaccessible to grazing animals. Many of the ash-woods of the Derbyshire Dales can be attributed to a relaxation of grazing pressure in the 19th century. Seedlings are susceptible to damping off in moist, shaded sites and to desiccation on shallow soils. These constraints on seedling establishment dictate that *F.e.* is most typically a seral species colonizing relatively open, little-grazed, base-rich sites. In historical times an increase in the abundance of *F.e.* was associated with forest clearance, with the result that in the Sheffield region *F.e.* is the most frequently recorded canopy species of limestone woodland, is the second most common hedgerow tree and is the third most commonly occurring solitary tree or bush. The species shows a bias towards N-facing slopes, and is virtually absent from acidic soils. In addition to its role in the formation of secondary woodland, *F.e.* is also a frequent constituent of the canopy of ancient woodland. However, although it attains its greatest size on fertile soils, *F.e.* reaches maximum abundance, and exerts maximum ecological impact, under conditions much less favourable for its growth, namely on relatively infertile calcareous soils where most individuals form stunted trees or shrubs. *F.e.* regenerates almost entirely by seed. Flowers are variously male, female and hermaphrodite, and may occur in various combinations on each tree. The proportion of male and female flowers can vary from year to year, but some trees, which do not fruit, are consistently male-flowered. Female trees have thinner trunks than males, apparently because of their greater allocation of resources to seed production. Fruits reach their full length by July and their maximum seed

size by August. However, embryos continue to develop until August or September and when the seed is shed (from September onwards) the embryo is still only half the length of the seed. Germination inhibitors are present both in endosperm and in embryo, and most fruits lie on the ground for two winters, during which time the embryo develops. In one Derbyshire ash-wood a majority of seeds were consumed by small mammals whereas others were predated by caterpillars. Germination of the few survivors occurs in April and May. Viable fruits may lie in the ground for up to six years, forming a shortly-persistent seed bank. This may facilitate rapid colonization following the formation of gaps in the woodland canopy. Seedlings and saplings expand a greater part of the leaf area before canopy expansion of the mature tree and, unlike the adults, which produce only sun-leaves, both sun- and shade-leaves are produced. Rapid growth appears to be dependent upon high rainfall in May and June. Roots do not penetrate below the level of the permanent water-table, but the species may persist in wetland habitats provided there is a shallow zone of well-drained soil for establishment. The canopy of *F.e.* produces a relatively light shade and the leaves, which are still green when shed, do not form persistent litter. *F.e.* combines relatively low palatability to the snail *Helix aspersa* with a low diversity of associated insects. As a forest crop *F.e.* is only economic under optimal soil conditions. In Germany, populations exploiting dry limestone soils are physiologically different from those exploiting moist fertile soils. A further ssp. has been recorded for SE Europe.
Current trends Regeneration of *F.e.* is favoured by disturbance, and the species may be increasing, at least on less-fertile soils.

Established strategy Between R and
  C–R.
Gregariousness Sparse.
Flowers Pink, purple or white,
  hermaphrodite, usually
  self-pollinated.
Regenerative strategies B$_s$.
Seed 4.83 mg.
British distribution Widespread (100%
  of VCs).
Commonest habitats Arable 13%, river
  banks 10%, hedgerows 5%, soil
  heaps 4%, acidic woodland 3%.
Spp. most similar in habitats *Myrrhis
  odorata* 73%, *Lapsana communis*
  69%, *Alliaria petiolata* 67%, *Galium
  aparine* 65%, *Petasites hybridus*
  63%.
Full Autecological Account *CPE* page
  288.

Synopsis The aggregate species is a tall
annual herb which exploits two main
types of disturbed ground: (a) arable
land and other open disturbed sites
and (b) moist, moderately shaded
habitats. In both types of habitat the
species usually functions as a summer-
annual, although there may be a
second period of germination in
autumn. The aggregate species consists
of two taxa, *G. bifida* and *G. tetrahit,
sensu stricto*. Extremely limited
observations suggest that *G.b.* is
restricted to arable and other similar
habitats and is thus, more overtly than
G.t., a follower of humans. *G.t.*
flourishes in open fertile ground, and
plants are considerably more robust in
broad-leaved crops and on fallow
ground than in cereal crops. In
woodlands *G.t.* occurs in clearings, and
in moist habitats exploits a niche
similar to that of *Impatiens glandulifera*
but, unlike this species, is not stand-
forming. In view of its tall stature, it is
not surprising to find that the species
does not occur in sites which are
regularly grazed, cut or trampled.
Usually only one generation per year is

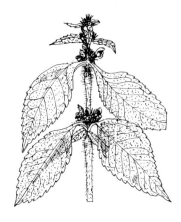

produced, and consequently the
species is vulnerable to the effects of
weeding and is seldom found in
gardens. *G.t.* can be toxic to stock. As
with other common ruderals, *G.t.* is
prolific (on average *c.* 2000 seeds per
plant), but the seeds are much larger
and are produced in smaller numbers
than those of other common weeds of
similar size and morphology (e.g.
Chenopodium album); this, coupled
with the strong seasonality of
germination, appears to reduce the
capacity of the species to colonize new
disturbed sites of transient duration.
Nutlets may float on water for two
days, and can survive ingestion by
stock and by birds. It is not known
whether a persistent seed bank is
formed in woodland, but one is
recorded from arable land, the
commonest habitat of the aggregate
species. In arable habitats, seed is also
effectively dispersed by humans since
some seed is still retained on the plant
when the crop is harvested and the
species may be an impurity in
commercial crop seed. However, it is
probably significant that, whereas seed
is now mainly found as an impurity in
harvested cereals, the species performs
best in association with broad-leaved
crops. *G.b.* and *G.t.* form a sterile
hybrid.

Current trends Possibly decreasing overall, especially on arable land. The high levels of disturbance associated with modern management are probably causing an increase of the species in its woodland habitats.

119 *Galium aparine* Goosegrass, Cleavers

Established strategy C–R.
Gregariousness Typically intermediate.
Flowers Greenish-white, hermaphrodite, protandrous, usually self-pollinated.
Regenerative strategies S, but often autumn- or winter-germinating.
Seed 7.25 mg.
British distribution Widespread (100% of VCs).
Commonest habitats Hedgerows 43%, river banks 35%, verges 26%, arable 17%, soil heaps 12%.
Spp. most similar in habitats *Stachys sylvatica* 89%, *Anthriscus sylvestris* 78%, *Glechoma hederacea* 77%, *Urtica dioica* 77%, *Myrrhis odorata* 73%.
Full Autecological Account *CPE* page 290.

Synopsis *G.a.* is an often tall, scrambling, facultatively winter- or summer-annual herb. The commonness of *G.a.* is related to an ability to occur in two different types of vegetation: tall-herb communities and ephemeral assemblages of disturbed land. In stands of tall herbs the large seed capital and ability to make appreciable growth during milder winters allow *G.a.* to keep pace in height growth with established perennials. The phenology of *G.a.* is semi-vernal, in that flowering commences in April, seed is set from June onwards and the plant dies in late summer or autumn, before the onset of frosty weather. *G.a.* flourishes under conditions of high soil fertility, and may form extensive monocultures on exceptionally enriched sites such as derelict sewage beds. The foliage of *G.a.* contains high concentrations of P and Ca and the species is sensitive to ferrous iron toxicity, which may explain the restriction of *G.a.* to the drier parts of wetland systems. The stems and leaves have backward-pointing prickles which provide adherence to and gain support from the shoots of canopy dominants. The stems also stick to clothing and fur, as do the mericarps, which are covered with hooked bristles. In this way *G.a.* is dispersed very effectively by humans and animals. Dispersal by water or after ingestion by birds or stock is also possible. Of the ruderal habitats with which *G.a.* is characteristically associated, the most important is arable land, where the species is classified as one of the world's worst weeds; it is the most frequent weed of winter wheat in Central S England. On arable land *G.a.* has been encouraged by the use of phenoxy-herbicides which are ineffective against the species. Control of *G.a.* is particularly difficult since the species is a contaminant of many species of crop seeds, and its adhesive seeds are readily dispersed in sacking, clothing and in mud. Seeds require a brief exposure to chilling, but are then capable of germination at low temperatures. Germination frequently occurs over a protracted period, and populations in spring may include both large, well-developed seedlings and others with the radicle just emergent. The stage of development is also dependent upon climate, with seedlings being more advanced in

spring after mild winters. *G.a.* forms a persistent seed bank, but seeds seldom persist beyond two years. Workers have identified obligately winter-, obligately summer- and facultatively polycarpic races within Europe, all with the same chromosome number ($2n = 64$). The distribution of other cytological races is not known.
Current trends Probably still increasing in all habitats except arable, where herbicides designed for specific use on *G.a.* have recently been introduced.

120   *Galium cruciata*   Crosswort

Formerly known as *Cruciata laevipes*

Established strategy C–S–R.
Gregariousness Occasionally patch-forming.
Flowers Yellow, insect-pollinated, terminal hermaphrodite, lateral male.
Regenerative strategies (V), S.
Seed 1.5 mm.
British distribution Uncommon in the N and W (63% of VCs).
Commonest habitats Limestone wasteland 7%, limestone pastures 4%, hedgerows/lead mine spoil/manure and sewage waste/river banks/verges 3%.
Spp. most similar in habitats Insufficient data.
Full Autecological Account *CPE* page 292.

Synopsis *G.c.* is a weakly scrambling winter-green herb on moist, often calcareous soils in vegetation of intermediate productivity. In lowland parts of the Sheffield region *G.c.* is mainly found at woodland margins and in hedgerows, while in upland areas it is characteristic of grasslands, especially on steep N-facing slopes subject to topographic shade. Thus, we suspect that *G.c.* is, like *Stellaria holostea*, a 'semi-shade' mesophyte. In common with *S. holostea*, the leaves

persist through the winter, becoming increasingly senescent, and are then replaced by a new flush of leaves in spring, followed by flowers in early summer. The canopy of *G.c.* is very low and, despite a capacity for lateral spread, only persists in sites where the growth of taller species is debilitated through the operation of disturbance factors such as occasional grazing. *G.c.* is susceptible to heavy grazing, and probably also to burning, and may become prominent for a number of years in grassland sites released from grazing. *G.c.* has a slender rhizome and subterranean stolons, and can form large clonal patches, particularly in shaded habitats. Detached shoots regenerate readily, and *G.c.* may spread along roadsides and river flood plains by this means. *G.c.* also regenerates by means of rather large seeds which germinate in spring and appear to be poorly dispersed. Lack of seed mobility may explain the failure of *G.c.* to colonize Ireland in Postglacial times. Seed production appears to be rather erratic; heavy fruiting in Derbyshire was observed in the unusually cool summer of 1985. No seed bank has yet been detected. *G.c.* is phenotypically plastic; tall scrambling colonies when grazed become prostrate with smaller leaves and shorter internodes.

Current trends Despite a limited capacity to colonize river banks and roadsides, *G.c.* is mainly restricted to older habitats and appears to be decreasing.

121 *Galium palustre*

Field records include sp. *elongatum*

Established strategy Between C–S–R and C–R.
Gregariousness Intermediate.
Flowers White, hermaphrodite, protandrous, insect-pollinated.
Regenerative strategies V, B$_s$.
Seed 0.91 mg.
British distribution Aggregate widespread (100% of VCs).
Commonest habitats Shaded mire/unshaded mire 16%, river banks 8%, soil heaps 2%, still aquatic/walls/acidic woodland 1%.
Spp. most similar in habitats *Myosotis scorpioides* 97%, *Cardamine pratensis* 89%, *Mentha aquatica* 88%, *Cardamine amara* 87%, *Juncus effusus* 86%.
Full Autecological Account *CPE* page 294.

Synopsis The aggregate species is a tall, scrambling wetland herb which is mainly restricted to sites where the vigour of potential dominants is to some extent suppressed (a) by shade from trees, (b) by disturbance factors such as light grazing or flooding, or (c) by a degree of nutrient stress. The aggregate species is found on a wide range of soil types, but is largely absent from calcareous strata. *G.p.*, in particular, is very frequent on mildly acidic peat and exhibits a moderate

Marsh Bedstraw

tolerance of ferrous iron. The aggregate species normally provides only a minor proportion of the total biomass, and frequently much of its foliage is in the lower layers of the leaf canopy. Survival appears to depend upon an opportunistic expansion into small gaps in the canopy of tall wetland vegetation. The capacity for lateral vegetative spread by means of a slender creeping stock and decumbent shoots may be important in this respect. As is the case for many other wetland species, detached cuttings root freely. This is in marked contrast with the response of cuttings of the grassland Bedstraws, *G. saxatile*, *G. sterneri* and *G. verum*, in which adventitious roots are poorly developed, even under controlled laboratory conditions. Detached fragments may be important in enabling the species to colonize more-distant sites beside water. The flowering shoots are taller and usually attain the height of the surrounding vegetation. Seed may survive ingestion by cattle, and may float for several weeks. Germination occurs mainly in spring and seeds may be incorporated into a persistent seed bank. The ecological differences between *G.e.* and *G.p.* have been little studied. *G.e.*

is the taller and is found in wetter sites, but the two are connected by intermediates. Both morphologically and ecologically, *G. uliginosum* shows close affinities with *G.p. sensu stricto*. Both occur in soligenous mire, but *G.u.* consistently occupies the drier sites. *G.u.* also differs from *G.p.* in its greater capacity to exploit highly calcareous mire, and its lower potential to form rooted cuttings.

Current trends Like most wetland species, probably decreasing as a result of habitat destruction.

122 *Galium saxatile*

Heath Bedstraw

Established strategy S.
Gregariousness Intermediate to patch-forming.
Flowers White, hermaphrodite, protandrous, self-incompatible, insect-pollinated.
Regenerative strategies V, B_s.
Seed 0.56 mg.
British distribution Widespread (99% of VCs).
Commonest habitats Acidic pastures 48%, limestone pastures 27%, acidic wasteland 19%, acidic quarries 17%, scrub/limestone wasteland 11%.
Spp. most similar in habitats *Vaccinium myrtillus* 94%, *Nardus stricta* 87%, *Deschampsia flexuosa Molinia caerulea* 74%.
Full Autecological Account *CPE* page 296.

Synopsis *G.s.* is a slow-growing, prostrate, mat-forming herb found in a wide range of rocky habitats on acidic soils. A classic calcifuge which, when planted on calcareous soils, grows poorly and exhibits root stunting and lime-chlorosis. The foliage of plants in the Sheffield region is high in Mn and Al. We suspect that the few field records at high soil pH represent

calcareous soils containing an undetected superficial leached horizon. *G.s.* appears to be moderately shade-tolerant and occurs in open woodland and as a diffuse understorey beneath grass or heather canopies. However, the low mats of *G.s.* tend to be suppressed in derelict grassland and heathland where they are often submerged beneath grass litter and low shrub canopies. As a result, *G.s.* tends to attain its highest frequencies in sheep pasture. The affinity for N-facing slopes suggests that in grassland the distribution of *G.s.* may be affected by

drought, an hypothesis which is consistent with known fluctuations in abundance corresponding to rainfall variation from year to year. The prostrate stems of *G.s.* readily form roots and are thus conducive to the development of large clonal patches. Unlike other species of similar growth form, *G.s.* has little capacity for regeneration from detached, rooted shoot tips. Seed is set regularly in unshaded sites and persistent seed banks may develop. However, regeneration by seed is probably infrequent. The significance of the seed bank in relation to the survival of *G.s.* in burned habitats requires investigation. Apart from the observed differences between individuals growing in shaded and unshaded habitats, little is known concerning the degree of phenotypic plasticity and ecotypic differentiation within *G.s.*, which is polyploid through most of its range. Slender, earlier-flowering diploid populations occur in Europe.
Current trends Destruction of habitats has led to a considerable decline in *G.s.* in many lowland regions, but *G.s.* is still common in the uplands.

123 *Galium sterneri* — Sterner's Bedstraw

Established strategy S.
Gregariousness Intermediate.
Flowers White, hermaphrodite, protandrous, self-incompatible, insect-pollinated.
Regenerative strategies V, ?B_s.
Seed 0.39 mg.
British distribution Confined to the N and W (25% of VCs).
Commonest habitats Scree 39%, limestone pastures 29%, lead mine spoil 25%, rock outcrops 8%, limestone quarries 6%.
Spp. most similar in habitats *Thymus praecox* 95%, *Helianthemum nummularium* 91%, *Campanula rotundifolia* 86%, *Teucrium scorodonia* 82%, *Carex flacca* 82%.
Full Autecological Account *CPE* page 298.

Synopsis Despite its rarity in Britain and Europe as a whole, *G.s.* is a common plant on the carboniferous limestone of N Derbyshire. It is a low-growing herb restricted to open grassland and rocky habitats on calcareous soils. It is a classic calcicole (and was formerly known as *G. sylvestre*), but in many other facets of its ecology *G.s.* is similar to the calcifuge *G. saxatile*, with which it forms a sterile hybrid in Wales and Scotland. In common with *G. saxatile*, *G.s.* tends to be absent from droughted sites, a feature perhaps related to the shallow root system and the delay of shoot growth until late spring. *G.s.* is also a poor competitor, and generally occurs as an inconspicuous component of the vegetation. *G.s.* is generally found in relatively open habitats, and appears to be less shade-tolerant than *G. saxatile*. *G.s.* is most conspicuous at the margin of screes and outcrops, where its shoots often form mats which extend over rocks and bared areas. However, these mats are less extensive than those often associated with *G. saxatile*, and their component shoots are less consistently rooted. Thus, regeneration by the seeds, which are commonly produced except where the plant is shaded, may assume a greater importance than in *G. saxatile*. A buried seed bank is suspected. Field studies are required to determine the phenology of seed germination and the circumstances conducive to seedling establishment. Genotypes differing in level of hairiness are often found close to each other, and some populations exploit lead-mine spoil. Otherwise, nothing is known of the genetic variability of populations and of the capacity of *G.s.* to form ecotypes. Derbyshire populations are tetraploid ($2n = 44$). Slender diploid populations ($2n = 22$) from the W coast of Britain and from Ireland also occur. These populations, although geographically and cytologically isolated, are morphologically indistinguishable from $4x$ plants. In addition, two closely related species are the octoploid *G. pumilum* ($2n = 88$) and *G. fleurotti*, found on calcareous soils in S Britain. Their ecological separation from the tetraploid *G.s.* is uncertain. All these species are part of a closely interrelated polyploid complex of 27 European species centred on *G. pusillum*.

Current trends Generally restricted to semi-natural or at least relatively ancient habitats. Changes in land management, such as the cessation of grazing in many of the limestone dales, are likely to have resulted in a decrease in the abundance of *G.s.*, but its survival on screes and rock outcrops seems assured.

---

124 *Galium verum* Lady's Bedstraw, Yellow Bedstraw

Established strategy Between C–S–R and S–C.
Gregariousness Normally intermediate.
Flowers Yellow, hermaphrodite, protandrous, insect-pollinated.
Regenerative strategies V, ?B$_s$.
Seed 0.4 mg.
British distribution Widespread (100% of VCs).
Commonest habitats Limestone pastures 16%, limestone wasteland 13%, rock outcrops 6%, limestone quarries/verges 2%.
Spp. most similar in habitats *Pimpinella saxifraga* 90%, *Stachys officinalis* 85%, *Avenula pubescens* 82%, *Carex caryophyllea* 80%, *Briza media* 77%.
Full Autecological Account *CPE* page 300.

Synopsis *G.v.* is an often low-growing, winter-green herb of unproductive grassland and, to a lesser extent, open rocky habitats. Within the Sheffield region, *G.v.* is particularly characteristic of soils of intermediate base-status over limestone, conditions which usually coincide with surface-leached calcareous soils. In Britain as a whole *G.v.* is also common on sandy soils, and here the unusually deep root system may be of critical significance. In turf microcosms, roots of this species became colonized by VA mycorrhizas with increases of up to ninefold in seedling yield. *G.v.* persists in dune grasslands in which it exhibits a limited ability to withstand burial. This may also explain the capacity to colonize anthills. Field observations suggest that *G.v.* has a midsummer peak in biomass. *G.v.* is relatively unpalatable to stock. The species is much more robust when released from grazing, and extensive mats of ascending shoots can develop in derelict grassland as a result of stolon growth. However, where grassland remains derelict and unburnt, *G.v.* is eventually suppressed by taller species. The seed ecology of *G.v.* has been studied little, but we suspect that regeneration by seed is rather rare and that a persistent seed bank is formed. Heavy-metal-tolerant populations have been identified within the flora of Derbyshire. In N Europe, populations of *G.v.* ssp. *verum* are 4*x* and in S Europe they are 2*x* (*FE4*). A large number of ecological and geographical races occur in S Europe, but the extent of genotypic variation within British populations appears not to have been investigated. The hybrid with *G. album* is frequent in S England.

Current trends An infrequent colonist of new habitats and, in common with most species of less-fertile habitats, *G.v.* is probably decreasing.

---

125 *Geranium molle*

Dove's-foot Cranesbill

Established strategy Between R and S–R.

Gregariousness Sparse.

Flowers Purple, hermaphrodite, insect-pollinated but probably often selfed.

Regenerative strategies S, B$_s$.

Seed 1.24 mg.

British distribution Widespread (100% of VCs).

Commonest habitats Rock outcrops 12%, soil heaps 4%, manure and sewage waste/scree/limestone wasteland 3%.

Spp. most similar in habitats *Saxifraga tridactylites* 81%, *Arabidopsis thaliana* 78%, *Myosotis ramosissima* 75%, *Sedum acre* 73%, *Trifolium dubium* 71%.

Full Autecological Account *CPE* page 302.

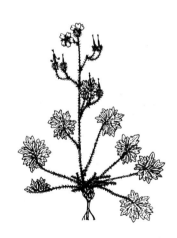

Synopsis *G.m.* is most typically a relatively robust, winter-annual herb exploiting open habitats where encroachment by perennials is checked annually by drought and their vigour reduced by nutrient stress. The species is associated with permanently open habitats, such as droughted rock outcrops and sandy wasteland, but occurs also in transient habitats, e.g. new soil heaps and localized areas of recently-bared ground in pasture and on road verges. Thus, its habitat range is remarkably similar to that of *Medicago lupulina*, with which *G.m.* is often associated in disturbed lawns and pastures. *G.m.* frequently extends into heavily-droughted habitats and, like the most ephemeral winter-annuals of these sites, flowers early and produces dormant seeds. However, in less-droughted sites *G.m.* often has an extended flowering period, perhaps because the long tap-root enables the species to exploit subsoil moisture. The possibility that the prolonged flowering may also be due to the production of more than one cohort of seedlings within a season (as in the case of *G. robertianum*) also merits investigation. *G.m.* is absent from woodland and wetland. Regeneration relies upon seeds, which exhibit hard-coat dormancy. A persistent seed bank may be formed. Large plants may produce 1500 seeds, but small plants may form < 100. Short-distance dispersal of seed (up to *c.* 6 m) is effected by means of explosive dehiscence of the fruit. Seeds survive ingestion by cattle and are an impurity in clover seed. The species of *Geranium* which are most closely related in ecology to *G.m.* are *G. dissectum* and *G. pusillum*. *G.d.* exploits moister, often impermanent, sites, while *G.p.* is more strictly a winter annual of dry lowland habitats.

Current trends Uncertain. Combines an attribute favourable to future expansion, namely the capacity for wide dispersal, with an unfavourable one, the dependence upon less-fertile habitats.

---

126 *Geranium robertianum*

Herb Robert

Established strategy Between R and C–S–R.

Gregariousness Sparse to intermediate.

Flowers Pink, hermaphrodite, protandrous, insect- or self-pollinated.

Regenerative strategies Exclusively by seed.

Seed 1.14 mg.

British distribution Widespread except in N Scotland (100% of VCs).

Commonest habitats Scree 67%, limestone woodland 19%, limestone quarries 18%, river banks 15%, lead mine spoil 11%.

Spp. most similar in habitats Insufficient data.

Full Autecological Account *CPE* page 304.

Synopsis *G.r.* is a much-branched, usually monocarpic herb, growing in base-rich sites where vegetation density is reduced by fairly frequent disturbance, coupled with either shade, low soil fertility, or both. The species occurs in such diverse habitats as shaded woodland beside rivers and streams, and slightly unstable limestone scree subject to intermittent disturbance, and is frequently the only ruderal in the habitats in which it occurs. Within the Sheffield region, suitable habitats are particularly well represented on limestone. In dry lowland areas *G.r.* is largely confined to moist shaded habitats. However, in damp upland climates *G.r.* occurs in a wide range of open and skeletal habitats and in maritime areas (where atmospheric moisture is also high) *G.r.* is also characteristic of shingle beaches. The species is relatively shade-tolerant, and flowers and fruits during the expanded phase of the woodland canopy. Foliage of *G.r.* sampled from the Sheffield region contains high concentrations of Ca, Na and Fe. When crushed, the foliage has an unpleasant smell; this characteristic may act as a deterrent to grazing. Throughout the summer months most populations contain a mixture of flowering individuals, juveniles and seedlings. Work in Poland suggests that this is because three cohorts of seedlings are often produced over the course of the year. Seedlings arising in spring flower and fruit the following summer, and have the highest survival rates and the highest reproductive potential. Thus, the spring cohort is the most important in the population dynamics of the species. The summer cohort of seedlings (July to August) overwinters as small plants which frequently do not flower until their third growing season. A few, flowering in the autumn of their second year, may also flower in their third, and are thus polycarpic. The autumn cohort is not always produced and few of the seedlings overwinter successfully. Nevertheless, such staggering of germination time appears likely to facilitate survival of a short-lived monocarpic species in environments subject to occasional, seasonally unpredictable disturbance. The seeds of the spring cohorts show the highest germination percentage. The seed, which exhibits hard-coat dormancy, is comparatively large and supports considerable height growth in the hypocotyl and cotyledon stalks. As in the case of the seedling of *Arrhenatherum elatius*, the resulting capacity for height growth in shade is important in the colonization of limestone screes. The presence of a persistent seed bank is predicted. The mericarps of *G.r.* float in water and are capable of attachment to animals and clothing. Both methods of dispersal may contribute to the relatively high colonizing ability of *G.r.*, both locally and within the secondary woodland of Lincolnshire. The species is variable and three ecological races were formerly separated as ssp. These races may hybridize with *G. purpureum*.

Current trends Not under threat, and perhaps even increasing in disturbed artificial habitats.

---

127 *Geum urbanum*   **Herb Bennet, Wood Avens**

Established strategy Between C–S–R and S.
Gregariousness Sparse to intermediate.
Flowers Yellow, hermaphrodite, self- or more rarely insect-pollinated.

Regenerative strategies S.
Seed 0.73 mg.
British distribution Widespread (100% of VCs).
Commonest habitats Limestone

woodland 12%, hedgerows 8%, scrub 5%, acidic woodland 4%, river banks 3%.

Spp. most similar in habitats *Arum maculatum* 89%, *Mercurialis perennis* 87%, *Melica uniflora* 87%, *Hedera helix* 86%, *Bromus ramosus* 83%.

Full Autecological Account *CPE* page 306.

Synopsis *G.u.* is a winter-green herb producing a rosette of ascending compound leaves. The species is frequent in situations of moderate shade in woodland, particularly on moist S-facing slopes, and is also widespread in the partial shade of hedgerows. Flowering and seed set are greatest in unshaded conditions. In coppiced woodland *G.u.* shows an intermediate degree of shade-tolerance comparable with that of *Cirsium palustre*. *G.u.* is perhaps best regarded as a 'semi-shade' species occurring in sites where the vigour of potential dominants of the herb layer is restricted by intermittent disturbance. *G.u.* is particularly frequent in hedgerows, in secondary woodland and along woodland rides. The species is virtually absent from waterlogged ground and from grazed habitats. *G.u.* has little capacity for vegetative spread. Like *Anthriscus sylvestris*, another hedgerow species of similar ecology, *G.u.* produces numerous seeds, sometimes > 1000 per plant, and apparently lacks a persistent seed bank. However, the achene has a strong hooked awn and is dispersed on clothing and fur. This mechanism may contribute to the ability of *G.u.* to occur widely in woodlands and hedgerows. *G. rivale*, a species of N-facing grassland, wet meadows and soligenous mire, forms hybrid swarms with *G.u.*, typically in open woodland. Usually introgression is towards *G.r.*, but in shaded, drier sites some hybrids may resemble *G.u.* more closely.

Current trends Favoured by the increasing level of disturbance of woodland and scrub, which may be compensating for losses of hedgerow habitats.

---

128 *Glechoma hederacea*

Ground Ivy

Established strategy C–S–R.
Gregariousness Patch-forming.
Flowers Violet, hermaphrodite (small female flowers also occur), insect-pollinated.
Regenerative strategies V, S.
Seed 0.69 mg.
British distribution Widespread except in the extreme W and N (97% of VCs).

Commonest habitats Hedgerows 20%, shaded mire 8%, soil heaps 7%, limestone woodland 6%, river banks 5%.
Spp. most similar in habitats *Stachys sylvatica* 85%, *Galium aparine* 77%, *Tamus communis* 74%, *Urtica dioica* 70%, *Stellaria holostea* 63%.
Full Autecological Account *CPE* page 308.

Synopsis *G.h.* is a carpet-forming herb most typical of fertile habitats where the vigour of potential dominants is restricted by a combination of shade and disturbance. *G.h.* is frequent in hedgerows, on shaded roadsides and in woodland, and tends to occur beside rides, on floodplains, beside rabbit scrapes or in clear-felled areas. The species is considered to be characteristic of secondary woodland. Although generally absent from pasture, *G.h.* is known to be toxic to stock. When growing in the open, *G.h.* produces more flowers and has a higher level of seed-set, forming leaves which are small relative to those seen in shaded sites. In common with many other species from shaded habitats, *G.h.* flowers early in the year; however, the leaf canopy reaches a maximum in summer. The species has a limited capacity to persist under tall herbs or beneath a continuous tree canopy, and is most characteristic of lightly or patchily shaded habitats. When released from canopy shade, *G.h.* often spreads rapidly and extensively by means of long creeping stems. Thus, at an early stage in colonization, the floor of a quarry adjacent to woodland has been seen

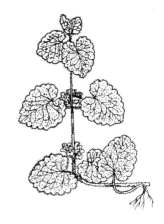

covered with *G.h.*. Effective regeneration appears to be primarily vegetative, and it is likely that, as in the case of *Stachys sylvatica*, detached shoot fragments are important for long-distance colonization. Seed-set is often extremely poor, and studies suggest that seed is only shortly persistent in the soil.

Current trends A regular colonist of secondary woodland, probably by vegetative means, and likely to remain relatively common in shaded habitats subject to occasional disturbance.

---

129 *Glyceria fluitans*

Flote-grass

Perhaps including field data for
Glyceria plicata

Established strategy C–R.
Gregariousness Sometimes
 patch-forming.
Flowers Green or purple,
 hermaphrodite, wind-pollinated.
Regenerative strategies V, B_s.
Seed 1.20 mg.
British distribution Widespread (100%
 of VCs).
Commonest habitats Running aquatic
 14%, unshaded mire 13%, still
 aquatic 11%, shaded mire 9%, river
 banks 5%.
Spp. most similar in habitats *Callitriche
 stagnalis* 94%, *Equisetum palustre*

92%, *Ranunculus flammula* 90%, *Apium nodiflorum* 89%.
Full Autecological Account *CPE* page 310.

Synopsis *G.f.* is a winter-green stand-forming grass of relatively fertile mire and semi-aquatic habitats. The species is usually rooted in mud, and frequently occurs with the shoots floating on shallow water at the edge of small ponds and drainage ditches. *G.f.* is one of the species which is most capable of exploiting sites where the water-table fluctuates extremely widely during the year. Although associated with a wide range of soil conditions, the species is absent from calcareous soils. Both in mud and in water *G.f.* reproduces vegetatively to form extensive patches by means of creeping shoots. Detached shoots readily re-root and may be transported by water and colonize new sites. Although populations of this clonal out-breeding species regularly set seed, regeneration by this means is probably not a regular occurrence. A persistent seed bank has been recorded. The ecological amplitude of *G.f.* is wide, and overlaps with that of a number of other *Glyceria* spp. However, *G.f.* tends to be replaced by the more calcicolous *G. plicata* in base-rich soligenous mire. Hybrids are found in flushes where the species co-occur. The hybrid *G.* × *pedicellata*, which is sterile but vegetatively vigorous, is locally common in the absence of its parents beside rivers on the upland carboniferous limestone. This may constitute a habitat which is intermediate in terms of hydrology and base status between those occupied by the parents. *G. declinata* is smaller, does not form extensive stands, and tends to replace *G.f.* in heavily trampled sites and base-poor soligenous mire. *G.d.* rarely hybridizes with *G.f.*. Unlike *G.f.*, both *G.d.* and *G.p.* are in-breeding, and each has smaller seeds than *G.f.*. In contrast with those of *G.d.* and *G.f.*, seeds of *G.p.* do not exhibit any dormancy. This may facilitate establishment in soligenous mire during autumn.
Current trends Activities such as ditch-making favour *G.f.*, and the species may be increasing in ponds, perhaps as a result of eutrophication. However, destruction of mire has been considerable and, in common with many other wetland plants, the future status of *G.f.* remains uncertain.

130   *Glyceria maxima*   Reed-grass

Established strategy C.
Gregariousness Often stand-forming.
Flowers Green, hermaphrodite, wind-pollinated.
Regenerative strategies V, B$_s$.
Seed 0.74 mg.
British distribution Widespread, especially in lowland areas (73% of VCs).
Commonest habitats Unshaded mire 8%, still aquatic 6%, river banks 5%, soil heaps 2%.
Spp. most similar in habitats *Eleocharis palustris* 98%, *Polygonum amphibium* 95%, *Hydrocotyle vulgaris* 93%, *Alisma plantago-aquatica* 91%, *Typha latifolia* 90%.
Full Autecological Account *CPE* page 312.

Synopsis *G.m.* is a tall robust grass producing a dense canopy and often forming a virtual monoculture as a result of expansion of an extensive and close-packed rhizome system. *G.m.* is highly aerenchymatous and is characteristic of waterlogged alluvial soils of high fertility, particularly in sites close to slow-flowing water. The species may also form floating rafts of prostrate stems which can choke canals. *G.m.* may compete successfully with *Phragmites australis* under productive conditions, since it produces a dense cover in spring before the development of *P.a.*, a phenomenon consistent with the higher nuclear DNA amount of *G.m.*. However, factors such as increased anaerobiosis may tilt the balance in favour of *P.a.*. *G.m.* is a lowland species mainly found below 150 m. The species is probably restricted in part by the scarcity and fragmented nature of suitable habitats in upland areas, but has been observed as a persistent introduction at 290 m. *G.m.* shows a low level of morphological variation and a modest ecological amplitude. *G.m.* is palatable to stock when young and was formerly cultivated as a fodder crop. It is tolerant of cutting, more so than *P.a.*, and its relative abundance may be increased by mowing. Seed-set is poor and establishment from the seeds which persist in the soil appears to be a rare event. Thus, *G.m.* is largely dependent upon vegetative means for regeneration. The species shows aggressive rhizome growth to form large clonal stands and detached rhizome pieces, which float, regenerate readily to form new colonies. By these means, whole water systems may be

colonized. However, such dependence upon large, water-borne vegetative propagules for regeneration severely limits the capacity of *G.m.* to colonize land-locked sites. Thus, *G.m.* may be abundant in a canal system but absent from nearby ponds, and we suspect that in many of its more isolated sites *G.m.* is an ornamental introduction.

Current trends Appears to have increased, apparently at the expense of *P.a.*, in the ditches beside arable land in lowland areas, and has choked a number of disused canals. However, since much wetland has been drained, on balance *G.m.* is probably decreasing.

---

131 *Hedera helix*

Ivy

For taxonomy see 'Synopsis'

Established strategy S–C.
Gregariousness Carpet-forming.
Flowers Yellow-green, hermaphrodite, sometimes protandrous, insect-pollinated.
Regenerative strategies V.
Seed 20.43 mg.
British distribution Aggregate widespread (100% of VCs).
Commonest habitats Limestone woodland 37%, hedgerows 27%, acidic woodland 16%, cliffs/walls 14%.
Spp. most similar in habitats *Bromus ramosus* 88%, *Geum urbanum* 86%, *Melica uniflora* 78%, *Elymus caninus* 74%, *Mercurialis perennis* 74%.
Full Autecological Account *CPE* page 314.

Synopsis *H.h.* is a long-lived (to 400 years), evergreen species. Its long woody stems grow either vertically up tree trunks, cliffs or walls, attached by means of numerous short roots, or horizontally to form continuous carpets on the woodland floor. Its capacity to extend over soil-less habitats from a base rooted in soil is unique within the British flora. The vertical stems are vital to the reproductive capacity of the species, since they alone bear flowering shoots and do so only in relatively unshaded sites. Plants in woodland or hedgerows seldom escape the shade of their accompanying trees and shrubs, and so rarely flower. Thus, like *Lonicera periclymenum* and *Tamus communis*, *H.h.* mainly sets seed in habitats which are less shaded than those in which it occurs most abundantly. The large, berried seeds are dispersed by birds, and germination probably occurs mainly in spring, though very young seedlings can be observed during winter. The seed is short-lived and does not form a persistent seed bank. Flowers which are produced in late autumn are susceptible to the effects of frost. Problems of initial colonization apppear to limit the woodland distribution of *H.h.*, and the species is characteristic of secondary rather than ancient woodland. Thus, colonies frequently radiate inwards from the woodland edge or are found on areas of former quarrying. It has been suggested that invasion is associated with hawthorn succession from open ground. The possibility that *H.h.* requires unvegetated, relatively unshaded sites for establishment from seed requires investigation. However, once established in woodland, *H.h.* can form extensive clonal patches by means of rooted horizontal stems. *H.h.* is also often abundant in woodland close to active limestone quarries, apparently due to its resistance to the effects of limestone dust. *H.h.* is heterophyllous with palmately lobed 'shade leaves' except on flowering

branches where larger, ovate sun-leaves occur. These two types of shoot have very different characteristics. 'Juvenile' shoots, with shade leaves, root readily but do not flower; in contrast, 'mature' shoots, with sun-leaves can be rooted only with great difficulty. Ivy is the only British member of the predominantly tropical family Araliaceae, and is susceptible to low winter temperatures. For this reason the species has been used by palaeobotanists as an indicator of mild winters. *H.h.* was gathered as fodder from Neolithic times until the 16th century. However, the plant is toxic if consumed in quantity and is rejected by gastropods. Foliage of *H.h.* appears to be subject to extremely low rates of herbivory, and provides excellent cover for some early-nesting birds. Two taxa are recognized, *H. helix* and *H. hibernica*. The latter, with $2n = 96$ $(4x)$, replaces *H.h.* in SW coastal regions of the British Isles, Ireland and the Iberian peninsula. Two further ssp. are recognized from S Europe.

Current trends Uncertain. Favoured by artificial disturbances, but some habitats, e.g. dead standing trees and old walls, are less frequent than formerly.

Established strategy S.
Gregariousness Intermediate.
Flowers Yellow, usually hermaphrodite, insect- or self-pollinated.
Regenerative strategies B_s.
Seed 1.38 mg.
British distribution Particularly in the S and SE (61% of VCs).
Commonest habitats Limestone pastures 27%, scree 23%, lead mine spoil 7%, rock outcrops/limestone wasteland 6%.
Spp. most similar in habitats *Avenula pratensis* 94%, *Galium sterneri* 92%, *Carex flacca* 90%, *Sanguisorba minor* 89%, *Polygala vulgaris* 89%.
Full Autecological Account *CPE* page 316.

Synopsis *H.n.* is a slow-growing, prostrate, evergreen undershrub most typically associated with species-rich, short turf on well-drained, infertile soils. *H.n.* is strongly calcicolous in SE England and in N Derbyshire. However, in E Scotland *H.n.* is characteristic of dry *Agrostis–Festuca* grasslands and may even occur with *Calluna vulgaris* and other calcifuges on soils of pH 3.8. This pattern of distribution with respect to soil reaction is similar to, though less extreme than, that shown by *Thymus praecox*. Leaves of plants established on calcareous soils in the Sheffield region have high Ca and Al concentrations. On the southern chalk, *H.n.* may be particularly frequent on anthills. The root systems of plants examined in chalk downland are described as intermediate in depth, but our own observations on fissured limestone suggest that the tap-root penetrates deeply and facilitates drought-avoidance. The species also exhibits some desiccation tolerance involving sclerophylly, and under extreme drought *H.n.* has the capacity

to shed leaves. The adult plant is also frost-resistant. *H.n.* can persist under moderate trampling and grazing, but cannot compete successfully in taller grasslands of the type which develops in response to dereliction or fertilizer application. Disturbance also creates the local areas of bare soil necessary for seedling establishment. This regenerative method is more important than the vegetative expansion which can occur by layering of old branches. Seeds have hard-coat dormancy, but further research is required to recognize the mechanisms which break dormancy under natural conditions and to examine the persistence of the seed bank. Germination tends to occur in spring, at least in N England, and seedlings are susceptible to droughting. Seeds, which are mucilaginous when moistened, appear to be poorly dispersed. As a result, *H.n.* is restricted to old or semi-natural habitats. This low dispersability may have contributed to the failure of *H.n.* to colonize Ireland and to its replacement in the Burren by *H. canum* which, elsewhere in Britain, is a species of xeric grassland. Some morphological variation exists and appears to be genetic in origin; a number of ecotypes have been identified. The existence of more-extreme morphological discontinuities allow the recognition of seven additional ssp. within the mountainous regions of C and S Europe. *H.n.* also hybridizes with *H. apenninum.*

Current trends Decreasing and close to extinction in some intensively-managed lowland areas, in which *H.n.* is becoming increasingly restricted to local refugia such as rock outcrops.

133 *Heracleum sphondylium* — Cow Parsnip, Hogweed, Keck

Established strategy C–R.
Gregariousness Reproductive
  individuals scattered and large.
Flowers White, protandrous,
  self-compatible, insect-pollinated.
Regenerative strategies S.
Seed 5.52 mg.
British distribution Widespread (100%
  of VCs).
Commonest habitats Limestone
  wasteland 41%, verges 38%,
  hedgerows 36%, meadows 35%,
  limestone quarries 31%.
Spp. most similar in habitats
  Anthriscus sylvestris 68%, *Galium
  aparine* 61%, *Stachys sylvatica* 59%,
  Dactylis glomerata 55%, *Poa
  pratensis* 54%.
Full Autecological Account *CPE* page
  318.

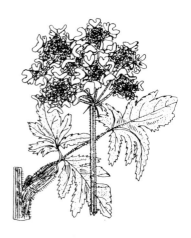

Synopsis *H.s.* is a robust, tap-rooted herb of rough grassland and wasteland, particularly on moist, fertile soils. The leaves are high in P, Ca and Fe, and when sampled in winter the plant contains unusually high concentrations of protein and fructose. *H.s.* is perhaps typically a short-lived perennial, but often persists for over five years. The species has no capacity for vegetative spread, and its ecology is limited by the need for regular establishment from seed. Following from this, *H.s.* is often found in sites within roadside and hedgerow vegetation where robust competitive perennials with a capacity for clonal expansion are debilitated by occasional cutting. *H.s.* also occurs in hay-meadows. The leaf canopy consists of a few large leaves which are vulnerable to heavy grazing. Consequently, *H.s.* is infrequent in pastures. *H.s.* has a low seedling relative growth rate primarily because at an early stage of seedling establishment photosynthate is allocated to the tap-root at the expense of new leaf development. However, the expansion of leaves and the growth of the stout stem in spring are relatively rapid in established plants due to a mobilization of these below-ground reserves. The tap-root may provide access to moisture in deep crevices, enabling *H.s.* to colonize rocky habitats. Like many ruderal species, *H.s.* has a protracted flowering period. *H.s.* is dependent upon seed for regeneration, and the mericarps, which are both winged and flattened, are initially scattered short distances by wind. Like many members of the Umbelliferae, seeds contain a poorly differentiated embryo whose development requires a period of chilling before germination can take place. Despite their large size and conspicuous presence on the soil surface over the winter, large numbers of the aromatic seeds escape predation and germinate in cohorts in spring. Populations often consist of a few flowering individuals, and young plants, which may persist for several years at the one-leafed stage, often lie partially hidden beneath litter and the established plants' canopy. Many of our field records refer to plants in this state, and the apparently wide habitat range of *H.s.* is due in large measure to the presence of seedlings and small plants in situations which will not be conducive to their further development. Like several other large-seeded species, e.g. *Anthriscus sylvestris* and *Galeopsis tetrahit*, there is circumstantial evidence that

seedlings of *H.s.* can establish, at least on a local scale, within closed perennial communities. Further research is required on this phenomenon. *H.s.* is phenotypically plastic in response to management treatment, and flowering stems developed after cutting are much shorter than those produced otherwise. *H.s.* is a very variable species, and nine geographical variants have been recognized within Europe. British material refers to ssp. *sphondylium* except in East Anglia, where ssp. *sibiricum* of NE and EC Europe also occurs. The distribution of *H.s.* overlaps with that of the introduced *H. mantegazzianum* and the two hybridize.

Current trends A common and probably increasing species. *H.s.* may become more important as an arable weed as a consequence of the increased popularity of minimum tillage.

---

134   *Hieracium pilosella*

Mouse-ear Hawkweed

See 'Synopsis' for taxonomy

Established strategy Between S and C–S–R.
Gregariousness Sometimes patch-forming.
Flowers Florets yellow, ligulate, hermaphrodite.
Regenerative strategies V, W.
Seed 0.15 mg.
British distribution Widespread except in Shetland (84% of VCs).
Commonest habitats Limestone pastures 33%, limestone quarries 31%, scree 28%, lead mine spoil/rock outcrops 21%.
Spp. most similar in habitats *Linum catharticum* 89%, *Leontodon hispidus* 85%, *Senecio jacobaea* 81%, *Lotus corniculatus* 81%, *Campanula rotundifolia* 81%.
Full Autecological Account *CPE* page 320.

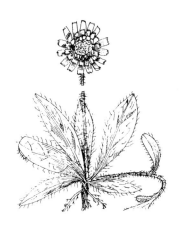

Synopsis *H.p.* is a low-growing stoloniferous herb of dry, often calcareous soils. The leaves of *H.p.* are palatable to a range of herbivores, and *H.p.* typically forms appressed rosettes in grazed grassland. Although the foliage may be held upright in taller vegetation, the low stature of the shoot causes *H.p.* to be vulnerable to the shade of taller plants when pastures become derelict. In the *PGE H.p.* is restricted to unproductive grassland, and in experimental plots in chalk grassland, production and flowering is stimulated by addition of nitrogen. The leafy stolons allow ramification over rocky and unstable substrata, and on rock outcrops daughter rosettes are sustained at a considerable distance from the points at which roots penetrate into soil crevices. However, *H.p.* has shallow roots and appears to show some susceptibility to drought. In turf microcosms, roots of this species became heavily colonized by VA mycorrhizas with increases of up to ninefold in seedling yield. *H.p.* regenerates largely by means of long stolons which give rise to rosettes and subsequently die back leaving unattached daughter plants. Stolons are produced only from rosettes which have produced a floral initial, even if this subsequently aborts. Rosettes are frequently short-lived, and considerable population fluctuations may occur as a result of the dynamics of stolon production and mortality.

H.p. also produces wind-dispersed seed, particularly in little-grazed sites. Seedlings have been observed in vegetation gaps mainly during autumn. A delay may occur between dispersal and germination but no persistent seed bank is developed. In established grassland, seedling establishment appears to be extremely rare, and Watt suggests that it takes place mainly where there is a low intensity of grazing and a wet spring. However, *H.p.* appears fairly widely on calcareous spoil, suggesting that regeneration from seed is effective in colonizing new habitats. 2*x* and 4*x* plants are sexual but higher ploidy levels are apomictic. Thus, in many cases reproduction by seed is an asexual process. *H.p.* has been subdivided into several ssp. differing in height and geographical distribution, and several occur in Britain. Data presented refer to the aggregate species, but both 4*x* (sexual) and 5*x* (apomictic) plants are recorded from the Sheffield region. The ecological differences between taxa and genotypes within the *H.p.* complex are poorly understood. A heavy-metal-tolerant race of *H.p.* has been described locally.

Current trends An effective colonist of artificial habitats such as quarries and railway banks, but perhaps decreasing due to the dependence upon infertile habitats.

135 *Hieracium*

Hawkweed

Subgenus *Hieracium*, for taxonomy see 'Synopsis'

Established strategy Varying between S and C–S–R.
Gregariousness Typically intermediate.
Flowers Yellow, ligulate, mostly apomictic.
Regenerative strategies W.
Seed (*H. exotericum* agg.) 0.40 mg.
British distribution Widespread (100% of VCs).

Commonest habitats Limestone quarries 57%, limestone wasteland 45%, cinder heaps 40%, rock outcrops 31%, coal mine waste 30%.
Spp. most similar in habitats *Medicago lupulina* 76%, *Leucanthemum vulgare* 73%, *Leontodon hispidus* 70%, *Crepis capillaris* 69%, *Hieracium pilosella* 68%.
Full Autecological Account *CPE* page 322.

Synopsis *H.* agg. is a perennial herb associated with a wide range of less-fertile, often rocky habitats. Habitat diversity coincides with the existence of many ecotypes within the grouping. *H.* agg. is a taxonomically difficult grouping and has been subdivided into a number of Sections, with over 240 microspecies. The Sections are often characterized by taxa of distinctive morphology and habitat range, but unfortunately it was not feasible to separate taxa during the fieldwork for this study. Over 66% of British taxa are either semi-rosette or rosette species and, consistent with this basal arrangement of the leaves, a majority are found in rocky (presumably little-vegetated and often desiccated) habitats. Taxa frequently have a deep tap-root, and appear able to exploit subsoil moisture in droughted periods. They are perhaps mainly stress-tolerators. In Britain, 'semi-rosette' species of *Hieracium* are most abundant on calcareous rocks, but some records are from siliceous strata. The stock may occasionally branch to give two rosettes, but effective regeneration is by seed which is produced in quantity and which is, in these apomictic species, a means of asexual reproduction. The plumed seed is widely dispersed and 'semi-rosette' species regularly colonize artificial habitats such as railway margins, road verges and old walls. Further, the presence of a number of alien species in these habitats suggests that the taxa are sometimes capable of dispersal over long distances, perhaps through human agency. Flowering stems are tall and the leaves relatively long-lived. Consequently, the taxa are

very susceptible to grazing and are seldom recorded from pasture. In a minority of taxa, persistent leaves only occur on the stem. Some, but not all, of such plants are associated with taller closed vegetation including the herb layers of open woodland. Taxa with leafy stems probably refer to *H. vagum* of Section Sabauda, which colonizes railway banks and coal mines and extends onto acidic soils. *H.v.* also has some capacity for lateral vegetative spread. Only one British species of subgenus *Hieracium*, *H. umbellatum*, is sexual, and hybrids between the two sspp. of the taxon have been reported. **Current trends** Uncertain. Reported to have decreased in the 20th century, perhaps as a result of sheep-grazing. However, the reduced levels of grazing in areas such as the Derbyshire dales, and the capacity to colonize artificial habitats, e.g. quarry heaps, suggests the potential for an increase in the abundance of certain taxa.

136   *Holcus lanatus*   Yorkshire Fog

Established strategy C–S–R.
Gregariousness Intermediate.
Flowers Greyish-green,
   wind-pollinated, self-incompatible.

Regenerative strategies S, B_s, V.
Seed 0.32 mg.
British distribution Widespread (100% of VCs).

Commonest habitats Meadows 72%,
coal mine waste/soil heaps 48%,
cinder heaps 46%, limestone
quarries 45%.

Spp. most similar in habitats *Cirsium
arvense* 81%, *Tussilago farfara* 59%,
Atriplex prostrata 58%, *Lathyrus
pratensis* 56%, *Dactylis glomerata*
55%.

Full Autecological Account *CPE* page
324.

Synopsis *H.l.* is a tufted, partially
winter-green grass. The ecological
range is unusually wide, but is limited
by five major attributes. First, *H.l.*
shows lax tillering and has relatively
few shoot buds; consequently, the
species is not tolerant of close grazing
or heavy trampling and tends to be
rather sparsely distributed in pasture.
Second, although present on soils of
widely different soil pH and fertility,
H.l. is most frequent on relatively
fertile soils of pH 5.0–6.0. In the *PGE*
it is reported that 'high nutrition
associated with soil acidity gives it
great encouragement'. Third, the
species achieves maximum vigour in
moist rather than waterlogged habitats,
although survival on anaerobic soils
can occur through the capacity of the
species to produce a network of fine
roots at the soil surface. Fourth, the
species has restricted tolerance of
shaded habitats. Fifth, *H.l.* is sensitive
to winter damage, and above-ground
biomass is reduced after severe
winters. Despite these constraints, *H.l.*
is an exceptionally versatile grass,
rivalled only by *Festuca rubra* and
Agrostis capillaris in its capacity to
exploit widely different soil conditions,
and forms of vegetation management.
Foliar levels of N, Ca and Fe are
relatively low in the Sheffield region,
but elevated concentrations of Na have
been found. Seed production, from the
plant's second year onward, is often
prolific and, since many seeds are shed
before harvesting, *H.l.* may be
abundant in hay-meadows. Caryopses
are released early and are capable of
rapid germination over a wide range of
temperatures. Seedling establishment
is the main source of population
expansion. Many seedlings appear in
autumn, but large numbers of
ungerminated seeds remain on the soil
surface and a persistent seed bank is
formed. Germination is stimulated by
the presence of light and by fluctuating
temperatures and is largely restricted
to gaps in the vegetation. The copious
seed production and wide seed
dispersal, often by human agency,
coupled with flexibility of germination
and high growth rate, makes *H.l.* a
highly efficient colonist of open
habitats and confers some persistence
under fluctuating types of land
management. There is a restricted
capacity for lateral spread through the
production of prostrate shoots which
root at the nodes. This phenomenon
probably only attains importance in
sites where seed set is prevented. *H.l.*
is possibly allelopathic. *H.l.* shows a
high degree of ecotypic differentiation,
and edaphic ecotypes are recorded.
H.l. also hybridizes with *H. mollis*.

Current trends *H.l.* has many
attributes which confer success in
common artificial habitats. The
species, which is a major problem in
herbage seed crops, and one of the
later colonists during sward
deterioration, is probably still
increasing. Further, although little

grown, its yield of herbage in field experiments compares favourably with those of sown cultivars of agricultural grasses; we predict that *H.l.* will be cultivated more frequently.

137 *Holcus mollis*

Creeping Soft-grass

Established strategy C.
Gregariousness Stand-forming.
Flowers Greyish-green, wind-pollinated, self-sterile except at increased polyploidy.
Regenerative strategies V.
Seed 0.32 mg.
British distribution Common except in the S Midlands and parts of Ireland (99% of VCs).
Commonest habitats Acidic woodland 36%, hedgerows 31%, broadleaved plantations/acidic wasteland 27%, shaded mire 26%.
Spp. most similar in habitats *Acer pseudoplatanus* (juv.) 80%, *Pteridium aquilinum* 73%, *Rubus fruticosus* 73%, *Silene dioica* 71%, *Sorbus aucuparia* (juv.) 69%.
Full Autecological Account *CPE* page 326.

Synopsis *H.m.* produces extensive and sometimes quite ancient clonal stands, or even monocultures, on acidic soils. In exceptional circumstances, old clones may form diffuse patches *c.* 1 km in diameter. Compared with most calcifuges, *H.m.* is capable of rapid dry-matter production and appears to be relatively nutrient-demanding. The distribution is generally restricted to brown earths and does not normally extend to strongly podsolized soils. Foliar concentrations of Ca are low. *H.m.* exploits a diversity of habitats from woodland to wetlands, grasslands and even arable land, but in many *H.m.* is subject to disturbance of one form or another. This is true even in woodland, its major habitat. *H.m.* is exceedingly vigorous in lightly shaded and open habitats in woodland, and may spread explosively after coppicing. However, where there is dense shade the expansion of the canopy in summer usually brings about etiolation and a premature decline in vigour and the shoots frequently succumb to fungal attack. It is thus difficult to understand why *H.m.* is not usurped from its woodland habitats by more-shade-adapted species. One possible reason may be the ability of *H.m.* to emerge through tree leaf litter and to complement the deeper rooting depth of *Hyacinthoides non-scripta*. *H.m.*, though frequent in grassland, does not persist in heavily grazed pasture, as its few robust shoot stems are eaten more quickly than they are replaced. But if grazing ceases, the erect growth form which the species adopts in unshaded habitats is competitively very advantageous. Most commonly, *H.m.* is 5*x* in W Europe and produces little seed. Plants in shade flower sparingly. Thus, the reproductive capacity of many populations of *H.m.* depends upon clonal expansion by means of long, spreading rhizomes and a facility for rapid regeneration from rhizome, and even shoot, fragments. The importance of regeneration from fragments in mechanically-disturbed sites is illustrated by this species' persistence in tree nurseries and, to a lesser extent, on arable land. There is no evidence of a persistent seed bank, but sites with 4x plants may not have been sampled. Greater vegetative vigour has been noted in 5x clones; 4x plants tend to occur on drier soils, but otherwise the ecological differences between the cytological races remain obscure. An additional ssp. is recorded from Spain. *H.m.* forms hybrids with *H. lanatus*. These resemble *H.m.*, and tetraploids of *H.m.* may have arisen as a result of past hybridization between *H.l.* and *H.m.*.

Current trends Uncertain. *H.m.* is able to exploit some disturbed conditions but is confined to moderately base-rich acidic soils.

138   *Hyacinthoides non-scripta*

Bluebell

Established strategy Between SR and C–S–R.
Gregariousness Intermediate or stand-forming.
Flowers Usually blue, hermaphrodite, partially self-compatible, insect-pollinated.
Regenerative strategies S, V.
Seed 6.17 mg.
British distribution Widespread (99% of VCs).
Commonest habitats Broadleaved plantations 50%, acidic woodland 35%, scrub 33%, limestone woodland 32%, coniferous plantations 26%.
Spp. most similar in habitats *Acer pseudoplatanus* (juv.) 91%, *Quercus* sp. (juv.) 83%, *Fagus sylvatica* (juv.) 81%, *Lonicera periclymenum* 80%, *Rubus fruticosus* 80%.
Full Autecological Account *CPE* page 328.

Synopsis *H.n-s.* is a woodland, bulb-forming, monocotyledonous herb with a vernal phenology, exploiting the light phase before the development of a full summer canopy, and restricted to sites where the light intensity does not fall below 10% of daylight between April and mid-June. This involves shoot expansion during late winter and early spring, a capacity consistent with its high nuclear DNA amount and the presence of exceptionally high concentrations of fructan and fructose in the bulb in winter. *H.n-s.* then undergoes a period of bulb aestivation prior to the expansion of a new set of roots in autumn. The number of leaves produced is predetermined and, since damaged leaves cannot be replaced, *H.n-s.* is very vulnerable to grazing, cutting or trampling. However, foliage contains toxic glycosides and, though eaten by cattle and sheep, is not grazed by rabbits. *H.n-s.* has a poorly-developed system for water conduction (vessels are confined to the roots), and seeds released in summer require exposure to warm moist conditions followed by chilling in order to become germinable. These attributes appear to be important in restricting *H.n-s.* to mesic sites, and perhaps also in confining this 'Atlantic' species to shaded habitats in areas of low rainfall. As a result of the action of contractile roots, the mature bulb of *H.n-s.* may be buried to a depth of 250 mm or more, rendering the species vulnerable to waterlogging and subject to replacement by the more superficial bulbs of *Allium ursinum* on alluvial terraces. Thus, *H.n-s.* is characteristic of moist, freely-drained environments. The location of the bulb in the subsoil may limit the ability of *H.n-s.* to occupy the woodlands of stabilized limestone scree. However, the deep root system often allows co-existence with strongly competitive woodland species. *H.n-s.* often exploits the subsoil below the rhizomatous growth of *Pteridium aquilinum.* Leaves are high in Na and Mn, but low in P and Mg. During the early phase of shoot emergence, the leaves form a compact spear-shoot, capable of penetrating through considerable depths of deciduous tree litter. In favourable situations, dense stands of robust individuals may exert local dominance of the ground flora during the vernal phase, and probably restrict the vigour of other species. Various studies consistently indicate that *H.n-s.* regenerates primarily by seeds, which germinate from October onwards, usually in microsites protected by a covering of leaf litter and within 0.4 m of the parent plant. A variable number of seeds remain ungerminated until spring. Seedlings suffer high mortality in close proximity to parent plants and in sites with deep tree litter. Plants do not flower for *c.* 5 years. At maturity they may produce daughter bulbs but most authorities have concluded that vegetative regeneration is of minor importance. Genotypic variation between and within populations has been reported, but few taxonomic variants have been described. *H.n-s.* forms hybrids with the related *H. hispanica.*

Current trends Uncertain. In moist, upland areas introduced *H.n-s.* is a frequent colonist of plantations, and may be among those woodland species which are less vulnerable to changes in forestry management. In some lowland areas, however, the species is largely restricted to ancient woodland, and may therefore be declining.

---

139   *Hydrocotyle vulgaris*

Pennywort, White-rot

Established strategy C–S–R.
Gregariousness Patch-forming.
Flowers Greenish-white tinged with pink, hermaphrodite, self-pollinated.
Regenerative strategies V.
Seed 0.31 mg.
British distribution Widespread (100% of VCs).
Commonest habitats Unshaded mire 7%, still aquatic 4%, walls 1%.
Spp. most similar in habitats Insufficient data.
Full Autecological Account *CPE* page 330.

Synopsis *H.v.* is a wetland species with creeping rooted stems bearing leaves held aloft on long petioles. The species is perhaps most characteristic of soligenous mire on peaty, mildly acidic soils where the growth of potential dominants is suppressed by an intermediate level of fertility and by occasional grazing. To a lesser extent, *H.v.* occurs in topogenous mire and plants rooted on, for example, ditch banks may extend into aquatic habitats. However, persistence and growth on reservoir margins down only to positions corresponding to the winter water level suggest an inability to exploit sites which are waterlogged throughout the year. The species may occur in local monocultures or in a turf with other species. *H.v.* may form an understorey to species such as *Juncus effusus*, but is usually absent from shaded mire. Canopy height is determined by the length of the petiole, and plants growing in dry or grazed habitats have much shorter petioles and smaller leaves than those in taller vegetation. Marked changes in morphology are often associated with plants growing beside water; individuals rooted under water often have a disproportionately long petiole and lack the hairs present on the leaves of terrestrial forms. *H.v.* regenerates vigorously by the creeping shoots which often form extensive patches. Thus, the species may be a particularly successful colonist of areas bared by disturbance. Detached fragments, which root readily, are probably important in the colonization of stream- and ditch-side sites following disturbance. Fruits, which float and may adhere to the feet of birds, may also play a role in the colonization of more-distant sites. However, regeneration by seed is probably of infrequent occurrence, particularly since *H.v.* appears rather shy-flowering when growing amongst tall vegetation. No persistent seed bank has been detected. The subfamily to which *H.v.* belongs, the Hydrocotyloideae (formerly treated as the separate family, Hydrocotylaceae), is mainly distributed in the S hemisphere, and *H.v.* is the only British member.
Current trends Decreasing as a result of habitat destruction. Not under any great threat at present in upland regions, but becoming rare in many lowland districts.

140 *Hypericum hirsutum*   Hairy St John's Wort

Established strategy Between C–S–R and S.
Gregariousness Isolated clumps.
Flowers Yellow, hermaphrodite, insect- or self-pollinated.
Regenerative strategies B_s, V.
Seed 0.07 mg.
British distribution Widely scattered (63% of VCs).
Commonest habitats Scree 6%, limestone pastures/scrub 4%, lead mine spoil 3%, limestone wasteland/limestone woodland 2%.
Spp. most similar in habitats Insufficient data.
Full Autecological Account *CPE* page 332.

Synopsis *H.h.* is a tall, tuft-forming herb characteristic of relatively unproductive wasteland on moist, often clayey, calcareous soils. Although tall, like *H. perforatum*, its capacity for dominance is restricted by the limited extent of lateral vegetative spread. Persistence of *H.h.* appears to be dependent upon infertility and occasional disturbance, factors which restrict encroachment by taller clonal species. *H.h.* exhibits a limited tolerance of burning. The species contains hypericin which, as in *H.p.*, confers some protection against herbivory. It is unusual to find extensive populations of *H.h.*, and there are grounds to suspect that the species is a plant of ecotones. In common with *Teucrium scorodonia*, *Fragaria vesca* and *Potentilla sterilis*, the species is most common at the boundary between shaded and unshaded habitats. *H.h.* is less overtly a colonizing species than *H.p.* and, although found on quarry spoil and shaded rock outcrops, *H.h.* is more typical of tall grassland, particularly at wood margins. Further contrasts with *H.p.* are evident in its greater restriction to calcareous strata and failure to exploit droughted habitats. *H.h.* produces numerous very small seeds which appear to require areas of bared ground for establishment. Seeds probably germinate mainly in spring after the inhibitor in the seed coat has been washed out. *H.h.* also forms a persistent seed bank, and this almost certainly plays an important role in allowing persistence during the 'closed' phases of coppice cycles. *H.h.* has only a limited capacity for vegetative reproduction, and detached shoots root much less readily than those of *H.p.*.

Current trends *H.h.* is restricted to a relatively narrow range of infertile habitats and, despite some capacity for persistence and dispersal, is probably decreasing.

---

141 *Hypericum perforatum*

Common St John's Wort

Established strategy Between C–R and C–S–R.
Gregariousness May form diffuse patches.
Flowers Yellow, hermaphrodite, usually apomictic.
Regenerative strategies B_s, V.
Seed 0.1 mg.
British distribution Common except in the N and W (94% of VCs).
Commonest habitats Limestone wasteland 12%, limestone quarries/rock outcrops 6%, arable/cinder heaps/limestone pastures/paths/soil heaps 2%.
Spp. most similar in habitats *Centaurea nigra* 82%, *Centaurium erythraea* 82%, *Brachypodium pinnatum* 81%, *Bromus erectus* 80%, *Inula conyza* 73%.
Full Autecological Account *CPE* page 334.

Synopsis *H.p.* is a tall herb characteristic of relatively unproductive wasteland on freely drained soils. Although a tall species, the potential to exercise dominance is restricted by the limited capacity to form dense stands. *H.p.* is most characteristic of dry rocky or sandy ground where, as a result of factors such as disturbance and low fertility, competition from more-robust species is restricted. In particular, *H.p.* is associated with chalk and limestone quarries and with railway ballast. *H.p.* is tolerant of, or even favoured by, fire and its grassland distribution is strongly biased towards burned sites. However, *H.p.* is ecologically wide-ranging and occurs on a diversity of soil types, and in association with widely contrasted levels of disturbance. It seems likely that the deep tap-root is of critical importance in allowing *H.p.* to sustain a summer peak in shoot biomass. In Australia and New Zealand, where the species is much more common than in Britain, the glandular shoots are a major hazard to farm animals, and heavy economic losses are experienced through poisoning of stock. This phenomenon is not common in Britain because *H.p.* is only a minor weed of grassland, and the major effects of hypericin toxicity are induced only in the presence of bright sunlight, an elusive factor during many British summers. Vegetative regeneration of *H.p.* takes two forms. First, many axillary shoots are produced from the decumbent rooting base of the clump; this enables a limited degree of lateral spread and tends to be particularly well developed in plants growing on deep soils. Second, and mainly in shallow soils, new clumps may develop from lateral roots at some distance from the parent shoots. This mode of regeneration is very similar to that of *Linaria vulgaris* with which *H.p.* is frequently associated on railway ballast. Detached shoot branches root fairly readily but in many of the habitats exploited by *H.p.* soil conditions are probably normally too dry to allow their establishment. Regeneration by seed, which is produced in large quantities (often > 30 000 per plant), is probably dependent upon the existence of relatively large areas of bare ground. Seeds are very small and tend to germinate mainly in spring, after an inhibitor in the seed coat has been washed out. *H.p.* regularly forms a persistent buried seed bank. The extent of ecotypic differentiation remains uncertain, but geographical races differing in leaf morphology have been identified and the existence of weed, woodland and sand-dune ecotypes is suspected. *H.p.* shows signs of hybrid origin and is sometimes male-sterile. Fertile hybrids between *H.p.* and *H.maculatum* ssp. *obtusiusculum* are recorded from Britain. *H.m.* tends to occur on moist, less calcareous soils, but its ecological distribution frequently overlaps with that of *H.p.*.

Current trends Possibly still increasing due to its capacity to exploit a number of artificial habitats, particularly those associated with railways.

142 *Hypochaeris radicata*   Cat's Ear

Often spelt *Hypochoeris*

Established strategy C–S–R.
Gregariousness Typically sparse to intermediate.
Flowers Florets yellow, ligulate, insect-pollinated; generally self-incompatible.

Regenerative strategies W, V.
Seed 0.96 mg.
British distribution Widespread (100% of VCs).
Commonest habitats Acidic quarries/limestone wasteland 15%,

rock outcrops 12%, limestone
quarries/meadows 11%.

Spp. most similar in habitats *Aira
praecox* 76%, *Agrostis capillaris*
68%, *Rumex acetosella* 67%,
Hieracium spp. 59%, *Calluna
vulgaris* 56%.

Full Autecological Account *CPE* page
336.

Synopsis *H.r.* is found in a wide range
of short grassland and open habitats,
but is particularly characteristic of dry,
sandy, slightly acidic soils. Though
highly palatable, *H.r.* often persists in
grazed turf, probably because the
leaves are restricted to a basal rosette
which is closely appressed to the
ground. In mown turf the tough,
springy flowering stems often evade
damage. *H.r.* is very tolerant of
drought. We suspect that this is
because the deep root system is able to
tap subsoil water during dry periods.
In much of Britain, *H.r.* overwinters
with a small rosette of leaves, but in
very cold regions the species is not
winter green. In heavily grazed or
mown turf, *H.r.* regenerates
vegetatively to form diffuse clonal
patches, whereas in lightly grazed sites
numerous, wind-dispersed seeds are
produced. Germination may occur
over much of the year, with peaks in
spring and autumn, but no persistent
seed bank is formed. Flexibility of
regeneration and transport in
commercial grass seed samples have
contributed to the ability of *H.r.* to
become an important weed of
grassland in many parts of the world.
This colonizing ability may have been
assisted by the capacity of plants under
favourable conditions to flower within
two months of establishment.
Considerable variation between
populations has also been reported.
Workers in Holland have suggested the
existence of two ssp., one from fertile
habitats and with long fruits, the other
from less-fertile, often acidic soils and
with smaller fruits and a more
prostrate growth habit. *H.r.* has a very
low chromosome number ($2n = 8$);
there are possible advantages of this
characteristic in facilitating rapid
evolution. The species hybridizes with
H. glabra.

Future trends Data suggest that *H.r.* is
probably still increasing worldwide.
However, *H.r.* is not an important
constituent of heavily fertilized
pasture, and the area of habitat
suitable for this species has declined in
response to modern methods of
grassland management. *H.r.* remains a
highly effective colonist of new
habitats, and its future as a common
plant appears assured.

143 *Impatiens glandulifera*   Policeman's Helmet

Established strategy C–R.
Gregariousness Robustly
   stand-forming.
Flowers Usually purplish-pink,
   hermaphrodite, protandrous, insect-
   pollinated.
Regenerative strategies S.
Seed 7.32 mg.

British distribution Locally common in
the N and W (39% of VCs).
Commonest habitats River banks 18%,
shaded mire 7%, bricks and
mortar/acidic woodland 3%,
limestone woodland 2%.
Spp. most similar in habitats *Angelica
sylvestris* 81%, *Festuca gigantea*
77%, *Petasites hybridus* 76%,
Alliaria petiolata 74%, *Myrrhis
odorata* 71%.
Full Autecological Account *CPE* page
338.

Synopsis *I.g.* is a tall, summer-annual
herb, typically forming linear stands
along river banks on fertile alluvial
soils. In these sites populations are re-
established annually from large seeds
which are dispersed explosively in
autumn. About 800 seeds are produced
by each medium-sized plant. Seeds
must experience chilling temperatures
in order to become capable of
germination. Germination of the entire
seed population occurs synchronously
in spring, and leads to the
establishment of dense even-aged
stands in which other species, including
perennials, may be submerged. River
banks colonized by *I.g.* frequently
have reduced floristic diversity. In the
British flora, ruderal dominance such
as that exercised by *I.g.* is relatively
uncommon except in weed-infested
arable land. *I.g.*, a native of the
Himalayas, is at 2 m the tallest annual
in the British flora. It is interesting to
note, too, that the tallest monocarpic
and polycarpic perennial herbs in the
flora (excluding twining plants),
respectively *Heracleum
mantegazzianum* (SW Asia, 3.5 m) and
Reynoutria sachalinensis (Sakhalin
Island, 3 m), are also aliens. The
capacity of *I.g.* to exploit riverside sites
appears to depend strongly upon
overwintering by seed. Thus, unlike
the perennials of alluvial terraces, *I.g.*

is relatively unaffected by winter
flooding, silt deposition and soil
erosion. *I.g.* is tolerant of moderate
shade, and is able to flower and set
seed in open woodland. We assume
that seeds are frequently transported
by water, since colonies on river banks
extend much more rapidly than others.
The species appears to be restricted to
very moist habitats; shoots wilt rapidly
after picking, and plants of drier sites
have short internodes and small leaves.
Like other British *Impatiens* spp., *I.g.*
has very brittle, hollow stems with a
high water content and a low
proportion of vascular tissue, and
relies upon buttressing by adventitious
roots (cf. *Zea mays*) to maintain its
erect stature. Flowers are
insect-pollinated. Such species often
experience problems with respect to
pollination when introduced into
foreign countries, and most other
introduced annual species are
self-pollinated. The degree of genetic
heterogeneity in populations is
uncertain, but flower colour frequently
varies within populations.
Current trends *I.g.*, which was first
recorded in the wild in Britain in 1848,
is still increasing.

Established strategy Between S and
S–R.
Gregariousness Sparse.
Flowers Yellow, central florets
hermaphrodite, marginal florets
female; visited by insects.
Regenerative strategies W.
Seed 0.26 mg.
British distribution Only in England
and Wales (43% of VCs).
Commonest habitats Limestone
quarries 14%, limestone wasteland
7%, limestone pastures 6%, rock
outcrops 5%, paths 2%.
Spp. most similar in habitats *Origanum
vulgare* 89%, *Erigeron acer* 81%,
Centaurium erythraea 74%,
Hypericum perforatum 73%,
Leontodon hispidus 69%.
Full Autecological Account *CPE* page
340.

Synopsis *I.c.* is a rosette-forming,
winter-green herb, mainly restricted to
dry, infertile, calcareous soils. The
ecology of the species is poorly
understood. *I.c.* appears to function as
a polycarpic perennial at the centre of
its distribution in S Europe, but
behaves as a monocarpic perennial at
sites in Holland. A consistent feature
of sites with *I.c.* in Britain is that they
are habitats with discontinuous
vegetation cover. This is doubtless
because regeneration is entirely
dependent upon small seeds. Seedlings
exploit vegetation gaps and appear
mainly during autumn. During
establishment, *I.c.* is a rosette species
with little capacity for lateral
vegetative spread. *I.c.* may be fatally
toxic to sheep and cattle and is little
grazed. However, as in Holland, we
suspect that the species is particularly
characteristic of dry wasteland at a
successional stage intermediate
between open ground and tall
grassland. *I.c.* is mainly restricted to S
Britain, and the distribution is biased
towards low altitudes and S-facing
slopes. Thus, despite an apparent
abundance of suitable habitats, *I.c.*
shows a restricted distribution on the
upland carboniferous limestone and
the species reaches its British
altitudinal limit on this substratum. As
in the case of *Centaurea scabiosa*,
therefore, there are indications that
some stages in the life-history may be
unusually thermophilous, and that the
geographical and ecological
distribution of *I.c.* is restricted by
climatic factors. Despite an association
with S-facing slopes, *I.c.* is largely
absent from the most heavily
droughted sites and we suspect that, in
dry situations, the deep root system
may sustain water supply during the
peak of growth in summer. A
persistent seed bank is suspected but
has not been demonstrated. The
relative size of the achene and pappus
appear well suited for long-distance
dispersal and *I.c.* has colonized a
number of foundry spoil tips situated
at a considerable distance from the
calcareous strata to which the species is
normally restricted.
Current trends *I.c.* is probably
increasing as a colonist of dry,
disturbed, usually calcareous habitats.

145 *Juncus articulatus*   **Jointed Rush**

Established strategy C–S–R.
Gregariousness Intermediate.
Flowers Brown, hermaphrodite,
protogynous, wind-pollinated.
Regenerative strategies V, B_s.
Seed 0.02 mg.
British distribution Widespread (100%
of VCs).
Commonest habitats Unshaded mire
13%, running aquatic 12%, river
banks 5%, still aquatic 4%, enclosed
pastures 3%.
Spp. most similar in habitats
Equisetum palustre 95%, *Glyceria
fluitans* 86%, *Juncus bulbosus* 84%,
Stellaria alsine 83%, *Apium
nodiflorum* 82%.
Full Autecological Account *CPE* page
342.

Synopsis *J.a.* is a morphologically
variable (low-growing to relatively
tall), winter-green rush which is most
typically found in base-rich mire in
situations where disturbance, and often
also a degree of soil infertility, limit
the growth of potential dominants. *J.a.*
is moderately resistant both to grazing
and to trampling, and is often found in
grazed mire, where it adopts a more
prostrate growth form and is usually
smaller than when growing in ungrazed
sites. *J.a.* is also found beside streams
in close proximity to fast-flowing
water, and may even behave as a semi-
aquatic, forming floating rafts in
upland rills or drainage ditches but
usually attached to the bank. In its
pronounced morphological plasticity
and wide ecological amplitude in
wetland habitats, *J.a.* shows ecological
affinities with the more shade-tolerant
Ranunculus flammula. *J.a.* is
edaphically wide-ranging, occurring on
mildly acidic to highly calcareous soils.
The leaves and stems have, like most
wetland species, well-developed air
spaces but *J.a.* exhibits only moderate
tolerance of ferrous iron. *J.a.*,
particularly in its prostrate forms, may
root at the nodes to form extensive
clonal patches. This is especially
evident close to water, and appears to
be important in the colonization of
stream margins. Even in grazed sites
plants produce large numbers of
minute seeds. These germinate mainly
during spring in vegetation gaps, or are
incorporated into a persistent seed
bank. The seeds, which are
mucilaginous and adhesive when
moist, are widely dispersed even
between landlocked sites, and are
probably transported in mud. The two
other frequent jointed rushes, *J.
acutiflorus* and *J. subnodulosus*, are
both taller, stand-forming dominants
of relatively undisturbed mire of
moderate productivity. Both are often
found on peat; the former on mildly
acidic, the latter on base-rich,
frequently calcareous substrata. *J.a.*
hybridizes in Britain with three other
rushes, the commonest being the
hybrid with *J. acutiflorus* which forms
clonal patches.
Current trends Although an effective
colonist, probably decreasing as a
result of wetland habitat destruction,
and perhaps also declining under the
impacts of eutrophication.

146 *Juncus bufonius*

Including *J. minutulus*

Established strategy R.
Gregariousness Sparse to intermediate.
Flowers Green, hermaphrodite, cleistogamous.
Regenerative strategies B$_s$.
Seed 0.02 mg.
British distribution Widespread (100% of VCs).
Commonest habitats Unshaded mire 9%, paths/river banks 8%, arable 5%, soil heaps 2%.
Spp. most similar in habitats Insufficient data.
Full Autecological Account *CPE* page 344.

Toad Rush

Synopsis The only common obligately annual, non-graminaceous monocotyledon in the British flora, *J.b.* is most typical of fertile sites which are submerged during winter and remain moist during summer. The species is low-growing, short-lived and without the capacity for vegetative expansion. Thus, *J.b.* is restricted to (a) continuously open sites such as exposed mud beside ponds and reservoirs, and (b) vegetated sites subject to moderate intermittent disturbance such as rutted cart-tracks and path margins. Plants may flower within *c.* 4 weeks and in dry habitats may be small and short-lived. However, in favourable sites where the level of competition is low, numerous flowering stems may be produced over a protracted period. *J.b.* regenerates by means of numerous minute seeds, and forms a persistent seed bank. Seeds can survive ingestion by cattle and horses, are mucilaginous and adhesive but do not float. However,

seedlings float and may be transported in water to new sites. Detached shoots, when young, are capable of rooting to form new plants; this capacity is probably significant mainly in enabling the plant to recover *in situ* following disturbance. The *J. bufonius* aggregate is represented in Europe by four near-diploids, differing in ecological and geographical distribution, and *J.b. sensu stricto*, which is *c.* 4x and 6x. It is suggested that *J.b.* has arisen as a result of hybridization amongst these or related species, a process which may be continuing, and *J.b.* is the most morphologically variable and ecologically diverse of the group. One of the near-diploids, *J. ambiguus* (*J. ranarius*) occurs in saline habitats and is allogamous. *J.a.* might be expected to occur on salted roadsides, but has not yet been recorded in this habitat.
Current trends A highly mobile species which exploits disturbed, fertile, artificial habitats. *J.b.* is one of the few wetland species which is clearly increasing.

Established strategy Between S–R and C–S–R.
Gregariousness Sparse to intermediate.
Flowers Reddish-brown, hermaphrodite, protogynous, wind-pollinated.
Regenerative strategies ?B_s, V.
Seed 0.03 mg.
British distribution Widespread (100% of VCs).
Commonest habitats Still aquatic/unshaded mire 5%, running aquatic 4%, river banks 3%.
Spp. most similar in habitats Insufficient data.
Full Autecological Account *CPE* page 346.

Synopsis *J.b.* is a low-growing rush of base-poor, often peaty wetland habitats where the growth of potential dominants is suppressed by the combined effects of infertility, relatively low pH and the various forms of wetland-habitat disturbances. Each plant consists of a tuft of basal leaves with axillary stems arising near the base. In mire, these stems are often short, whereas in aquatic systems they may reach 1 m. Leaves are also longer on submerged shoots, and the aquatic and mire forms are morphologically so dissimilar that they are scarcely recognizable as the same species. This phenotypic plasticity appears to render a single genotype capable of exploiting two different habitats (mire and aquatic) and may be a reflection of fluctuating water-table in many of the habitats with which *J.b.* is associated. Thus, *J.b.* often occurs in sites subject to seasonal variation in water depth (shallow pools, ditches, moorland reservoirs), and even in soligenous mire. *J.b.* is often found in close proximity to running water. Populations may vary markedly in size from year to year and it seems likely that, as in other species of *Juncus*, populations are buffered by the presence of a persistent seed bank. However, *J.b.* occurs in several apparently more-stable habitats such as mountain lakes, and many aspects of its ecology are poorly understood. The leaves and stems, like those of most wetland species, have well-developed air spaces. In aquatic systems, *J.b.* forms large floating patches, and detached portions of shoot readily regenerate and may even be carried some distance in the water. Unlike individuals growing in mire, aquatic forms do not flower and semi-aquatic plants are often viviparous. No information is available concerning the existence of ecotypes and *J. kochii*, which was formerly treated as a separate taxon, is no longer recognized.
Current trends Appears relatively mobile, with a capacity to colonize even land-locked sites; it may be encouraged by habitat disturbance. However, since *J.b.* is associated with infertile wetlands, it is decreasing rapidly in many lowland areas. In upland areas its status is more uncertain, and the species may be increasing in response to the acidification of water systems by 'acid rain' (as suggested in Holland).

Established strategy Between C and
S–C.
Gregariousness Patch-forming.
Flowers Brown, hermaphrodite,
protogynous, wind-pollinated or
cleistogamous.
Regenerative strategies V, B_s.
Seed 0.01 mg.
British distribution Widespread (100%
of VCs).
Commonest habitats Unshaded mire
31%, river banks 18%, shaded
mire/acidic pastures 11%, still
aquatic 6%.
Spp. most similar in habitats
Cardamine pratensis 90%,
Epilobium palustre 87%, *Stellaria
alsine* 86%, *Galium palustre* 85%,
Polygonum amphibium 83%.
Full Autecological Account *CPE* page
348.

Synopsis *J.e.* is a tall perennial with
photosynthetic stems and scale-like
basal leaves. *J.e.* is one of the most
abundant and wide-ranging of wetland
species, and is by far the most common
British rush. *J.e.* is found on a wide
range of moist or waterlogged soils,
and is particularly widespread on the
more-acidic mineral soils. Levels of N,
P and Ca in the leaves are low, but
those of Na and Fe are relatively high.
J.e. is frequent in grazed habitats,
where it tends to be avoided by sheep
or cattle, and is moderately tolerant of
annual cutting and trampling. *J.e.* also
occurs in some shaded habitats, often
without flowering. The success of *J.e.*
in wetland appears to be related to (a)
the dense basal structure of the
tussocks and persistent litter, both of
which tend to prevent the
establishment of other species, (b) the
extreme tolerance of ferrous iron
toxicity, and (c) a regeneration
mechanism which allows seed
persistence in the soil and rapid
colonization of disturbed habitats.
Over 13 000 seeds may be produced by
a single inflorescence. Many are
incorporated *in situ* into a buried seed
bank, and small seed banks have been
detected at sites where there is no
historical record of *J.e.*, suggesting that
there is effective dispersal of the seeds
through animals or some other agency.
Thus, *J.e.* is, in many mire habitats,
among the first species to establish on
soil bared by disturbance. Germination
occurs particularly in spring. Seeds,
which are mucilaginous when moist,
sink in water but colonization of mud
at the edge of drainage ditches and
larger water bodies can arise through
'beaching' of seedlings which may be
free-floating. However, seedlings do
not appear to establish under water.
Thus, *J.e.* normally colonizes only
those sites which are above the water
table for part of the year. *J.e.* will
survive in shallow water if established
plants are subject to a change in water-
table, e.g. as a result of land
subsidence. Once established,
regeneration by seed is much less
important. *J.e.* is strongly rhizomatous
and may form extensive clonal patches;
breakdown of the rhizome results in
the isolation of daughter plants.
Several genotypes have been
recognized, including var. *compactus*,
which is a variant normally found on

upland acidic soils. *J.e.* is morphologically similar to *J. conglomeratus* and *J. inflexus*, neither of which is common in the more waterlogged sites exploited by *J.e.*. In comparison, *J.e.* is less robust, flowers earlier and is particularly characteristic of sites of intermediate soil pH. It is most abundant in base-rich, particularly calcareous conditions and is much less tolerant of ferrous iron toxicity. *J.e.* hybridizes both with *J.c.* and with *J.i.*.

Current trends One of the most rapid colonists of artificial wetland habitats and moist disturbed ground, and probably increasing.

149   *Juncus squarrosus*

Heath Rush

Established strategy S.
Gregariousness Intermediate, but can form patches.
Regenerative strategies V, B$_s$.
Seed 0.05 mg.
British distribution Widespread, especially in the N and W (97% of VCs).
Commonest habitats Acidic pastures/paths 4%, unshaded mire/rock outcrops 3%, acidic wasteland 2%.
Spp. most similar in habitats *Empetrum nigrum* 79%, *Nardus stricta* 71%, *Molinia caerulea* 68%, *Carex pilulifera* 61%, *Vaccinium myrtillus* 61%.
Full Autecological Account *CPE* page 350.

Synopsis *J.s.* is a winter-green clump-forming rush developing flattened rosettes and exhibiting a rather restricted ecological and geographical range. The species is virtually confined to infertile acidic soils, and seedlings from sites of low pH grow very poorly and may be highly chlorotic when planted on calcareous soils. Growth of *J.s.* is strongly inhibited by high external concentrations of calcium. Leaves are rich in Na but have low Ca content. *J.s.* is occasionally eaten by beef cattle and horses and, in winter and early spring when food is scarce, by sheep. However, *J.s.* is generally avoided by stock, and is increasing in those upland areas which are now only grazed by sheep in summer. Thus, in the upland areas of N and W Britain, *J.s.* exploits a niche similar to, but wetter than, that of *Nardus stricta*. *J.s.* grows slowly and tends to be a very subordinate component of plant communities, particularly since its foliage is held close to the ground surface. *J.s.* is also intolerant of shade and is dependent upon the creation of short turf or bare ground by grazing or trampling. *J.s.* is tolerant of occasional burning. Like *N.s.*, *J.s.* regenerates in established communities mainly by rhizomatous growth, to form larger clumps. This is particularly effective on moist soils, and die-back of older rhizomes in the centre of a patch results in the isolation of daughter plants. *J.s.* may release up to 10 000 seeds m^{-2}, which germinate in vegetation gaps during spring or are incorporated into a persistent seed bank. *J.s.* can expand locally within sites bared by fire or other disturbances. Seeds are mucilaginous and adhesive, and the species shows some potential for long-distance colonization of new sites. Regeneration by seeds is probably infrequent in established vegetation. The species exhibits little morphological variation.

Current trends *J.s.* is relatively mobile and appears to have colonized a number of acidic artificial habitats from a long distance. However, the rate of appearance of these new habitats is much lower than the rate of destruction of heathland. Consequently *J.s.* is decreasing in many lowland areas. However, current management regimes in upland areas (i.e. reduction in the extent of winter and early spring sheep grazing) probably favour *J.s.*.

150 *Koeleria macrantha*   Crested Hair-grass

Established strategy S.
Gregariousness Intermediate.
Flowers Green, hermaphrodite, homogamous, wind-pollinated.
Regenerative strategies S.
Seed 0.3 mg.
British distribution General (86% of VCs).
Commonest habitats Limestone pastures 59%, scree 20%, rock outcrops 16%, lead mine spoil 15%, limestone wasteland 11%.
Spp. most similar in habitats
  Sanguisorba minor 96%, *Avenula pratensis* 93%, *Carex caryophyllea* 89%, *Helianthemum nummularium* 87%, *Campanula rotundifolia* 83%.
Full Autecological Account *CPE* page 352.

Synopsis *K.m.* is a slow-growing, loosely tufted, usually short, winter-green grass most characteristic of infertile calcareous grassland and rock outcrops. *K.m.* shares several attributes with calcareous races of *Festuca ovina* including (a) high leaf-extension rates in early spring, (b) a short, early flowering period occurring before the onset of summer drought, (c) a dependence for regeneration on seed, which germinates soon after release, resulting in the appearance of cohorts of seedlings during autumn,

(d) intermediate seed size, (e) a wide temperature range for seed germination and (f) the absence of a persistent seed bank. However, unlike *F.o.*, *K.m.* is never a dominant contributor to the shoot biomass. It seems likely that this may be related to the many aspects of its biology in which *K.m.* exhibits an intermediate degree of specialization compared with the various dominants with which it is able to coexist. Thus, *K.m.* is less xeromorphic than *F.o.* and appears more palatable to sheep, but is less productive and resilient under close-grazing than *Agrostis capillaris*. *K.m.* is also less robust than *Avenula pratensis*. Hence, because *K.m.* is an 'intermediate', its populations may be expected, under fluctuating conditions of climate, soil fertility and management, to show a greater degree of homeostasis than those of potential dominants. In this respect the ecology of *K.m.* in calcareous pasture is somewhat analogous to that of *Trisetum flavescens* in meadows. In some areas *K.m.* is less strongly calcicolous and may occur, for example, on sea cliffs in essentially base-poor soil which has been nutritionally enriched by sea spray. Plants with various ploidy levels are recorded from Europe, but cytological studies have not been carried out in Britain. Ecological comparisons of *K.m.* with the newly separated *K. glauca* have not yet been effected. As a result, the geographical and ecological distribution of *K.m.* in Britain requires clarification. Although *K.m.* has a relatively narrow ecological range in the Sheffield region, the existence of ecotypic differences between local populations has been demonstrated. Heavy-metal-tolerant populations have been recorded and *K.m.* appears to be an 'excluder'. Hybridizes with *K. vallesiana*.

Current trends Mainly restricted to semi-natural vegetation, and hence decreasing as a result of habitat destruction. Now close to extinction in some lowland areas.

151   *Lamiastrum galeobdolon*   Yellow Archangel

Ssp. *montanum*

Established strategy Between S and S–C.
Gregariousness Intermediate.
Flowers Yellow, hermaphrodite, homogamous, insect-pollinated, rarely self-fertilized.
Regenerative strategies V, S.
Seed 1.98 mg.
British distribution Common only in the S and E (43% of VCs).
Commonest habitats Limestone woodland 33%, scrub 20%, acidic woodland 13%, riverbanks/verges 3%.
Spp. most similar in habitats *Arum maculatum* 88%, *Mercurialis perennis* 84%, *Anemone nemorosa* 84%, *Sanicula europaea* 83%, *Fraxinus excelsior* (juv.) 83%.
Full Autecological Account *CPE* page 354.

Synopsis *L.g.* is a slow-growing herb restricted to moist woodland and other shaded habitats. *L.g.*, which is winter green and produces both sun and shade leaves, is tolerant of deep shade but under these conditions flowering is suppressed. *L.g.* is morphologically well-suited to exploit deep accumulations of tree litter in which it may form large, often diffuse stands as a result of stolon growth. *L.g.* is cold-tolerant but susceptible to drought, and although found on a wide range of soils is scarce on sandy substrata. The greater abundance in N-facing woodland probably relates both to the sensitivity to desiccation and to the shade tolerance exhibited by *L.g.*. *L.g.* is susceptible to trampling. The mineral nutrition of *L.g.* deserves attention since, despite a low growth rate, *L.g.* is regarded by some ecologists as a plant of relatively nutrient-rich soils. Leaves contain high concentrations of N and Fe, but appear to be low in P and Mg. *L.g.* produces long spreading stolons which root at the nodes and may form large clonal patches. Breakdown of the internodes results in the formation of daughter plants. We suspect that shoot pieces may be transported along woodland streams to form new colonies. Also regenerates by means of seeds, which germinate in spring. The balance between seed production and stolon development appears to be regulated by climate. In drier, open sites, and in

unusually warm summers, more seed is set and stolons are shorter. Thus, *L.g.* may be more dependent upon vegetative regeneration towards the N edge of its range, particularly in upland areas, than in S England. Seeds of *L.g.* have elaiosomes and are dispersed by ants. Long-distance dispersal is generally poor and *L.g.* has been identified as an indicator species of older woodlands. Ssp. *galeobdolon*, which is less vigorous and flowers earlier, occurs in N Lincolnshire. Ssp. *m.*, the main British taxon, is believed to have originated through hybridization between ssp. *g.* and ssp. *flavidum*, and occupies an intermediate geographical range.

Current trends There appears to be no significant recruitment of *L.g.* into the wild from garden populations, and *L.g.* is presumably decreasing as a result of habitat destruction.

---

152   *Lamium album*   **White Dead-nettle**

Established strategy C–R.
Gregariousness Intermediate.
Flowers White, hermaphrodite, homogamous, insect-pollinated.
Regenerative strategies V.
Seed 1.43 mg.
British distribution Common except in the W and N (82% of VCs).
Commonest habitats Soil heaps 17%, verges 9%, bricks and mortar/walls 6%, hedgerows 5%.
Spp. most similar in habitats *Rumex obtusifolius* 58%, *Elymus repens* 57%, *Arctium minus* 52%, *Poa trivialis* 51%, *Urtica dioica* 50%.
Full Autecological Account *CPE* page 356.

Synopsis *L.a.* is a sprawling, relatively low-growing rhizomatous, winter-green herb of fertile, well-drained and often lightly-shaded sites, where the vigour of taller potential dominants is suppressed by occasional disturbance.

Thus, *L.a.* is particularly characteristic of hedgerows and road verges which are infrequently disturbed by verge cutting and hedge trimming and where, as a result, species such as *Urtica dioica* can only achieve incomplete dominance. The species flowers in early summer, at which time it is a conspicuous component of the flora. Subsequently, the growth of taller species usually renders *L.a.* a subordinate member of the community. Although attaining maximum shoot biomass in summer, the phenology of *L.a.* involves much shoot growth in early spring. This may in part explain the restriction of *L.a.* to lowland sites in the British Isles. Effective regeneration appears to be vegetative; *L.a.* forms loose clonal patches by means of rhizome growth. Seed production is occasionally prolific (up to 2400 per plant), but regeneration by seed has been little studied and no persistent seed bank has yet been detected. The seed has an elaiosome and may be dependent upon ants for dispersal. Long-distance seed dispersal during Postglacial colonizing episodes appears to have been slow, and may account for the absence of *L.a.* from the native Irish flora. However, *L.a.* is described as a colonist of secondary woodland, and the possibility must be considered that some long-distance dispersal of seeds or detached portions of the vegetative plant is taking place.

Current trends Colonizes roadsides and hedgerows, and present in some secondary woodlands. Favoured both by fertile conditions and by disturbance; probably increasing.

153 *Lamium purpureum*   Red Dead-nettle

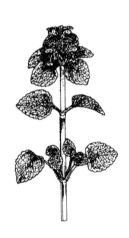

Established strategy R.
Gregariousness Sparse.
Flowers Purple, hermaphrodite, insect- or self-pollinated.
Regenerative strategies B_s, (V).
Seed 0.70 mg.
British distribution Widespread (100% of VCs).
Commonest habitats Arable/bricks and mortar/soil heaps 6%, cinder heaps 2%.
Spp. most similar in habitats *Urtica urens* 81%, *Tripleurospermum inodorum* 78%, *Capsella bursa-pastoris* 76%, *Papaver rhoeas* 76%, *Sinapis arvensis* 71%.
Full Autecological Account *CPE* page 358.

Synopsis *L.p.*, sometimes regarded as doubtfully native, is a short-lived annual herb which exploits gardens and a wide range of other disturbed, relatively fertile habitats. In some sites *L.p.* is facultatively winter-annual. Thus, the species occurs on rock outcrops liable to summer drought and on steep or sandy hedgebanks, where seed set precedes the major growth period of associated woody and perennial herbaceous species. In winter barley, too, *L.p.* sets seed early, before the crop has developed a dense canopy. Important controls upon the phenology of *L.p.* are exercised by its germination biology. Summer- and winter-annual forms with different requirements for germination may occur and most local populations appear facultatively winter- or summer-annual. In Britain, the germination period in the field is mainly from May to October, both on arable land and in a vegetable garden population examined in N Derbyshire, where two generations regularly occur in a season. In some N American populations, seed is dormant when shed in early summer, and remains so until autumn. Any residual seeds are returned to a state of secondary dormancy by low temperatures in winter. *L.p.* often produces > 1000 seeds per plant which show some

persistence in the soil. The species also has a restricted capacity for vegetative spread. During the colder months *L.p.* may produce short, rooted, prostrate shoots, forming a clump.

Non-flowering shoot tips readily re-root, and plants can establish and set seed following spring rotavation. Little is known about the dispersal of *L.p.*, but the species appears to be mainly dependent for its long-range transport upon movement of seeds contained in soil, although some local dispersal by ants has been observed.

Current trends Not under threat, and perhaps increasing locally in lowland areas.

---

154 *Lapsana communis*   **Nipplewort**

Established strategy Between R and C–R.

Gregariousness Sparse.

Flowers Florets yellow, ligulate, hermaphrodite, self- or insect-pollinated.

Regenerative strategies S, B$_s$.

Seed 1.27 mg.

British distribution Widespread except in N Scotland (100% of VCs).

Commonest habitats Arable 14%, soil heaps 7%, hedgerows 5%, manure and sewage waste/river banks 3%.

Spp. most similar in habitats
Polygonum persicaria 78%, *Veronica persica* 78%, *Myosotis arvensis* 78%, *Spergula arvensis* 77%, *Fallopia convolvulus* 77%.

Full Autecological Account *CPE* page 360.

Synopsis *L.c.* is a tall winter- or summer-annual herb which exploits fertile habitats, e.g. (a) arable land and soil heaps where the growth of perennial herbs has been checked or prevented by major disturbance before the growing season or (b) where the growth of perennials is restricted by light-to-moderate shade, and by

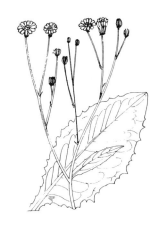

disturbances such as soil creep (e.g. on hedgebanks) or the creation of woodland clearings. *L.c.* occurs in a range of habitats similar to that exploited by *Galeopsis tetrahit*, although *L.c.* is less frequent in wetter sites. The two species are of similar height and, under favourable conditions, both become very tall plants capable of dominating neighbours, including smaller perennials. In autumn each produces relatively large seeds, which may contaminate crop seeds and may become incorporated into a shortly persistent seed bank. On average, plants produce *c.* 1000 seeds. Seeds appear to germinate with equal facility in autumn and spring and this, coupled with the lesser vulnerability to weeding of the non-flowering juvenile rosette (compared with that of the erect leafy *G.t.*), may contribute to the persistence of *L.c.* in shaded gardens and explain the absence of *G.t.* from this habitat. The semi-rosette growth form may also prevent *L.c.*, which is found mainly in cereals, from exploiting sites where taller dicotyledonous crops (e.g. *Brassica* spp. and potatoes) are grown. The outer achenes in each capitulum are much longer than the inner. Whether they also differ in germination characteristics is not known. The extent to which ecotypes have developed in this wide-ranging species, which exploits both fugacious and semi-stable habitats, is uncertain, but three further ssp. have been recorded from Europe. One, ssp. *intermedia* from SE Europe, is introduced into Britain. Ssp. *i.* is an annual to polycarpic perennial and has a larger involucre (and presumably also larger achenes).

Current trends Perhaps increasing in shaded habitats, as a result of the greater level of mechanized disturbance in land management, but probably decreasing on arable land.

---

155 *Lathyrus montanus*   Bitter Vetch

Established strategy Between C–S–R and S.
Gregariousness Intermediate.
Flowers Purple, hermaphrodite, insect-pollinated, never selfed.
Regenerative strategies V, ?B$_s$.
Seed 12.49 mg.
British distribution Widespread, especially in the W and N (95% of VCs).
Commonest habitats Limestone wasteland 11%, limestone pastures 8%, lead mine spoil 3%, enclosed pastures 1%.
Spp. most similar in habitats *Stachys officinalis* 90%, *Avenula pubescens* 88%, *Trifolium medium* 83%, *Pimpinella saxifraga* 80%, *Bromus erectus* 79%.
Full Autecological Account *CPE* page 362.

Synopsis *L.m.* is a slow-growing, rhizomatous legume of unproductive grassland, usually on soils of intermediate pH. Nitrogen-fixing root nodules are formed in conjunction with *Rhizobium leguminosarum*. The species appears to be restricted to continuously moist soils and is characteristic of neutral grasslands dominated by *Agrostis capillaris*. In these circumstances *L.m.* is often associated with a well-defined group of species of similar ecology (*Stachys officinalis*, *Hypericum pulchrum*,

Potentilla erecta and *Viola lutea*), all of which resemble *L.m.* in sensitivity to moisture stress and in susceptibility to lime-chlorosis. *L.m.* is observed at woodland margins in lowland areas. In common with species such as *Ophrys insectifera* and *Primula vulgaris*, the species is predominantly found in shaded sites in lowland Britain and in more-open localities in upland areas. Occurrence in woodlands appears to be correlated with the fact that the phenology of *L.m.* is 'semi-vernal'. Consistent with its high nuclear DNA value, *L.m.* produces a rapid flush of growth in early spring by mobilizing the reserves in its tuberous rhizome. Flowering commences in April, and little further shoot growth takes place during summer. The distribution of *L.m.* within pastures is strongly affected by its morphology and phenology. *L.m.* has an erect stature (150–400 mm) and is confined to · lightly-grazed sites. The species appears to be absent from sites which are regularly grazed during the early spring. *L.m.* persists through summer with short vegetative shoots at, or a little below, the canopy height of the pasture. Effective regeneration is mainly by means of rhizomatous growth to form diffuse patches. Usually the large seed of *L.m.* is released with a hard seed coat which requires scarification before germination is possible. However, under wet summer conditions seeds have been observed to germinate in pods which are still attached to the parent plant. Germination in autumn or spring is hypogeal and the seedling is presumably dependent upon its seed reserves for a considerable period during establishment. A species of low colonizing ability, considered to be characteristic of ancient woodland in Lincolnshire.

Current trends *L.m.* is rather narrowly restricted to an increasingly scarce set of conditions of soil type and management. A further reduction in abundance seems inevitable.

156 *Lathyrus pratensis*   Meadow Vetchling

Established strategy C–S–R.
Gregariousness Intermediate.
Flowers Yellow, hermaphrodite, insect-pollinated.
Regenerative strategies V, ?B$_s$.
Seed 12.85 mg.
British distribution Widespread except in N Scotland (100% of VCs).
Commonest habitats Meadows 27%, limestone wasteland 13%, verges 12%, lead mine spoil 9%, coal mine waste 7%.
Spp. most similar in habitats *Ranunculus acris* 67%, *Festuca pratensis* 65%, *Ranunculus bulbosus* 63%, *Rumex acetosa* 63%, *Trifolium pratense* 61%.
Full Autecological Account *CPE* page 364.

Synopsis *L.p.* is a rhizomatous legume supported on its leaf tendrils by the surrounding vegetation. It is most typical of meadows, roadsides and other tall vegetation on soils of moderate fertility. In the *PGE*, *L.p.* benefited from treatments involving applications of minerals alone or of sodium nitrate and occasional farmyard manure. The species was suppressed by 'starvation, ammonium salts or sodium nitrate alone'. *L.p.* is common on calcareous soils, and much more resistant to lime-chlorosis than *Lathyrus montanus*. N-fixing root nodules are formed in association with *Rhizobium leguminosarum*. As a consequence of its ascending growth form, the occurrences of *L.p.* in pastures are restricted to lightly grazed situations. *L.p.* is often tall and has the capacity for lateral vegetative spread, but tends to exercise only localized and temporary dominance in undisturbed vegetation. The species tends to be restricted to sites where the vigour of potential dominants, e.g. *Arrhenatherum elatius*, is debilitated by occasional mowing or by some other form of infrequent disturbance. The capacity of *L.p.* to behave as a dominant appears to be restricted by (a) its scrambling habit and dependence upon support from other vegetation and (b) its capacity to fix nitrogen which, we suspect, may ameliorate soil infertility to an extent that faster-growing species may dominate. Regeneration is predominantly vegetative; *L.p.* can produce rhizomes up to 7 m in length. The importance of vegetative regeneration is particularly evident in habitats such as pasture, where grazing often prevents flowering and seed-set, and in hay meadows which are often cut before seeds are ripe. This vigorous vegetative reproduction is coupled with rather modest seed-set (usually 10–12 ovules, 3–6 seeds). The large seeds are generally shed close to the plant and so the species has only limited colonizing ability. However, *L.p.* is widespread and in the past seed may have been dispersed within hay-crops. Seed can also survive ingestion by animals. Many seeds germinate in spring. Seed-bank studies have not been undertaken. More recently, mechanized disturbance resulting in the movement of soil containing rhizomes may have allowed the spread of *L.p.* in disturbed habitats. Two main cytotypes occur in Europe; $2n = 14$ ($2x$), which is widespread in Europe but with a predominantly E distribution, and $2n = 28$ ($4x$) in W and C Europe. This apparent autotetraploid appears to be the predominant type in Britain; despite the slightly heavier seeds and lower pollen fertility, it is morphologically indistinguishable from diploid plants, and has a similar habitat range. However, the two are genetically isolated. Other closely related but ecologically separated species occur in Europe.

Current trends Remains common, but is decreasing on farmland.

157 *Lemna minor*

Field records may include *L. gibba*

Established strategy C–R.
Gregariousness Stand-forming.
Flowers Minute, with no well-defined pollination mechanism.
Regenerative strategies V.
Seed *c.* 0.6 mg.
British distribution Widespread except

Duckweed, Duck's-meat

in the Scottish Highlands (98% of VCs).
Commonest habitats Still aquatic 54%, shaded mire 21%, unshaded mire 15%, running aquatic 6%, walls 3%.
Spp. most similar in habitats Insufficient data.
Full Autecological Account *CPE* page 366.

Synopsis *L.m.* is the most widespread free-floating vascular plant in the world and is by far the most common aquatic in Britain. *L.m.* is absent from base-poor, and perhaps also highly calcareous, waters but elsewhere is frequent on still waters where these are of sufficiently small extent to escape the major effects of wave action and wind blow. Despite the poorly protected epidermis and a water content of *c.* 96%, beached thalli are capable of active growth. *L.m.* persists, albeit with reduced vigour, on water or on mud shaded by taller species. Each plant is minute, consisting of a single green, flat thallus up to 4 mm in diameter and a single unbranched root. The vascular tissue lacks vessels. Plants contain oxalate raphides, but it is not clear to what extent these deter predation. Vegetative buds are produced from two pouches on each frond, and new fronds may separate or remain joined to the parent. Each thallus may survive for 5–6 weeks during summer. By this means the water surface may be quickly covered, and in exceptional circumstances watercourses may become choked. Seed is rarely set, but high rates of flowering occurred during Postglacial, Boreal and Atlantic periods, when the climate was warmer and milder. Whole plants may adhere to the feathers or feet of birds and *L.m.* is an effective colonist of still water. In Europe, *L.m.* is predominantly tetraploid and more rarely pentaploid, but diploid, triploid and octaploid counts have been recorded within the extra-European range of the species. The distribution of two other related free-floating species overlaps with that of *L.m.*: *L. gibba* and *Spirodela polyrhiza*. *L.g.*, which does not normally sink in winter and is thus vulnerable to the effects of low air temperatures, is largely confined to canals and ditches in lowland areas. *L.g.* exhibits a high phosphorus requirement and a higher relative growth rate, and in mixed populations at high density may

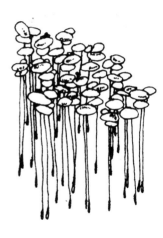

exclude *L.m.* through the ability of the buoyant thalli to occupy a superior position in the canopy. Gibbosity is however perhaps primarily a specialization for water-retention, of value in more-arid climates in the event of the water body drying up. The character is often poorly developed in British populations, leading to confusion with *L.m.*. Indeed, some field records of *L.m.* may include *L.g.* or even consist entirely of this species. (Records may also include some data for *L. minuscula*, which is also readily mistaken for *L. minor*, but this species has not yet spread to N England.) The larger *S.p.* has a high nitrogen and phosphorus requirement and forms turions. *S.p.* survives winter well but, in contrast with *L.m.* and *L.g.*, has a high minimum temperature for growth. Perhaps for this reason, within the Sheffield region *S.p.* is mainly found in shallow water, which may be expected to warm up more rapidly.

Current trends A colonist of eutrophic water, *L.m.* may still be increasing despite the continued destruction of aquatic habitats. However, it has been suggested that *L.g.* is the potential dominant in warmer eutrophic waters. In lowland regions eutrophication may be creating habitats more suitable for the latter species.

Established strategy Between C–S–R and R.
Flowers Florets yellow, ligulate, hermaphrodite, self-incompatible, pollinated by insects.
Regenerative strategies W, B_s.
Seed 0.70 mg.
British distribution Widespread (100% of VCs).
Commonest habitats Lead mine spoil 21%, bricks and mortar 13%, cinder heaps 10%, paths/verges 8%.
Spp. most similar in habitats Insufficient data.
Full Autecological Account *CPE* page 368.

Synopsis *L.a.* is a rosette-forming herb of short turf and open ground. Found under a range of edaphic conditions, but particularly characteristic of road verges and pasture on moist fertile soil. *L.a.* is highly plastic in morphology, and is capable of exploiting sites subject to periodic disturbance by grazing, cutting or trampling. In short vegetation the leaves, which are palatable to stock, tend to be held close to the soil and may be pinnatifid. In taller grassland the leaves are generally long, ascending and relatively undissected. *L.a.* is absent from woodland and tall herbaceous vegetation, and appears to be intolerant of shading. The shoot phenology contrasts markedly with that of many of the grasses, such as *Lolium perenne*, with which it grows, since there is a pronounced peak in leaf area during summer. *L.a.* has a limited capacity for vegetative spread, and may produce a branched stock, with each branch terminating in a rosette. There is, however, no evidence of the development of extensive clones. *L.a.* regenerates by

means of wind-dispersed seeds. Germination is dependent upon warm temperatures, and is partially inhibited by leaf canopies. Some seedlings appear in vegetation gaps during autumn, but there is evidence, for the Sheffield region and elsewhere, that *L.a.* forms a transient reserve of surface-lying or superficially-buried seeds. The pappus remains stiff when wet, an unusual feature among wind-dispersed composites; this appears to provide anchorage during seedling establishment. Fruit production is often extremely limited in summer-grazed sites. As well as being phenotypically plastic, *L.a.* is also out-breeding and genetically variable. Populations of shorter, few-flowered plants have been separated recently as ssp. *pratensis* and occur mainly in montane areas.
Current trends *L.a.* exploits some common artificial habitats, is relatively mobile and appears to be increasing, perhaps most conspicuously in continuous bands close to the metalled surface of many major roads and motorways.

Established strategy S.
Gregariousness Intermediate.
Flowers Florets yellow, ligulate,
  hermaphrodite, self-incompatible,
  insect-pollinated.
Regenerative strategies W, V, ?B_s.
Seed 0.85 mg.
British distribution Widespread except
  in the N (77% of VCs).
Commonest habitats Limestone
  quarries 47%, scree 26%, limestone
  pastures 23%, lead mine spoil 21%,
  limestone wasteland 16%.
Spp. most similar in habitats *Linum
  catharticum* 89%, *Scabiosa
  columbaria* 86%, *Hieracium pilosella*
  85%, *Senecio jacobaea* 85%,
  Solidago virgaurea 77%.
Full Autecological Account *CPE* page
  370.

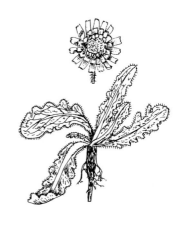

Synopsis *L.h.* is a slow-growing
rosette-forming species which grows in
a range of unproductive grasslands and
open habitats, particularly on
calcareous soils. It is moderately
palatable to sheep and tends to
subtend its leaves close to the ground
surface in grazed turf. In the *PGE*, the
species is mainly confined to
unproductive plots. In turf
microcosms, roots became heavily
colonized by VA mycorrhizas with
increases of up to 14-fold in seedling
yield. The marked capacity to
regenerate after close grazing and
cutting is related to the ability to re-
sprout from buds on the stock. In tall
grassland, leaves are more upright but
L.h. is suppressed by larger species in
rank vegetation. Leaf production in
spring is considerably delayed
compared with that of many of the
grasses with which the species grows.
The tap root penetrates to a
considerable depth and in dry habitats
appears to sustain water supply during
the peak of shoot growth, which occurs
in summer. We suspect that in dry
habitats the dependence upon water

provided via the tap-root severely
limits the capacity of *L.h.* for
vegetative spread, since in moist
habitats *L.h.* can form clonal patches
through the production of a branching
stock with adventitious roots. These
clones may reach 200 mm in diameter.
Otherwise, as in quarries, *L.h.*
regenerates mainly by means of wind-
dispersed seeds which germinate in
vegetation gaps. Ungerminated seeds
of *L.h.* have a limited capacity for
persistence on the soil surface.
However, seed-set is often low in
summer-grazed turf and establishment
from seed is usually slow although
seedlings invading large vegetation
gaps in Derbyshire have been observed
to reach the flowering stage two years
after germination. The field
distribution of *L.h.* reveals that the
species often behaves as a colonizer.
L.h. is particularly common in open
communities on unstable calcareous
substrata, e.g. recently abandoned
quarry heaps; this capacity for invasion
is clearly related to the production of
wind-dispersed seeds. *L.h.* is
subdivided into six subspecies in
Europe and is essentially outbreeding.
L.h. hybridizes with *L. taraxacoides*.

Current trends Uncertain, since although it is mobile and exploits artificial habitats such as quarries and railway banks, *L.h.* is largely restricted to less-fertile habitats. As a result of the ability to colonize derelict land, it is nevertheless more secure than the majority of species characteristic of species-rich pastures.

160 *Leucanthemum vulgare*

Does not include *L. maximum*, a garden escapee

Ox-eye Daisy, Moon Daisy, Marguerite

Established strategy Between C–S–R and C–R.
Gregariousness Intermediate.
Flowers Central florets tubular, yellow and hermaphrodite, outer florets ligulate, white and female; insect-pollinated or selfed.
Regenerative strategies S, B$_s$, V.
Seed 0.33 mg.
British distribution Widespread but less frequent in the N (100% of VCs).
Commonest habitats Limestone quarries 28%, meadows 19%, limestone wasteland 15%, cinder heaps 14%, rock outcrops 9%.
Spp. most similar in habitats *Hieracium* spp. 73%, *Crepis capillaris* 72%, *Medicago lupulina* 72%, *Origanum vulgare* 70%, *Dactylis glomerata* 68%.
Full Autecological Account *CPE* page 372.

Synopsis *L.v.* is a loosely-tufted winter-green herb of grassy and open habitats where the growth of potential dominants is restricted by a degree of soil infertility and disturbance factors such as cutting for hay and light grazing. The leaves of *L.v.* are basal and situated close to the ground over the winter period but, associated with a high value for nuclear DNA, there is a rapid shoot expansion in spring. In ungrazed sites during the summer, functional leaves are largely confined to the flowering stems. These features, together with the production of seeds in early summer, enables *L.v.* to occur as a weed of hay meadows. In common with many arable weeds, *L.v.* retains many seeds on the plant at harvest time, is capable of germinating in autumn or spring, and forms a persistent seed bank. *L.v.* also exploits pasture grazed by horses, sheep and goats, and was formerly eaten as a salad plant, but is said to be generally avoided by cows and pigs. Although tolerant of grazing and trampling, *L.v.* is generally less abundant in pastures; this may be a consequence of the limited capacity for clonal spread (patches usually < 50 mm in diameter) and low seed-set under grazed conditions. Ungrazed plants may produce up to 4000 achenes, which may explain why *L.v.* is often an effective colonist of areas of discontinuous vegetation cover on spoil heaps, roadsides and railway banks.

L.v. is considered drought tolerant; in outdoor pot experiments, substantial seedling emergence occurred between March and September and the characteristics of the achene allow rapid germination at the surface of relatively dry soils. L.v. forms a polyploid series and numerous ecotypes have been recognized. Tetraploids replace diploids in N and E Britain, while in S and W they tend to occur only in sites subject to heavy-metal pollution. 2x plants are also commonest in N and W Europe, with 4x plants most frequent in the N and E, and in the S on mountains. Both occur in C Europe. Drought or low N supply can induce leaf succulence, a character under genetic control in maritime populations.

Current trends The wide distribution of L.v. owes much to human activities. However, L.v. is uncommon in highly fertilized sites, and appears to be decreasing in meadows and pastures. As a result L.v. is becoming more characteristic of areas such as railway and motorway banks, and less common in farmland.

161 *Linaria vulgaris* Yellow Toadflax

Established strategy C–R.
Gregariousness Sparse to intermediate.
Flowers Yellow, hermaphrodite, self-incompatible, insect-pollinated.
Regenerative strategies (V), B_s.
Seed 0.14 mg.
British distribution Mainly England, Wales and S Scotland (83% of VCs).
Commonest habitats Cinder heaps 11%, coal mine waste 9%, bricks and mortar/hedgerows 3%, walls 1%.
Spp. most similar in habitats *Senecio viscosus* 78%, *Chaenorhinum minus* 65%, *Artemisia vulgaris* 64%, *Atriplex prostrata* 61%, *Chamerion angustifolium* 61%.
Full Autecological Account *CPE* page 374.

Synopsis L.v. is a tall herb mainly restricted to open, artificial habitats on dry soils. L.v. is particularly common in situations where deep mineral soils are overlain by coarse debris, and in this respect the species resembles *Tussilago farfara* and *Equisetum arvense*. Although tall and with an extensive root system, L.v. has only a limited capacity to spread laterally and to develop a dense leaf canopy; consequently the species is generally restricted to sites where the growth of more-robust perennials is debilitated. Thus, L.v. is most frequent in sites of relatively low annual productivity and some disturbance, and is particularly characteristic of open cinders and ballast beside railways, where the vigour of other potentially fast-growing, colonizing species is limited by the low nutritional status of the substratum. Persistence beside the permanent way may also arise from a resistance to selective herbicides, and we suspect that L.v. is also tolerant of burning. L.v. has a relatively deep root system which may sustain water supply during the peak of growth, which occurs during summer. The foliage is said to be toxic to stock but *L.v.* is rarely present in pasture. Like *Hypericum perforatum*, with which it often grows, *L.v.* regenerates by means of adventitious buds produced on the roots, and forms large but diffuse clonal patches by this means. Disturbance appears to have a deleterious effect on vegetative spread. Regeneration by seed is also important and each inflorescence has the potential to produce over 1000 small, winged seeds, but *L.v.* is self-incompatible and seed-set is often poor. Seeds may persist for several years in the soil and germinate in spring and early summer. *L.v.* is part of a taxonomically complex grouping centred on S Russia and hybridizes in Britain with *L. repens*.

Current trends Probably increasing due to a capacity to exploit railways, industrial spoil and other artificial habitats.

162 *Linum catharticum*   Purging Flax

Established strategy S–R.

Gregariousness [H]Intermediate, but can develop high population densities.

Flowers White, hermaphrodite, pollinated by various insects or selfed.

Regenerative strategies S, B_s.

Seed 0.15 mg.

British distribution Widespread (100% of VCs).

Commonest habitats Lead mine spoil/limestone quarries 31%, limestone pastures 27%, rock outcrops 22%, scree 20%.

Spp. most similar in habitats *Hieracium pilosella* 89%, *Leontodon hispidus* 89%, *Campanula rotundifolia* 85%, *Senecio jacobaea* 82%, *Thymus praecox* 80%.

Full Autecological Account *CPE* page 376.

Synopsis *L.c.* is a diminutive, slow-growing, biennial or annual herb most characteristic of open, infertile, calcareous grassland and damp skeletal habitats. The root system of *L.c.* is both delicate and shallow. The occurrences of *L.c.* within the *PGE* are restricted to unproductive plots. In grassland, *L.c.* is associated with open turf, but germination and seedling survival is greatest in microsites providing some cover by perennial plants. *L.c.* reaches greater abundance on N-facing slopes and can develop large populations in wet skeletal habitats, such as quarry floors, and in open calcareous mire. *L.c.* usually overwinters as small plants, flowering in the following summer. In winter the tissues contain high concentrations of fructan. The tendency to exploit the margins of transient vegetation gaps and to persist beneath the turf canopy suggest both a vulnerability to moisture stress and a degree of shade tolerance. *L.c.* is potentially toxic to stock but is predated by slugs. Emergence through the sward and effective display of flowers is facilitated by the erect growth form of the shoot. *L.c.* regenerates entirely by means of seeds, which are released in summer or autumn and require chilling at *c.* 5°C for 6–8 weeks before germination is possible. Recent investigations in England and Holland reveal that some seeds of *L.c.* persist in the habitat and revert to a dormant state during the warmer summer months (cf. *Polygonum aviculare*). Seedlings appear in large populations mainly in spring but occasionally in autumn.

There are high mortalities. The timing of spring germination has been found to be relatively constant from year to year, and it has been suggested that a daylength cue may be implicated. In Dutch grassland, biennial plants produce *c.* 75 seeds per plant and annuals only 20–30. The annual forms may represent a different ecotype.

Current trends *L.c.* exploits infertile artificial habitats such as quarry spoil and railway banks. It is not clear to what extent this compensates for the losses occurring through destruction of more-ancient grassland habitats.

163   *Lolium perenne*

Rye-grass, Ray-grass, Perennial Rye-grass

Ssp. *perenne*

Established strategy Between C–R and C–S–R.
Gregariousness Intermediate.
Flowers Green, hermaphrodite, wind-pollinated.
Regenerative strategies S.
Seed 1.79 mg.
British distribution Widespread (100% of VCs).
Commonest habitats Meadows 87%, paths 42%, verges 38%, bricks and mortar/manure and sewage waste 36%.
Spp. most similar in habitats
   Cynosurus cristatus 79%, *Phleum pratense* 76%, *Trifolium repens* 76%, *Bellis perennis* 69%, *Plantago major* 66%.
Full Autecological Account *CPE* page 378.

depends upon the introduction and maintenance of a sward of *L.p.*. There is a current annual sowing of more than 20 000 t of *L.p. sensu lato*, including ssp. *multiflorum* (*L.multiflorum*) and ssp. *multiflorum* × ssp. *perenne*. This is almost ten times the amount sown of any other grass. Many genotypes have been bred or identified, but *L.p.* may be conveniently subdivided into two

Synopsis *L.p.* has been cultivated since at least the 17th century and is the most economically important forage grass in Britain. The management of productive grasslands, for both agriculture and amenity, at present

groups: early-heading types, which tend to be short-lived and to have few tillers and an erect growth form, and late-heading types, which are often longer-lived and with many tillers. Both are extensively sown. The former are considered in historical terms to have been 'followers of man', while the latter are characteristic of old fattening pasture and may have entered Britain before humans. Leaf extension commences under cool conditions in the early spring and, consistent with this phenology, the tissues of *L.p.* are rich in fructan. *L.p.* has a high nutrient requirement, and is highly productive on fertile soils, very tolerant of trampling, mowing and grazing, and palatable to stock. Foliage has a high content of P but, in common with most grasses, tends to be low in Ca. Despite the high densities of seed sown, *L.p.* is rapidly replaced by other grasses unless carefully managed. The low persistence in many pastures relates to the transient nature of the seed bank and to ineffective vegetative spread. Both of these seriously reduce the cost-effectiveness of *L.p.* for agricultural and amenity use, and often necessitate frequent re-sowing.

Heavy-metal-tolerant populations have been identified in Derbyshire and, apart from hybrids with ssp. *m.*, *L.p.* also crosses with various species of *Festuca*.

Current trends *L.p.* is abundant, and perhaps increasing since vast amounts of seed are sown annually, both for agriculture and for amenity. Seed is also shed on tracks and roadsides during harvesting. Commercially, however, there is an over-reliance upon *L.p.*, and when the ecological features of other grass species are better understood by agronomists, *L.p.* will be less-consistently recommended as a grassland crop. Also, the advantages of using slower-growing, less nutrient-demanding and more-persistent species on roadsides and in some amenity grasslands are now being more generally appreciated. In the near future the abundance of *L.p.* is thus likely to decrease.

164 *Lonicera periclymenum*   Honeysuckle

Established strategy S–R.
Gregariousness Forms stands or patches even at low frequencies.
Flowers Creamy-yellow, hermaphrodite, usually insect-pollinated.
Regenerative strategies V, S.
Seed 5.21 mg.
British distribution Widespread (100% of VCs).
Commonest habitats Acidic woodland/limestone woodland 4%, hedgerows/broadleaved plantations/scrub 3%.
Spp. most similar in habitats *Milium effusum* 81%, *Hyacinthoides non-scripta* 80%, *Geum urbanum* 80%, *Arum maculatum* 77%, *Rubus fruticosus* 77%.
Full Autecological Account *CPE* page 380.

Synopsis *L.p.*, which has an sub-Atlantic distribution, is represented by two main phenotypes: (a) a tall (often > 3 m), woody, freely-flowering, deciduous climber of hedgerows and other more-open habitats and (b) a prostrate, non-flowering undershrub of woodland, dependent upon the creation of gaps in the canopy by tree-fall or woodland clearance before flowering can take place. In these respects the species resembles several other climbing plants of woodland habitats, e.g. *Hedera helix*, *Tamus communis* and *Clematis vitalba*. In N England the species exhibits a bias towards S-facing slopes in shaded habitats and a tendency to occur on N-facing slopes in the open. In shoot phenology, shade-tolerance and reproductive biology *L.p.* resembles *Rubus fruticosus*, and the two species have similar ecologies in both woodlands and hedgerows. However, the distribution of *L.p.* is strongly biased towards less-fertile acidic soils, and *R.f.* has a greater potential to dominate herbaceous vegetation. Also, of course, *L.p.* is without spines and,

judging by its rarity in grazed sites, is more vulnerable to grazing. In woodland sites the prostrate stems, which root at irregular intervals to form large clonal patches, are the only means of regeneration. In hedgerow sites, where plants are erect, vegetative reproduction is absent, or at best restricted, and regeneration appears to be mainly by seeds, which germinate in spring. There is no information on the presence or otherwise of a bank of persistent seed, or any direct evidence concerning the frequency of establishment by seed. However, *L.p.* is an effective colonist of secondary woodland, both in the Sheffield region and within Lincolnshire, presumably because its berries are widely dispersed by birds. It is not known whether stems of *L.p.* cut during, for example, hedge-trimming, regenerate to form new plants. A further European ssp. is recorded.

Current trends Uncertain. A frequent colonist of broad-leaved plantations on freely-drained acidic soils, but not an early recruit in hedgerows, where most seed is set.

---

165   *Lotus corniculatus*

Birdsfoot-trefoil, Eggs-and-Bacon

Established strategy Between S and C–S–R.
Gregariousness Intermediate.
Flowers Yellow, hermaphrodite, insect-pollinated.
Regenerative strategies B_s.
Seed 1.67 mg.
British distribution Widespread (100% of VCs).
Commonest habitats Limestone pastures 42%, limestone wasteland 40%, limestone quarries 32%, lead mine spoil 26%, cinder heaps 14%.
Spp. most similar in habitats *Pimpinella saxifraga* 82%, *Hieracium pilosella* 81%, *Linum catharticum* 74%, *Campanula rotundifolia* 73%, *Briza media* 72%.
Full Autecological Account *CPE* page 382.

Synopsis *L.c.*, which forms nitrogen-fixing nodules (in conjunction with *Rhizobium lupini*), is the commonest legume of unproductive semi-natural grassland, and is perhaps the most ecologically wide-ranging legume in Britain. The species extends from maritime to montane environments, from moderately low to high soil pH, from infertile to moderately fertile soils, and from spoil and open habitats to grassland. *L.c.* is restricted to unproductive plots of the *PGE* and is strongly suppressed by nitrogenous manures. *L.c.* is generally regarded as relatively tolerant of grazing, is palatable to stock, and attains agricultural importance in less-productive grasslands. It has even been suggested that *L.c.* is the second most important forage legume in N America. Some populations of *L.c.* produce cyanogenic glucosides, which are an effective deterrent against small herbivores and may also be toxic to stock. The presence of *L.c.* produces local enrichment of soil nitrogen. This can result in invasion and dominance of the patch by more-N-demanding species, e.g. *Festuca rubra*. *L.c.* is not a drought-tolerant species and the establishment and persistence of *L.c.* on desiccated, often S-facing limestone soils, is probably related to its relatively large seed and the capacity of its roots to penetrate fissures deeply. In ungrazed sites, up to 100 seeds may be produced in each inflorescence, and this moderate seed output is apparently sufficient to allow *L.c.* to colonize successfully many infertile habitats, e.g. quarry spoil heaps, where an ability to fix nitrogen may be important. However, in many grazed habitats seeds are produced infrequently, and population expansion may be dependent upon the persistent seed bank and the limited capacity for clonal expansion. Populations from S England frequently have a greater potential for vegetative regeneration than those from N upland sites. *L.c.* can also regenerate from root or stem cuttings, but this is probably only rarely important under natural conditions. *L.c.* is genotypically and probably to a lesser extent phenotypically variable. Fourteen varieties are identified from Europe, where such variability has been crucial in determining the ability of *L.c.* to exploit a range of different habitats. A heavy-metal-tolerant race of *L.c.* has been described within the Sheffield area. As a result of its N-fixing ability *L.c.* is potentially a valuable species for the reclamation of derelict land. In Britain, seed of *L.c.* is presently sown for agricultural and amenity use at the rate of < 5 t year^{-1}. More-erect, alien genotypes, with larger leaflets, have been introduced in seed mixtures sown onto new road verges; these are capable of persistence in relatively productive swards. *L.c.* and *L. uliginosus* are part of a taxonomically complex group which are represented by at least 12 spp. in Europe.

Current trends In improved pasture *L.c.* is rapidly suppressed, and the species is a poor competitor in the tall grasslands which result from the cessation of grazing in marginal land. Thus, although *L.c.* exploits a number of new artificial habitats, e.g. limestone and cinder spoil, a further decline in its abundance seems inevitable.

---

166 *Lotus uliginosus*

Large Birdsfoot-trefoil

Established strategy S–C.
Gregariousness Intermediate, but occasionally stand-forming.

Flowers Yellow, hermaphrodite, protandrous, insect-pollinated, self-incompatible.

Regenerative strategies V, B$_s$.
Seed 0.4 mg.
British distribution Widespread except in the extreme N (98% of VCs).
Commonest habitats Limestone wasteland 7%, unshaded mire 5%, paths 2%, still aquatic/enclosed pastures 1%.
Spp. most similar in habitats Insufficient data.
Full Autecological Account *CPE* page 384.

Synopsis *L.u.* is a relatively tall, patch-forming legume of the drier parts of topogenous mire and of soligenous mire. The species is the only common wetland legume in the British flora, and forms nitrogen-fixing nodules in association with *Rhizobium lupini*. The general absence of legumes from wetlands may be a reflection of the conflict between a requirement for oxygen during nitrogen-fixation and a possible requirement for oxygen diffusion from the root as part of the detoxification of the anaerobic soil environment. Consistent with these constraints, *L.u.* is, within topogenous mire, largely restricted to drier habitats, and is fairly sensitive to ferrous iron toxicity. *L.u.* occurs both as a colonist of infertile habitats (e.g. on stony ground beside reservoirs and in gravel pits) but in tall wet grassland and on soligenous mire the species appears to be a component of relatively stable and moderately productive vegetation. We suspect that *L.u.* is restricted to sites where the growth of potential dominants is limited to some extent by soil infertility. However, the extent to which the capacity to fix nitrogen affects the status and persistence of the species in communities is not known. *L.u.* is tolerant of light grazing and may be cut for hay. The species is, like other wetland species, sensitive to drought but is moderately salt-tolerant.

L.u. can form clonal patches by means of spreading stolons, and breakdown of stolons results in the isolation of daughter colonies. Shoot pieces root relatively slowly compared with those of many wetland species, and regeneration by means of shoot fragments is probably uncommon. Reproduces also by seeds which appear to germinate in spring and a persistent seed bank is reported. Despite the fact that seeds are relatively large and lacking in any obvious dispersal mechanism, *L.u.* is a not-infrequent colonist of moist, disturbed sites. The degree of ecotypic differentiation within this out-breeding species is not known. *L.u.* is part of the *Lotus corniculatus* complex, which consists of 12 species in Europe.
Current trends *L.u.* has only a restricted colonizing ability, and is decreasing through the destruction of wetland habitats, particularly in lowland areas. Considered agriculturally superior to *L.c.* due to its higher protein content, but not at present used in any great quantity for agricultural purposes or in landscape reclamation.

167 *Luzula campestris*

Field records include *L. multiflora*

Established strategy Between S and C–S–R.
Gregariousness Intermediate.
Flowers Chestnut-brown, hermaphrodite, wind-pollinated.
Regenerative strategies V, B_s.
Seed 0.64 mg.
British distribution Widespread (100% of VCs).
Commonest habitats Limestone pastures 53%, limestone wasteland 20%, lead mine spoil 17%, meadows 16%, verges 15%.
Spp. most similar in habitats
Anthoxanthum odoratum 76%, *Danthonia decumbens* 74%, *Briza media* 73%, *Carex caryophyllea* 69%, *Agrostis capillaris* 69%.
Full Autecological Account *CPE* page 386.

Synopsis *L.c.* is a winter-green, patch-forming species which is in need of further study. Although occurring in association with a relatively wide range of habitats and soil types, *L.c.* typically occupies the lower stratum in short turf on moist, less fertile, often mildly acidic soils. The species is primarily a pasture plant and, despite its delayed leaf growth, is one of the earliest-flowering of the common pasture species. In short turf *L.c.* often forms conspicuous dense patches as a result of stolonifery. Although absent from most of the taller types of herbaceous vegetation, *L.c.* displays a marked ability to persist under a discontinuous canopy of pasture grasses, and occasionally occurs as non-flowering

Field Woodrush

colonies in scrub. This strongly suggests that the species is shade-tolerant. Palatability to stock is similar to that of *Festuca ovina* and *Agrostis* spp. Seed has a low temperature range for germination, and the timing of germination is uncertain. A persistent seed bank is also formed, but regeneration by seed appears to be infrequent. The mucilaginous seeds have an elaiosome and may be dispersed by ants. *L. multiflora* is a densely-tufted taller plant, virtually restricted to acidic peaty and sandy soils. In Britain, *L.m.* includes sspp. *multiflora* and *congesta*.
Current trends Uncertain. Most common in less-fertile habitats, but capable of persisting as a rather inconspicuous component of some relatively productive, artificial habitats such as lawns and road verges.

168 *Luzula pilosa*

Established strategy S.
Gregariousness Sparse to intermediate.
Flowers Brown, hermaphrodite, wind-pollinated.

Hairy Woodrush

Regenerative strategies S, B_s, V.
Seed 0.81 mg.
British distribution Widespread except in Ireland (91% of VCs).

Commonest habitats Coniferous
plantations 5%, acidic woodland
4%, cliffs/broadleaved
plantations/scrub 3%.
Spp. most similar in habitats
Dryopteris dilatata 89%, *Pteridium
aquilinum* 81%, *Quercus* spp. (juv.)
79%, *Sorbus aucuparia* (juv.) 77%,
Hyacinthoides non-scripta 73%.
Full Autecological Account *CPE* page
388.

Synopsis *L.p.* is a tufted, winter-green
species most typically associated with
situations in woodlands where the
vigour of potential dominants is
restricted by shade and by soil
infertility. *L.p.* is particularly
characteristic of moist acidic soils. The
species is widespread on clay but
virtually absent from the freely-drained
acidic soils, such as those of the Bunter
sandstone of S Yorkshire and N
Nottinghamshire. In lowland parts of
Britain, *L.p.* is largely confined to
broad-leaved woodland; in
Lincolnshire the plant is characteristic
of ancient woodland. In upland areas
the species is ecologically more wide-
ranging; here *L.p.* may occur in old
pastures, often with *Oxalis acetosella*
and is a frequent colonist of gritstone
quarries, particularly where these are
adjacent to woodland. Presumably
because of its shade tolerance, *L.p.* is
also a common constituent of upland
conifer plantations in the Sheffield
region, and is recorded more
frequently from this habitat than from
any other. The short stolons confer
only limited capacity for vegetative
spread, and regeneration by seed is
important in population expansion and
persistence. On average a plant

produces *c.* 200 seeds year^{-1}; these
have a well-developed elaiosome and
are frequently dispersed by ants. Even
under heavy shade, *L.p.* produces
seed, but it is in open areas after forest
fire or clear felling that seed set is
greatest. These are also the
circumstances most favourable for
seedling establishment, either from
freshly-shed seed or from the long-
persistent seed bank. On this basis,
infrequent periods of catastrophic
disturbance may play an essential role
in the maintenance of populations of
L.p. in woodland habitats. This
regenerative strategy is remarkably
similar to that associated with *Carex
pilulifera* on burned heathland. *L.p.*
forms hybrids with *L. forsteri*, and the
distributions of the two species overlap
in S England.
Current trends Probably decreasing in
lowland areas, where it shows a low
capacity to colonize new sites, but with
perhaps a stable or even an increasing
distribution in forested upland areas.

---

169 *Matricaria matricarioides*

Also known as *Chamomilla suaveolens*

Established strategy R.
Gregariousness Intermediate.
Flowers Florets greenish-yellow,
hermaphrodite, tubular, seldom

Pineapple Weed, Rayless Mayweed

visited by insects.
Regenerative strategies B$_s$.
Seed 0.08 mg.
British distribution Widely naturalized
(86% of VCs).

Commonest habitats Arable 50%, bricks and mortar 46%, paths 34%, soil heaps 21%, manure and sewage waste 18%.

Spp. most similar in habitats *Capsella bursa-pastoris* 96%, *Poa annua* 93%, *Polygonum aviculare* 91%, *Tripleurospermum inodorum* 79%, *Plantago major* 78%.

Full Autecological Account *CPE* page 390.

Synopsis A native of NE Asia and perhaps also of NW America, first recorded in Britain in 1871. Strongly aromatic when crushed. *M.m.* is facultatively summer or winter annual, and exploits compacted soils where dominance by perennial species is prevented by trampling and by vehicular pressure. When flowering, the species is upright with leafy stems, and at this stage is potentially more vulnerable to damage from trampling than more-decumbent species such as *Poa annua*; thus, *M.m.* tends to be abundant only on road margins and in the central grassy areas of cart-tracks, and scarce on the most heavily trampled areas of paths. The success of the species is related in part to the production of numerous seeds. Although a long-day plant, seed may be set within 40–50 days in spring-germinating plants, a much shorter period than that required by *Tripleurospermum inodorum*. A plant from little-disturbed sites produces on average over 6000 small seeds; an exceptionally large capitulum may release over 400 achenes. Even small plants with a single flower head, such as may be found in heavily trampled sites, may produce over 50 seeds. Seeds may be carried in mud on tyres, and the rapid spread of *M.m.* at the beginning of the 20th century relates to the increased use of motor cars. Seeds may also survive ingestion by horses. The spread of the species to distant sites in England, Ireland and Scotland in the 19th century illustrates the capacity of *M.m.* for long-distance dispersal. *M.m.* forms a persistent seed bank and, although spring and autumn peaks are evident, has the capacity to germinate over most of the year, which undoubtedly enables populations to survive repeated disturbance. The occurrence of *M.m.* on bare areas nearest to the metalled road surface suggest that the species possesses some salt tolerance. On the basis of its association with sites subject to soil compaction, a modest tolerance of waterlogging is predicted, but this hypothesis has not been tested. The high frequency with which *M.m.* is recorded from arable fields in the Sheffield region probably results from sampling bias since, to avoid crop damage, most records were collected at field margins. *M.m.* is particularly abundant in gateways and in other access areas for tractors, but is largely absent from gardens and the centres of arable fields.

Current trends Remains abundant on fertile paths, tracks and road margins at low to moderate altitudes. Perhaps still increasing in range and abundance.

Established strategy Between S–R and R.
Gregariousness Intermediate.
Flowers Yellow, hermaphrodite, self- or insect-pollinated.
Regenerative strategies B_s.
Seed 2.01 mg.
British distribution Widespread except in Scotland and Ireland (99% of VCs).
Commonest habitats Rock outcrops 19%, cinder heaps/limestone quarries 16%, limestone wasteland 9%, arable 7%.
Spp. most similar in habitats *Myosotis ramosissima* 78%, *Arenaria serpyllifolia* 78%, *Hieracium* spp. 76%, *Leucanthemum vulgare* 72%, *Saxifraga tridactylites* 69%.
Full Autecological Account *CPE* page 392.

Synopsis *M.l.* is a short-lived legume forming nitrogen-fixing nodules in conjunction with *Rhizobium meliloti*, and is characteristic of relatively disturbed, infertile sites, particularly on calcareous soils. *M.l.* produces a long tap-root and frequently exploits grasslands on S-facing slopes and other habitats where summer droughting of perennial vegetation creates areas of bare ground. The species has a predominantly lowland distribution in Britain, extending only to 400 m. *M.l.* is grazed by sheep but is not very palatable to cattle. *M.l.* can withstand repeated cutting and may persist in prostrate form as a lawn weed. In short turf, fruits are borne close to the ground and are thus protected to some extent from predation. Regeneration is entirely by seed, which is often set over a prolonged period. Under moisture stress *M.l.* tends to sustain seed production at the expense of vegetative development, and in droughted habitats *M.l.* usually functions as a winter annual. In moist fertile sites a life-span of at least four years is probably common and robust plants may produce over 2000 seeds annually. Some genetic differences exist with respect to life-history; diploid and tetraploid races have been recorded. *M.l.* may set seed within nine weeks of germination, forms a persistent buried seed bank, and is essentially a colonist of bare areas. Because of its capacity to fix nitrogen, *M.l.* is sometimes recommended for use in land reclamation.
Current trends Uncertain. An effective colonist of disturbed habitats and spoil, but potential range is limited by the association with infertile soils.

171 *Melica uniflora*

Established strategy Between S and
S–C.
Gregariousness Stand-forming.
Flowers Green or brown,
hermaphrodite, wind-pollinated.
Regenerative strategies V, S.
Seed 2.78 mg.
British distribution Absent from N
Scotland (91% of VCs).
Commonest habitats Limestone
woodland 16%, scrub 6%, cliffs 4%,
hedgerows/broadleaved plantations
3%.
Spp. most similar in habitats
Mercurialis perennis 92%,
Moehringia trinervia 89%,
Brachypodium sylvaticum 89%,
Sanicula europaea 88%, *Geum
urbanum* 87%.
Full Autecological Account *CPE* page
394.

Synopsis *M.u.* is a stand-forming
woodland grass more rarely found in
hedgerows and restricted to
moderately base-rich but relatively
unproductive sites where the vigour of
taller potential dominants is reduced
both by shade and by low soil fertility.
The annual increase in shoot biomass
continues into the shaded phase but,
unusually amongst woodland grasses,
M.u. is not winter green. *M.u.* is
characteristic of freely-drained sites;
frequent occurrence on shaded rock
ledges and the absence of the species
from mire suggest intolerance of
waterlogging. Although associated
with a wide range of soil types, the
species is recorded most frequently

from calcareous strata. *M.u.*
regenerates *in situ* by means of
rhizomatous growth to form large
clonal patches. Stands are often
extremely localized, suggesting that
seedling colonization is infrequent
either because the comparatively large
seeds are underdispersed or because
they are heavily predated. Seeds
germinate in spring and a persistent
seed bank has been detected. Spikelets
of *M.u.* consist of one fertile floret and
2–3 reduced to form a club-shaped
mass. This appendage encourages
dispersal by ants. *M.u.* is only rarely
among the recruits to recent artificial
habitats, and within the context of
Lincolnshire is regarded as a species of
ancient woodland. This neglected
species is in urgent need of ecological
investigation.
Current trends Probably decreasing.

172 *Mentha aquatica*

Established strategy Between C and
C–R.
Gregariousness Intermediate, but
sometimes stand-forming.

Flowers Lilac, gynomonoecious,
protandrous, insect-pollinated, self-
compatible.
Regenerative strategies V, B_s.

Seed 0.14 mg.
British distribution Widespread, but rarer in N Scotland (100% of VCs).
Commonest habitats Shaded mire 22%, unshaded mire 10%, still aquatic 4%, walls 1%.
Spp. most similar in habitats *Caltha palustris* 96%, *Myosotis scorpioides* 94%, *Solanum dulcamara* 88%, *Galium palustre* 88%, *Filipendula ulmaria* 85%.
Full Autecological Account *CPE* page 396.

Synopsis *M.a.* is a rhizomatous, aromatic, stand-forming herb typically associated with mire adjacent to open water but more common in mid-marsh positions than those close to the water-line. Less frequently, *M.a.* may occur wholly or partially submerged in ditches. The species appears to be restricted to situations where the vigour of taller waterside dominants is reduced by a degree of soil infertility. *M.a.* is also found in moderately shaded habitats. Because of its erect stature, *M.a.* is sensitive to trampling and grazing and, despite the apparent toxicity of *Mentha* spp. to stock, the species is suppressed in grazed marshes. *M.a.* is found in a wide range of edaphic conditions, and is particularly frequent in soligenous mire on calcareous soils. The species forms extensive patches by means of creeping rhizomes, breakdown of which often results in the formation of daughter clones. Rhizome or shoot fragments detached as a result of disturbance during, for example, flooding readily regenerate and *M.a.* may be transported in water to colonize distant sites. Various related sterile hybrid mints have colonized water systems extensively in this way. Regeneration by seeds, which may float on water for long periods, is probably infrequent. Seeds normally germinate in vegetation gaps during spring or after incorporation into a persistent seed bank. *M.a.* varies greatly in habit, leaf shape and degree of hairiness, both as a result of genetic differences and phenotypic plasticity. In drier sites where the water-table fluctuates widely, e.g. woodland rides and reservoir margins, *M.a.* is frequently replaced by *M. arvensis*. The usually sterile but vegetatively vigorous hybrid between the two (*M.* × *verticillata*) is also frequent, often in the absence of one or both parents, and usually in slightly drier habitats than those exploited by *M.a.*. Two further hybrids involving *M.a.* occur in Britain; these are often of horticultural origin.
Current trends Still common, but perhaps decreasing as a result of eutrophication and habitat destruction.

173 *Mercurialis perennis*   Dog's Mercury

Established strategy S–C.
Gregariousness Stand-forming.
Flowers Green, usually dioecious, wind-pollinated.
Regenerative strategies V, S.
Seed 2.15 mg.

British distribution Widespread except in N Scotland and Ireland (80% of VCs).

Commonest habitats Limestone woodland 70%, scrub 29%, hedgerows 20%, broadleaved plantations 15%, limestone pastures/scree 12%.

Spp. most similar in habitats *Brachypodium sylvaticum* 97%, *Arum maculatum* 94%, *Anemone nemorosa* 94%, *Melica uniflora* 92%, *Sanicula europaea* 90%.

Full Autecological Account *CPE* page 398.

Synopsis *M.p.* is a rhizomatous, stand-forming woodland herb exploiting relatively unproductive base-rich sites where the growth of more-robust dominants is suppressed. In lowland habitats the species is largely restricted to shaded sites. However, in the uplands *M.p.* is found in unshaded herbaceous vegetation, particularly on moist N-facing slopes; in some populations established in the open, *M.p.* forms the upper layer of the canopy and may suffer considerable loss of chlorophyll. It is quite common, however, for *M.p.* to occupy the shaded zone beneath the canopy of a taller herb, e.g. *Filipendula ulmaria*. *M.p.* is relatively deeply rooted (normally 100–150 mm) and is very susceptible to waterlogging and to the effects of Fe^{2+} toxicity. On wet soils roots are shallower and *M.p.* is susceptible to summer drought. The foliage of *M.p.* contains high concentrations of N and Ca, but levels of P are low. The leaves have an unpleasant odour when bruised and rarely show signs of extensive predation. They contain, amongst other substances, toxic volatile oils, and are highly poisonous both to stock and to humans. The species combines winter green and vernal aspects in its rather unusual phenology. As with many vernal species, new shoots are formed in the autumn and expand in late winter (February–March). The emergent shoot is bent over at the tip, facilitating emergence through leaf litter. The plant flowers before the leaves are fully expanded. *M.p.* reaches a peak in biomass at the end of May at the termination of the light phase. However, unlike vernal species, most shoots and leaves persist through the summer and gradually senesce during winter. Regeneration is mainly by vegetative means, and *M.p.* forms extensive long-lived clones which may extend by 100–150 mm year^{-1}. The species, which can survive at very low light intensities, responds little to the additional light present during clear felling or coppicing, but survives beneath other species and returns to prominence after the regrowth of the tree canopy. Although normally dioecious, female plants usually set seed (up to 300 per plant), but many may be non-viable and seedling establishment is uncommon. No persistent seed bank is formed. Male and female plants appear to differ in their ecological distribution. Male plants have been recorded more frequently than females in well-illuminated conditions and at high pH. *M.p.* is a common species of ancient woodland in Lincolnshire, and its large seeds, which have an elaiosome and are said to be specialized for dispersal by ants, appear to be poorly dispersed. *M.p.* shows some morphological variations which have been given varietal rank. The species is also phenotypically plastic, and can form sun or shade leaves which differ in morphology and anatomy.

Current trends Shows poor mobility and is probably decreasing with the destruction of older woodland.

174 *Milium effusum*   Wood Millet

Established strategy Between C–S–R and S.
Gregariousness Intermediate.
Flowers Green, hermaphrodite, wind-pollinated.
Regenerative strategies B_s, V.
Seed 1.2 mg.
British distribution More frequent in England (79% of VCs).
Commonest habitats Scrub 10%, acidic woodland 8%, limestone woodland 4%, hedgerows 3%, paths/coniferous plantations 2%.
Spp. most similar in habitats *Sorbus aucuparia* (juv.) 81%, *Lonicera periclymenum* 81%, *Rubus fruticosus* 80%, *Lamiastrum galeobdolon* 79%, *Oxalis acetosella* 78%.
Full Autecological Account *CPE* page 400.

Synopsis *M.e.* is a loosely-tufted woodland grass occurring in sites where the vigour of potential dominants is reduced both by shade and by low site fertility. The species is winter green, with an annual increase in shoot biomass beginning in spring and continuing into the shaded phase. The robust shoots are relatively efficient at penetrating layers of deciduous tree litter. *M.e.* is restricted to moist soils, particularly clays, and is uncommon on sandy soils. Within the Sheffield region, *M.e.* is often recorded from mildly acidic soils. However, elsewhere *M.e.* may be characteristic of calcareous sites. The species has only a limited capacity for vegetative spread, but numerous seeds are produced. These show low rates of germination when first transferred to the laboratory. *M.e.* exhibits variation in germination biology over its geographical range. At southern stations, *M.e.* flowers early and seeds germinate predominantly in autumn. Nearer to its northern limit, seed-set is later and at least some germination may be delayed until spring. A similar pattern has been reported for *Festuca ovina*. *M.e.* forms a persistent seed bank, which suggests that the species may have the capacity to exploit sites where periodic disturbance occurs in the form of either tree-felling or fire. 'Seeds', which can survive ingestion by some animals, are shed without an attached lemma. They show no obvious specialization for dispersal, and in Lincolnshire *M.e.* is regarded as a species of ancient woodland. However, *M.e.* has colonized more-recent woodland in some upland sites. *M.e.* was formerly planted for

235

ornament and its seeds provided food for game birds, but there is no evidence that *M.e.* has been introduced in the Sheffield region.

Current trends Uncertain, but probably decreasing, particularly in lowland areas.

175 *Minuartia verna*

Ssp. *verna*

Vernal Sandwort

Established strategy S.
Gregariousness Forms small patches.
Flowers White, hermaphrodite, insect-pollinated or self-fertilized.
Regenerative strategies V, B_s.
Seed 0.08 mg.
British distribution Scattered throughout the W and N (19% of VCs).
Commonest habitats Lead mine spoil 51%, limestone pastures 2%, unshaded mire 1%.
Spp. most similar in habitats *Euphrasia officinalis* 93%, *Rumex acetosa* 67%, *Carlina vulgaris* 67%, *Leontodon autumnalis* 63%, *Thymus praecox* 56%.
Full Autecological Account *CPE* page 402.

Synopsis *M.v.* is a low-growing, cushion-forming herb, which is relatively uncommon and has a very disjunct distribution in Britain. The species may perhaps best be regarded as a late glacial relic. Nearly all populations in N Derbyshire occur on lead-mine spoil; the remainder, in natural rock outcrops and limestone quarries, also occur on contaminated soils. *M.v.* also exploits serpentine soils. In rocky montane sites, the other main habitat type for the species, heavy-metal concentrations are assumed to be low. The wide edaphic tolerance of *M.v.* is further illustrated by its capacity to grow on mine spoil on acidic soils in addition to its more typical calcareous habitats. *M.v.* is, after *Thlaspi alpestre*, the most heavy-metal-tolerant species in the Sheffield flora and is an accumulator of heavy metals. This may be a reflection of the absence of mycorrhizas in the root system, since all the other heavy-metal-tolerant plants mentioned in this volume are mycorrhizal and behave as excluders. The foliage of plants sampled from lead mine heaps in Derbyshire contains unusually low concentrations of N but high levels of Ca and Fe. *M.v.* is found over a wide range of climatic conditions in Britain, extending from sites near sea-level which are virtually frost-free to montane areas where the number of days with frost may exceed 150. The distribution is mainly restricted to sites with at least 1000 mm of rainfall each year. It has been suggested that summer humidity may be important in the survival of this oceanic montane species. This is consistent with local observations of heavy fatalities during summers of exceptional drought. Many Derbyshire sites for *M.v.* are highly toxic and, as a result, have an incomplete vegetation cover. The species is particularly associated with

236

sites liable to soil erosion; plants detached as a result of soil disturbance are frequently observed in the field. *M.v.* has an unusually long flowering period and produces numerous very small seeds. The presence of a persistent seed bank is probably critical in enabling the species to recover after drought. Under more-favourable conditions *M.v.* also spreads by clonal expansion but field observations and recent experiments have shown that here the species is exceedingly vulnerable to competition from metal-tolerant races of *Festuca rubra*. *M.v.* is polymorphic and five further ssp. have been identified in Europe. Some European populations are tetraploid but their recognition is to some extent confounded by the interaction between genetic diversity and phenotypic plasticity. Thus, individuals of the same biotype may appear very different in contrasted environments whereas genetically distinct races may be indistinguishable through the effects of climate. Crosses between diploid populations from widely differing geographical areas are successful, but diploids are genetically isolated from tetraploids.

Current trends Reported as decreasing in S Scotland, and in Derbyshire also, as a result of habitat destruction.

176 *Moehringia trinervia*   **Three-nerved Sandwort**

Established strategy S.
Gregariousness Individuals usually scattered.
Flowers White, hermaphrodite, visited by insects and automatically self-pollinated.
Regenerative strategies B_s.
Seed 0.22 mg.
British distribution Common except in N Scotland and Ireland (95% of VCs).
Commonest habitats Limestone woodland 8%, hedgerows/broadleaved plantations/scrub 3%, cliffs/coniferous plantations/rock outcrops/soil heaps 2%.
Spp. most similar in habitats *Melica uniflora* 89%, *Mercurialis perennis* 85%, *Arum maculatum* 84%, *Brachypodium sylvaticum* 82%, *Geum urbanum* 81%.
Full Autecological Account *CPE* page 404.

Synopsis *M.t.* is one of only two relatively common annual species which are exclusive to British woodlands, the other being *Corydalis claviculata*. As with *C.c.*, *M.t.* is a summer-annual. In a majority of sites with *M.t.*, the cover of tree litter is less than 25%, and it seems likely that the presence of large amounts is prejudicial to establishment; indeed, the general absence from woodland of winter annuals and other autumn-germinating species may be due to problems of seedling establishment arising from the impacts of leaf-fall and litter persistence. More rarely, and perhaps only in light shade, *M.t.* may behave as a short-lived, winter-green polycarpic perennial. The species exploits disturbed ground, e.g. rabbit scrapes or sites of tree-fall, and in appearance and in ecology *M.t.* is 'the *Stellaria media* of woodland'. The distribution of *M.t.* shows a slight bias towards acidic soils, and the slight tendency for the species to occur on S-facing slopes in woodland suggests that, in common with many other ephemerals, *M.t.* requires well-insolated, relatively warm sites for completion of the life-cycle. The average output per plant is said to be *c.* 2500 seeds, and, although this estimate appears rather high, *M.t.* is undoubtedly a prolific seed producer. Germination is inhibited by high temperatures and is delayed until spring, and, unusually for a woodland

species (and typically for species exploiting unpredictably disturbed habitats), a persistent seed bank is formed. In this latter specialization an obvious parallel exists with another woodland opportunist, *Digitalis purpurea*. The seed has an oily elaiosome and dispersal is often influenced by ants although, judging by the distribution of the species, dispersal by humans is much more important.

Current trends *M.t.* is regarded as a characteristic species of secondary woodland. It is favoured by the high level of mechanized disturbance associated with modern forestry, and is probably increasing.

177 *Molinia caerulea* — Purple Moor-grass

Established strategy S–C.
Gregariousness Intermediate to patch-forming.
Flowers Green or purplish, hermaphrodite, wind-pollinated.
Regenerative strategies V, ?B_s.
Seed 0.53 mg.
British distribution Widespread (100% of VCs).
Commonest habitats Acidic pastures 9%, unshaded mire/river banks 3%, shaded mire/acidic wasteland 2%.
Spp. most similar in habitats Insufficient data.
Full Autecological Account *CPE* page 406.

Synopsis *M.c.* is a tufted, or turf-forming, deciduous grass which, like *Festuca ovina*, has a bimodal pH distribution with peaks of abundance on both highly acidic and calcareous soils. At both high and low pH *M.c.* tends to be associated either with moist grassland or with soligenous mire possessing a well-oxygenated soil profile, but it is tolerant of ferrous iron. The lack of tolerance of

waterlogging may in part stem from the deep root system (down to at least 800 mm) and in areas with a relatively sharp transition between wetland and dryland *M.c.* often occupies the transition zone. In drier moorland habitats *M.c.* gives way to *Nardus stricta*, whereas in waterlogged habitats it is replaced by species such as *Juncus effusus* and *Eriophorum vaginatum*. *M.c.* is tolerant of grazing and burning, and is particularly important in acidic upland areas where the species has a grazing value intermediate between that of *Agrostis–Festuca* grassland and impoverished *Nardus* grassland. In *Molinia*-dominated communities there are high rates of nutrient turnover. This contrasts with the behaviour of *Eriophorum vaginatum*, which exploits impoverished soils and appears to show more-efficient mineral retention and recycling (see *E.v.*). *M.c.* litter was formerly harvested for animal bedding. The species persists in moderate shade, but tends to produce few flowers under these conditions. Leaves are deciduous with an abscission zone, but new growth, fuelled by below-ground carbohydrate reserves, begins in April or May. *M.c.* is vulnerable to damage in cold springs, particularly in upland areas. In Derbyshire, *M.c.* has a tussock growth form in moorland areas, but forms more-uniform carpets in moist calcareous grassland and in soligenous mire with a pronounced lateral water flow. This behaviour is consistent with findings which suggest that the height of tussocks is greatest when both the growth of *M.c.* and the rate of decomposition of its dead remains is slow. This combination of conditions may result when there is a fluctuating water-table. *M.c.* regenerates *in situ* through lateral vegetative spread, particularly in habitats favouring the non-tufted growth form. However, tussocks may include more than one genotype. Break-up of patches of *M.c.* is generally associated with patch degeneration rather than with regeneration, and is often accompanied by invasions by other species, e.g. bryophytes and *Deschampsia flexuosa*, on top of the tussocks. *M.c.* produces much seed which germinates in spring and colonizes bare ground. A persistent seed bank has been recorded. In common with various grasses and sedges of wet northerly habitats (e.g. *E.v.*), there is a high-temperature requirement for seed germination. This characteristic is likely to reduce the risk of excessive damage to seedlings by spring frosts. Taller plants with larger 'seeds' have been separated as ssp. *arundinacea*, but many intermediates occur. Ssp. *a.* is recorded from rather base-rich mineral soils with a fluctuating water-table and has (as *M. litoralis*) been recorded from S England and Wales.

Current trends Although no longer of agricultural importance, *M.c.* continues to be widespread and abundant in upland regions. The species is declining in many lowland areas as a result of habitat destruction.

---

178 *Mycelis muralis*

Wall Lettuce

Established strategy C–S–R.
Gregariousness Sparse.
Flowers Florets yellow, ligulate, hermaphrodite.
Regenerative strategies W.
Seed 0.34 mg.
British distribution Common except in Scotland and Ireland (62% of VCs).
Commonest habitats Cliffs/rock outcrops/limestone woodland 6%, walls 5%, limestone quarries 3%.
Spp. most similar in habitats *Athyrium filix-femina* 88%, *Asplenium* *ruta-muraria* 86%, *Asplenium trichomanes* 86%, *Elymus caninus* 83%, *Cystopteris fragilis* 82%.
Full Autecological Account *CPE* page 408.

Synopsis *M.m.* is a tall winter-green herb. In the Sheffield flora, four of the five species most similar in habitat range are ferns. This highlights the distinctive ecology of the species, which arises from the unusual combination (among British herbs) of shade tolerance with effective seed dispersal by wind. These two attributes undoubtedly confer a characteristic ability to exploit shaded but relatively inaccessible situations on cliffs, walls and rock outcrops. Despite its tolerance of shade, *M.m.* is not a common constituent of woodland herb layers; this may be related to sensitivity to dominance by more-robust species and submergence of the basal rosette by tree litter. The species is highly calcicolous within the Sheffield flora, occurring mainly on limestone rocks or mortared walls. *M.m.* is a drought-avoiding species whose roots penetrate fissures in the rock or stonework. Seed is produced throughout late summer and autumn, with only five seeds in each capitulum, a remarkably low number for a member of the Compositae. The species lacks effective vegetative spread and is restricted to small rocky or disturbed microsites which are large enough to sustain the established plant

but too small to allow the vegetative expansion of more-dominant species. It is not known whether the species forms a persistent seed bank.

Current trends Uncertain, but not under threat. A frequent associate of artificial habitats, but some populations, particularly on old walls, are decreasing in abundance.

179 *Myosotis arvensis*   Common Forget-me-not

Established strategy Between R and S–R.

Gregariousness Sparse.

Flowers Blue, hermaphrodite, usually self-pollinated.

Regenerative strategies S, B_s.

Seed 0.29 mg.

British distribution Widespread (100% of VCs).

Commonest habitats Arable 15%, limestone pastures/soil heaps 4%, acidic wasteland 1%.

Spp. most similar in habitats *Spergula arvensis* 95%, *Veronica persica* 95%, *Fallopia convolvulus* 93%, *Papaver rhoeas* 91%, *Anagallis arvensis* 87%.

Full Autecological Account *CPE* page 410.

Synopsis *M.a.* is a relatively tall, annual herb which occurs in disturbed artificial habitats and exploits a wide range of soil types. The species tends to occur on relatively shallow calcareous soils or dry sandy soils in sites where, as a result of summer drought and often also some physical disturbance of the soil, the dominance of polycarpic perennials is incomplete. *M.a.* usually occurs as scattered individuals, and tends to occupy sites which have both deeper soils and are more transient than those of the majority of small winter annuals. However, the species does normally behave as a winter annual, and germination of the seeds, which are shed in summer, is usually delayed until autumn. Other transient habitats exploited by *M.a.* include waste ground and woodland clearings. The species appears to be highly mobile. The seeds can survive ingestion by cattle and horses, and may be dispersed also by humans or in the fur of animals, enclosed in the hispid spiky calyx. Seeds are also incorporated into a persistent seed bank, which is particularly important in the main habitat, arable land. Here the species may behave to a limited extent as a summer annual, but *M.a.* is more frequent in winter wheat and winter barley than in spring barley. Responses to agricultural management have been little studied, but *M.a.* has been observed to re-sprout and flower following defoliation. *M.a.* varies greatly in habit according to environment, but the role of ecotypic differentiation in determining the wide ecological amplitude of *M.a.* is uncertain. Two ssp. occur in Europe. It seems likely that local populations belong to ssp. *arvensis* ($2n = 52$) rather than the larger fruited ssp. *umbrata* with $2n = 66$.

Current trends Uncertain and strongly dependent upon developments in agricultural practice.

180 *Myosotis ramosissima*   **Early Forget-me-not**

Established strategy S–R.
Gregariousness Often loosely clumped.
Flowers Blue, hermaphrodite, perhaps usually self-pollinated.
Regenerative strategies S, ?B_s.
Seed 0.11 mg.
British distribution Absent from most of the N and W (74% of VCs).
Commonest habitats Rock outcrops 11%, limestone quarries 5%, cinder heaps/limestone pastures 4%, bricks and mortar/limestone wasteland 3%.
Spp. most similar in habitats *Saxifraga tridactylites* 93%, *Arenaria serpyllifolia* 85%, *Arabidopsis thaliana* 79%, *Medicago lupulina* 78%, *Trifolium dubium* 75%.
Full Autecological Account *CPE* page 412.

Synopsis *M.r.* is a small, winter-annual herb restricted to shallow organic soils on limestone, or to dry sandy soils, where, in both, the cover of perennials is restricted by summer drought. *M.r.* may also occur on anthills. Seeds are released during summer, and immediate germination is prevented by an after-ripening requirement. Seeds germinate in autumn and plants, which overwinter as rosettes, require vernalization before flowering. The flowering period commences in early spring and continues until the onset of drought. *M.r.* is much more common in long-established, winter-annual communities than in fugacious sites. The species shows a similar restriction to more-predictably open sites in Holland, where its complex germination biology is regarded as more conservative and less opportunistic than that of *Veronica arvensis*. The existence of a persistent seed bank is suspected. Common species of similar ecology in Britain are *Desmazeria rigida*, *Erophila verna* and *Saxifraga tridactylites*. The former is later-flowering and is presumably associated with less drought-prone sites. *E.v.* is earlier-flowering. In addition, *E.v.* and (particularly) *S.t.*, produce numerous small seeds, while *M.r.* forms fewer (often < 100), larger seeds. The reasons for the restriction of *M.r.* to predictably open sites are unknown. However, there is no evidence implicating low mobility and one local colony on railway cinders in N Derbyshire lies at least 10 km from the next known site. Populations from the limestone strata of N England appear to refer to ssp. *ramosissima*. In sandy places, especially near the sea in S England, the species is represented by ssp. *globularis*.

Current trends As a colonist of quarries, sand pits and railway banks, *M.r.* has probably increased in the recent past. However, current trends in land management are destroying many of its sites, and the species appears to be decreasing.

181 *Myosotis scorpioides* — Water Forget-me-not

Established strategy C.

Gregariousness Intermediate to stand-forming.

Flowers Blue, hermaphrodite, insect-pollinated, largely self-incompatible.

Regenerative strategies V, ?B$_s$.

Seed 0.28 mg.

British distribution Widespread (97% of VCs).

Commonest habitats Shaded mire 17%, unshaded mire 15%, still aquatic/river banks 3%, running aquatic/soil heaps 2%.

Spp. most similar in habitats *Galium palustre* 97%, *Mentha aquatica* 94%, *Cardamine amara* 89%, *Cardamine pratensis* 87%, *Solanum dulcamara* 86%.

Full Autecological Account *CPE* page 414.

Synopsis *M.s.* is a low-growing, winter-green herb of fertile mire close to open water, usually in situations where the growth of potential dominants is restricted by disturbance associated with winter flooding. Thus, *M.s.* is characteristic of ditch, pond and river margins, and may sometimes form floating rafts. The species is perhaps more tolerant of a fluctuating water-table and of submergence than the ecologically similar but smaller-seeded *Veronica beccabunga*. *M.s.* occurs in shaded sites, and may also persist at the edges of stands of taller dominant species. *M.s.* forms clonal patches by means of rhizomes and runners. Older stems break down to leave isolated daughter plants. Shoot fragments washed away by flooding readily regenerate to form new colonies, and may be important in colonizing water systems. Reproduction by seed which germinate in vegetation gaps during spring is probably of lesser importance in established colonies. However, the seeds do not float, and we suspect that they are incorporated into a persistent seed bank. The germination requirements of *M.s.* have been studied little and there is a need to ascertain whether, as in many other wetland species of similar ecology, germination is promoted by fluctuating temperatures. The extent of ecotypic differentiation in this out-breeding species is uncertain, but the progeny from selfed plants may show in-breeding depression. *M.s.* has a wide edaphic range, extending from calcareous to mildly acidic soils. However, beside nutrient- and base-poor water, generally in upland areas, *M.s.* is often replaced by the morphologically similar polycarpic perennial, *M. secunda*. The annual *M. laxa* ssp. *caespitosa*, replaces *M.s.* in annual communities on bare mud, and the hybrid between these two species is perennial and vegetatively vigorous.

Current trends Tolerant of eutrophic waters, and relatively mobile. Likely to remain common.

---

182 *Myrrhis odorata*　　Sweet Cicely

Established strategy Between C and C–S–R.
Gregariousness Can form stands even at low frequencies.
Flowers White, insect-pollinated.
Regenerative strategies S.
Seed 35 mg.
British distribution Largely restricted to the N (58% of VCs).
Commonest habitats River banks 8%, hedgerows/verges 3%, acidic woodland 2%, arable/limestone woodland 1%.
Spp. most similar in habitats *Alliaria petiolata* 90%, *Galium aparine* 73%, *Petasites hybridus* 73%, *Galeopsis tetrahit* 73%, *Impatiens glandulifera* 71%.
Full Autecological Account *CPE* page 416.

Synopsis *M.o.* is a robust garden escapee, smelling of aniseed and formerly cultivated as a pot-herb. The species is now thoroughly naturalized, particularly on shaded river banks, and has been established in semi-natural vegetation for so long that it was formerly regarded as native. The species lacks the capacity for lateral vegetative spread and forms stands in moist, fertile sites only, where the growth of dominants is restricted by factors such as occasional flooding and light shade. *M.o.* is usually associated with deep mineral soils, and the foliage is characterized by high concentrations of nitrogen and phosphorus. Despite its association with riversides, *M.o.* is not a wetland species, tending to occur on freely-draining soils at the tops of river banks. Where *M.o.* and *Impatiens glandulifera* occur together, *M.o.* occupies the areas further from the water's edge. On riverbanks the species benefits from the stout tap-root which affords very effective anchorage. Following floods, it is not unusual to observe several centimetres of exposed root in stream-side populations. *M.o.*

is frequently found on roadsides, where it survives occasional cutting (but not grazing). The ecology of the species in these sites may be similar to that of *Rumex alpinus*, also a garden escapee, which may persist near its site of introduction but shows little capacity for dispersal. In Britain, *M.o.* is rare in lowland areas, a pattern matching the montane distribution in C and S Europe, where it is native. *M.o.* appears to be entirely dependent upon seed for regeneration. When shed, the seed, which is the largest in the herbaceous flora of the British Isles, has a poorly-differentiated embryo which occupies only 10% of the seed length. The embryo grows while the seed is in a cold, imbibed condition during the winter and, as a result, seed germination is delayed until spring. The large fruits float when dry, but when imbibed generally sink within 12 h. Apart from an evident capacity to spread along river systems, *M.o.* is a poor colonist.

Current trends Uncertain. Does not appear immediately threatened, at least within upland Britain.

183 *Nardus stricta*

Mat-grass

Established strategy S.
Gregariousness Intermediate.
Flowers Green or purplish, hermaphrodite, apomictic.
Regenerative strategies V, S.
Seed 0.38 mg.
British distribution Particularly abundant in the N and W (99% of VCs).
Commonest habitats Acidic pastures 52%, acidic wasteland 10%, acidic quarries 9%, limestone pastures 8%, unshaded mire 7%.
Spp. most similar in habitats *Molinia caerulea* 90%, *Galium saxatile* 87%, *Empetrum nigrum* 82%, *Vaccinium myrtillus* 81%, *Carex pilulifera* 75%.
Full Autecological Account *CPE* page 418.

Synopsis *N.s.* forms long-lived, spreading tussocks, and is frequently a major constituent of unproductive grassland and heath vegetation on relatively free-draining acidic soils. *N.s.* is particularly abundant in upland areas of high rainfall. The foliage is low in Ca and is unpalatable to sheep, though it may be eaten in winter when other food is scarce. *N.s.* may re-colonize fired areas by seed, but adult plants are sensitive to burning and to cutting. *N.s.* is, however, tolerant of trampling, and may adopt a flattened growth form beside paths. Regeneration is effected principally through a limited growth and branching of the rhizome system followed by decay of older rhizomes, producing daughter plants. Tussocks may expand at a rate of *c.* 20 mm year^{-1}. *N.s.* also regenerates from detached pieces of rhizome. Regeneration by seed appears to be less important, particularly in closed vegetation, but may play a role in colonization of bare ground where incursions by other species have been delayed. Seed-set may be low in poor summers, particularly at high altitudes, and the majority of caryopses are dormant when freshly shed. However, a clump 200 mm in diameter may produce 2000 florets under favourable conditions. Thus, under some circumstances *N.s.* may be an effective local colonist. There are isolated records suggesting that *N.s.* has a persistent seed bank, but negative results have been obtained in other investigations and there is no evidence of seed persistence in the Sheffield region. Since most seed is shed close to the parent plant, *N.s.* is unlikely to be an effective long-distance colonist of new habitats. In parts of continental Europe, *N.s.* is less-exclusively calcifuge and populations from acidic soil have been shown to grow successfully on highly calcareous soils following the addition of calcium hypophosphate. *N.s.* is morphologically relatively constant and apomictic.

Current trends In many upland areas of Britain the growth of *N.s.* was checked by winter grazing. Now sheep are usually released onto the hills only during the summer months and, being selective feeders, they choose more-palatable species, which are often present in abundance. Thus, *N.s.* has increased dramatically in parts of N Britain and is replacing *Agrostis–Festuca* grassland on some sites in N Derbyshire. In lowland areas habitat destruction for agriculture, forestry and urban development has diminished *N.s.*.

184   *Nasturtium officinale*

Includes *N. microphyllum*, *N. officinale* and their hybrid

Established strategy C–R.
Gregariousness Often forms patches.
Flowers White, hermaphrodite, homogamous, insect-pollinated, self-compatible.
Regenerative strategies V, ?B$_s$.
Seed *N.m.*: 0.13 mg.
British distribution Aggregate common throughout British Isles.
Commonest habitats Running aquatic 31%, shaded mire 13%, still aquatic 8%, unshaded mire 6%.

Water-cress

Spp. most similar in habitats *Apium nodiflorum* 99%, *Veronica beccabunga* 97%, *Ranunculus penicillatus* 91%, *Equisetum palustre* 89%, *Epilobium parviflorum* 88%.
Full Autecological Account *CPE* page 420.

Synopsis Despite belonging to a different family, *N.o.* agg. is ecologically and morphologically very similar to *Apium nodiflorum* but extends further N in Britain. Young shoots are eaten by cows (although this may result in tainted milk) and by water-fowl. The taxon survives heavy grazing. As in the case of *A.n.*, procumbent stems root freely at the soil surface and give rise to large clones. Detached stem pieces have a remarkable ability to root and form new plants. *N.m.* × *o.* appears to have colonized whole water courses by this means. Seeds may be important for initial establishment, particularly in landlocked sites. Although seeds lack dormancy, germination probably takes place in the spring. *N.o.* agg. is apparently more ruderal than *A.n.* in the following respects: (a) roots are produced even more freely on stem fragments (often within 24 h); even the shoot of a seedling at the cotyledon stage and detached inflorescences are capable of rooting; (b) stems are exceedingly short-lived; (c) the numerous small seeds are capable of either immediate or delayed germination, and a persistent seed bank is predicted on the basis of laboratory characteristics of the seed. *N.o.* and *N.m.* × *o.* are cultivated as

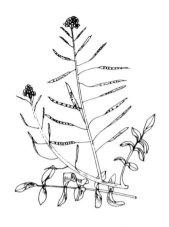

salad plants in the S and both were formerly used as medicinal herbs; in some sites the two taxa may have been introduced for this purpose.
Future trends Still common, but decreasing in aquatic habitats as a result of water pollution. Probably lost from some mire sites through eutrophication and following invasion by taller wetland species.

---

185 *Origanum vulgare*   Wild Majoram

Established strategy Between C–S–R and S–C.
Gregariousness Intermediate.
Flowers Purple, gynodioecious, not self-fertile, insect-pollinated.
Regenerative strategies B_s, V.
Seed 0.1 mg.
British distribution Widespread, but local in Scotland and N Ireland (88% of VCs).
Commonest habitats Limestone quarries 24%, limestone wasteland 13%, limestone pastures 4%, scree 3%, cliffs/rock outcrops/soil heaps 2%.
Spp. most similar in habitats *Inula conyza* 89%, *Leontodon hispidus* 72%, *Erigeron acer* 72%, *Leucanthemum vulgare* 70%,

Scabiosa columbaria 64%.
Full Autecological Account *CPE* page 422.

Synopsis *O.v.* is a tall, tufted aromatic herb of relatively dry and infertile, usually calcareous, soils. The species grows most vigorously at high external concentrations of Ca, and at low levels of supply exhibits Ca deficiency. *O.v.* exhibits only a limited capacity for lateral spread and is unable to coexist with taller, fast-growing species. As a result of its stature *O.v.* is vulnerable to grazing and is infrequent in pasture. The species tends to occur either in grassland vegetation subject to periodic burning or on rock outcrops. *O.v.* has a relatively deep root system which often bears very long and numerous root hairs and almost certainly allows the species to exploit subsoil water during periods of drought. The species is essentially of lowland distribution and differences in phenology are observed at high and low altitude. Observations in the Sheffield region suggest that in lowland sites young green shoots overwinter, while during severe winters shoot growth in upland sites may be delayed until spring. *O.v.* produces numerous small seeds which germinate in vegetation gaps during spring. *O.v.* also forms very large buried seed banks. In samples from W Europe 1–62% of individuals are male-sterile plants. Populations from disturbed vegetation show a higher incidence of male-sterility than those from more-stable sites. It would seem that some populations have experienced high selection pressures for out-breeding, and that this has led to the formation of the many variants and ecotypes of *O.v.* which have been recorded. However, it should be pointed out that, at least in the British Isles, *O.v.* has a very narrow ecological range.
Current trends The species is capable of exploiting disused quarries and other artificial calcareous habitats, and has extended its range locally along roadsides. However, *O.v.* is restricted to rather infertile sites and may be expected to decline as these habitats diminish.

186 *Oxalis acetosella*

Wood Sorrel

Established strategy Between C–S–R and S.
Gregariousness Intermediate to patch-forming.
Flowers Solitary, long-stalked, white, hermaphrodite, sparingly pollinated by insects.
Regenerative strategies V, S, ?B_s.
Seed 1.01 mg.
British distribution Common except in E England and C Ireland (100% of VCs).
Commonest habitats Scree 23%, acidic woodland 19%, scrub/ limestone woodland 10%, coniferous plantations 8%.
Spp. most similar in habitats *Sorbus aucuparia* (juv.) 84%, *Milium effusum* 78%, *Rubus fruticosus* 70%,

Hyacinthoides non-scripta 67%,
Lonicera periclymenum 66%.
Full Autecological Account *CPE* page
424.

Synopsis *O.a.* is a rhizomatous, patch-forming woodland herb with a superficial resemblance to *Trifolium repens* both in habit and in the nastic 'sleep' movements of its trifoliate leaves. *O.a.* has a slow growth rate and the leaf canopy is lower than that of most other members of the woodland ground flora. Thus, *O.a.* tends to play a subordinate role in herbaceous communities. The species is suppressed by heavy depositions of deciduous tree litter, and for this reason is often restricted to raised areas of the woodland floor such as tree bases and small hummocks. *O.a.* is, however, very shade-tolerant, surviving in dimly-lit rock crevices and maintaining dry matter production and setting seed under moderate shade. *O.a.* is also physiologically attuned to sites of low illumination and relatively low temperatures. The rhizome of *O.a.* lies on the soil surface, and the root system is typically shallow. Thus, *O.a.* exploits a relatively thin layer of soil and is largely restricted to continuously moist habitats. *O.a.* is found on a wide range of soil types from highly calcareous to acidic, but is more frequent on the latter. The leaves are unusually low in P, Ca, Mg and Fe. It is suspected that in some calcareous habitats the niche exploited by *O.a.* is occupied by species such as *Hedera helix*; similarly, *O.a.* may be replaced by *Deschampsia flexuosa* in woodlands on the poorest acidic soils. *O.a.* also occurs in semi-natural grassland, usually on N-facing slopes, but tends to show less winter growth in such habitats than in more-sheltered woodland sites. *O.a.* appears to be little grazed, perhaps because its leaves contain oxalic acid and are poisonous to livestock. *O.a.* regenerates mainly as a result of clonal growth, rhizome extension in beech woods averaging 100 mm year^{-1}, and through the production of cleistogamous flowers. A persistent seed bank is reported. *O.a.* is phenotypically plastic with respect to size and the proportion of resources allocated to rhizome and root. The leaves of plants from open sites differ in anatomy, colour and longevity from those of shaded sites.
Future trends *O.a.* is one of the few woodland species which exploits coniferous plantations. However, *O.a.* is most characteristic of long-established woodland sites, and appears to be decreasing in abundance.

187 *Papaver rhoeas* Field Poppy

Established strategy R.
Gregariousness Sparse to intermediate.
Flowers Usually red, hermaphrodite, insect-pollinated, self-incompatible.
Regenerative strategies B$_s$.
Seed 0.09 mg.
British distribution Widespread except in N Scotland (94% of VCs).
Commonest habitats Arable 10%, soil heaps 4%, cinder heaps/ paths/verges 2%.
Spp. most similar in habitats *Sinapis arvensis* 92%, *Myosotis arvensis* 91%, *Spergula arvensis* 89%, *Veronica persica* 89%, *Fallopia convolvulus* 86%.
Full Autecological Account *CPE* page 426.

Synopsis *P.r.* is a summer-annual restricted to disturbed artificial habitats (particularly arable land), and may have been introduced from S Europe along with agricultural crops. Seed remains have been identified which date from the Late Bronze Age. *P.r.* is more frequent in cereal than in root crops. With the advent of selective weed-killers, *P.r.* is now mainly restricted to fallow ground and to disturbed sites not managed for agriculture. *P.r.* is infrequent in grazed habitats. Nevertheless, the species contains toxic alkaloids and is unpalatable to livestock. Seeds also contain toxins. *P.r.* is sensitive to trampling, but may occasionally survive harvesting of the crop and produce new flowering shoots. Regeneration is entirely by seed which is rapidly shed from a many-seeded capsule by a censer mechanism. Seedlings originating in autumn may survive mild winters, but most-effective regeneration is in spring. *P.r.* is capable of survival in an unpredictably disturbed landscape, and dispersal in time is effected by the formation of a buried seed bank. Thus, the former boundaries of an arable field which had been grassed for over 20 years have been recognized by the presence of poppies derived from a persistent seed bank. *P.r.* may be dispersed considerable distances and occasionally occurs as an alien introduction from abroad. *P.r.* shows considerable plasticity, both in form and in number of seeds set. Under unfavourable conditions *P.r.* may produce from a single small flower as few as four seeds. Robust plants are much branched, perhaps with over 400

flowers; the mean number of seeds per capsule has been given as 1360. The species hybridizes with *P. dubium*.
Current trends Formerly a common plant in lowland areas, often forming large populations in arable fields. Still common, but decreasing both in number of sites and in size of populations. *P.r.* is beginning to become characteristic of waste places rather than of arable fields. However, unlike the related *P. dubium*, which has recently increased by colonizing dry, sandy and cindery railway banks, *P.r.* has remained strictly confined to productive ruderal sites. As *P.r.* becomes more restricted to small populations, we predict selection for increased in-breeding.

188 *Petasites hybridus* Butterbur

Established strategy C.
Gregariousness Stand-forming at low frequencies.
Flowers Florets pinkish-violet, dioecious, insect-pollinated.
Regenerative strategies V, W.
Seed 0.26 mg.
British distribution Widespread except in N Scotland (99% of VCs).

Commonest habitats River banks 10%, walls/limestone woodland 3%, still aquatic/shaded mire/soil heaps/acidic woodland 2%.
Spp. most similar in habitats *Impatiens glandulifera* 76%, *Myrrhis odorata* 73%, *Alnus glutinosa* (juv.) 70%, *Alliaria petiolata* 69%, *Festuca gigantea* 67%.

Full Autecological Account *CPE* page
428.

Synopsis *P.h.* is a robust rhizomatous
herb of open or partially shaded river
and stream terraces, usually growing
on moist, fertile alluvial soils. The
leaves, which may approach 1 m in
diameter, are supported by petioles up
to 2 m in length. They are by a
considerable margin the most massive
leaves found within the native British
flora, and are in marked contrast with
those of other competitive dominants
(e.g. *Epilobium hirsutum*) which create
dense shade by means of a large
number of small leaves supported on
an erect stem. Mobilization of reserves
in the stout rhizome allows a dense leaf
canopy to be produced rapidly during
late spring and early summer. In many
habitats *P.h.* is associated with species
such as *Ranunculus ficaria* which have
complementary vernal phenologies.
Because of its morphology, *P.h.* is
particularly vulnerable to grazing,
trampling, cutting and other
disturbances during summer and
autumn. The roots of *P.h.* abscise at
the end of the growing season and are
replaced each spring. *P.h.* regenerates
mainly by means of rhizome growth to
form extensive patches. Older
rhizomes break down to leave isolated
daughter plants, and rhizome pieces
may be transported to other sites
during flooding of river banks and
periods of erosion. *P.h.* is dispersed
effectively along waterways in this
manner and, since in some drainage
systems all colonies are of the same
sex, the possibility of a common
vegetative parentage seems likely. The
tendency for *P.h.* to be restricted to
river-sides may arise from the poor
mobility of its rhizome fragments.
P.h., which flowers in early spring,
before leaf expansion, is dioecious.
Where male and female plants coexist,
abundant seed is often produced. The
seeds are short-lived and
wind-dispersed, and germinate
immediately after shedding in spring or
in early summer. *P.h.* does not form a
persistent seed bank, and
establishment of new colonies by seed
is probably rare. Indeed, since the
female plant has a restricted
geographical distribution, regeneration
by seed is unlikely over much of
Britain. Female plants are also
infrequent in parts of N Europe.
Where only male plants of *P.h.* are
found, they may have been introduced
to safeguard a good early source of
nectar for bees. A further ssp. is
recorded from Europe.
Current trends Although an occasional
colonist of roadsides and railway
banks, *P.h.* has rather specific habitat
requirements, and seems likely to
remain only locally abundant.

Established strategy C.
Gregariousness Stand-forming.
Flowers Green, hermaphrodite, wind-pollinated, self-incompatible.
Regenerative strategies V, ?B$_s$.
Seed 0.67 mg.
British distribution Widespread (100% of VCs).
Commonest habitats River banks 13%, shaded mire 9%, running aquatic/unshaded mire 8%, still aquatic/walls 3%.
Spp. most similar in habitats Insufficient data.
Full Autecological Account *CPE* page 430.

Synopsis *P.a.* is a tall, strongly rhizomatous grass which forms extensive stands in mire and along water-courses. This potential dominant produces a dense canopy of broad leaves in summer, and is highly productive in unmanaged, undisturbed sites. The growth of other flowering plants may be suppressed also by the considerable quantities of litter which accumulate when the shoots die back each autumn. *P.a.* is a flood-tolerant species, and an increased amount of aerenchyma is produced in the roots after inundation. However, the species is deeply rooted and is most characteristic of habitats which are incompletely waterlogged in summer, e.g. river banks and the upper drier parts of topogenous mire. The species tends to be confined to drier sites than, for example, *Glyceria maxima* and *Phragmites australis*, and in aquatic habitats is mainly recorded from sites with flowing water. Although normally found in unmanaged sites, *P.a.* is tolerant of intermittent cutting. The young shoots, which are palatable to stock, may be exploited for hay and for grazing on land subject to flooding. However, the plant contains alkaloids and may be highly toxic to sheep. The species regenerates mainly by rhizomes to form large clonal patches. Nodal portions of vegetative shoots are also capable of regeneration, and fragments which float offer a potential means of colonizing river systems. Despite being self-incompatible, seed production is often prolific, but there is frequent infection of the inflorescence by ergot (*Claviceps purpurea*). Distribution by means of floating seeds is probably only important in the colonization of new sites. Germination, which in the field occurs in spring, may be accelerated by scarification, dry storage, stratification or fluctuating temperatures, and the formation of a persistent seed bank is suspected. The less-robust ssp. *rotgesii* also occurs in Europe.

Current trends Uncertain. Not particularly imperilled by current patterns of land-use.

Includes sspp. *pratense* and *bertolonii*

Established strategy *p.* between C–S–R
and C–R; *b.* between S and C–S–R.
Gregariousness Sparse to intermediate.
Flowers Green, hermaphrodite, wind-
and typically cross-pollinated.
Regenerative strategies S.
Seed *p.*: 0.45 mg.
British distribution Aggregate
widespread (100% of VCs).
Commonest habitats Meadows 53%,
arable/enclosed pastures 10%,
manure and sewage waste/verges
8%.
Spp. most similar in habitats (Relating
mainly to *p.*) *Cynosurus cristatus*
87%, *Bellis perennis* 85%,
Alopecurus pratensis 84%, *Festuca
pratensis* 80%, *Rhinanthus minor*
79%.
Full Autecological Account *CPE* page
432.

Synopsis Ssp. *p.* is a tall, tufted,
winter-green grass now close to
extinction in what are presumably its
native habitats (water meadows and
other low-lying moist grassland), but
widely cultivated in fertile hay
meadows and pastures, from which it
escapes frequently. The species is very
cold-tolerant and some growth may
occur during winter, with tillers being
produced in spring and, to a lesser
extent, in autumn. Because of this
phenology ssp. *p.* can withstand some
grazing during winter and early spring.
Like other grasses producing tall nodal
stems in summer, ssp. *p.* grows poorly
if subjected to heavy summer grazing,
and is perhaps best suited to regimes
involving only occasional cutting or
grazing. Ssp. *p.* is typical of moist
habitats, and can survive short periods
of waterlogging in winter. This taxon is
intolerant of trampling. Although
individual plants may survive for at
least 20 years, observations in the
Sheffield region suggest that ssp. *p.* is
only a transient constituent of
meadows and pastures. This is

probably related to the absence of a
buried seed bank and to the virtual
absence of vegetative spread in this
laxly tufted plant, both features which
limit the capacity to colonize gaps in
vegetation. Ssp. *p.* flowers during its
first summer and, like *Arrhenatherum
elatius*, is unusual among British
pasture grasses in not having a chilling
requirement for floral induction. Ssp.
b. is a much smaller, relatively long-
lived plant which, within the Sheffield
region, is particularly characteristic of
sandy soils and rocky ground in
calcareous pasture. In many respects
its ecology is similar to that of ssp. *p.*,
but ssp. *b.* is generally found in less-
fertile and drier habitats. Although
sometimes cultivated, ssp. *b.* is native
in many sites and shows some capacity
for vegetative spread. It has been
claimed that seed may persist in the
soil for 20 years but this is at variance
with data for ssp. *p.* and is in need of
confirmation.
Current trends Ssp. *p.*: nine cultivars
are recommended and over 1400 t of
seed are purchased each year for use in
short-term grassland and, to a lesser
extent, in turf. A decrease in the
amount sown would drastically reduce
the abundance of ssp. *p.*.

Approximately 20 t of seed of ssp. *b.* are supplied annually for use in permanent pasture or turf. It is not known to what extent the sowing of these cultivated strains compensates for the loss of native populations in semi-natural habitats.

191 *Phragmites australis*   Reed, Common Reed

Established strategy C.
Gregariousness Forms extensive stands.
Flowers Purplish, hermaphrodite, wind-pollinated.
Regenerative strategies V, W.
Seed 0.12 mg.
British distribution Widespread, particularly in the S and E (100% of VCs).
Commonest habitats Coal mine waste/river banks 3%, unshaded mire/acidic wasteland 2%, still aquatic 1%.
Spp. most similar in habitats Insufficient data.
Full Autecological Account *CPE* page 434.

Synopsis *P.a.* is a long-lived (perhaps to 1000 years), wetland grass with annual bamboo-like stems and rhizomes which may remain functional for *c.* 5 years. *P.a.* is the tallest non-woody species in the British flora (to over 3 m), forms a dense canopy with, in places, > 100 shoots m^{-2} and shows extensive rhizome growth. A covering of persistent litter may also accumulate on the soil surface except in sites liable to cutting or flooding. As a result, *P.a.* forms virtual monocultures by competitive exclusion in more fertile sites. *P.a.* is susceptible to intensive grazing, cutting or trampling. Often damaged by ground frost in spring and showing its most vigorous growth in warmer climates. *P.a.* has well-developed aerenchyma and shows slightly greater persistence in flowing water than *Glyceria maxima*. *P.a.* can regenerate after occasional ploughing. *P.a.* spreads vigorously by means of rhizomes, and up to 25% of shoots may produce an inflorescence containing hundreds of spikelets. However, seed-set is usually very low and the young seedlings are very vulnerable to flooding. Shoots detached as a result of disturbance (e.g. maintenance work on drainage ditches and winter flooding) float readily and are capable of regenerating. Thus, dispersal and establishment often involves fragmentation rather than the wind-dispersal of seeds. However, effective long-distance dispersal, apparently by wind, does occur. *P.a.* was one of the first colonists of Krakatau and is frequently an early colonist of spoilage in the S Yorks coal-field. Existence of a persistent seed bank remains uncertain. Apart from being one of the world's most widely distributed angiosperms, *P.a.* is also one of the world's worst weeds, capable of blocking up waterways and drainage canals. The species is also of considerable value to humankind. In Britain the persistent stem litter was formerly extensively cropped during winter for thatching, and is still harvested locally. *P.a.* has a high protein content and the species is used for fodder, fuel, fertilizer and paper-making. Although ecologically wide-ranging and to some extent phenotypically plastic and existing as a number of cytotypes, *P.a.* shows comparatively little morphological variability in Britain. In the Netherlands a peatland ecotype with short shoots and high shoot density has been distinguished from a river margin form which has fewer larger shoots and is more tolerant of salt. These genotypes of *P.a.* with long stems replace other tall riverside species beside brackish water. Within the Sheffield region the only areas where

P.a. is more abundant than *Typha latifolia* and *Glyceria maxima* are low-lying 'warp' lands formerly flooded with brackish water, colliery flashes, where Na and Cl concentrations may also be high, and on the magnesian limestone, where abnormally high levels of Mg in the water may be predicted.

Current trends *P.a.* is an effective colonist of some new habitats, particularly colliery waste, and may even have been planted in some of its extant sites. Severe losses of *P.a.* have occurred within recent historical times through the widespread drainage of wetlands. Also, in some areas *P.a.* appears to have been replaced by *Glyceria maxima*, which may perhaps exploit better than *P.a.* the lowered water-table associated with increased drainage. A continuing decline in the abundance of *P.a.* is predicted.

192 *Pimpinella saxifraga*   Burnet Saxifrage

Established strategy S.

Gregariousness Intermediate.

Flowers White, mostly hermaphrodite, insect-pollinated.

Regenerative strategies S.

Seed 1.2 mg.

British distribution Most areas except N Scotland and N Ireland (94% of VCs).

Commonest habitats Limestone pastures 21%, limestone wasteland 19%, limestone quarries 8%, lead mine spoil/rock outcrops/scree 3%.

Spp. most similar in habitats *Galium verum* 90%, *Avenula pubescens* 88%, *Stachys officinalis* 86%, *Briza media* 86%, *Carex caryophyllea* 85%.

Full Autecological Account *CPE* page 436.

Synopsis *P.s.* is a rosette-forming, winter-green herb of short turf and rocky habitats on base-rich,

unproductive soils. *P.s.* is low-growing and tends to be suppressed in derelict or fertilized grassland. Restricted to unproductive plots in the *PGE*, *P.s.* exhibits a capacity for sustained growth during summer, and the distribution extends into dry, but not severely droughted, habitats, where the long tap-root provides access to moisture in soil crevices. In non-flowering plants there is a basal rosette consisting of a small number of leaves, but by the time flowers are produced these have normally withered and have been replaced by a small quantity of stem foliage. Thus, even where *P.s.* attains high population densities, the contribution to the total shoot biomass of the community is small. The capacity of *P.s.* to produce offsets is less than that of many other small tap-rooted species, e.g. *Sanguisorba minor* and *Scabiosa columbaria*. Consequently, *P.s.* is even more dependent than these species upon seedling establishment for the maintenance of populations and the colonization of new sites. In common with those of *Succisa pratensis*, fruits develop late in the growing season and, in Britain, many are in a green state at the onset of winter. The seeds have the capacity to survive ingestion

by cattle, and this may be important in pasture sites. As with many other members of the Umbelliferae, seeds contain a small embryo and have a chilling requirement for germination. In the field, germination occurs in a synchronous burst in spring. Seedlings examined on a N-facing slope in N Derbyshire appeared at similar densities in closed turf and in gaps, and showed comparable mortalities (*c.* 50%) over the initial two years. *P.s.* has a low colonizing ability, and is regarded as an indicator species of old grassland. Although mainly found over calcareous strata, *P.s.* has been observed occasionally in a number of grassland habitats on non-calcareous rocks. *P.s.* varies in size, pubescence, leaf shape and flower colour. The relationship, if any, between this polymorphism and the ecological distribution of *P.s.* is not known.
Current trends Decreasing, at least in many lowland areas. The fate of this species will be closely related to that of old calcareous grasslands.

193 *Plantago lanceolata*   Ribwort

Established strategy C–S–R.
Gregariousness Intermediate.
Flowers Brown, protogynous, hermaphrodite or gynodioecious, self-incompatible, wind-pollinated and visited by insects.
Regenerative strategies B_s, V.
Seed 1.9 mg.
British distribution Widespread (100% of VCs).
Commonest habitats Meadows 51%, rock outcrops/limestone pastures 40%, lead mine spoil/limestone quarries 37%.
Spp. most similar in habitats *Trisetum flavescens* 85%, *Festuca rubra* 75%, *Cerastium fontanum* 74%, *Linum catharticum* 73%, *Dactylis glomerata* 67%.
Full Autecological Account *CPE* page 438.

Synopsis *P.l.* is a frost-tolerant, winter-green, rosette-forming herb capable of flowering in its first year and, in some genotypes, exhibiting a potential life-span of > 12 years. Although largely absent from woodland and wetlands, *P.l.* is found in a very wide range of habitats and the shoot is extremely plastic. There are ascending linear lanceolate leaves in dense vegetation, such as hay-meadows, and short ovate leaves forming prostrate rosettes in exposed or closely grazed sites. Although many roots are superficial, some penetrate deeply and appear to confer drought avoidance in dry grassland and on S-facing slopes. In the *PGE*, *P.l.* is 'chiefly associated with poor, exhausted soils' and is scarce in plots treated heavily with nitrogen. The foliage of plants collected from the Sheffield region is low in N, P and Fe, but high in Na and Ca and contains sorbitol. In turf microcosms, roots of *P.l.* were heavily colonized by VA mycorrhizas, and infection caused increases of up to 12-fold in seedling yield. *P.l.* is palatable to sheep, but because of its appressed rosettes in short turf the foliage is not readily eaten in cattle pastures. The flowering heads are often heavily predated. *P.l.* has been described as one of the world's 12 most successful colonizing species, and as one of the world's most successful weeds. This status is achieved despite the fact that *P.l.* is seldom a principal weed of any one crop. The production of buds from the stem affords a method of vegetative regeneration in some established colonies. Connections between ramets subsequently decay. *P.l.* is also capable of regeneration from root fragments, but it is uncertain to what extent this is of importance in the field. Thus, *P.l.* has a limited capacity for vegetative spread and much flexibility in regeneration by seed. Some seed is shed in summer and some may overwinter on the parent plant. In one study, a decline in mean seed weight was noted in response to trampling and soil compaction. Regeneration from seeds freshly dispersed into bare areas, such as those occurring in overgrazed grassland, may occur in spring and autumn. Seeds have no well-defined dispersal mechanism, but may survive ingestion by animals; seeds also produce a sticky mucilage when wet. *P.l.* is mainly transported as a result of man's activities. The species was recorded among the six most common roadside plants in a large-scale survey in Denmark, and seed may be a contaminant of hay crops. *P.l.* also forms a buried seed bank. In addition to the high level of morphogenetic plasticity, *P.l.* is also extremely variable genetically. Heavy-metal-tolerant populations occur in Derbyshire, and ecotypes differing in longevity, rosette number, leaf number and size, inflorescence shape and height, seed size and seed dormancy have been distinguished in comparisons between dune grassland and hay-meadow populations. Many other population studies have been carried out, and the evolution of ecotypes is potentially rapid. However, as the result of a large and detailed demographic study, it has been concluded that phenotypic variation is much more important than genotypic differences in determining the nature of grassland populations of *P.l.*. The extremely wide range of morphological variation within this out-breeding species has resisted taxonomic classification.

Current trends *P.l.* has long been associated with human activities, and its abundance increased dramatically after early deforestation. *P.l.* is still very common, and appears capable of widespread persistence despite current changes in land-use.

194　*Plantago major*

Rat-tail Plantain

Ssp. *major*

Established strategy Between R and
　C–S–R.
Gregariousness Intermediate.
Flowers Green, hermaphrodite, wind-
　pollinated, protogynous but capable
　of full self-fertilization.
Regenerative strategies B_s.
Seed 0.24 mg.
British distribution Widespread (100%
　of VCs).
Commonest habitats Paths 67%, bricks
　and mortar 38%, soil heaps 24%,
　arable 21%, cinder heaps 20%.
Spp. most similar in habitats *Poa
　annua* 89%, *Matricaria
　matricarioides* 78%, *Capsella bursa-
　pastoris* 68%, *Lolium perenne* 66%,
　Polygonum aviculare 59%.
Full Autecological Account *CPE* page
　440.

Synopsis *P.m.* is a rosette-forming herb
found in a wide range of relatively
fertile, disturbed habitats. It is
exceptionally resistant to trampling
and is characteristic of paths, tracks
and gateways, where it usually forms
appressed rosettes and is often
associated with *Matricaria
matricarioides* and *Poa annua*. In
closed vegetation the foliage is held
erect. *P.m.* is frequent on compacted
soils with impeded drainage, but is
generally absent from sites subject to
prolonged summer waterlogging.
These two growth forms are to some
extent under genotypic control. Erect
genotypes are less tolerant of mowing
and trampling than prostrate genotypes
from lawns, and appear to be better
attuned to tall vegetation. *P.m.* is less
frequent on dry soils than *P.
lanceolata*, and the foliage appears to
be frost-sensitive. In grasslands *P.m.* is
largely confined to short, heavily
trampled turf, and in this habitat as
elsewhere is strongly reliant upon seed
for regeneration. *P.m.* germinates and
shows effective root penetration in
compacted soil but often fails to

establish on loose or sandy soil.
Establishment occurs in bare areas
resulting from trampling or
overgrazing in spring, but seedlings
also appear sporadically during
summer; *P.m.* often develops a large
persistent seed bank. Although the
leaves may be eaten by stock, the
fruiting heads are not consumed. Seed
production may be prodigious, and an
exceptionally large plant may release
14 000 seeds. Under optimal
conditions seeds may be produced
within six weeks. In one study an
increase in average seed size was
detected in plants subjected to
trampling. Plants may produce
daughter rosettes by means of lateral
buds, and plants with several rosettes
often have a higher seed output than
those with a single rosette. On
occasion, the production of side-shoots
follows mechanical damage, but
vegetative regeneration is less
important than in *P. lanceolata*. It is
uncertain whether *P.m.* is capable of
resprouting from shoot fragments
created by ploughing of arable land or
by heavy trampling on paths. Seeds,
which are mucilaginous when wet, are
usually retained over winter on the
parent plant and are probably
dispersed by humans, sometimes in

crops. *P.m.* has colonized many parts of the world and, as a result, is identified as one of the world's most successful weeds. Two other ssp. of *P.m.* are described in Europe, including ssp. *intermedia*, which grows in damp, usually saline, sites.

Current trends Further increases in abundance and ecological range may be expected.

195 *Poa annua* — Annual Poa, Annual Meadow-grass

Established strategy Typically R.
Gregariousness Intermediate.
Flowers Green, hermaphrodite, homogamous, wind-pollinated; self-compatible and usually self-pollinated.
Regenerative strategies S, B_s and V in some biotypes.
Seed 0.26 mg.
British distribution Widespread (100% of VCs).
Commonest habitats Paths 79%, arable/bricks and mortar 73%, manure and sewage waste 48%, soil heaps 45%.
Spp. most similar in habitats *Matricaria matricarioides* 93%, *Plantago major* 89%, *Capsella bursa-pastoris* 88%, *Polygonum aviculare* 84%, *Atriplex patula* 73%.
Full Autecological Account *CPE* page 442.

Synopsis *P.a.* is regarded by some as only doubtfully native to the British Isles. It is arguably the most successful ruderal species in the British flora, with a habitat range which includes arable fields and agricultural and urban grassland. The species is among the first colonists during 'sward deterioration' of sown grassland and is a common colonist of both winter- and summer-sown arable crops. *P.a.* also occurs in the second phase of colonization of new polders in Holland following desalination. Leaves of *P.a.* have high concentrations of N and P but are low in Ca. Many factors contribute to the wide ecological amplitude, including: (a) Considerable genotypic variation and phenotypic plasticity. Most characteristically *P.a.* is a small annual plant of disturbed ground and in extreme circumstances may complete its life-cycle within six weeks. In contrast, perennial genotypes tend to occur in less disturbed, usually grassland, sites. Here they may not flower until their second year and can show stoloniferous growth. In some wetland sites *P.a.* may even approach *Catabrosa aquatica* in size. Twofold intraspecific variation in nuclear DNA amount (at constant ploidy) has been recorded within one population. (b) Flexibility of regeneration. This has several components, including the capacity for immediate germination and the formation of a buried seed bank, germination throughout most of the year, often with a peak in autumn, and survival of seedlings and young plants in winter conditions. Early-flowering individuals under unfavourable conditions may produce as few as ten seeds, only 0.05% of the seed production possible under favourable conditions. (c) Tolerance of grazing, trampling and herbicides. Unlike most ruderals, *P.a.* is tolerant of severe defoliation and can set seed when cut regularly to as low as 6.5 mm. Also resistant to trampling,

P.a. is the most common species on paths within the Sheffield region. *P.a.* is known to have developed resistance to some herbicides.

Current trends *P.a.* appears to be increasing. It combines success as a weed with palatability as a pasture grass. Now sown for amenity purposes.

196 *Poa pratensis* — Smooth-stalked Meadow-grass

Field data may include *P. angustifolia* and *P. subcaerulea*

Established strategy C–S–R.
Gregariousness Typically intermediate.
Flowers Green, hermaphrodite, wind-pollinated or apomictic.
Regenerative strategies V, ?B$_s$.
Seed 0.25 mg.
British distribution (Including *P.a.* and *P.s.*). The British Isles (100% of VCs).
Commonest habitats Limestone wasteland 73%, verges 68%, limestone quarries 53%, cinder heaps 37%, bricks and mortar 33%.
Spp. most similar in habitats *Dactylis glomerata* 78%, *Festuca rubra* 72%, *Taraxacum* agg. 67%, *Achillea millefolium* 59%, *Plantago lanceolata* 56%.
Full Autecological Account *CPE* page 444.

Synopsis *P.p.* is a wide-ranging grassland species exploiting both fertile and relatively infertile habitats, and occurring on moist or on dry soils. *P.p.* is frequent in pastures, is tolerant of grazing and trampling, and is potentially high-yielding. The species rarely dominates vegetation, but is widespread as a minor component in grassland. In the *PGE*, *P.p.* is 'present on most plots and is tenacious of its position in spite of the very small amounts that usually occur'. The attached tiller rhizome system may be of benefit in these managed sites since, after cutting, the nutritional independence of the tillers may end

following a reinstatement of the shoot/rhizome system. Another important feature of the rhizomes is their ability to penetrate deeply into the soil. This may explain in part the capacity of *P.p.* to exploit dry soils and sites with uneven soil depth, e.g. rock outcrops, quarry heaps and cinder tips. *P.p.* can produce exceedingly long leaves in tall vegetation. This is apparent in productive tall herb communities on wasteland and unmanaged road verges, where *P.p.* frequently persists as a minor component dominated by, e.g., *Arrhenatherum elatius* and *Elymus repens*. *P.p.* is cold-tolerant and grows appreciably during winter. The '*Poa pratensis* group' includes a number of ecologically and morphologically separable complexes which may be treated as species. In Britain these are *P.p.*, *P.a.* and *P.s.*. Each has arisen as a result of 'hybridization of various ancestral species followed by an increase in chromosome number and the disruption of the sexual breeding system' and inter-crossing between them is probably rare. Each 'species' has the capacity for extensive rhizomatous spread to form large clonal patches and is represented by an aneuploid series. No morphological separation of cytotypes of British species has been attempted and plants are in any event phenotypically plastic. Different cytotypes may cohabit in the same site and *P.a.* and *P.s.* may sometimes coexist with *P.p.*. Seed production is largely apomictic. A persistent seed bank has been reported, but numerous observations in the Sheffield region have failed to confirm this finding. *P.p.* is by far the most common segregate, and the only one commercially sown (at a rate of *c.* 400 t year^{-1} for permanent pasture or turf-grass). *P.a.* is found in dry, less-productive, mainly lowland sites, usually growing on calcareous soils with *Brachypodium pinnatum*, often on freely-drained railway banks. The

period of spring growth is short, and within the Sheffield region the species is absent from grazed sites, though tolerant of burning. *P.s.* is found primarily in two moist but unproductive types of habitat, namely montane grassland and dune slacks. *P.s.* is probably at its S geographical limit in Britain. *P.s.* produces fewer larger seeds than *P.p.*.

Current trends *P.p.* has a modest colonizing ability, but because it is only sown in small quantities and not generally recommended for agricultural use, the taxon is possibly decreasing. *P.a.* and *P.s.* are mainly restricted to older habitats and may also be decreasing. Changes in land-use which decrease the area of infertile grassland are likely to reduce the degree of geographical isolation between populations of *P.p.*, *P.a.* and *P.s.*. This brings the prospect of hybridization, with some breakdown of the ecological, genetic and morphological discontinuities between the three species. In some sites where *P.a.* and *P.p.* occur together the presence of intermediates is already suspected.

197 *Poa trivialis*

Rough Meadow-grass, Rough-stalked Meadow-grass

Established strategy Between C–S–R and C–R.
Gregariousness Intermediate.
Flowers Green, hermaphrodite, wind-pollinated.
Regenerative strategies V, B$_s$.
Seed 0.09 mg.
British distribution Widespread (100% of VCs).
Commonest habitats Meadows 93%, river banks 71%, soil heaps 67%, verges 66%, manure and sewage waste 53%.
Spp. most similar in habitats *Urtica dioica* 74%, *Ranunculus repens* 63%, *Alnus glutinosa* (juv.) 58%,

Cardamine flexuosa 57%, *Epilobium hirsutum* 53%.

Full Autecological Account *CPE* page 446.

Synopsis *P.t.* is a relatively fast-growing, winter-green grass with low-growing stolons, found in a wide variety of fertile habitats including wetland, woods, grassland and arable fields. *P.t.* is a palatable grass exploited for hay, although it is now little sown. Despite low persistence under close mowing and susceptibility to trampling, the species is a useful turf-grass and about 40 t of seed are sown annually. *P.t.* appears to be moderately shade-tolerant, typically forming an understorey below taller vegetation, and also occurring in woodland. In this latter habitat, the low stature and weak physical structure are ill-suited for emergence through tree litter, which exerts a strong negative influence on *P.t.*. *P.t.* also has a very shallow root system and is drought-sensitive. Foliage persists longer, and stolons are more robust, in unshaded but permanently moist habitats. The shoot phenology (75% of its annual yield is produced before June) is complementary to that of tall herbs such as *Urtica dioica*, with which *P.t.* often coexists. In pasture *P.t.* usually occupies a stratum below the dominant grasses. Despite its cold-tolerance, *P.t.* exhibits little winter growth; in accordance with its relatively low nuclear DNA amount, little growth occurs before April. *P.t.* exhibits a degree of heterophylly, the culm leaves being much larger than those of the low-growing stolons. There is evidence of a light requirement for germination, and natural seed populations are least dormant in spring and autumn. However, seeds generally mature early and show little dormancy. Rapid germination is possible across a wide temperature range and huge cohorts of seedlings appear on bare soil each autumn. Individuals originating in autumn produce more-numerous and heavier seeds than those in spring. Seeds which become buried in the soil are capable of persistence. In lawns, and in other situations where seeding is prevented, *P.t.* frequently spreads vegetatively by means of stolons. Plants can withstand shallow burial, and regeneration from shoot fragments occurs on arable land, though here (and elsewhere) regeneration by seed is important. *P.t.* is a weed of winter cereals, is a major problem in herbage seed crops and is often one of the first colonists during the 'sward deterioration' of cultivated grassland which often has a deleterious effect on rye-grass yield. Although *P.t.* is rarely if ever a dominant, the species is arguably the most successful subordinate constituent of plant communities in the British Isles. At present it is unclear to what extent this is due to the evolution of ecotypes, to phenotypic plasticity or to the capacity of *P.t.* to occupy similar niches in a wide range of vegetation types. Seeds of arable populations appear to be markedly more dormant than seeds collected from grassland. Seeds from open arable habitats germinate poorly when exposed to a ratio of far-red/red irradiance similar to that found in natural shade, while seed from closed vegetation is less inhibited. Another ssp. is recorded from Europe.

Current trends Apparently increasing in grassland and arable land and a frequent colonist of artificial habitats.

Established strategy S.
Gregariousness Sparse.
Flowers Blue, purple or white,
   hermaphrodite, self- or
   insect-pollinated.
Regenerative strategies ?S.
Seed 1.72 mg.
British distribution Widespread,
   especially in the S (97% of VCs).
Commonest habitats Limestone
   pastures 16%, scree 9%, limestone
   wasteland 7%, lead mine spoil 5%,
   limestone quarries 2%.
Spp. most similar in habitats *Potentilla
   sterilis* 93%, *Helianthemum
   nummularium* 89%, *Avenula
   pratensis* 89%, *Carex caryophyllea*
   88%, *Primula veris* 87%.
Full Autecological Account *CPE* page
   448.

Synopsis *P.v.* is a low-growing herb
with a woody base, found in short,
unproductive calcareous turf, where
dominance by more-robust species is
restricted by mineral-nutrient stress
and grazing. The species is also found
occasionally on base-rich soils
associated with non-calcareous strata.
In turf, *P.v.* tends to occupy the lower
levels of the leaf canopy, and we
suspect that it is relatively
shade-tolerant. However, *P.v.* reaches
highest frequencies in more-open sites
at the stabilized margins of limestone
screes. The species forms small patches
but the branches, although prostrate,
do not root and there is no evidence of
vegetative regeneration. Thus,
effective regeneration occurs by seeds
which germinate mainly in spring. No
persistent seed bank has been
detected. The seed is large, with an
elaiosome and is dispersed by ants.
However, as in the case of many other

species with this dispersal mechanism,
seed mobility appears to be localized.
Although *P.v.* is found in many old
quarries and on lime-enriched road
verges across acidic moorland, the
species seldom colonizes recent
artificial habitats and is regarded as an
indicator species of ancient grassland.
P.v. is the most common and
ecologically wide-ranging species of the
genus in Europe, and British material
is genetically and biochemically
diverse. *P.v.* is also morphologically
variable and has been subdivided
recently into four ssp., represented in
the British Isles by ssp. *collina*. *P.v.*
hybridizes with two other *Polygala*
species in S England. On acidic soils
P.v. is replaced by *P. serpyllifolia*, and
in only one site in the Sheffield region,
a species-rich, non-calcareous pasture
of intermediate pH, has the ecological
range of the two been observed to
overlap.
Current trends Decreasing,
particularly in lowland areas as a result
of habitat destruction and the
relaxation of grazing in old grassland.

Established strategy C–R.
Gregariousness Stand-forming.
Flowers Pink, hermaphrodite, insect-pollinated.
Regenerative strategies (V), ?B_s.
Seed 4.48 mg.
British distribution Widespread, especially in the C and S (99% of VCs).
Commonest habitats Unshaded mire 10%, still aquatic 6%, shaded mire/river banks 3%.
Spp. most similar in habitats *Alisma plantago-aquatica* 98%, *Eleocharis palustris* 97%, *Glyceria maxima* 95%, *Hydrocotyle vulgaris* 94%, *Typha latifolia* 92%.
Full Autecological Account *CPE* page 450.

Synopsis *P.a.* is a robust, rhizomatous herb capable of exploiting both aquatic and terrestrial habitats apparently by virtue of its extreme phenotypic plasticity. The species behaves as an aquatic with glabrous, floating leaves with functional stomata restricted to the upper surface. In mire or on ditch and river banks, however, the plant has hairy leaves with a much higher stomatal frequency on the lower surface. *P.a.* exploits fertile habitats, and the rhizomatous shoot system may extend to 13 m and spread laterally at a rate of > 50 mm day^{-1}. However, in terrestrial habitats *P.a.* reaches only 750 mm in height and only achieves abundance where the vigour of potential dominants is suppressed. Thus, *P.a.* is particularly abundant along reservoir margins and, to a lesser extent, river and ditch banks, where water levels are subject to wide fluctuations. The species is also frequent in areas of mining subsidence, which are now more waterlogged than formerly. Aquatic populations are less common than those of the terrestrial habitats, and there is circumstantial evidence that many, if not most, have been secondarily derived from

populations of the land form following flooding. Although primarily a wetland species, *P.a.* may be long-persistent in drained habitats and has been observed as a weed at the edge of potato fields in the Sheffield region. The species has not been recorded from calcareous soils. *P.a.* is very frost-sensitive and is perhaps the first wetland species to die back in autumn; this may explain the restriction of *P.a.* to lowland sites < 230 m. *P.a.* regenerates mainly by vegetative means, forming extensive patches by rhizome extension. Detached vegetative shoot fragments root within two days in the laboratory, and under field conditions this form of regeneration is an important mechanism in the founding of new colonies; long-distance colonization along water courses may be effected by this means. *P.a.* also regenerates, perhaps only infrequently, by seeds which germinate during spring. A persistent seed bank is predicted but has not yet been demonstrated. Aquatic populations tend to flower moderately freely; terrestrial plants are shy-flowering.
Current trends *P.a.* exploits disturbed,

263

fertile, moist conditions, but is not generally an early colonist of new sites. Despite a number of characteristics which make it potentially an aggressive weed, *P.a.* appears to be decreasing.

200 *Polygonum aviculare*

Knotgrass

Data also include *P. arenastrum*

Established strategy R.
Gregariousness Intermediate.
Flowers Pink or white, hermaphrodite, probably always self-fertilized.
Regenerative strategies B_s.
Seed 1.45 mg.
British distribution Probably in 100% of VCs.
Commonest habitats Arable 74%, bricks and mortar 31%, paths 24%, soil heaps 23%, manure and sewage waste 21%.
Spp. most similar in habitats *Capsella bursa-pastoris* 96%, *Matricaria matricarioides* 91%, *Polygonum persicaria* 86%, *Tripleurospermum inodorum* 85%, *Stellaria media* 84%.
Full Autecological Account *CPE* page 452.

Synopsis The aggregate species, which exploits a wide range of open, disturbed, usually fertile, artificial habitats, is a summer-annual herb often with a deep tap-root. It is represented primarily by *P.are.* and *P.avi.*, both of which are very common. All other species in Section *Polygonum* are rare in Britain. *P.are.* is tetraploid, smaller in its parts, more usually prostrate, very tolerant of trampling and more frequent in drier sites. *P.avi.* is particularly common in cereal crops, whereas *P.are.* predominates on tracks and roadsides, waste ground and spoil and, to some extent, in broad-leaved arable crops. However, there is a considerable overlap in habitats between the two, and mixed populations are not uncommon. The geographical distribution of the two is similar. The aggregate species shares a number of attributes with other successful ruderals. (a) A relatively high potential for seed production; a robust plant may produce over 1000 relatively large seeds whereas, under conditions of stress, fewer than 10 may be produced. (b) Effective mechanisms of seed dispersal by human or other agencies. Seeds occur both as an impurity in the harvested crop and as a contaminant of sown seed. They may be dispersed in mud on footwear or tyre treads, and can survive ingestion by stock or by birds. (c) Genetic variability. Both are polymorphic; morphologically separable distinct ecotypes of *P.are.* can be identified. Seeds germinate only in spring and are returned to a state of secondary dormancy during the rise in late spring temperatures. Consequently, the species produces only one generation each year, and is ill-equipped to exploit bared sites created by disturbance in summer. The major habitats of each (cereal fields in the case of *P.avi.*, trampled areas in the case of *P.are.*) are subject to a rather predictable pattern of disturbance. As with many other tap-rooted species, the aggregate is largely absent from wetlands. Both are also sensitive to shading, but can survive defoliation. Indeed, *P.are.* is frequent in mown or grazed grassland subject to trampling. Both, particularly *P.are.*, show a degree of seed polymorphism, the ecological significance of which requires study. Each occurs by the sea and on road verges close to the metalled road surface, and a degree of salt tolerance is suspected in some populations.

Current trends An effective colonist of artificial habitats, and probably still increasing in range.

201 *Polygonum persicaria* — Redshank, Persicaria

Established strategy R.

Gregariousness Individuals typically scattered.

Flowers Pink or rarely white, hermaphrodite, self- or insect-pollinated, sometimes cleistogamous.

Regenerative strategies B_s.

Seed 2.12 mg.

British distribution Widespread (100% of VCs).

Commonest habitats Arable 51%, soil heaps 14%, manure and sewage waste 13%, coal mine waste/unshaded mire/river banks 3%.

Spp. most similar in habitats *Fallopia convolvulus* 94%, *Veronica persica* 92%, *Spergula arvensis* 92%, *Stellaria media* 88%, *Myosotis arvensis* 87%.

Full Autecological Account *CPE* page 454.

Synopsis *P.p.* is an erect summer annual, exploiting disturbed fertile soils. *P.p.* is essentially a 'follower of humans' and is particularly characteristic of arable land in both cereal and dicotyledonous crops, although usually more vigorous and more frequent in the latter. *P.p.* is found on a wide range of moist soils, and occurs on exposed mud beside ponds, where it may be a native of long standing. Although there are no British records of poisoning, *P.p.* contains oxalate and is potentially toxic. *P.p.* resembles many other arable weeds, e.g. *Sinapis arvensis*, in three respects: (a) seed is retained on the plant, leading to contamination of harvested crop seed, (b) a persistent seed bank is formed and (c) seed may be ingested and dispersed by birds and animals. Seeds are also polymorphic, and there is variation in seed weight and chilling requirement between populations and even individual plants. Seeds in the soil have the potential to germinate each spring, but are returned to a state of secondary dormancy during the annual rise in late spring temperatures. This behaviour restricts the capacity of *P.p.* to exploit areas which are disturbed within the summer and to autumn. Cuttings of *P.p.* readily root at the node, and this facility may allow rapid re-establishment of plants during disturbance. *P.p.* is phenotypically very plastic. Thus, *P.p.* may adopt a prostrate form on paths and, when grazed by water-fowl, can produce short, much-branched plants bearing seed. In unfavourable habitats, as few as three flowers may be formed, whereas large specimens can produce 1200. The species shows a high level of in-breeding, and there are a large number of morphological differences between populations. *P.p.* is very similar to *P. lapathifolium*, both in form and in ecology. *P.l.* tends to be more robust, exhibits a more southerly distribution in Britain and is less common, but only a few ecological differences between the two species have been established. *P.p.* hybridizes with *P.l.* and three other species in Britain.

Current trends Apparently decreasing on arable land, particularly in cereal crops. Future status on wasteland uncertain.

202 *Potamogeton crispus*

Curled Pondweed

Established strategy C–R.
Gregariousness Occasionally patch-forming.
Flowers Green, hermaphrodite, wind-pollinated.
Regenerative strategies S, ?B_s.
Seed 4.0 × 2.5 mm.
British distribution Widespread except in upland areas (94% of VCs).
Commonest habitats Still aquatic 7%, running aquatic 2%.
Spp. most similar in habitats *Elodea canadensis* 98%, *Potamogeton natans* 93%, *Sparganium erectum* 90%, *Equisetum fluviatile* 89%, *Lemna minor* 87%.
Full Autecological Account *CPE* page 456.

Synopsis *P.c.* is a submerged aquatic of relatively still waters. As a consequence of its winter-annual phenology, *P.c.* experiences little competition from summer-green aquatics, and may even exert early season dominance (cf. *Poa trivialis* in some terrestrial communities). The production of summer foliage, characterized by strongly undulate, serrate leaves, is followed (about June or July) by the formation of fruiting spikes and dormant apices, the latter appearing in response to long days. The plant dies back during July and the dormant apices, now black and thorny, sink to the bottom where they aestivate. When the water body cools down in autumn, each germinates to give rise to a plant with small uncrisped leaves. Typically this overwintering stage is deeply submerged. Sites with *P.c.* may develop a surface layer of ice in winter. In the following spring the small winter leaves die and are replaced by new summer foliage. A single plant may produce 900 dormant apices in a season. Vegetative regeneration of this type is extremely effective, as illustrated by the fact that sterile, hybrid taxa may nevertheless colonize large rivers or canal systems. Detached vegetative shoots do not develop roots readily, but dormant apices are formed irrespective of photoperiod. Many fewer fruits are produced than dormant apices, and regeneration by seed, which appears to be stimulated by unusually low winter temperatures, is probably a rare event. The exact conditions for optimal germination of the drupaceous fruit, which exhibits a slight degree of hard-coat dormancy, remain to be elucidated. Fruits may survive ingestion by water-fowl and float in water for long periods. The shoots of *P.c.* are eaten by wildfowl, and they also support a large assemblage of aquatic invertebrates. *P.c.* occurs in calcareous, brackish and relatively eutrophic environments, but the species is poorly represented in peaty water. In highly fertile ponds *P.c.* is often suppressed, apparently by the dominant effect of a floating raft of species such as *Lemna minor*. *P.c.* is not recorded from very stony substrata, and is usually rooted in silt. *P.c.* does not exhibit aquatic acid metabolism to any appreciable extent (see *Ranunculus flammula*). The species normally utilizes carbon dioxide for photosynthesis, but is capable of using the bicarbonate ion. As with other submerged aquatics, the cuticle of the leaf is very thin (0.05 μm), allowing foliar absorption of inorganic nutrients. However, vascular tissue in the stem is better developed than that of, for example, *P. pectinatus*. As a result *P.c.* can take up more nutrients from the rooting substrate than *P.p.*, and can extend into less-eutrophic waters than *P.p.*. *P.c.* is highly plastic in morphology, and the varieties described by early workers appear to have been phenotypic rather than genetic in origin. Hybridizes in Britain with six other species, some broad- and others narrow-leaved.

Current trends A mobile species, but succumbs to dominance by *Lemna minor* and filamentous green algae under highly eutrophic conditions. Apparently decreasing.

---

203   *Potamogeton natans*   **Broad-leaved Pondweed**

Established strategy Possibly between
  C and S–C.
Gregariousness Patch-forming.
Flowers Green, hermaphrodite, wind-
  pollinated.
Regenerative strategies (V).
Seed 4.25 mg.
British distribution Widespread (100%
  of VCs).
Commonest habitats Still aquatic 22%,
  unshaded mire 1%.
Spp. most similar in habitats
  Equisetum fluviatile 96%, *Lemna
  minor* 94%, *Potamogeton crispus*
  93%, *Elodea canadensis* 92%,
  Sparganium erectum 87%.
Full Autecological Account *CPE* page
  458.

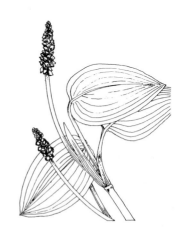

Synopsis *P.n.* is a large, heterophyllous aquatic with long-stalked, leathery, floating leaves and narrow, translucent, inconspicuous, often short-lived, submerged leaves. The species is widespread in ditches, ponds, canals and sheltered lake-sides and is characteristic of relatively deep, stagnant water (exceptionally up to 2m). However, *P.n.* is absent from very deep water, and light may be an important factor controlling the zonation. In the Sheffield region *P.n.* is absent from reservoirs, suggesting that the species is vulnerable to wave action or has a limited tolerance of fluctuating water levels. In some areas of disused canal, the growth of *P.n.* appears to be suppressed by a layer of *Lemna minor* and *L. gibba* at the water surface. Land forms are reported but, unlike *P. polygonifolius*, *P.n.* is of transient occurrence in mire. In running or deep water, phenotypes with leaves wholly submerged may occur. The species overwinters by means of dormant buds formed at the stem apex. Some stems, unlike those of *P. crispus*, live for more than one season, and *P.n.*, in common with other *Potamogeton* species, is without vessels, the vascular tissue being less condensed than that of *P.c.*. *P.n.* is unpalatable to the grass carp (*Ctenopharyngodon idella*), and may form extensive rhizomatous patches in fisheries. Breakdown of older rhizomes isolates daughter plants. Detached pieces of shoot (and presumably also rhizomes) are capable of rooting. These may either remain *in situ* or are carried some distance by water. Regeneration by the seeds, which show hard-coat dormancy, is probably infrequent, even though seed is produced in abundance. It is suspected that germination occurs during spring. No information is available concerning the persistence of any seed bank. Seeds can survive ingestion by water-fowl and may be dispersed also in water, where they may float for up to one year. *P.n.* is an effective colonist, even of land-locked sites such as pools in quarries. However, it is uncertain whether such sites are colonized by seed or by transported vegetative fragments. The species extends from peaty, somewhat acidic sites to relatively base-rich conditions, and appears to exploit both nutrient-rich and more-nutrient-poor conditions. However, at high levels of acidity, *P.n.* is replaced by *P.p.*, and in strongly calcareous waters *P. coloratus* becomes important. *P.n.* hybridizes, albeit rarely, with four other species in Britain.

Current trends Despite being an effective colonist, *P.n.* appears to be decreasing through the destruction, and perhaps eutrophication, of aquatic habitats.

---

204   *Potentilla erecta*   Common Tormentil

Established strategy Between S and C–S–R.
Gregariousness Intermediate.
Flowers Yellow, hermaphrodite, insect-pollinated, self-incompatible.
Regenerative strategies V, B$_s$.
Seed 0.58 mg.
British distribution Widespread (100% of VCs).
Commonest habitats Limestone pastures 31%, limestone wasteland 21%, acidic pastures 12%, unshaded mire/verges 8%.
Spp. most similar in habitats Insufficient data.
Full Autecological Account *CPE* page 460.

Synopsis *P.e.* is a slow-growing, tufted herb typically found in a wide range of grassland and heathland habitats on acidic soils. *P.e.* reaches highest frequency and abundance in pastures of intermediate fertility and is, for example, particularly common on brown earths and incipient podsols developed on loessic soils over limestone. The distribution extends into calcareous grasslands, but is restricted here to soils of high moisture status such as those occurring on many N-facing dalesides of the Derbyshire limestone. Foliage of *P.e.* is low in P but unusually high in Fe, Mn and Al. Field measurements and laboratory experiments show that vegetative growth is much more vigorous on non-calcareous soil. On calcareous soils it seems likely that *P.e.* may be excluded from dry S-facing sites where shallow root development renders seedlings vulnerable to summer drought; the same phenomenon could explain the apparently wider edaphic range of *P.e.* in the more humid climate of N and W Britain. The shallow root system of *P.e.* is also consistent with the occurrences of *P.e.* in soligenous mire. Flowering shoots of *P.e.* die back completely over winter, new ones being produced rapidly in the late spring by mobilization of reserves from the stout basal stock. This phenological pattern is very different from that of many of the species with which *P.e.* occurs on acidic soils, and would perhaps not represent a viable strategy in an infertile grazed habitat but for the fact that *P.e.* is very unpalatable to stock. *P.e.* is self-incompatible and has a long flowering period. Achenes are often slow to ripen and many are released late in the summer, often in a green condition. Rapid germination is

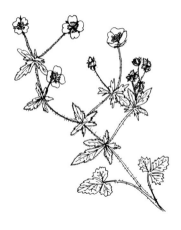

usually achieved only after warm moist incubation and there is no evidence of germination in autumn when many seeds become incorporated into a buried seed bank. Populations have been shown to differ genetically according to habitat and geographical location. Hybrids occur with *P. anglica* and *P. reptans*, both of which regenerate vegetatively by producing long-rooted runners. The former is apparently an allo-octoploid hybrid between *P.e.* and *P.r.. P.a.*, which is self-compatible, tends to occur on soils of intermediate pH, whereas *P.r.*, which is self-incompatible, is confined to base-rich soils.

Current trends As a result of habitat destruction *P.e.* is decreasing in many lowland areas. The species remains abundant, however, in much of upland Britain.

205   *Potentilla sterilis*

Barren Strawberry

Established strategy S.
Gregariousness Sparse.
Flowers White, hermaphrodite, insect-pollinated or selfed.
Regenerative strategies V.
Seed 0.58 mg.
British distribution Widespread except in N Scotland and around the Wash (98% of VCs).
Commonest habitats Limestone pastures 14%, scree 6%, lead mine spoil/verges/limestone wasteland 3%.
Spp. most similar in habitats

Insufficient data.
Full Autecological Account *CPE* page 462.

Synopsis *P.s.* is a tufted, low-growing winter-green herb of relatively infertile hedgebanks and pastures. *P.s.*, which has a relatively shallow root system, appears to be restricted to relatively moist base-rich soils or those of intermediate pH, and is absent from both waterlogged and droughted sites. In common with a number of other winter-green forbs, such as *Viola riviniana*, *Fragaria vesca* and *Ajuga reptans*, *P.s.* tends to be restricted to partially shaded habitats in lowland areas, but extends into more-open habitats in upland regions. Foliage persists throughout the summer and, though clearly shade-tolerant, *P.s.* is normally absent from deep shade. The occurrence of *P.s.* in pasture may in part be a reflection of the low palatability of the leaves to stock and the ability of the foliage to exploit shaded layers low in the turf canopy. *P.s.* flowers and sets seed remarkably early, and appears capable of sustaining growth at low temperatures. It is not known whether a persistent seed bank is formed, and the season of seed germination remains uncertain. The species is often found in lightly disturbed sites such as steep banks subject to a degree of soil creep and the margin of woodland rides, and compared with other species of shaded habitats the achenes are relatively small. We predict that regeneration by seed is usually associated with disturbance events. *P.s.* forms extensive, usually diffuse, patches through the growth of rooted stolons. In shaded habitats, plants tend to have larger leaves and longer petioles and a more erect habit than those from open sites. The extent of ecotypic differentiation in *P.s.* has not been investigated.

Current trends Regarded as a plant of ancient woodland with a low colonizing ability. *P.s.* spreads relatively slowly on roadsides, and has a tendency to be restricted to ancient habitats. Probably decreasing.

206 *Primula veris*   Cowslip, Paigle

Established strategy S.
Gregariousness Sparse.
Flowers Yellow, hermaphrodite, insect-pollinated.
Regenerative strategies V.
Seed 0.69 mg.
British distribution Widespread but local outside England (92% of VCs),
Commonest habitats Limestone pastures 10%, limestone wasteland 7%, scree 6%, limestone quarries 2%, enclosed pastures 1%.
Spp. most similar in habitats *Polygala vulgaris* 87%, *Avenula pubescens* 82%, *Danthonia decumbens* 82%, *Carex caryophyllea* 82%, *Briza media* 81%.
Full Autecological Account *CPE* page 464.

Synopsis *P.v.* is a spring-flowering, long-lived, winter-green, rosette-forming herb particularly characteristic of short, species-rich turf and only rarely occurring in shaded habitats. *P.v.* occurs predominantly on moist calcareous soils, but can be found occasionally on dry soils over non-calcareous strata. The species has an unusual phenology, in that it commences growth during winter and flowers in spring but achieves peak biomass in summer. This differs from, but overlaps, the phenology of summer-growing herbs such as *Leontodon hispidus*. The leaves, which are often held appressed to the ground, tend to be little predated by stock. *P.v.* has a very limited capacity for vegetative replication through branching of the rhizome, although this may be the most common mechanism of regeneration in stable communities. Elaborate mechanisms, including heterostyly, reduce the amount of self-fertilization. Nevertheless, *P.v.* produces considerable quantities of seed and in disturbed areas may establish relatively rapidly to form extensive populations. The timing of germination and nature of the seed bank is uncertain. *P.v.* forms four ssp. in Europe, of which only one, ssp. *veris*, occurs in Britain. The extent of genotypic variation and phenotypic plasticity between British populations is uncertain, though even within populations individual plants are known to vary considerably in leaf size and number of flowers per umbel.

Current trends *P.v.* has greatly decreased through the ploughing of old pastures and the relaxation of grazing pressure in semi-natural grassland. Although becoming ever rarer, *P.v.* is showing some resurgence along the verges of some motorways and trunk roads. Plants of the 'pin' form, with the long stigma, are able to self in the absence of 'thrum' plants, with the short stigma. Normally the proportion of 'pin' types is about 60%, but in small isolated populations a gradual increase in the proportion of 'pin' plants may be expected and, in unstable sites and newly colonized areas, the greater heterozygosity associated with a high proportion of 'thrum' plants may be advantageous.

---

207 *Prunella vulgaris*

Self-heal

Established strategy C–S–R.
Gregariousness Sparse to intermediate.
Flowers Violet, hermaphrodite, self-sterile, insect-pollinated.
Regenerative strategies (V), ?B_s.
Seed 0.73 mg.
British distribution Widespread (100% of VCs).

Commonest habitats Meadows 29%, limestone pastures 17%, rock outcrops 12%, verges 11%, unshaded mire/limestone quarries 8%.
Spp. most similar in habitats *Bellis perennis* 85%, *Cynosurus cristatus* 75%, *Cerastium fontanum* 70%,

Trifolium repens 67%, *Ranunculus bulbosus* 65%.
Full Autecological Account *CPE* page 466.

Synopsis *P.v.* is a short, rather slow-growing, winter-green, patch-forming herb of grassland, typically associated with moist, moderately fertile soils. Like *Bellis perennis*, with which it often grows, *P.v.* is easily dominated by taller herbs and is abundant only in short turf, particularly in lawns and permanent pasture, where large clones may develop by vegetative expansion. *P.v.* is favoured by close grazing, presumably because of its creeping growth-form, moderate resistance to trampling and rather low palatability. *P.v.* is infrequent in burned sites. The species exhibits some shade-tolerance and frequently occurs along woodland rides. It persists in meadows if grazing occurs after harvest. *P.v.* has a marked summer peak in above-ground biomass. Its parts are short-lived and, like *Trifolium repens*, *P.v.* is relatively mobile, expanding temporarily into gaps within turf before being replaced by more-robust species. The species forms loose clonal patches and connections with daughter ramets may decay in less than one year. We suspect that *P.v.* may regenerate from shoot fragments detached as a result of trampling, or other forms of damage. Regeneration also occurs by means of seeds, which germinate mainly in spring. Establishment from seed can occur in comparatively dry exposed microsites. Plants developed from small lengths of shoot closely resemble seedlings in appearance, and the relative importance of seedlings and vegetative fragmentation during the re-colonization of disturbed ground requires investigation. Accounts vary as to whether or not any seed bank is formed. The relative importance of regeneration by seed and by vegetative means varies greatly between populations, and in Europe both annual and perennial races occur. *P.v.* is phenotypically plastic, and considerable change in morphology can be induced by temperature and by the intensity of grazing or mowing. Seed size is also variable; some plants from shaded sites have been observed to produce heavier seeds. Populations also vary in morphology and in growth rate according to habitat and geographical area. There is clear ecological importance of both phenotypic plasticity and genetic diversity in the widespread success of the species.
Current trends Common, and probably increasing.

208 *Pteridium aquilinum*   Bracken

Established strategy C.
Gregariousness Stand-forming even at low frond density.
Sporangia In sori on underside of frond.

Regenerative strategies V, W, ?B$_s$.
Spore 33 × 28 μm.
British distribution Widespread (100% of VCs).
Commonest habitats Coniferous

plantations 37%, scrub 31%, acidic woodland 22%, acidic pastures 17%, acidic wasteland 13%.
Spp. most similar in habitats *Luzula pilosa* 81%, *Sorbus aucuparia* (juv.) 80%, *Rubus fruticosus* 80%, *Milium effusum* 73%, *Holcus mollis* 73%.
Full Autecological Account *CPE* page 468.

Synopsis *P.a.* is found in a wide range of shaded and unshaded habitats, particularly on deep acidic soils. In lowland areas *P.a.* is markedly more frequent in shaded habitats, while in upland areas it is equally, if not more, widespread in open sites. Whether this difference is brought about by changing climatic or land-use factors is not known. Unlike other common British ferns, *P.a.* possesses an extensive system of underground rhizomes and, in the context of the British flora, *P.a.* is a uniquely competitive and aggressively invasive fern. These attributes are associated with the presence of xylem vessels in the rhizome, an unusual feature in ferns. Although *P.a.* is most characteristic of acidic strata, the species is usually restricted to deeper soils, and shows maximum vigour on productive brown earths. Over calcareous strata, *P.a.* is often restricted to situations where there is overlying drift or loess. In vegetation subject to run-off from calcareous road foundations *P.a.* is often chlorotic and stunted. Leaves have low levels of Ca, but contain relatively high concentrations of P and the species is very efficient in recycling nutrients within the plant. The whole plant is heavily defended chemically from mammalian and insect predators, and is toxic to livestock and to humans. *P.a.* is also considered allelopathic by some biologists. Young fronds produce extrafloral nectaries. These provide food for ants which may in turn rid *P.a.* of some of its potential insect predators. Young shoots of *P.a.*, the first of which emerge in April or May,

are very sensitive to frost and trampling. *P.a.* was originally a woodland plant before the deforestations. However, spore production and growth of the prothallus is greater in unshaded habitats. In open sites, dominance by *P.a.* may be largely due to the presence of much surface-lying and standing litter. Litter accumulation may even suppress *P.a.* and this can lead eventually to re-invasion by other species. In woodland, breakdown of litter is more rapid; here populations of *P.a.* are often more stable. Once established, *P.a.* regenerates vegetatively to form large clonal patches, bypassing the vulnerable prothallial stage that follows spore germination. Individual parts of the rhizome system may live for > 50 years and their presence in the subsoil confers resistance to burning. Up to 30 million spores may be produced by a single fertile frond and, though initially wind-dispersed, viable spores are often found in abundance within the soil profile, and a buried spore bank is suspected. Spores retain their viability for up to ten years and may even survive ingestion by insects. Although establishment of the sporeling is rapid, regeneration from spores is mainly confined to areas of disturbed or burnt ground and may occur in spring. Thus, spores are probably only important in the colonization of new sites. The species shows genetic variation and, for example, exhibits polymorphism for cyanogenesis. Formerly of economic importance as a source of fuel, thatch, litter, compost, food and potash, *P.a.* has some potential as a possible energy resource, either as a fuel or in the production of methanol or methane.

Current trends Now a weed, primarily of grasslands but also of forestry.

Increasing in upland areas and difficult to eradicate, despite recent efforts to reclaim grasslands by application of asulam and other mechanical or chemical means. In lowland Britain, many of its habitats have been destroyed as a result of modern land-usage, and *P.a.* may be decreasing in some areas. Nevertheless, *P.a.* remains common and is still an invasive plant in many lowland areas.

209   *Quercus*

Oak

Includes *Q. petraea*, *Q. robur* and their hybrid

Established strategy S–C.
Gregariousness Typically sparse.
Flowers Green, monoecious, wind-pollinated.
Regenerative strategies S.
Seed *Q.p. c.* 2500 mg.
British distribution *Q.p.* more common in the N and W (80% of VCs); *Q.r.* more in SE and C England (81% of VCs).
Commonest habitats Broadleaved plantations/acidic woodland 8%, acidic quarries/limestone quarries 6%, scrub 4%.
Spp. most similar in habitats *Hyacinthoides non-scripta* 83%, *Acer pseudoplatanus* (juv.) 81%, *Luzula pilosa* 79%, *Lonicera periclymenum* 75%, *Sorbus aucuparia* (juv.) 72%.
Full Autecological Account *CPE* page 470.

Synopsis Tall but slow-growing, deciduous, timber trees, with ring porous wood and a late-emerging canopy. *Q.* spp. grow to 30 m, may live for > 500 years and can persist in relatively infertile sites. Oak is the commonest tree of British broad-leaved woodlands (315 000 ha, 44% of total) except in Scotland, where birch is even more common. The concept of oak woodland as the climax vegetation of the British Isles

appears unfounded. The production of new large xylem vessels in spring precedes bud-break by *c.* 1 week. Shoot growth and leaf expansion occur in April and May and subsequent shoot growth is intermittent with a resting period of 4–6 weeks. The genus *Quercus* has a predominantly warm-temperate distribution, and the two northern outlying species described here are susceptible to frost damage to the foliage in spring and show other signs of maladaptation to cool, temperate climates. *Q.p.* is the more common species in ancient woodland and is found primarily on soils of low pH, particularly in the uplands of N and W Britain. In contrast, *Q.r.* is more generally associated with mixed woodland and hedgerows, and extends more regularly onto soils of higher pH or of a moist clayey consistency. *Q.r.* has larger acorns and the faster initial growth. Ecological distinctions between the two species have been blurred, first as a result of hybridization, particularly in the N, and second, because of the widespread planting of *Q.r.* for timber (*Q.p.* is now regarded as superior for this purpose). The two species are highly variable and show much overlap in their characteristics and were not separated during fieldwork. Regeneration is entirely by seed from *c.* 40 years old, often slightly earlier in *Q.r.*, and seed production reaches a maximum at between 80 and 120 years. Mast years, with up to 90 000 acorns per tree, only occur every 6–7 years, and even moderate yields may occur only every 3–4 years. Good mast years are more frequent in S England and acorns are larger and tend to have higher viability under these circumstances. Production of acorns may be erratic from one site to another. A single tree may produce several million acorns in its lifetime. Acorns are shed from September to November, but those falling before mid-October are usually non-viable. Acorns are killed by excessive water-loss and are susceptible to frost damage. Those of *Q.p.* are susceptible to waterlogging. A covering of litter provides a suitably moist environment for seedling establishment. The synchronization of fruit and leaf fall may aid this process and also reduce the accessibility of acorns to predators. Seeds are heavily predated by insects, birds and mammals. However, birds and mammals may aid dispersal by burying seed at some distance from the parent, and acorns appear to be transported to greater distances than are beech nuts. Maximum growth is likely to occur at 20–40% of full daylight. Acorns show epicotyl dormancy, which is broken by chilling. Germination is hypogeal and the seedling has a large tap-root which

may constitute 75% of the weight of the plant. These reserves enable *Q.* spp. to colonize closed grassland (other common tree species are restricted to gaps) and confer resilience should the leading shoot be removed by grazing. The foliage is toxic and has an unpleasant taste; trees are attacked by squirrels and rodents to a lesser extent than are *Fagus sylvatica* or *Acer pseudoplatanus*. Although probably more insects predate *Q.* spp. than any other British tree, relatively few have a drastic effect and foliage is unpalatable to the snail *Helix aspersa*. However, two species of Lepidoptera may cause severe defoliation to mature trees during early summer and establishment of juveniles is often poor under oak, perhaps in part because defoliating caterpillars falling from the canopy predate the juveniles present. In long-lived species such as oak, with thousands of apical meristems, mutations may result in some branches being genetically distinct from the next. The ecological significance of a 'genetic mosaic' produced in this way remains to be assessed but, in theory, interactions with pests and pathogens may well be modified. The ability of *Q.* spp. to regenerate appears to have diminished greatly in the last 150 years. In the case of *Q.p.*, which appears the more shade-tolerant, effective regeneration may be restored in many upland sites simply by excluding grazing animals from woodland. It has been suggested that the combined effects of shade and predation by insects, mammals, birds and oak mildew may be too great to allow survival, and points to the fact that *Q.r.* frequently becomes established in unshaded habitats. In lowland areas the cessation of coppicing and pollarding, and thus the removal of a light phase, shows some correlation with the decline of oak regeneration.

Current trends Becoming less important as a timber tree and the bark is no longer utilized in the tanning industry. However, *Q.r.* appears to be amongst the less vulnerable of our native trees. *Q.r.*, and in many upland regions *Q.p.*, shows some capacity for colonizing new, unshaded habitats.

210 *Ranunculus acris* — Meadow Buttercup

Established strategy C–S–R.
Gregariousness Intermediate.
Flowers Yellow, typically hermaphrodite and protogynous, often self-sterile.
Regenerative strategies B_s, V.
Seed 1.9 mg.
British distribution Widespread (100% of VCs).
Commonest habitats Meadows 61%, lead mine spoil 26%, limestone pastures 17%, verges 12%, enclosed pastures 9%.
Spp. most similar in habitats *Rumex acetosa* 79%, *Cerastium fontanum* 73%, *Alopecurus pratensis* 73%, *Phleum pratense* 72%, *Trifolium repens* 71%.
Full Autecological Account *CPE* page 472.

Synopsis *R.a.* is, like the other common buttercups (*R. bulbosus* and *R. repens*), a winter-green, frost-tolerant herb of grazed or mown grassland. *R.a.* is a plant of moist, but not waterlogged, habitats. In some ridge-and-furrow grasslands the three buttercups occupy different positions, with *R.b.* found on the more freely-drained ridge top, *R.a.* on the moist ridge side and *R.r.* in the wetter furrows. *R.a.* is also taller than the other two and can compete effectively in moderately productive meadows. Detailed demographic studies have been carried out on all three species. Though often disadvantaged by being in full flower when hay is cut, *R.a.* is more common in meadows than in any other habitat but, even here, its range is limited to relatively unproductive sites and its yield can be suppressed by treatment with ammonium salts. Like *R.b.*, *R.a.* is unpalatable and it is avoided by stock, in which it reduces yield. The claim that *R.a.*, which contains protoanemonin, is also highly toxic may be exaggerated. Overgrazing produces areas of bare soil suitable for seedling establishment, but the seed has no obvious mechanism for dispersal. However, some vegetative regeneration occurs. With respect both to habitat range and to dependence upon seed for regeneration, the three common buttercups may be ordered *R.b.* < *R.a.* << *R.r.*. Seed of *R.a.*, which may be incorporated into a persistent seed bank, is produced in much greater numbers than is the case in *R.r.*. *R.a.* is a variable and polymorphic species in need of taxonomic study in Britain and elsewhere. Two of the five ssp. tentatively recognized from Europe are recorded in Britain; ssp. *acris* and the montane ssp. *borealis*. Studies on the origin and ecological significance of this variation are required.

276

Current trends *R.a.* exploits overgrazed, moderately fertile pastures. Since it is easily controlled, has poor dispersal and is slow to establish, *R.a.* is now mainly confined to permanent pastures subject to lax management. *R.a.* has almost certainly passed its peak of abundance and may be expected to decline further.

211   *Ranunculus bulbosus*   Bulbous Buttercup

Established strategy Between S–R and C–S–R.
Gregariousness Sparse.
Flowers Yellow, hermaphrodite or more rarely gynodioecious, insect-pollinated.
Regenerative strategies B_s.
Seed 3.1 mg.
British distribution Widespread (99% of VCs).
Commonest habitats Meadows 29%, limestone pastures 21%, limestone wasteland 6%, lead mine spoil/enclosed pastures/limestone quarries/rock outcrops 3%.
Spp. most similar in habitats *Anthoxanthum odoratum* 74%, *Rhinanthus minor* 71%, *Ranunculus acris* 70%, *Phleum pratense* 68%, *Bellis perennis* 67%.
Full Autecological Account *CPE* page 474.

Synopsis *R.b.* is a small rosette plant possessing a more-limited capacity for growth in height and in lateral spread than the other two common buttercups, *R. acris* and *R. repens*. Individual organs are replaced annually but the plant itself is relatively long-lived. *R.b.* is particularly common in older, often species-rich, permanent pastures on freely-drained soils in sites where shading is prevented by heavy grazing or, in unproductive turf, by low soil fertility. The leaves of *R.b.* are mainly epimastic in a basal rosette, and the species tends to be suppressed in hay meadows. *R.b.* appears to be particularly characteristic of S and W slopes in sites where summer drought limits the growth of more-robust species. *R.b.* exhibits an unusual phenology in which flowering and seed set are earlier than in *R.a.* and *R.r.* and the plant usually aestivates as a below-ground 'corm' from mid-July. This phenology affords a possible mechanism of drought-avoidance in drier sites and may also reduce sensitivity to the impacts of summer dominants in more-fertile sites. *R.b.* is encouraged by overgrazing, almost certainly because seedling establishment, which is critical for population increase, occurs on the bare soil created by trampling. No well-defined mechanism of dispersal has been described, and seeds are often shed close to the plant. Seeds of this species, and of *R. acris* are produced in greater numbers than is the case with *R.r.* but the seed bank of *R.b.* is less persistent. *R.b.* is unpalatable and, together with other herbage in its vicinity, is generally avoided by stock. As a result, agricultural productivity of pasture is reduced by infestations of *R.b.* and there are records of stock,

and humans, being poisoned.
However, the level of seed predation is higher than that experienced by *R.a.*, though lower than in the case of *R.r.*. Ecotypes have been described and 6 ssp., including two found in Britain, have been tentatively recognized within Europe.

Current trends *R.b.* is mainly restricted to older areas of permanent pasture. The species reduces grassland production and is easily controlled by weed-killers. A continuing decrease in its abundance is therefore inevitable.

212 *Ranunculus ficaria*

Includes sspp. *ficaria* and *bulbifer*

Lesser Celandine

Established strategy Between R and S–R.
Gregariousness Intermediate.
Flowers Yellow, usually hermaphrodite, protandrous, insect- or self-pollinated.
Regenerative strategies S.
Seed Ssp. *f.* achene, 1.04 mg.
British distribution Common except in Scotland and Ireland (100% of VCs).
Commonest habitats River banks 13%, limestone woodland 11%, acidic woodland 10%, shaded mire/scrub 8%.
Spp. most similar in habitats *Silene dioica* 91%, *Festuca gigantea* 89%, *Allium ursinum* 80%, *Lamiastrum galeobdolon* 79%, *Circaea lutetiana* 78%.
Full Autecological Account *CPE* page 476.

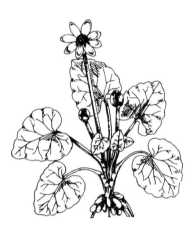

Synopsis *R.f.* is a tuber-forming herb which minimizes competition with summer-growing herbs, grasses and trees by virtue of its vernal phenology. The species grows in woods and, to a lesser extent, tall-herb and grassland communities. *R.f.* often also occurs on path sides and in grazed or mown habitats. Because of the vernal phenology, the growth of *R.f.* is often near completion in these sites before trampling, grazing or mowing reach any great intensity. Like other vernal geophytes, *R.f.* has a high nuclear DNA amount. *R.f.* is very cold-resistant, and is the earliest common species to emerge, flower and

senesce. As in *Hyacinthoides non-scripta*, unshaded sites are more favourable for growth, despite the fact that *R.f.* is frequent in woodland. *R.f.* is common on moist, relatively fertile soils in sites with little accumulation of tree litter. The leaves of *R.f.* contain relatively high concentrations of Na. They are toxic to stock (the poisonous principle is protoanemonin) and tend to escape predation. The seedling, like that of *Conopodium majus*, is monocotyledonous. *R.f.* exists as two main ecotypes. Ssp. *b.* is more vernal and more strictly a woodland plant, and regenerates both by means of bulbils (up to 24 per plant) borne in the axils of the lower leaves and by means of root tubers if these possess a bud. However, ssp. *b.* sets little seed. In contrast, ssp. *f.* produces seed but not bulbils. Thus, although the two overlap ecologically and can occur together, ssp. *b*. tends to be found in the more disturbed habitats, and may even be a garden weed. The two occasionally hybridize, and two further ssp. occur in Europe.

Current trends Ssp. *f*., which tends to be found in less-disturbed sites, may be decreasing. Ssp. *b*. is probably more mobile, both within and between woods, and the proportion of localities suitable for ssp. *b*. is likely to be increasing.

213 *Ranunculus flammula*

Lesser Spearwort

Established strategy Between C–R and C–S–R.

Gregariousness Intermediate.

Flowers Yellow, hermaphrodite, insect-pollinated, largely self-incompatible.

Regenerative strategies V, B_s.

Seed 0.37 mg.

British distribution Widespread (100% of VCs).

Commonest habitats Shaded mire 6%, unshaded mire 5%, running aquatic 4%, limestone wasteland 3%, still aquatic 2%.

Spp. most similar in habitats Insufficient data.

Full Autecological Account *CPE* page 478.

Synopsis *R.f.* is found in a variety of wetland habitats where the growth of potential dominants is suppressed by disturbance and, in some sites, by infertility. The species is characteristic of mildly acidic, often peaty, soils, but has also been recorded from highly calcareous sites within the Sheffield region. Much of the wide tolerance of wetland conditions coincides with phenotypic plasticity. Thus, under aquatic conditions *R.f.* exhibits nodal rooting; terrestrial and aquatic forms also differ in leaf shape and prostrate, rooted phenotypes are erect in cultivation. In N. America, some genotypes are more heterophyllous than others. These are adaptable both to submergence and to desiccation, and are found in fluctuating habitats such as lake margins. Less-plastic genotypes are associated with more-stable conditions. However, despite the extent of phenotypic plasticity, single populations of this essentially out-breeding species may include morphologically-distinct genotypes. Apart from a tendency to exploit disturbed habitats such as the stony edges of oligotrophic lakes and areas of soligenous mire, *R.f.* also occurs in vegetation kept low by grazing. Palatability has not been studied extensively, but the species contains protoanemonin and is toxic to stock. *R.f.* is found in shaded mire and may persist in taller unshaded vegetation, where it adopts a more upright growth form. Aquatic acid metabolism confers advantages where the concentration of CO_2 in the aquatic environment is low and diffusion rates are slow. Several aquatic species of oligotrophic waters, e.g. *Littorella uniflora*, possess this adaptation, but *R.f.* is not one of them. The species shows some tolerance of ferrous iron. Prostrate stems root freely and diffuse clonal patches may be formed, particularly in open habitats. Shoots detached as a result of disturbance readily regenerate to form new plants; this may be an important mode of regeneration in lake- and stream-side populations. Regeneration by seed is probably less important. Seeds may be incorporated into a persistent seed bank and tend to germinate mainly in spring. Rapid germination can be induced if seeds are alternately frozen and thawed, but the significance of this and other possible germination cues remains uncertain. Seeds have been shown to survive ingestion by horses. Long-distance transport in water seems unlikely since the seeds lack buoyancy. Three ssp. are recorded from Britain, but only ssp. *flammula* occurs in most of the British Isles. The other two, ssp. *minimus* and ssp. *scoticus*, are endemic to Scotland and Ireland. The former has a maritime and the latter a lake-side distribution. *R.f.* shows introgression with *R. reptans*, which is now believed to be extinct in Britain. **Current trends** Occasionally colonizes new habitats, but probably decreasing as a result of wetland destruction and eutrophication.

---

214 *Ranunculus peltatus*   Common Water-crowfoot

Established strategy Perhaps between R and C–S–R.
Gregariousness Sparse to intermediate.
Flowers White, hermaphrodite, insect-pollinated or selfed.
Regenerative strategies V, ?B_s.
Seed 0.25 mg.
British distribution Widespread except in Scotland.
Commonest habitats Still aquatic 4%, unshaded mire 3%, shaded mire 2%.
Spp. most similar in habitats *Typha latifolia* 96%, *Equisetum fluviatile* 93%, *Lemna minor* 92%, *Sparganium erectum* 91%, *Potamogeton natans* 87%.
Full Autecological Account *CPE* page 480.

Synopsis *R.p.* is a still-water aquatic also capable of growing on wet mud and exploiting sites in which the fluctuation of water level prevents dominance by either obligate aquatics or by emergent 'semi-aquatic' and mire species. The species shows well-developed heterophylly; aquatic

phenotypes usually have finely divided submerged leaves which collapse in air, and floating leaves with a well-defined blade. Terrestrial individuals usually possess rigid leaves with short, rigid, filiform segments. Plants are susceptible to frost if not submerged, but can withstand being frozen in ice. In *R. aquatilis*, with which *R.p.* has close affinities, the type of leaf produced is dependent upon photoperiod and whether the shoot is submerged or terrestrial, and the same appears to hold for *R.p.*. This heterophylly appears to be of critical importance in allowing exploitation of water bodies subject to fluctuations in water level. In particular, the capacity of laminar leaves to intercept light at the water surface may be important in competitive interactions with other aquatic species. In its terrestrial form *R.p.* may occur beneath robust annuals, reflecting some capacity to tolerate reduced irradiance. Three other species of water-crowfoot are frequent in still water in N England; their distribution appears to be partially determined by hydrology. The non-heterophyllous *R. circinatus* is virtually confined to aquatic systems, often occurring in relatively deep water and showing little capacity to exploit bare mud. The two most common species in ditches and small pools, habitats liable to dry out completely during drought, are *R.a.* and the non-heterophyllous *R. trichophyllus* (although *R.t.*, and to a lesser extent *R.a.*, are also found in deep water). In contrast, *R.p.* is typical of aquatic systems with fluctuating levels of water, and is by far the most common water-crowfoot in reservoirs. The distribution of *R.p.* is biased towards mildly acidic (often peaty) habitats. The species is absent from calcareous waters, where it appears to be replaced by *R.t.*, which is also a species of 'brackish' waters associated with colliery workings. The regenerative strategies of *R.p.* are incompletely understood. Seedlings have been observed during autumn and spring, and mature plants may overwinter. Seeds float for short periods. No evidence of a persistent seed bank is available, but its existence is predicted from the ecology of the species. Vegetative shoot fragments root freely to form new plants, and this may be important in the dispersal of *R.p.* during the summer. The hybrid with *R. fluitans* has spread extensively along one river system, and *R.p.* also hybridizes with *R.t.* and with three further species.

Current trends Although apparently an effective colonist of recently created habitats, like most aquatics *R.p.* is probably now decreasing as a result of habitat destruction.

215 *Ranunculus penicillatus*

Formerly known as *R. pseudofluitans*

Established strategy Probably C.
Gregariousness Patch-forming even at
  low frequency.
Flowers White, hermaphrodite, insect-
  pollinated or selfed.
Regenerative strategies V.
Seed 0.39 mg.
British distribution Not frequent in the
  extreme N.
Commonest habitats Running aquatic
  27%, still aquatic/walls 1%.
Spp. most similar in habitats
  Insufficient data.
Full Autecological Account *CPE* page
  482.

Common Water-crowfoot

Synopsis *R.p.* is a dominant and productive submerged aquatic herb forming dense and extensive monocultures in rivers and streams but developing a standing crop which is usually much less than those of dominant emergent species such as *Typha latifolia*. Peak biomass is attained in summer about one month after flowering, but there is also some photosynthetic activity during winter. In water, *R.p.* produces long, submerged, finely-dissected leaves which offer little mechanical resistance to the strong currents often experienced. Shoots produced in spring have thicker, hollow stems, produce flowers and float at or near the water surface. Photosynthesis is greater at faster flow rates and is possibly limiting in slow-moving water, as perhaps are nutrient uptake and the intake of oxygen for respiration. Beached plants have the same dissected leaf form shown by *R. peltatus*. The control of *R.p.* by cutting in spring is ineffective and simply encourages further vegetative growth. In many river systems *R.p.* is the only aquatic macrophyte present, but in some regions, as in the Derbyshire Peak District, *R. fluitans* may also occur. Stretches of water with a rocky or gravelly bottom are characterized by *R.p.*, which is replaced by *R.f.* where the substratum is more silty. This distribution appears unrelated to water chemistry, since both species occur on the base-poor millstone grit and on the carboniferous limestone. However, *R.p.* produces roots throughout the year and can thus remain attached to an unstable rocky substrate, whereas *R.f.* roots only during winter and is restricted to streams with a more stable silty bottom. Regeneration is mainly vegetative. Stems may grow up to 4 m in the direction of the water flow, and single plants can form large clonal patches. Vegetative fragments root readily, and establishment in this manner has been observed. *R.p.* often produces abundant seed, which floats for only a short period. It is not known when seeds normally germinate. No persistent seed bank has so far been detected. It is likely that establishment by seed is rare. Populations in N England refer mostly to var. *calcareus*, which occurs in slow-flowing water subject to frequent flooding. Var. *penicillatus*, which may form floating laminar leaves, is found in fast-flowing water, and var. *vertumnus* in clear, slow-flowing water, including canals not subject to regular flooding. There is much additional variation within the species group, mostly ecotypic in origin. The *R.p.* group is believed to have resulted from the hybridization of *R.f.* with *R. aquatilis*, *R. trichophyllus* and possibly *R. peltatus*. The first hybrid, which is quite sterile, has been confirmed cytologically for Derbyshire, and some field records for this hybrid may in error have been assigned to *R.p.*.

Current trends Much less common than formerly, particularly in lowland areas, presumably as a result of increased eutrophication and/or industrial pollution.

---

216 *Ranunculus repens*

Creeping Buttercup, Crowfoot

Established strategy C–R.
Gregariousness Can form patches, more usually interspersed with other spp.
Flowers Yellow, normally hermaphrodite, insect-pollinated.
Regenerative strategies (V), B_s.

Seed 2.32 mg.
British distribution Widespread (100% of VCs).
Commonest habitats Meadows 53%, river banks 32%, soil heaps 31%, shaded mire 27%, unshaded mire 24%.

Spp. most similar in habitats *Galium palustre* 68%, *Cardamine pratensis* 68%, *Juncus effusus* 65%, *Myosotis scorpioides* 64%, *Stellaria alsine* 63%.

Full Autecological Account *CPE* page 484.

Synopsis *R.r.* is a stoloniferous, winter-green herb of fertile and disturbed habitats, particularly on poorly-drained soils, and in areas of high rainfall. *R.r.* flourishes in mire, grassland and a range of disturbed habitats, and displays a wider habitat range than *R. acris* or *R. bulbosus*. *R.r.* is frequent in meadows, though less so than *R.a.*, and can survive infrequent digging or ploughing. The occurrence of *R.r.* is generally indicative of present or past human disturbance. *R.r.* appears to be less-effectively defended against predation than *R.a.* or *R.b.*, and it is frequently eaten by stock, apparently without ill-effect. Seeds, too, are more heavily predated than those of other common buttercups. *R.r.* is more sensitive to translocated herbicides, and its distribution relative to *R.a.* and *R.b.* may be restricted accordingly. Demographic studies have been conducted on pasture populations. Like other buttercups, *R.r.* is dependent upon bare areas for regeneration by seed; open areas also encourage vigorous proliferation of stolons. There is a rapid turnover of ramets which may allow survival and spread of the species in sites subject to moderate disturbance. *R.r.* is the common buttercup most dependent upon vegetative regeneration. Some population biologists term the vegetative spread of *R.r.* (by means of widely-spaced offsets borne on stolons) 'the guerrilla strategy', as opposed to 'the phalanx strategy' of tufted plants, where shoots are closer together. In close turf, stolons may serve primarily to replace the parent by a daughter, but in the absence of competition on fertile soils stolons up to 1.5 m in length are formed, producing several widely spaced ramets. Colonization of new sites appears to be effected both by seed and by detached plantlets, probably dispersed through the agency of humans or domestic animals. Although produced in smaller numbers, seeds of *R.r.* are more persistent in the soil than those of *R.a.* or *R.b.*. The wide range of genetic variation within *R.r.* requires taxonomic treatment.

Current trends *R.r.* colonizes and persists in a wide range of artificial habitats. A further increase in abundance is predicted

---

217 *Ranunculus sceleratus*   Celery-leaved Crowfoot

Established strategy R.
Gregariousness Sparse.
Flowers Yellow, hermaphrodite, insect-pollinated, self-incompatible.
Regenerative strategies B$_s$.

Seed 0.16 mg.
British distribution Common except in N Scotland (94% of VCs).
Commonest habitats Unshaded mire 7%, manure and sewage waste 6%,

shaded mire 4%, river banks 3%, soil heaps 2%.
Spp. most similar in habitats Insufficient data.
Full Autecological Account *CPE* page 486.

Synopsis *R.s.* is winter- or, more commonly, summer-annual and is characteristic of bare, fertile mud, particularly at the edges of ponds and ditches. *R.s.* and *Rorippa palustris* share a number of attributes which facilitate exploitation of transient waterside habitats. Under productive conditions, each produces large quantities of small seeds (up to 56 000 per plant in the case of *R.s.*). Like *R.p.*, *R.s.* forms a persistent seed bank and submerged seeds remain viable for many years. In both species, germination is enhanced by fluctuating temperatures and by exposure to sunlight. Achenes are adhesive to animals and may float for several days. Both species are capable of rooting from detached stem pieces, although this does not appear to constitute a common mechanism of regeneration. The major differences between the two species are as follows. (a) In *R.s.* a proportion of the freshly dispersed seeds are readily germinable, and seedlings may become established in the field during autumn as well as in spring and summer. Seedlings are frost-tolerant. (b) *R.s.* has determinate growth and plants often set seed and die in *c.* 2 months. Thus, *R.s.* may produce two generations within a growing season. In contrast, the shorter *R.p.* tends to maintain flower and fruit production throughout the growing season if conditions remain favourable. (c) *R.s.* is more succulent and less woody than *R.p.*, and appears to be less persistent in drier sites. These differences accord with the tendency of *R.s.* to exploit fugacious sites close to the water's edge, whereas *R.p.* is particularly associated with areas of more-predictably exposed mud beside reservoirs. Plants of *R.s.* which have become submerged have been observed to develop floating leaves. Modified submerged leaves have also been recorded. Like *R.p.*, *R.s.* has a southern and lowland distribution, and its sites seldom exceed 300 m. *R.s.* is potentially one of the more troublesome species of *Ranunculus* to farmers, since it may often be eaten by stock as a result of its succulent texture; the poisonous principle is protoanemonin. A further ssp. is recorded from Europe.
Current trends An effective colonist of disturbed, fertile wetlands, which may be increasing. Perhaps also spreading to higher altitudes as a result of human activities.

218   *Reynoutria japonica*

Var. *japonica*

Established strategy C.
Gregariousness Stand-forming even at low frequencies.

Japanese Knotweed

Flowers Greenish-white, functionally dioecious, insect-pollinated.
Regenerative strategies V.

Seed 0.61 mg.
British distribution Extensively
naturalized (99% of VCs).
Commonest habitats Soil heaps 4%,
river banks 3%, coal mine waste
2%, acidic woodland 1%.
Spp. most similar in habitats
Insufficient data.
Full Autecological Account *CPE* page
488.

Synopsis Probably the most aggressive
of the relatively common herbaceous
dominants of the British flora. It is
now an offence to allow *R.j.* to escape
into the wild. *R.j.* is a native of Japan,
Taiwan and N China, and was first
introduced into Britain in 1825 as an
ornamental plant. It was first recorded
as a naturalized alien in 1886. In
Japan, *R.j.* is often an early colonist of
volcanic soils, and may occur on soils
of pH < 4.0. In Britain the species is a
widespread colonist of habitats such as
river banks, road verges and railway
banks, with coal-mine spoil and cinder
tips being perhaps the nearest
equivalent of its native habitat. Many
of its habitats are undoubtedly
productive but coal-mine sites, some of
which have a low pH, are likely to be
less fertile. Moreover, *R.j.* has a well-
defined capacity to grow on soils of low
N status in Japan. The concentration
of *R.j.* in urban areas in Britain is a
reflection of its horticultural origin.
R.j. is an exceedingly tall (to 2 m),
polycarpic perennial, and a dense
canopy develops each summer beneath
which few species are capable of
surviving. Coexistence with *R.j.* is
restricted further by the persistent
stem litter which accumulates within
established stands. The shoots die back
each autumn, and early growth of new
shoots in the spring is sustained by
reserves in the thick fleshy rhizome.
R.j. is toxic to at least some species of
livestock. The species is functionally
dioecious, and only female flowers
(with reduced anthers) have been
observed within the Sheffield region. A
small amount of seed may be set, and
some has been induced to germinate in
the laboratory. However, no seedlings
have been recorded in the field in
Britain. Thus, effective regeneration
appears to be entirely by means of the
extensive and rapidly-growing rhizome
system. Clones many metres in
diameter are common. Where *R.j.*
occurs adjacent to rivers it may be
widely dispersed by natural means, but
on most forms of wasteland
long-distance dispersal appears to be
sporadic and dependent upon human
disturbance, such as the transport and
tipping of soil contaminated with
rhizomes. Apart from a geographical
restriction resulting from low mobility,
the spread of *R.j.* may be constrained
further by climatic factors. Shoots are
vulnerable both to late frosts and to
summer drought, and *R.j.* has spread
most rapidly in regions where these
phenomena are less severe,
particularly in the W of Britain and in
relatively frost-free urban areas. If it
were not for these limiting factors,
R.j., which is the tallest common
polycarpic herbaceous species
(excluding twining plants) in the
British flora, would be even more
common, particularly since the species
is resistant to a variety of herbicides.
The robust shoot apices can penetrate
asphalt. Var. *compacta*, a native of
India with stems < 700 mm tall and

$2n = 44$ (4x), is less frequent in Britain. Plants with $2n = 66$ (6x) may be referable to the hybrid between *R.j.* ($2n = 88$) and the more robust (to 3 m) and less-frequently naturalized *R.*

sachalinensis ($2n = 44$).

Current trends An invasive, highly competitive, polycarpic perennial weed, which is likely to become even more troublesome.

219 *Rhinanthus minor*   Yellow-rattle

Established strategy Between R and S–R.
Gregariousness Frequently forms dense populations.
Flowers Yellow, hermaphrodite, insect-pollinated or selfed.
Regenerative strategies S.
Seed 2.84 mg.
British distribution Widespread (100% of VCs).
Commonest habitats Meadows 16%, limestone pastures/rock outcrops 2%, enclosed pastures 1%.
Spp. most similar in habitats *Bromus hordeaceus* 93%, *Festuca pratensis* 82%, *Phleum pratense* 79%, *Alopecurus pratensis* 78%, *Ranunculus bulbosus* 71%.
Full Autecological Account *CPE* page 490.

Synopsis *R.m.* is a hemi-parasitic, summer-annual herb found in a wide range of grassland habitats on soils of moderate to low fertility. The species is most typical of hay meadows, where survival of *R.m.* depends critically upon the capacity to shed seed before harvest. *R.m.* is potentially toxic to stock, and is extremely sensitive to the effects of heavy grazing, reflecting the vulnerability of the flowering shoots to predation and the total dependence upon seed production each year for regeneration and persistence of populations. *R.m.* is absent from droughted and from shaded habitats. Although capable of limited autotrophic growth, *R.m.* appears to be an obligate hemi-parasite in the field. *R.m.* has a wide host-range, particularly amongst the Gramineae,

but grows considerably larger when attached to certain species, most notably *Trifolium repens*. *R.m.* receives carbohydrates, and presumably also water and mineral nutrients, from its host. The possibility that the relatively high levels of mannitol present in the plant plays an important osmotic role in the maintenance of hemi-parasitism deserves investigation. *R.m.* produces moderately large winged seeds. These have a poorly differentiated embryo, and chilling is required to break dormancy. Germination occurs in spring and no persistent seed bank is produced. *R.m.* is polymorphic, and the ecotypic variation present is not readily subdivided into taxa. The two commonest ecotypes are var. *stenophyllus*, which is found mainly in moist grassland, has a northern bias and flowers from July to August, and var. *minor*, which is associated with drier sites, particularly in the S, and flowers from May to July. Ecotypes from montane habitats and with an intermediate flowering period have also been separated. The origins of these variants, which are thought to have evolved recently, are uncertain. **Current trends** Decreasing as a result of habitat destruction and the re-sowing and application of fertilizers to meadows.

220 *Rorippa palustris*   Marsh Yellow-cress

Formerly included in *R. islandica*

Established strategy R.
Gregariousness Sparse.
Flowers Yellow, hermaphrodite, autogamous.
Regenerative strategies B$_s$, S, (V).
Seed 0.07 mg.
British distribution Widespread except in N Scotland (86% of VCs).
Commonest habitats Unshaded mire 7%, manure and sewage waste 6%, walls 3%, arable 1%.
Spp. most similar in habitats
Ranunculus sceleratus 76%, *Epilobium obscurum* 75%, *Epilobium palustre* 74%, *Alopecurus geniculatus* 70%.
Full Autecological Account *CPE* page 492.

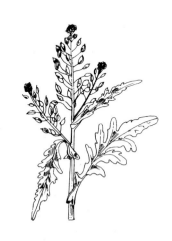

Synopsis *R.p.* is a summer-annual which is characteristic of bare, fertile mud which becomes exposed in summer at the edge of reservoirs and, to a lesser extent, ponds and ditches. Also rarely recorded from moist arable land and as a weed of horticulture. In drier sites, where the mud is exposed early, plants are frequently small and may produce < 200 seeds, whereas *c.* 130 000 may be formed on a plant of average size. *R.p.* has a southern and lowland distribution in Britain, and does not extend above 320 m. The species is associated with a wide range of edaphic conditions from mildly acidic silt flooded by acidic peaty water to exposed river beds in limestone dales. *R.p.* is absent from sites grazed by stock. Regeneration is almost entirely by seed, and involves the formation of a persistent seed bank. The germination of seeds is promoted by fluctuating temperatures and by light. Thus, seeds close to the surface of exposed mud may germinate, but those which are in submerged soil will not. Detached portions of vegetative shoots rapidly re-root. This capacity often causes the plant to re-anchor following flooding. Exceptionally, main stems dislodged onto the surface of the mud may produce roots and shoots along their entire length, to form small clonal patches. Seeds float in water for several days and adhere to birds. They may also be transported by humans and *R.p.* is an early colonist of wetland sites. *R.p.* is ecologically very similar to *Ranunculus sceleratus*; differences between the two are described under *R.s.*. Hybrids between *R.p.* and *R. amphibia* have been reported. *R.p.* is tetraploid ($2n = 32$); the closely related diploid ($2n = 16$) *R. islandica sensu stricto* has been

recognized. *R.i.* has a very restricted northern coastal distribution in Britain and Europe, with outlying stations in the Alps and Pyrenees.
Current trends Although the

distribution of *R.p.* is restricted at present, the association with fertile artificial habitats suggests the potential to expand into disturbed fertile wetlands; probably increasing.

221 *Rubus fruticosus* **Blackberry, Bramble**

Includes *Rubus* subgenus *Rubus*,
  except *R. caesius*

Established strategy S–C.
Gregariousness Stand-forming.
Flowers White or pink, hermaphrodite, cross-pollinated by insects, or apomictic.
Regenerative strategies V, B_s.
Seed 2.49 mg.
British distribution Widespread (100% of VCs).
Commonest habitats Hedgerows 46%, acidic woodland 34%, scrub/limestone woodland 33%, coniferous plantations 29%.
Spp. most similar in habitats *Acer pseudoplatanus* (juv.) 80%, *Milium effusum* 80%, *Pteridium aquilinum* 80%, *Hyacinthoides non-scripta* 80%, *Sorbus aucuparia* (juv.) 78%.
Full Autecological Account *CPE* page 494.

Synopsis *R.f.* is a robust, spiny shrub or undershrub, capable of local dominance of herbaceous vegetation. Some suggest that *R.f.* cannot achieve or maintain dominance in unshaded habitats. This is consistent with the predominantly woodland and hedgerow distribution, but takes insufficient account of the capacity of some taxa to form extensive thickets in unshaded sites or on derelict land and, in particular, on railway property. The greater frequency of *R.f.* on acidic soils is confirmed by our own data for the Sheffield region. However, *R.f.* is remarkably wide-ranging and occurs in most habitat types. *R.f.* regenerates mainly by vegetative means. During the short days of autumn, stem apices become positively geotropic and, when

they establish contact with the soil, produce roots and a resting bud. More rarely, adventitious shoots (suckers) may be formed on the roots from a depth of up to 0.5 m. *R.f.* may also reproduce from root fragments. The woody stems usually persist for only 2–3 years, but stands do not degenerate in the centre as in the case of species such as *Pteridium aquilinum*. In their second year, stems may produce flowers and seed production, often the result of apomixis, may reach 40 000 per 'plant' in open situations. However, in very shaded sites flower and fruit production is low. Much of the seed is non-viable, and germination does not occur until the second spring after berries ripen. This is because there is double dormancy, imposed by a hard seed coat coupled with a chilling requirement. Seedlings appear to be more vulnerable than new vegetative shoots to dominance by established plants. Although regeneration by seed in dense perennial communities is infrequent, the berries are palatable to birds, animals and humans and seed is probably important in the colonization of new sites. Plants originating from seed do not flower until they are at least three years old. *R.f.* is plastic, in growth form and morphology, to factors such as irradiance level, which makes some taxa difficult to identify. However, the most important variations within *R.f.* are of genetic origin. Thus, *R.f.*, as defined here, is a complex group encompassing over 386 British microspecies, classified into seven sections. Most are apomictic, and many probably arose during the Pleistocene through the combined effects of hybridization and facultative apomixis. This speciation is still continuing. The wide habitat and strategic range of *R.f.* is in part a reflection of ecological differences between segregates. Thus, the commonest, and the only British diploid, *R. inermis* is usually sexual, has a many-flowered, compound, late-flowering inflorescence, has stems rooting at the apex, and is found on chalk and on heavy clay soils. In contrast, members of Sect. Suberecti have a simple, early-flowering inflorescence, stems which do not root at their ends and are associated with acidic soils. Most-recent treatments of the genus remove *R. caesius*, in the Sheffield region typically a plant of calcareous woodland, to subgenus *Glaucobatus* and place Sect. Triviales, a group which has arisen as a result of hybridization with *R.c.*, outside the aggregate altogether. However, because of problems of field identification the broader, but taxonomically incorrect, treatment which includes Sect. T. in *R.f.*, has been adopted here. Species with late leaf-fall are restricted to areas where extreme winter conditions are short-lived; they have a smaller distributional area within Sweden. Wide-ranging taxa in Sweden tend to have heavier seeds, form fewer seeds per berry, have early ripening of fruits and possess seeds with high germinability.

Current trends Uncertain. Reduction of populations has occurred through, for example, the removal of hedgerows but the species is encouraged by the increased disturbance associated with modern forestry.

---

222 *Rumex acetosa*　　　　**Common Sorrel**

Established strategy C–S–R.
Gregariousness Intermediate.
Flowers Greenish, dioecious, wind-pollinated.
Regenerative strategies S, V.
Seed 0.74 mg.
British distribution Widespread (100% of VCs).
Commonest habitats Lead mine spoil 65%, meadows 64%, limestone pastures 31%, limestone wasteland 29%, verges 18%.
Spp. most similar in habitats
　Ranunculus acris 79%, *Cerastium fontanum* 75%, *Minuartia verna* 67%, *Euphrasia officinalis* 66%, *Anthoxanthum odoratum* 65%.

Full Autecological Account *CPE* page 496.

Synopsis *R.a.* is found in a wide variety of habitats, but is particularly characteristic of meadows and pastures. The rosette of ascending leaves and the tall, flowering stem, coupled with the capacity for early seed set, facilitate the exploitation of hay-meadows. Under grazing, *R.a.* persists as a small, low-growing rosette and sets seed less regularly. This plasticity of response can operate within a single season, apparently enabling *R.a.* to exploit both the mown and the grazed phases of the 'hay and aftermath' management regime applied to many upland meadows. Much of the additional morphological variation within the species is also phenotypic in origin. *R.a.* extends occasionally onto calcareous soils, and is locally abundant on lead-mine spoil, where heavy-metal-tolerant races occur. In Europe, races differing in shade tolerance are recorded. The species most consistently achieves abundance on mildly acidic brown earths. The roots are not subject to heavy colonization by VA mycorrhizas, and in turf microcosms no consistent benefit from inoculation occurred in circumstances where other species, e.g. *Scabiosa columbaria*, were strongly promoted. The leaves contain oxalates and are potentially toxic to stock if eaten in large quantities. Regeneration is mainly by means of seed, which is set abundantly despite the fact that *R.a.* is dioecious. No persistent seed bank is formed. Seeds can survive ingestion by cows. However, *R.a.* is probably mainly dispersed by human activity as an impurity within hay and seed crops. *R.a.* exhibits a limited capacity to produce daughter ramets, and this may be of local significance in grazed and trampled habitats. Populations normally have about twice as many female as male ramets, despite the fact that the sex ratio of seed is *c.* 1:1. Female inflorescences are formed later, but on taller stems, and leaves of female ramets are located at a greater height and are maintained later into the summer. On the basis of these results, it has been suggested that male and female plants occupy different niches. A number of related species occur in Europe. But, in contrast, with subgenus *Rumex* (including *R. crispus* and *R. obtusifolius*), hybridization is not common.

Current trends Since *R.a.* is most characteristic of mildly acidic soils, is dioecious, has little capacity for effective vegetative spread, and has no persistent seed bank, we predict that it will decline rapidly in areas converted to intensive agricultural management.

223   *Rumex acetosella*

Field data for the aggregate species only

Established strategy Between C–S–R and S–R.
Gregariousness Often patch-forming.
Flowers Green, dioecious, wind-pollinated.
Regenerative strategies V, B$_s$.
Seed 0.4 mg.
British distribution Widespread (100% of VCs).
Commonest habitats Coal mine waste

Sheep's Sorrel

25%, acidic wasteland 20%, acidic quarries 18%, acidic pastures 14%, rock outcrops 11%.
Spp. most similar in habitats *Agrostis capillaris* 79%, *Aira praecox* 78%, *Hypochaeris radicata* 67%, *Deschampsia flexuosa* 61%.
Full Autecological Account *CPE* page 498.

Synopsis *R.a.* is a patch-forming herb characteristic of relatively infertile sandy or peaty soils. Because of its low growth habit, *R.a.* is readily dominated by taller herbs and grasses. Although capable of persistence in a non-flowering condition in open scrub and other shaded habitats, *R.a.* tends to be restricted to areas where, as a result of mechanical disturbance, fire or a rocky or sandy terrain, the vegetation is very open. *R.a.* also persists in poor pasture over acidic strata. The relatively deep root system affords access to subsoil water during summer and enables *R.a.* to coexist on dry sandy soils with winter annuals such as *Aira praecox*. The foliage is little grazed, and the plant contains oxalates which are toxic to stock. *R.a.* is characteristically calcifuge; seedlings grow very poorly and are highly chlorotic when grown on calcareous soils. Among the species exploiting acidic soils, *R.a.*, like *Holcus mollis*, exhibits an unusually high potential relative growth rate. Perhaps for this reason both *R.a.* and *H.m.* are sometimes important weeds of ornamental or forestry tree nurseries. The species regenerates vegetatively, forming extensive patches by producing adventitious buds on horizontal roots. The subsequent breakdown or mechanical severing of these roots results in the formation of daughter colonies. *R.a.* also regenerates by seeds, which germinate in spring and form a persistent bank of buried seed. Although dioecious, most populations appear to set seed and, though varying considerably, tend to show a 1:1 ratio between males and females. Some suggest that males and females differ in response to water-stress, and there is evidence suggesting that males allocate more resources to vegetative regeneration and are

therefore more persistent in closed vegetation. Male inflorescences are produced earlier than female inflorescences. The dispersal of *R.a.* is frequently aided by humans. Seeds occur as a commercially important contaminant of horticultural peat. Seeds can survive ingestion by birds, cattle, horses and pigs, and in some experiments germination has been found to be stimulated by nitrate. Most, and possibly all, of our field records refer to *R.a. sensu stricto*. The aggregate species consists of a polyploid series (*R. angiocarpus*, 2*x*, which may not occur in Britain; *R. tenuifolius*, 4*x*, and *R. acetosella*, 6*x*). In N England, forms resembling *R.t.* are found on some of the most acidic sandy soils of the Bunter sandstone, but many intermediate with *R.a.* occur and the two are not readily separable. **Current trends** Population expansion can occur rapidly on disturbed acidic soils, and the species seems likely to remain locally common in upland and lowland Britain.

---

224   *Rumex crispus*

Curled Dock

Established strategy C–R or R.
Gregariousness Mainly occurring as
scattered individuals.

Flowers Green, usually hermaphrodite
and wind-pollinated.
Regenerative strategies B$_s$, (V).

Seed 1.33 mg.
British distribution Widespread (100% of VCs).
Commonest habitats Cinder heaps 17%, bricks and mortar 16%, soil heaps 14%, coal mine waste 9%, lead mine spoil/unshaded mire 7%.
Spp. most similar in habitats *Senecio viscosus* 74%, *Atriplex prostrata* 70%, *Artemisia vulgaris* 70%, *Tussilago farfara* 64%, *Senecio squalidus* 63%.
Full Autecological Account *CPE* page 500.

Synopsis *R.c.* is a short-lived perennial or, more rarely, annual herb, with a tap-root and one to several clumped stems. *R.c.* is capable of flowering in its first year, and is most frequently associated with sites that are both disturbed and fertile. As well as being one of the world's most widely distributed angiosperms, *R.c.* is also considered to be one of the world's worst weeds and is recorded as important in 37 countries in 16 arable crops and in pastures. *R.c.* is toxic to stock, and in Britain its control is required under the 1957 Weeds Act. In common with many species exploiting agricultural land and other artificial habitats, *R.c.* displays high phenotypic plasticity and forms well-defined ecotypes. Seeds, which are borne in numbers ranging from less than 100 to over 40 000 per plant, are polymorphic with respect both to size and to germination characteristics; seed characters may differ according to ecotype. The germination is complex. The probability of flowering is correlated with rosette size during the previous autumn. *R.c.* is capable of regeneration from root fragments, but effective reproduction appears to be almost exclusively by seed, though establishment of seedlings is often rather slow. A persistent seed bank is formed and seeds retained viability in soil for 80 years in one experiment. Fruits survive ingestion by some birds and *R.c.* may be dispersed in this way.

The most noteworthy ecotype is var. *littoreus*, which tolerates immersion in sea water and is often abundant on maritime shingle. This variety lacks the polymorphism with respect to seed dormancy of typical ruderal populations and may not form a persistent seed bank. In addition this ecotype has three tubercles on its perianth segments (most inland races and typical British *R. obtusifolius* have one), and this characteristic, which is shared by many European docks that grow near water, may aid buoyancy and enable the dispersule to float in sea water for many months. The fruits of other forms of *R.c.* also float, but for a lesser period. In a cultivation experiment involving both European and N American populations, it was found that a majority of variation was associated with phenotypic plasticity, suggesting that populations of *R.c.* were mainly in-breeding. Indeed, in cultivation, populations from N America and the British Isles allocate a similar proportion of their resources to reproductive parts, but some differences occur, e.g. in seed size, and populations from ruderal sites, maritime habitats and tidal mud of rivers appear genetically distinct in a number of ecologically important respects. Ten hybrids involving *R.c.* are recorded for Britain.
Current trends *R.c.* is an effective

colonist of many artificial habitats and may even be increasing as a result of the continuing high levels of disturbance. Since it is now uncommon on arable land, *R.c.* may be expected to show an increasing tendency towards perenniality.

225 *Rumex obtusifolius*

Ssp. *obtusifolius*

Established strategy C–R.
Gregariousness Scattered individuals.
Flowers Green, usually hermaphrodite, pollination mainly by wind, self-fertile.
Regenerative strategies B_s.
Seed 1.1 mg.
British distribution Aggregate widespread (100% of VCs).
Commonest habitats Soil heaps 43%, bricks and mortar 33%, river banks 20%, arable 18%, cinder heaps 13%.
Spp. most similar in habitats *Artemisia vulgaris* 72%, *Tripleurospermum inodorum* 72%, *Senecio vulgaris* 69%, *Lamium purpureum* 63%, *Epilobium ciliatum* 62%.
Full Autecological Account *CPE* page 502.

Synopsis *R.o.* is essentially similar to *R. crispus,* both in form and in ecology and the two frequently hybridize. *R.o.* is rather less ruderal, typically more robust, exclusively perennial and usually does not flower until its second year. *R.o.* also exhibits a greater capacity for vegetation dominance. Although classified as one of the commonest weeds, *R.o.* is clearly much less important than *R.c.* worldwide, and has the narrower geographical range. Also, *R.o.* is probably less wide-ranging ecologically. In Britain the species is designated under the 1959 Weeds Act, and is probably now more troublesome than *R.c.* as a weed of pastures, particularly around field edges and in trampled gateways. It is also more common than *R.c.* in the infrequently cut grass swards of roadsides. *R.o.* is

Broad-leaved Dock

generally refused by cattle, sheep and horses. However, seed number and weight are affected by the grazing of the beetle *Gastrophysa viridula* and *R.c.* is even more susceptible. Established plants of *R.o.* survive mowing and show a greater tendency than those of *R.c.* to produce two crops of seed. For both *R.c.* and *R.o.*, stems cut down 14 days after anthesis produce seed with a germination percentage not significantly different from that of controls. Seeds of *R.o.* mature less rapidly than those of the more ruderal *R.c.*. In addition to the mechanisms for dispersal already described, the spiny teeth on the perianth enclosing the fruit facilitate dispersal in animal fur and in clothing. Seed germinates mainly in spring and forms a persistent bank of buried seeds. *R.o.* often forms compact patches, but not of sufficient size to permit effective vegetative spread. However, pieces of underground stem and, in spring, root fragments readily regenerate after ploughing. Seeds are polymorphic with respect to germination behaviour. *R.o.* frequently grows close to *Urtica dioica* and juice from the leaves is customarily applied to the skin to relieve nettle stings. A further three ssp. are recorded from Europe including ssp. *sylvestris* and ssp. *transiens*. These are naturalized in S England. Within the British Isles hybrids are recorded with 11 other species.

Current trends Well established in a wide range of artificial habitats, and perhaps increasing.

226 *Sagina procumbens*

Procumbent Pearlwort

Established strategy Between R and C–S–R.
Gregariousness Intermediate.
Flowers Green, hermaphrodite, homogamous, often self-pollinated.
Regenerative strategies B_s, (V).
Seed 0.02 mg.
British distribution Widespread (100% of VCs).
Commonest habitats Cinder heaps 25%, paths 12%, walls 9%, verges 8%, bricks and mortar 6%.
Spp. most similar in habitats
Chaenorhinum minus 64%, *Salix caprea/cinerea* (juv.) 58%, *Plantago major* 54%, *Taraxacum* agg. 53%, *Epilobium ciliatum* 52%.
Full Autecological Account *CPE* page 504.

Synopsis *S.p.* is a herb of appressed growth habit and moss-like appearance, found in a wide range of moist, disturbed, moderately fertile habitats. As a consequence of the low growth form, *S.p.* is susceptible to competition from the majority of other pasture species, and is particularly abundant in lawns, where the growth of other species is debilitated by the effects of close and frequent mowing. *S.p.* is equally frequent in sites with disturbed soil, such as spoil heaps with high exposure of bare ground. In pasture *S.p.* is little grazed by stock. *S.p.* is tolerant of trampling and colonizes cracks between paving stones. *S.p.* is also a common garden weed, often occurring in lightly shaded situations. Regeneration by a combination of seedling establishment and vegetative means occasionally results in the formation of local monocultures of up to 0.25 m². Vegetative regeneration involves the rooting of prostrate stems, which we suspect may form new plants if detached. As with many other ruderal species, the season over which flowering and seed set may occur is long. Seeds are numerous, minute and highly mobile, and may germinate in autumn or in spring. A large bank of buried seed may be formed. The longevity of established plants of *S.p.* varies. In highly disturbed sites *S.p.* may be short-lived, and behaves as a summer annual on reservoir margins. However, the species appears to be relatively long-lived in some grassland sites, and plants have been observed which appear to persist for *c.* 8 years in a lawn. A change, for no apparent reason, has also been noted from a long- to a short-life duration within a lawn population of *S.p.*. Plants adopt a weakly ascending habit in light shade. We suspect that, like several other lawn weeds, *S.p.* is both phenotypically plastic and genotypically variable; the species is very variable in Europe. Although mainly self-pollinated, hybrids between *S.p.* and other *Sagina* spp. have been reported, and *S.p.* tends to be replaced by other species of *Sagina* in dry, maritime, heathland and montane environments.

Current trends The seed of *S.p.* is both persistent in the soil and readily dispersed by human activities. *S.p.* is a common and probably increasing colonist of artificial habitats.

227   *Salix*   Sallow

Field records include
   seedlings/saplings; spp. include
   caprea, cinerea and related hybrids

Established strategy C.
Gregariousness Usually sparse to
   intermediate.
Flowers Usually dioecious, mainly
   insect-pollinated, some
   wind-pollination possible, visited by
   birds.
Regenerative strategies W, (V).
Seed 0.09 mg.
British distribution Widespread (*S.cap.*
   99% of VCs).
Commonest habitats Cinder heaps
   10%, bricks and mortar/limestone
   quarries 8%, unshaded mire 7%,
   manure and sewage waste/walls 6%.
Spp. most similar in habitats *Rorippa
   palustris* 63%, *Epilobium hirsutum*
   61%, *Sagina procumbens* 58%,
   Epilobium ciliatum 55%, *Alnus
   glutinosa* (juv.) 53%.
Full Autecological Account *CPE* page
   506.

Synopsis *S.* spp. is a deciduous shrub or small tree with diffuse-porous wood and early canopy emergence. As a seedling or young sapling *S.* spp. occurs on bared ground in a wide range of moist but little-grazed habitats. This broad distribution in disturbed habitats stems from the capacity of *S.* spp., despite being dioecious, to produce an abundance of minute, wind-dispersed seeds. These are short-lived (germinating exceedingly rapidly over a wide range of temperatures in sunlight and in darkness) and do not form a persistent seed bank. In *S.cap.*, male and female bushes tend to occur in equal numbers; in *S.cin.* there is a bias towards females. *S.* spp. are tolerant of cutting, and the capacity of detached twigs to re-root and sprout, forming new plants, also may be of importance. It is not known whether, like *S. fragilis*, *S.* spp. have primordial adventitious roots in their twigs. The ecological limits of the mature plants are much narrower than those of juveniles. Thus, as canopy components, *S.* spp. are typically associated with moist soils and are more frequent on N-facing slopes. *S. spp.* tends to occur on relatively fertile soils, and produces a dense canopy. Litter is generally only shortly persistent. Seedlings have a high growth rate relative to other woody species, and establishment from seed appears restricted to unshaded sites. Plants are also capable of growing from a coppiced stump at the rate of 50 mm day^{-1} and may attain a height of 3–4 m in a single season. The mature plants occur mainly as bushes or as open scrub on wasteland, spoil or in mire, and they are much more frequent as canopy species in scrub than in woodlands. Other studies illustrate that the invasion by *S.* spp. may represent the first stage in the development of secondary woodland or of fen carr, and the distribution of *S.* spp. outlined above is consistent with the premise that *S.* spp. is a seral species of developing woodland or carr. However, as is the case for *Crataegus monogyna*, the plant may also form long-persistent thickets, particularly in upland areas with well-established scrub. Leaves of *S.cin.* are relatively unpalatable to the snail *Helix*

aspersa and willows in general harbour a large number of insects. *S.cin.* ssp. *oleifolia* (*S. atrocinerea*) is probably the most common willow in N England, and is found on stream-sides, marshes, woodland edges and hedgerows. Outside Britain the taxon appears restricted to France, Spain and Portugal. *S.cap.* is also widespread and shows some tendency to exploit drier, more-calcareous sites, including demolition sites. Hybrids between the two taxa are apparently common, and both segregates form ecotypes. Thus, *S.cin.* ssp. *cinerea* is found in base-rich mire, particularly in East Anglia, and *S.cap.* var. *sphacelata* is a shrub of the Scottish Highlands, extending to 850 m.

Current trends The two segregates of *S.* spp. are effective colonists of bared ground, and are probably increasing. The high level of disturbance in the contemporary landscape is presumably also tending further to erode the ecological barrier between *S.cap.* and *S.cin.*, promoting additional hybridization between the two taxa.

---

228 *Sambucus nigra*

(Field records as seedlings)

Elder, Bourtree

Established strategy C.
Gregariousness Sparse.
Flowers White, hermaphrodite, usually insect-pollinated.
Regenerative strategies S.
Seed 3.4 mg.
British distribution Widespread (100% of VCs).
Commonest habitats Bricks and mortar/hedgerows/river banks 8%, manure and sewage waste 6%, scrub 5%.
Spp. most similar in habitats *Rubus fruticosus* 59%, *Urtica dioica* 58%, *Digitalis purpurea* 57%, *Galium aparine* 57%, *Bromus sterilis* 57%.
Full Autecological Account *CPE* page 508.

Synopsis *S.n.* is a short-lived shrub, or small tree to 10 m, with a rough bark. The wood of *S.n.* is diffuse-porous, and bud-break is early. Indeed, leaves may be formed during warmer spells during winter, and are often subsequently killed by severe frosts. *S.n.* has a Sub-Atlantic distribution. In unshaded habitats the species often produces an abundant crop of berries which are avidly removed by birds as soon as they become ripe. Consequently, seeds are widely distributed and seedlings are recorded from a greater diversity of habitats than the adult plant. The occurrence of seedlings and the establishment of saplings appear to be consistently associated with bared ground. Seedlings appear to be sensitive to competition from polycarpic perennial herbs and to grazing by stock. Once established, *S.n.* is seldom grazed; the species contains cyanogenic glycosides, and tissue has an unpleasant odour when crushed. The leaves are toxic to mammals, and stands of *S.n.* are characteristic of rabbit warrens. It has also been suggested that *S.n.* possesses inducible defences against foliar predation by insects. Litter is only shortly persistent. The species is characteristic of well-drained, base-rich sites of high fertility, and is often found with *Urtica dioica* and *Galium aparine* in farmyards and in other sites close to human habitation.

Long-established colonies may occur on rabbit warrens and other fertile derelict land without any evidence of replacement by taller species. Elsewhere, succession tends to lead to the formation of woodland. *S.n.* is characteristic of more-disturbed and fertile woodlands. Thus, although not one of the most frequent woodland species, *S.n.* is recorded as the third most common volunteer species both in broad-leaved and in coniferous plantations in the Sheffield region. Like *Crataegus monogyna*, the shrub may persist in shaded sites, but few flowers and fruits are formed. Thus, the regeneration of the species is heavily dependent upon plants growing in woodland margins and clearings. No seed bank has been reported. The species is very resilient when coppiced, and cuttings appear to root readily. *S.n.* is a common plant of recent, species-poor hedgerows.

Current trends A common and increasing colonist of fertile derelict land.

229 *Sanguisorba minor*

Salad Burnet

Ssp. *minor*

Established strategy S.
Gregariousness Intermediate.
Flowers Greenish, upper often female, middle and lower hermaphrodite, wind-pollinated.
Regenerative strategies ?S.
Seed 1.8 mg.
British distribution Mainly in the S, rarer in the N and W (64% of VCs).
Commonest habitats Limestone pastures 31%, scree 12%, lead mine spoil 11%, limestone wasteland 9%, rock outcrops 8%.
Spp. most similar in habitats *Koeleria macrantha* 96%, *Avenula pratensis* 93%, *Helianthemum nummularium* 89%, *Carex caryophyllea* 87%,

Campanula rotundifolia 86%.
Full Autecological Account *CPE* page
510.

Synopsis *S.m.* is a tap-rooted, rosette-forming herb particularly characteristic of species-rich pastures on both dry and moist, relatively infertile, calcareous soils. In the *PGE*, *S.m.* only occurs on unfertilized, unproductive plots. Calcium and magnesium contents of the leaves are unusually high. *S.m.* maintains a small rosette of overwintering leaves, and expands the shoot biomass slowly in the spring, reaching a peak in summer. This phenological pattern, which contrasts with species such as *Festuca ovina* with which *S.m.* often grows, may relate to the capacity of the long tap-root to provide access to moisture in deep soil crevices during the period of summer growth. In turf microcosms roots of *S.m.* became heavily colonized by VA mycorrhizas and infection caused up to fivefold increases in seedling yield. Where calcareous soil accumulates a surface layer of acidic humus, larger, older, individuals of *S. minor* with roots mainly exploiting the calcareous subsoil may be observed in situations where seedling establishment is no longer possible. *S.m.* survives grazing both as a seedling by gastropods and as a mature plant by sheep and rabbits. Leaves were formerly used in salads. Because of the low canopy and lack of lateral vegetative spread, *S.m.* is a poor competitor. Thus, although *S.m.* is occasionally represented by large individuals in abandoned pastures, the species is not persistent in tall turf.

S.m. appears to be susceptible to burial and is infrequent on anthills. Shoots regenerate from buds situated on the stock, and it is not unusual to find small multiple rosettes arising. In view of the extended longevity of *S.m.* it seems likely that occasional vegetative offspring and recruitments from seed may be sufficient to sustain population levels (cf. *Primula veris*). Seeds germinate in spring. Seed-set is relatively low, and often as few as four fruits per capitulum contain viable seed. It is not known whether a persistent seed bank is formed. Fruits are large. Thus, seedlings have some capacity to etiolate in response to shading, and at the second true leaf stage the radicle may attain 60 mm in length. Nevertheless, fatalities are high during seedling establishment. Although they may survive ingestion by rabbits, *S.m.* has an extremely limited capacity to disperse to new sites. However, the species may be found occasionally on railway banks on non-calcareous strata. The extent to which *S.m.* forms ecotypes is uncertain, but the difference between low-growing and taller populations has some genetic basis, and five other ssp. have been recognized in Europe. One variant formerly grown for fodder and a native of S Europe, ssp. *muricata*, is established locally in Britain and, in particular, may be observed in relatively tall calcareous grassland on railway banks. Ssp. *m.* is taller and has larger fruits (dispersule 13.0 mg) than ssp. *minor*.
Current trends Mainly restricted to species-rich, calcareous pastures, and decreasing, particularly in lowland areas.

230 *Sanicula europaea*   Sanicle

Established strategy S.
Gregariousness Sparse to intermediate.
Flowers White, insect-pollinated, in umbels of male and hermaphrodite

flowers.
Regenerative strategies S, V.
Seed 2.97 mg.
British distribution Widespread except

in Scotland (98% of VCs).
Commonest habitats Limestone
woodland 13%, scrub 4%, paths
2%, broadleaved
plantations/limestone woodland 1%.
Spp. most similar in habitats
Brachypodium sylvaticum 91%,
Anemone nemorosa 90%,
Mercurialis perennis 90%, *Arum
maculatum* 89%, *Melica uniflora*
88%.
Full Autecological Account *CPE* page
512.

Synopsis *S.e.* is a slow-growing, winter-
green, perennial herb, characteristic of
sites where the growth of potential
dominants is restricted by heavy shade
and probably also by nutrient stress.
The species is long-lived, with a half-
life of between 59 and 360 years. Thus,
S.e. may live for as long as the trees
above it. The species is very shade-
tolerant, a fact illustrated by the N-
facing bias of its woodland sites and its
ability to persist under the canopy of
Taxus baccata and *Fagus sylvatica*. The
species is relatively sensitive to burial
by beech litter. *S.e.* is susceptible to
summer drought. Consistent with this,
the species is characteristic of clayey
calcareous soils, and in N England is
particularly common on the heavier
soils of the magnesian limestone and
Keuper marl. Species with a similar
edaphic restriction include *Carex
sylvatica* and *Viola reichenbachiana*.
Leaves are long-lived and of low
palatability to unspecialized
herbivores; each year the older leaves
are replaced during the spring flush of
growth. The species has only a limited
capacity to produce daughter rosettes
but, because of the low mortality of
mature ramets, bifurcation of the
rhizome may play an important role in
regeneration. Indeed, even though the
rhizome gradually breaks down and no
extensive rhizome system is formed,

S.e. is in many ways comparable with a
Trifolium repens 'growing in slow
motion'. Plants do not flower until
8–16 years old and often produce less
than 100 seeds year^{-1}, which
germinate in spring. No persistent seed
bank has been recorded in field studies
or in pot experiments. The pioneering
studies of plant demography conducted
on permanent plots in woodland in
Sweden confirm the existence of a
bank of persistent seedlings in this
species. The mericarps of *S.e.* are
covered in hooked bristles and are
dispersed by humans and animals.
Within upland Britain, where *S.e.* is
scarce, there is evidence of
long-distance dispersal and, perhaps
rather surprisingly in view of the
population biology of the species, it
has been classified as a fast-colonizer
of Lincolnshire secondary woodland. It
has been confirmed that *S.e.* has some
colonizing ability in secondary
woodland in E England. *S.e.* is one of
the few British representatives of the
subfamily Hydrocotyloideae.
Current trends Uncertain, but perhaps
decreasing through the destruction of
broad-leaved woodland habitats.

Established strategy S–R.
Gregariousness Sparse.
Flowers White, hermaphrodite, usually
    self-pollinated.
Regenerative strategies S, B$_s$.
Seed 0.01 mg.
British distribution Lowland (82% of
    VCs).
Commonest habitats Rock outcrops
    11%, limestone quarries 5%, cinder
    heaps 2%.
Spp. most similar in habitats *Myosotis
    ramosissima* 93%, *Arabidopsis
    thaliana* 88%, *Arenaria serpyllifolia*
    88%, *Trifolium dubium* 86%, *Sedum
    acre* 82%.
Full Autecological Account *CPE* page
    514.

Synopsis *S.t.* is a diminutive, winter-
annual herb of dry, nutrient-deficient,
calcareous, rocky or sandy habitats in
which the cover of perennials is
restricted by summer drought. The
species is narrowly restricted to 'semi-
permanent' winter-annual communities
on rock outcrops, crevices in walls and
sand dunes, and in habitat and
phenology closely resembles *Erophila
verna*. Both *S.t.* and *E.v.* are late-
germinating and early-flowering, and
are presumably associated with the
most drought-prone microsites. *S.t.*
differs primarily in its much smaller
seeds, which are released in summer;
these are prevented from immediate
germination by an after-ripening
requirement and germinate in autumn.
Seedlings perennate as very small
rosettes. Flowering is induced by
vernalization, but there is no absolute
requirement for long days and
flowering occurs rapidly in early
spring. A persistent seed bank is
formed. Seed appears to be mainly
dispersed close to the parent plant.
However, the species is capable of
wider dispersal, e.g. along railway
tracks, and the frequent occurrence of
S.t. on walls suggests that the minute
seeds are capable of movement in air
currents. A perennial hybrid between
S.t. and *S. hypnoides* has been
recorded once for the British Isles.
Current trends Not under immediate
threat, but infrequent as a colonist of
artificial habitats. Confined to local
refugia and probably decreasing
slowly.

232 *Scabiosa columbaria*   **Small Scabious**

Established strategy Between S and
    S–R.
Gregariousness Sparse to intermediate.
Flowers Florets lilac, homogamous to
    protandrous, insect-pollinated.

Regenerative strategies S.
Seed 1.32 mg.
British distribution Absent from
    Ireland and most of Scotland (48%
    of VCs).

Commonest habitats Limestone pastures/scree 12%, limestone quarries 11%, limestone wasteland 9%, cliffs 5%.

Spp. most similar in habitats
Leontodon hispidus 86%,
Campanula rotundifolia 76%,
Hieracium pilosella 74%, *Teucrium scorodonia* 73%, *Arabis hirsuta* 70%.

Full Autecological Account *CPE* page 516.

Synopsis *S.c.* is a winter-green, rosette-forming herb largely restricted to short turf and rocky habitats on dry or moist infertile calcareous soils. *S.c.* has been described as an 'hapaxanth'. Populations under garden cultivation are also short-lived (*c.* 5 years), and floriferous if cultivated on fertile soils, but field observations suggest that plants develop much more slowly and often persist for > 10 years in the Derbyshire dales. Low stature and lack of lateral vegetative spread restrict the capacity of *S.c.* to exploit tall or productive communities. *S.c.* occurs both in grazed and in burned sites. The species is particularly associated with microsites of lower vegetation density in calcareous grassland, and this capacity for sustained growth during summer may be related to the long tap-root which provides access to moisture in deep soil crevices. In turf microcosms, *S.c.* became colonized by VA mycorrhizas and seedlings showed up to fivefold increases in yield as a result of infection. In Derbyshire populations, several rosettes frequently arise from the basal stock, but *S.c.* forms only very small compact patches. The association of *S.c.* with open vegetation appears to be linked to its dependence upon seed for regeneration. Although visited more rarely than *Centaurea* spp. by insects, in ungrazed sites *S.c.* produces abundant seed which germinates mainly in autumn. However, flowering and fruiting heads are often heavily grazed by rabbits, in which case the average number of seeds per plant may only be in the order of 20–60. Mortality rates of seedlings are low, but seed predation may be high. A persistent seed bank has been reported, but on the basis of laboratory germination characteristics we suspect that seed longevity is short. Fruits vary greatly in size within the same capitulum. Seedlings have a relatively similar growth rate irrespective of seed size. Fruits are spiny and have been described as specialized for wind dispersal; this interpretation is not confirmed by our own field observations, and *S.c.* is an ineffective colonist of new sites. The absence of *S.c.* from Ireland may relate to poor dispersal during Postglacial colonizing episodes. The species exhibits much variation in shape and degree of dissection of the leaves, and variation also occurs in life-history. Heavy-metal-tolerant races have been recorded in Derbyshire populations. *S.c.* is part of a taxonomically complex grouping, and two further ssp. are recognized within Europe.

Current trends Largely restricted to old or semi-natural vegetation. Decreasing and near extinction in some lowland areas.

Established strategy S–C.
Gregariousness Typically sparse to
intermediate.
Flowers Yellow, hermaphrodite,
protandrous, insect-pollinated,
sometimes selfed.
Regenerative strategies V, ?B_s.
Seed 0.03 mg.
British distribution Widespread (99%
of VCs).
Commonest habitats Rock outcrops
14%, scree 6%, limestone quarries
5%, lead mine spoil/walls 3%.
Spp. most similar in habitats *Arenaria
serpyllifolia* 86%, *Saxifraga
tridactylites* 82%, *Arabidopsis
thaliana* 78%, *Arabis hirsuta* 76%,
Myosotis ramosissima 75%.
Full Autecological Account *CPE* page
518.

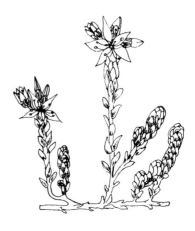

Synopsis *S.a.* is a slow-growing, leaf
succulent forming tight cushions which
expand vegetatively over shallow
infertile soil and bare rock. *S.a.* is the
most widespread British succulent.
Under droughted conditions carbon
fixation shifts from the normal C_3 type
to crassulacean acid metabolism
(CAM). However, the extent of the
carbon gain resulting from CAM
appears slight and the importance of
CAM to the drought resistance
exhibited by *S.a.* is uncertain.
Notwithstanding, the species is
particularly characteristic of dry sand
dunes and steeply sloping, S-facing
exposures of rock. On sand dunes *S.a.*
can survive a limited degree of burial.
S.a. is strongly protected against
herbivory. The foliage is highly acrid
and the plant also contains alkaloids
which are toxic to stock. *S.a.* is almost
exclusively calcicolous in the Sheffield
region. In W Britain a comparable
ecological niche on more-acidic soils is
occupied by the morphologically
similar *S. anglicum*. In addition to
patch formation by creeping stems,
vegetative spread may occur from
detached shoots, or even leaves, which
under suitably moist conditions root to
form new colonies. Numerous small
seeds are produced during summer.
These show increased germination
following laboratory storage, and
probably germinate in autumn. *S.
album* is known to form a persistent
seed bank and the same is predicted
for *S.a.*. This is a morphologically
variable species, but the extent of
ecotypic differentiation between native
populations is not clear. The species
exhibits occasional escapes from
cultivation on rockeries, walls and
cottage roofs, and may be observed
locally in situations such as the cinders
adjacent to disused railway stations. It
is not known to what extent these
'domestic' populations differ
genetically and ecologically from the
native stock. *S.a.* is part of a complex
in which 11 species have been
recognized.
Current trends Despite its frequent
cultivation in gardens and the fact that
the seeds are produced in large
numbers in suitable habitats, the
species is not an early colonist of
artificial habitats, and may be
decreasing.

Established strategy Between R and
 C–R.
Gregariousness Mainly scattered.
Flowers Florets yellow, in capitula,
 disc hermaphrodite/ray female,
 insect-pollinated.
Regenerative strategies W, B_s, (V).
Seed 0.05 mg.
British distribution Widespread (100%
 of VCs).
Commonest habitats Limestone
 quarries 45%, scree 37%, limestone
 pastures 31%, rock outcrops 25%,
 verges 18%.
Spp. most similar in habitats
 Leontodon hispidus 85%, *Linum
 catharticum* 82%, *Hieracium
 pilosella* 81%, *Festuca rubra* 76%,
 Scabiosa columbaria 70%.
Full Autecological Account *CPE* page
 520.

Synopsis *S.j.* is typically a biennial, but
often takes more than two years to
flower. Flowering individuals have
significant carbohydrate reserves in the
stem bases and roots. It has been
suggested that this may allow the
species to behave also as a facultative
perennial. Up to 30 000 achenes may
be produced by a single plant and the
distribution of *S.j.* is limited by a
dependence upon vegetation gaps for
colonization by seed. Thus, *S.j.* is
characteristic of open habitats and is a
noxious weed of overgrazed grassland.
Germination occurs mainly during
autumn, with subsidiary cohorts of
seedlings appearing in spring and
summer. *S.j.* forms a buried seed bank
in some circumstances but seed is only
briefly persistent. Most seed is
dispersed only short distances from the
parent plant, but field evidence
suggests that *S.j.* is highly mobile. The
seed survives ingestion by sheep but
not birds. Survival in relatively
undisturbed habitats may be effected
by vigorous recolonization in bared
areas formerly occupied by individuals

of *S.j.*. Plants also have the capacity to
regenerate from root fragments, which
may be important in disturbed
habitats. New basal branching may
occur, particularly following damage,
to give several rosettes which may
become independent. *S.j.* is highly
toxic to cattle and horses, and is
avoided if alternative food is available.
The poisonous principles are
pyrrolizidine alkaloids which are not
destroyed by drying or storage. The
species is also unpalatable to rabbits
and may be prominent near rabbit
warrens. *S.j.* is grazed by sheep,
sometimes with harmful effects, and
often persists in sheep pastures as non-
flowering rosettes for many years.
Shoots of *S.j.* are often defoliated and
flowering is suppressed by the
caterpillars of the cinnabar moth
(*Tyria jacobaeae*), which has been used
in the biological control of *S.j.* in
countries where *S.j.* has become a
serious weed. *S.j.* shows a number of
morphologically distinct variants.
There is seed polymorphism and a
degree of heterophylly (stem leaves are
much more dissected than those of
rosettes). The existence of ecotypes
has been recognized. Taxonomists
have recently separated certain rayless,
low-growing coastal populations in the

Scottish and Irish coasts as ssp. *dunensis*, and other populations from dunes and sea cliffs with swollen stem bases as ssp. *jacobaea* var. *condensatus*. In wetter sites *S.j.* is replaced by *S. aquaticus*, with which it hybridizes.

Current trends *S.j.* undoubtedly owes its present abundance to an ability to exploit artificial habitats. Management to eliminate, or at least contain, *S.j.* in pastures is a legal requirement under the Weeds Act of 1959 and, although still common, *S.j.* is now only abundant in poor pasture and wasteland, particularly on sandy or other freely drained soils where the presence of many gaps in the vegetation allows the establishment of seedlings. *S.j.* is thus probably decreasing in many, but not all, areas.

235 *Senecio squalidus*

Oxford Ragwort

Established strategy Between C–S–R and R.
Gregariousness Sparse.
Flowers Florets yellow, self-incompatible, insect-pollinated.
Regenerative strategies W.
Seed 0.21 mg.
British distribution Throughout the S, local elsewhere (39% of VCs).
Commonest habitats Bricks and mortar 46%, cinder heaps 23%, soil heaps 21%, coal mine waste 16%, rock outcrops 9%.
Spp. most similar in habitats *Senecio viscosus* 82%, *Sonchus oleraceus* 80%, *Artemisia absinthium* 78%, *Artemisia vulgaris* 75%, *Chamerion angustifolium* 70%.
Full Autecological Account *CPE* page 522.

Synopsis *S.s.* is typically a short-lived perennial, fast-growing, bushy, winter-green herb which is a native of disturbed areas in C and S Europe. The species was first recorded from Britain as an established alien on walls at Oxford in the late 18th century. Its subsequent spread, particularly since reaching the railway at Oxford in *c.* 1879, has been rapid, and the plumed seeds of this species have been widely dispersed, particularly in the gusts formed by passing trains and through the long-distance transport of railway ballast. Related in part to its habitat range (which includes demolition sites, cinders and ballast beside the railway and cracks at the edge of pavements), *S.s.* has a lowland distribution. We suspect that the distribution in Britain is also restricted by climatic factors; persistence may be favoured by the warmer winter temperatures associated with certain urban habitats. Thus, *S.s.* has become the most characteristic species of areas of urban dereliction in lowland S and C Britain. *S.s.* also colonizes some naturally-occurring rocky habitats. In all of these sites the lateral spread of perennials is restricted by the rocky nature of the substrata. The species shows an association with soils of high pH, a reflection in part of the large proportion of records from sites contaminated by cement and mortar. *S.s.* is absent from grazed, regularly mown, woodland and wetland sites, and is restricted to lowland regions. The species regenerates entirely by seed, and plants on average produce *c.* 10 000 seeds year^{-1}. The rapid geographical spread of this species is possibly related to its capacity to produce abundant flowers and seeds over much of the year. Seeds exhibit only limited persistence in the soil. Plants vary markedly in leaf shape, even within single populations, and the level of this variation is much greater than that of the native plants of C Europe. *S.s.* hybridizes with *S. viscosus* and *S. vulgaris*, and is one parent of the allohexaploid *S. cambrensis* (see account for *S. vulgaris*).

Current trends Probably still increasing, both in range and in abundance.

236   *Senecio viscosus*

Stinking Groundsel, 'Sticky Groundsel'

Established strategy R.
Gregariousness Intermediate.
Flowers Florets yellow, with a high selfing ability.
Regenerative strategies W.
Seed 0.60 mg.
British distribution Lowland (58% of VCs).
Commonest habitats Bricks and mortar/cinder heaps 16%, coal mine waste 7%, soil heaps 4%, manure and sewage waste 3%.
Spp. most similar in habitats *Senecio squalidus* 82%, *Linaria vulgaris* 78%, *Rumex crispus* 74%, *Artemisia vulgaris* 73%, *Tripleurospermum inodorum* 71%.
Full Autecological Account *CPE* page 524.

Synopsis *S.v.* is an erect, summer-annual herb which is introduced in Ireland and is only doubtfully native in Britain. The only semi-natural habitat with which *S.v.* is frequently associated is maritime shingle. The species is most frequently recorded from urban spoil, particularly cinder tips, railway ballast and demolition sites, all providing substrates on which consolidation by large perennials is restricted. Occurrences on more-natural soils are less frequent and, within the Sheffield region, they are largely restricted to the freely-drained Bunter sandstone. *S.v.* is unrecorded from grazed, shaded or waterlogged habitats. The species is foetid and clothed in viscid hairs. These characteristics tend to deter herbivory by insects and are presumably effective against some other predators. *S.v.* regenerates entirely by means of wind-dispersed seeds, which germinate in spring, and an average plant may produce 6000 seeds. Seeds produced late in the season are frequently non-viable, and the species has a lowland distribution, with a limit at 310 m in the British Isles. The degree of persistence of seed in the soil requires investigation. The dispersules (achene and pappus) are less buoyant in air than those of many members of the Compositae. However, the species is an effective colonist of disturbed artificial habitats, and may be dispersed by air currents along roadsides and railways. Transport by humans is implicated by the fact that new sites have been reported up to 360 km from the nearest known localities. Genetic differences between populations have been detected, and a dwarf, early-flowering, form occurs in some shingle sites. The hybrid with *S. squalidus*, which is sterile and annual, is occasionally observed on wasteland, demolition sites and on refuse and cinder tips. The hybrid with *S. sylvaticus* L. has also been reported.

Current trends *S.v.*, which was first recorded in Britain in 1666, has spread dramatically in this century. From a distribution centred on SE England in 1900 the species has extended over much of the UK. Dispersal was perhaps facilitated by the development of a modern network of roads and railways. *S.v.* is probably still increasing.

237   *Senecio vulgaris*

Records include var. *hibernicus*

Groundsel

Established strategy R.
Gregariousness Intermediate.
Flowers Florets yellow, tubular, hermaphrodite, usually self-fertilized.
Regenerative strategies W, B_s.
Seed 0.25 mg.
British distribution Abundant except in the Highlands of Scotland (100% of VCs).
Commonest habitats Bricks and mortar 43%, manure and sewage waste 41%, soil heaps 33%, arable 20%, cinder heaps 16%.
Spp. most similar in habitats *Atriplex patula* 92%, *Chenopodium album* 88%, *Capsella bursa-pastoris* 73%, *Matricaria matricarioides* 72%, *Poa annua* 72%.
Full Autecological Account *CPE* page 526.

Synopsis *S.v.* is an annual herb which exploits impermanent sites such as soil and manure heaps, as well as disturbed habitats with greater continuity, such as garden plots and arable fields (particularly those with broad-leaved crops). *S.v.* frequently occurs in mixed populations with *Poa annua*. However, unlike *P.a.*, *S.v.* is a relatively tall plant, and is unable to exploit regularly grazed, mown and trampled sites. *S.v.* is also obligately annual. The success of *S.v.* in exploiting fertile, unshaded, disturbed habitats appears to be related to the association between a short life-span and the capacity to germinate over much of the year; seeds may be produced within six weeks, and up to three generations may be formed each year. The species displays a high potential for seed production and an average plant produces over 1000 seeds. Although *S.v.* thrives on soils of high mineral-nutrient status, the species is also capable of regenerating under conditions of nutrient stress and drought; exceptionally stunted plants on skeletal habitats such as rock outcrops or cinder tips may produce a single capitulum and set < 40 seeds. Although there is flexibility in seed production, reproductive effort expressed as the percentage of the total weight of the plant allocated to seeds tends to be maintained at a constant level. Seeds of *S.v.* are effectively dispersed by wind, by humans and by other agencies. The small plumed seeds are buoyant in air currents, and the hairs of the pappus are sufficiently viscid to adhere to animals or to clothing. Achenes can also survive ingestion by birds. The seeds exhibit limited persistence in the soil and are strongly responsive to changes in light intensity and quality. Although *S.v.* is essentially in-breeding, ecotypic differentiation has occurred. Herbicide resistance, coupled with reduced growth rate, has been reported, and some populations from industrial areas have been reported to be more resistant to 'acid rain', while those from road verges are more salt-tolerant. Var. *hibernicus*, which may have arisen from the introgression of *S. squalidus* into *S.v.*, appears to be increasing. This taxon shows a higher degree of out-breeding than typical *S.v.*, produces more seed and shows greater restriction to urban spoil and to roadside habitats than typical forms. In addition, it is particularly conspicuous in towns and villages at upland locations above the altitudinal range of *S.s.*. Another rayed form (ssp. *denticulatus*) is a winter annual, has a lower reproductive potential, shows marked seed dormancy and, except in Mediterranean regions where it occurs on mountains, has a maritime distribution. *S.v.* occasionally forms sterile hybrids with *S.s.*, and the fertile allohexaploid between the two, *S. cambrensis*, is endemic to Britain.

Current trends An effective colonist of a wide range of fertile artificial habitats. Remains exceedingly common.

238   *Silene dioica*

Red Campion

Established strategy C–S–R.
Gregariousness Sparse to intermediate.
Flowers Rose-coloured, dioecious, insect-pollinated.
Regenerative strategies B_s.
Seed 0.69 mg.
British distribution Widespread except in Ireland (96% of VCs).
Commonest habitats River banks 15%, shaded mire/limestone woodland 13%, hedgerows/broadleaved plantations 8%.
Spp. most similar in habitats *Ranunculus ficaria* 91%, *Festuca gigantea* 86%, *Fraxinus excelsior* (juv.) 79%, *Allium ursinum* 79%, *Circaea lutetiana* 78%.
Full Autecological Account *CPE* page 528.

Synopsis *S.d.* is a potentially fast-growing, winter-green herb in which the leaf canopy is situated close to the ground. The species is most typically associated with relatively fertile sites where the vigour of potential dominants is reduced by shade and by the occasional disturbance which results from factors such as soil creep on steep slopes or flooding on river banks. *S.d.* is perhaps typically a short-lived polycarpic perennial which may flower in its first year. The species is a mesophytic long-day plant, is largely confined to lightly shaded habitats and can survive in a non-flowering condition in deep shade. However, the species is most prominent during the open phase of the coppice cycle and flowers most profusely at woodland margins. In moist upland and maritime environments *S.d.* often grows on rock ledges and in other sites which are inaccessible to stock. In these open sites *S.d.* is frequently found on highly calcareous soils. However, in calcareous woodland *Myosotis sylvatica*, a species of closely similar life-history, morphology and ecology, may partially replace *S.d.*. In lowland Britain the distribution of *S.d.* is biased towards non-calcareous strata. The foliage of *S.d.* contains high concentrations of P. The species is susceptible to frost damage and to both drought and waterlogging. Vegetative spread by means of short stolons is largely ineffectual. In contrast, seed-set is frequently good. Flowers are often infected by the smut *Ustilago violacea*. Individuals are said to produce 7000 seeds, on average, but our own studies in N England suggest that less than half this number may be more realistic. Seed production is nevertheless prodigious in comparison with that of most other perennial woodland herbs. Pot experiments strongly indicate tendencies for seed persistence and early spring germination. At a relatively fine scale, plants of the two sexes appear to differ in their ecology and distribution. Male plants normally produce more leaves and flowers than females, and tend to occur in greater numbers within the population. However, female plants grown adjacent to male plants often produce more shoots and seeds, while the size of the male plant is reduced. Further, they occupy slightly different temporal niches. Thus, females are more prominent in the June canopy while by August, male and female plants contribute equally. Seed size is very variable within populations, and it has been suggested that the production of many small, relatively disadvantaged seeds may serve to reduce the level of insect predation, which for *S.d.* is high. Ecotypic differentiation in plant form, in heavy-metal tolerance, and in germination characteristics have been described for European populations, and in Britain a montane variety with seeds almost three times as heavy as normal was formerly described as ssp. *zetlandica*. Hybrid swarms with *S. alba*, a short-lived species of disturbed wasteland, occur in disturbed lowland habitats where the two species overlap.

Current trends As a result of deforestation and agriculture, *S.d.* is a decreasing species in many lowland areas and is showing increasing hybridization with *S. alba*. However, *S.d.* is not under immediate threat in upland habitats.

---

239 *Sinapis arvensis*

Charlock, Wild Mustard

Established strategy R.
Gregariousness Usually as isolated individuals.

Flowers Yellow, hermaphrodite, self- and cross-pollinated by insects.
Regenerative strategies B_s.

Seed 1.15 mg.
British distribution Widespread (100% of VCs).
Commonest habitats Arable 11%, cinder heaps/soil heaps 4%, coal mine waste 3%, paths/limestone quarries 2%.
Spp. most similar in habitats *Papaver rhoeas* 92%, *Spergula arvensis* 86%, *Veronica persica* 86%, *Myosotis arvensis* 85%, *Fallopia convolvulus* 84%.
Full Autecological Account *CPE* page 530.

Synopsis *S.a.* is a ruderal and arable weed, probably originating from the Mediterranean region. In keeping with the Mediterranean origins, seed-set in *S.a.* appears to be greatly influenced by climatic factors. *S.a.* is regarded as palatable to stock and has been included in human diets, but seeds are very poisonous to livestock. The species is sensitive to mechanical damage and is seldom recorded from grazed habitats. *S.a.* takes a minimum of six weeks to flower and, due to restricted seed output and mobility, does not exploit sites of transient disturbance as effectively as species such as *Poa annua* and *Senecio vulgaris*. Seeds germinate in spring and to a lesser extent in autumn and winter, and germination of seed populations may be partially inhibited by a leaf canopy. Colonization of much of Europe coincident with the spread of agriculture has probably been assisted by a number of features of its biology. Some seeds are usually still in the siliqua at harvest time, and *S.a.* has been spread as an impurity along with seeds of crop plants. Seeds may pass undamaged through the intestines of birds, which are important as dispersal agents. The long-lived seed bank enables survival in sites subject to infrequent cultivation and disturbance. It has been estimated that at certain sites one reproductive phase every 11 years is sufficient to maintain the seed bank. Like other ruderals, *S.a.* shows marked phenotypic plasticity. In unfavourable habitats an individual may yield as few as 16 seeds, while in robust plants up to 25 000 may be produced. A number of varieties have been recognized but the extent of ecotypic differentiation in *S.a.* requires investigation.

Current trends Formerly *S.a.* was a serious weed, particularly of cereals. This is no longer the case and *S.a.* is recorded more frequently from broad-leaved than from cereal crops in the Sheffield region. This decrease, like that of many arable weeds, stems from the widespread use of weed-killers and, to some extent, from the improved purity of crop seeds. A further reduction in the abundance of *S.a.* is predicted.

---

240 *Solanum dulcamara*

Bittersweet, Woody Nightshade

Established strategy Between C–S–R and C.

Gregariousness Stand-forming even at low frequencies.

Flowers Purple, hermaphrodite, cross-pollinated by insects.
Regenerative strategies (V), S.
Seed 1.49 mg.
British distribution Common except in Scotland and Ireland (92% of VCs).
Commonest habitats Shaded mire 21%, river banks 15%, unshaded mire 9%, still aquatic 6%, hedgerows 5%.
Spp. most similar in habitats *Caltha palustris* 92%, *Filipendula ulmaria* 91%, *Cardamine amara* 91%, *Mentha aquatica* 88%, *Myosotis scorpioides* 86%.
Full Autecological Account *CPE* page 532.

Synopsis *S.d.* is a scrambling woody plant of fertile habitats, many situated in wetland. The species, which lives for at least 20 years, is capable of dominance within the herb layer of vegetation shaded by trees, but in unshaded habitats it may be suppressed by taller-growing herbaceous species. The Solanaceae show their greatest diversity in the tropics and, consistent with this trend, *S.d.* is largely confined to southern and lowland Britain and extends only to altitudes of 310 m. *S.d.* is strongly defended against herbivory through the presence of glycoalkaloids, and only the fully ripe berry is not highly toxic. Regeneration *in situ* is mainly by vegetative means. Stems in contact with the ground root readily, and large clonal patches may result. Pieces of stem are also capable of regeneration, and the frequency of *S.d.* on river banks may be a reflection of its ability to form new plants in this way. Plants growing in shade often produce fewer flowers and fruits but are reported to bear a mean value of 1400 seeds per plant. Seeds are dispersed by birds within many-seeded berries and are probably only of importance in colonizing new sites. Seedlings emerge in spring from down to 50 mm in the soil; on this basis and from more recent pot experiments, it appears that *S.d.* does not form a persistent seed bank. Prostrate, fleshy-leaved forms may constitute a maritime ecotype, and outside Britain sun and shade ecotypes have been identified. Races also occur in Europe, differing in the alkaloid composition of their above-ground parts and in their geographical distribution.
Current trends Uncertain. Exploits fertile habitats but appears to be a poor colonist of new sites.

241 *Solidago virgaurea*   Golden-rod

Established strategy S.
Gregariousness Sparse.
Flowers Florets yellow, insect-pollinated, self-incompatible.
Regenerative strategies W.
Seed 0.52 mg.

British distribution Widespread, but rare in C and E England (99% of VCs).
Commonest habitats Limestone quarries 8%, coal mine waste 7%, bricks and mortar/scree/limestone

wasteland 3%.

Full Autecological Account *CPE* page
534.

Synopsis *S.v.* is a semi-rosette herb
with a stout stock. *S.v.* is mainly
restricted to infertile habitats, and
achieves local abundance on cliffs and
outcrops. Plants produce on average *c.*
4000 small, wind-dispersed seeds. This
feature has contributed significantly to
the ability of *S.v.* to colonize quarries
and railway cuttings. At the other
extreme of its ecological range *S.v.* is
also found at wood margins; here the
species flowers less regularly. *S.v.* does
not persist, however, in tall herb
communities. Regeneration is by seed,
and the rarity of *S.v.* in pasture is
presumably a reflection of the
vulnerability to grazing of the long
flowering spike. The species occurs
across a wide range in soil pH values.
In lowland parts of the region *S.v.*
tends to be calcifuge, whereas in the
uplands the species occurs on both
highly calcareous and highly acidic
soils. Consistent with this trend, *S.v.*
shows a predominantly calcifuge
distribution in SE England. The wide
ecological range of this out-breeding
species appears to depend to a
considerable extent on its capacity to
form ecotypes. Both British and
European populations are highly
polymorphic and, for example, dwarf
montane populations with large
capitula may merit taxonomic
separation. Many ecotypes have also
been described for Europe. In
particular, ecotypes from shaded and
unshaded habitats differ markedly in
their response to light intensity and
exposure to high irradiance causes a
differential amount of
chlorophyll-bleaching in shade
ecotypes. *S.v.* hybridizes with the N
American garden escapee, *S.
canadensis*.

Current trends *S.v.* is essentially
restricted to infertile habitats, and its
capacity to colonize new artificial
habitats is limited, despite the large
number of wind-dispersed propagules
which each plant may produce. As a
result of habitat destruction *S.v.* is now
close to extinction in many lowland
areas. However, the species is under
no immediate threat in its upland
habitats.

---

242 *Sonchus asper*

Spiny Milk- or Sow-Thistle

Established strategy Between R and
C–R.
Gregariousness Sparse.
Flowers Florets yellow, ligulate, visited
by insects, self-compatible.

Regenerative strategies W, ?B$_s$.
Seed 0.32 mg.
British distribution Lowland (100% of
VCs).
Commonest habitats Limestone

quarries 9%, soil heaps 7%, coal mine waste 5%, manure and sewage waste 3%, cinder heaps/rock outcrops/paths/limestone wasteland 2%.

Spp. most similar in habitats
Desmazeria rigida 62%, *Crepis capillaris* 60%, *Artemisia vulgaris* 58%, *Erigeron acer* 58%, *Tussilago farfara* 50%.

Full Autecological Account *CPE* page 536.

Synopsis *S.a.* is a ruderal species with numerous wind-dispersed seeds (over 1500 on a robust plant). The species is particularly common in transient artificial habitats, where it exhibits considerable variation in size. *S.a.* often develops a rosette of leaves and a tap-root in autumn, and produces a flowering stem during the following summer, but spring-germinating individuals also occur and may function as summer annuals. *S.a.* reaches a larger size than many strict ruderals and takes longer to set seed (*c.* 10 weeks in spring-germinating plants), and is rarely an important arable weed. Despite their prickles, the leaves of *S.a.* are palatable to grazing stock, and *S.a.* seldom reaches maturity in heavily grazed areas. *S.a.* regenerates solely by means of seeds, which germinate on open ground during autumn or remain dormant until the following year. A shortly persistent seed bank is suspected. The plumed achenes are buoyant in air and are widely dispersed. Thus, a high proportion of the occurrences in the Sheffield region refer to isolated seedlings originating from seed dispersed into habitats where there is low probability of survival to flowering. This may explain the heterogeneity of the list of species 'similar in habitats'. The ecological significance, if any, of variation within the species remains to be elucidated. The ecology and geographical distribution of *S.a.* is remarkably similar to that of *S. oleraceus*, with which it hybridizes. The slight differences between the two are described under *S.o.*. A further biennial subspecies of *S.a.* is recorded within Continental Europe.

Current trends *S.a.* was formerly regarded as a troublesome arable weed, even more so than *S.o.*. Although generally less abundant than *S.o.*, *S.a.* appears to be increasing. The species is now less common in farmland, but remains an effective colonist and is apparently still extending its geographical range in other recent artificial habitats, particularly those in urban areas.

243 *Sonchus oleraceus*

Milk- or Sow-Thistle

Established strategy Between R and C–R.
Gregariousness Usually as isolated individuals.

Flowers Florets yellow, ligulate, visited by insects, mainly self-fertilized.
Regenerative strategies W, B_s.
Seed 0.27 mg.

British distribution Lowland (100% of VCs).

Commonest habitats Bricks and mortar 38%, soil heaps 17%, limestone quarries 16%, arable 11%, manure and sewage waste/verges 8%.

Spp. most similar in habitats *Senecio squalidus* 80%, *Artemisia absinthium* 76%, *Senecio vulgaris* 67%, *Tripleurospermum inodorum* 65%, *Atriplex patula* 62%.

Full Autecological Account *CPE* page 538.

Synopsis The distinctive features of *S.o.* are essentially the same as those of *S. asper*. *S.o.* differs in that it has a slightly more restricted geographical distribution, tending to be replaced by *S.a.* at the N edge of its range. The species is strongly confined to impermanent habitats and, in the Sheffield region, does not share the tendency of *S.a.* to occur in some winter-annual communities on limestone outcrops and dry sandy soils. In *S.o.* the pappus is usually more securely attached to the ripe achene, which is slightly longer and thinner. There appears to be a greater capacity in *S.o.* for ungerminated viable seeds to persist on or near the soil surface. *S.o.* is widely distributed, and as a result is classified as one of the world's worst weeds. Like *S.a.*, *S.o.* is palatable but lacks spiny leaves and was formerly eaten as salad. *S.o.* (4x) is believed to have arisen through hybridization between *S.a.* (2x) and the S European *S. tenerrimus* (2x). Some of the variability of the species may relate to its hybrid origin. The existence of ecotypes is suspected.

Current trends Although perhaps less frequent on arable land than formerly, *S.o.* is a common plant of many other artificial habitats, and appears to be increasing.

---

244 *Sorbus aucuparia*

(Field records as seedlings)

Established strategy S–C.
Gregariousness Sparse.
Flowers White, hermaphrodite, insect-pollinated or selfed.
Regenerative strategies S.
Seed 4.0 × 2.0 mm.
British distribution Widespread, though rare in some C and E counties (99% of VCs).
Commonest habitats Acidic woodland 9%, coniferous plantations 5%, scrub 4%, lead mine spoil/limestone woodland 3%.
Spp. most similar in habitats *Oxalis acetosella* 84%, *Milium effusum*

Rowan, Mountain Ash

81%, *Pteridium aquilinum* 80%, *Rubus fruticosus* 78%, *Luzula pilosa* 77%.

Full Autecological Account *CPE* page 540.

Synopsis *S.a.* is a small deciduous tree or shrub with diffuse porous wood, 15–20 m in height and living a maximum of 150 years. Bud-break occurs relatively early in spring. Seedlings and mature individuals are concentrated mainly in skeletal habitats, e.g. crevices in rock outcrops, and in woodland. *S.a.* extends on rocky ground to over 900 m, higher than any of the other British trees. The buds of montane populations show a considerable resistance to desiccation. In wooded habitats the species occurs more frequently in woodland than in scrub, suggesting that establishment by seed and the subsequent development of the tree usually take place under shaded conditions. The small seedling is considered to be exceptionally shade-tolerant. The species is wide-ranging with respect to soil pH, but is more frequent on acidic soils, often occurring rather sparsely with oak or birch. *S.a.* is characteristic of well-drained soils, and does not accumulate persistent litter. Indeed, leaves show a high palatability to the snail *Helix aspersa*, and *S.a.* supports a relatively species-poor insect fauna. However, *S.a.* appears to possess a mechanism of inducible defence against foliar predation by insects. Establishment in woodland, and presumably also in skeletal habitats, appears to be adversely affected by the presence of grazing stock. *S.a.* regenerates entirely by seed, which decreases in viability with altitude. Seeds germinate in spring, have a chilling requirement for germination, and are dispersed within a red fleshy fruit by birds and mammals. Where suitable habitats are adjacent to mountain streams they are also dispersed by water. No persistent seed bank has been detected within the Sheffield region, but it has been suggested that seeds have considerable longevity in the soil. If confirmed, this would make *S.a.* the only tree in the British flora characterized by a persistent seed bank. The species is

perhaps most similar in ecological distribution to *Betula* spp., since both are relatively short-lived trees most characteristic of acidic soils and extending to high altitudes. In regenerative biology, however, the two clearly differ. *B.* spp. produce vast numbers of small, wind-dispersed seeds and colonize relatively unshaded areas, whereas *S.a.* is capable of regeneration in more-shaded sites and produces fewer, larger, bird-dispersed seeds. *S.a.* is represented by a further four ssp. in Continental Europe, and hybridization between *S.a.* and other taxa has greatly increased the taxonomic complexity of the genus *Sorbus* in Britain.

Current trends Uncertain. Not under any immediate threat, and probably increasing in some lowland areas as a result of a widespread use for amenity or in gardens (and formerly for fruit or 'to frustrate witches').

---

245 *Sparganium erectum*

Bur-reed

Established strategy Between C and C–R.
Gregariousness Stand-forming.
Flowers Greenish, monoecious, protogynous, wind-pollinated.

Regenerative strategies (V), ?S_s.
Seed 18.71 mg.
British distribution Widespread but scarce in uplands (99% of VCs).
Commonest habitats Still aquatic 19%,

unshaded mire 10%, running aquatic 8%, shaded mire 3%.

Spp. most similar in habitats *Elodea canadensis* 94%, *Lemna minor* 93%, *Equisetum fluviatile* 93%, *Juncus bulbosus* 93%, *Callitriche stagnalis* 92%.

Full Autecological Account *CPE* page 542.

Synopsis *S.e.* is a long-lived, robust, stand-forming, emergent aquatic or marsh species resembling a diminutive *Typha latifolia* in form, phenology and habitat range. Each plant is capable of extending 2 m laterally each year. *S.e.* grows most vigorously in shallow water (*c.* 100 mm), but is rarely found in fast-flowing sites and is intolerant of wave action. *S.e.* is typically rooted in loose, often anaerobic, silt and has well-developed air spaces in its roots, stems and leaves. *S.e.* is also found growing less vigorously in mire beside open water, but the species is sensitive to drought and is confined to situations where the water-table is above the level of the roots (*c.* 100 mm below the surface) for most of the year. New shoots examined in early spring were found to contain high concentrations of fructose. *S.e.* occurs on a wide range of soils and can survive low levels of salinity. *S.e.* grows poorly in shade and produces few flowering stems. The species appears unable to compete effectively with the taller *Phragmites australis* and *T.l.*. Despite forming extensive and persistent stands, the clones are rapidly fragmented and each ramet is short-lived. Rhizomes survive only one growing season and corms, which give rise either to vegetative or to flowering shoots and are monocarpic, survive for a maximum of three years. Corms in winter and rhizome pieces in summer may be transported in water, allowing new sites to be colonized. Apart from vegetative fragmentation *S.e.* also regenerates, albeit rarely, by seeds produced at the rate of up to 720 per flowering head. The fruit has a spongy exocarp and may float in water for one year. The endocarp is also hard and the micropile is plugged, and seeds pass undamaged through the gut of waterfowl. Fruits may also be transported externally by birds. Seedlings establish under water or on wet mud, and are sensitive to competition, even from algae. Establishment is not observed in turbid water, and seedlings, unlike their parents, tend not to survive in eutrophic habitats. It seems likely that regeneration by seed is of primary significance in the colonization of new sites. Fruits are relatively long-lived under a variety of conditions and the possibility of a 'floating seed bank' requires investigation. Three subspecies are recognized, differing in fruit characters and geographical distribution. They do not appear to be ecologically distinct and have not been separated here. Hybrids between two of the subspecies have also been reported.

Current trends Uncertain. Although *S.e.* has the capacity to exploit highly eutrophic aquatic systems, colonization of new land-locked sites is slow, presumably because dispersal by birds, on which *S.e.* appears to rely, is less effective than the wind-borne dispersal of species such as *T.l.*.

Established strategy R.
Gregariousness Intermediate.
Flowers White, hermaphrodite, mostly self-pollinated.
Regenerative strategies B_s.
Seed 0.42 mg.
British distribution Common (92% of VCs).
Commonest habitats Arable 31%, soil heaps 6%, bricks and mortar 3%, paths 2%.
Spp. most similar in habitats *Veronica persica* 99%, *Fallopia convolvulus* 98%, *Myosotis arvensis* 95%, *Polygonum persicaria* 92%, *Anagallis arvensis* 91%.
Full Autecological Account *CPE* page 544.

Synopsis *S.a.* is a slender summer-annual herb found in Britain on arable land and in other fertile, disturbed habitats. *S.a.* is thought to have been a common weed of flax from the Iron Age onwards. Whether its form and life-history are best suited to this crop is uncertain. Today, *S.a.* is cosmopolitan in distribution, and is classified as one of the world's worst weeds, particularly of cereals. *S.a.* is regarded as an important weed in 25 crops and in 33 countries. Unlike most other common British arable weeds, the occurrences of *S.a.* are most frequent on surface-leached sandy soils of pH 6.0 or less. *S.a.* responds to defoliation by producing new flowering shoots from the base, but is not tolerant of trampling. *S.a.* is palatable to sheep, cattle and poultry, and has been grown for fodder. Seeds have, in historical times, also been eaten by humans. Regeneration is entirely by seed, which germinates mainly in spring. Seed production is both rapid and prolific. Robust plants may produce as many as 7500 seeds; these may be released within ten weeks of

germination, and two generations may be accommodated within a growing season. Germination may be inhibited by a leaf canopy. Like many other arable species, *S.a.* forms a large persistent seed bank, which may reach 23 million seeds ha^{-1}. Seeds appear to be capable of extended survival in the soil and can withstand ingestion by many mammals and birds. The species is often transported to new sites on agricultural machinery. In common with most ruderals, *S.a.* is phenotypically highly plastic. Genotypic differences are also apparent. Seeds with papillae tend to germinate at higher temperatures and lower moisture tension, and tend to be replaced by non-papillate forms towards the N and W. Hairiness of the shoot varies in a similar manner.
Current trends Since arable soils with pH values below 6.5 are now infrequent, *S.a.* is probably decreasing, despite remaining a locally common plant of disturbed sandy soils. The proportion of densely hairy plants may be increasing and this may be related to changes in climate.

Established strategy S.
Gregariousness Individuals typically
  scattered.
Flowers Purple, hermaphrodite, insect-
  pollinated.
Regenerative strategies V, S.
Seed 1.37 mg.
British distribution Widespread except
  in Scotland and Ireland (63% of
  VCs).
Commonest habitats Limestone
  pastures 12%, limestone wasteland
  10%.
Spp. most similar in habitats *Lathyrus
  montanus* 90%, *Viola hirta* 89%,
  Succisa pratensis 87%, *Pimpinella
  saxifraga* 86%, *Galium verum* 85%.
Full Autecological Account *CPE* page
  546.

Synopsis *S.o.* is a slow-growing,
rosette-forming herb which is mainly
restricted to short turf on infertile soils
where potential dominants are
restricted in vigour either by a low
intensity of grazing or cutting or by
burning. Like *Lathyrus montanus*, with
which it often grows, *S.o.* is most
commonly found on soils which are
mildly acidic (pH 5.0–6.5), at least in
the superficial horizons. In laboratory
experiments *S.o.* exhibits high
sensitivity to lime-chlorosis on
calcareous soils. *S.o.* tends to be
associated with relatively deep clayey
soils, and is usually absent from
waterlogged and from droughted
situations. In a field experiment
individuals of *S.o.* transplanted onto a
shallow calcareous soil in Derbyshire
suffered irreversible wilting during
summer droughts, and had much lower
survivorship than transplants onto a
neighbouring brown earth naturally
colonized by the species. At low
altitudes *S.o.* persists at woodland
margins, and experimental studies also
indicate a degree of shade-tolerance.
S.o. shows a pronounced summer peak
in both biomass and leaf canopy and
thus has a somewhat different
phenology from that of many of the
grasses with which it is associated.
Plants grown continuously at low
temperatures (< 15°C) produce dark
green leaves with small laminae; these
are quite unlike the large leaves
produced at warm temperatures. We
suspect that *S.o.* is long-lived, and that
vegetative spread, although restricted
in extent, is more important in the
development of populations than
regeneration by seed. Seed matures
late in the year and is released slowly.
In a recent experiment it was found
that germination was delayed until
spring, and was not associated with the
formation of a persistent seed bank.
The seed appears to be poorly
dispersed, and *S.o.* is virtually
restricted to old or semi-natural
vegetation. Lack of mobility during a
Postglacial colonizing episode may be
partly responsible for the present rarity
of *S.o.* in Ireland. Various extracts of
the leaves were formerly used in home
remedies for coughs, stomach upsets
and kidney, bladder and spleen
complaints.
Current trends Decreasing,
particularly in lowland areas.

Established strategy Between C and
  C–R.
Gregariousness Can form patches.
Flowers Purple, hermaphrodite, insect-
  pollinated, self-compatible.
Regenerative strategies (V), B_s.
Seed 1.4 mg.
British distribution Common except in
  N Scotland and Ireland (99% of
  VCs).
Commonest habitats Hedgerows 20%,
  verges 15%, river banks 13%, soil
  heaps 11%, scrub 8%.
Spp. most similar in habitats *Galium
  aparine* 89%, *Glechoma hederacea*
  85%, *Anthriscus sylvestris* 77%,
  Urtica dioica 73%, *Tamus communis*
  68%.
Full Autecological Account *CPE* page
  548.

Synopsis *S.s.* is a rhizomatous, stand-
forming herb of moist fertile sites,
where the vigour of taller potential
dominants is reduced by disturbance
and often also by the presence of light-
to-moderate shade. Thus, *S.s.* is
particularly characteristic of
hedgerows, road verges, river banks,
the edge of woodland rides and
floodplains. The species also occurs on
infrequently-cut roadsides, but despite
the fact that its leaves are glandular
and foetid when crushed, it is absent
from pasture. Individual parts of the
rhizome system are short-lived and
shoots are, in addition, proliferous.
Colonies of *S.s.* are perhaps similar to
those of *Ranunculus repens* in terms of
the dynamic turnover of ramets. This
capacity for intensive vegetative
reproduction ensures high competitive
ability and facilitates the active
foraging for optimal sites within a
changing environment. Effective
regeneration is probably mainly by
vegetative means, and *S.s.* forms large
clonal patches as a result of rhizome
growth. The species also regenerates

freely from rhizome pieces. Thus, *S.s.*
can persist in vegetable gardens
provided that the weeding regime is
not too severe. Long-distance dispersal
of the largely sterile *S.* × *ambigua*, (*S.
palustris* × *S.s.*) occurs regularly as a
result of human activities, and
introductions of *S.s.* by similar means
are probable. *S.s.* is a rapid colonizer
of secondary woodland. Although *S.s.*
exhibits a degree of shade tolerance,
the species has a peak of biomass in
summer, and lacks both overwintering
shoots and pronounced vernal growth.
This leads us to suspect that the
capacity for effective dispersal and
early establishment (often low in many
woodland species) are more critical to
the success of *S.s.* in exploiting shaded
habitats than phenological
specialization. The conditions required
for establishment by seed have been
little studied. However, seeds which
form a persistent seed bank are
produced more consistently in
unshaded sites, and appear to lack any
well-defined mechanism for
long-distance dispersal.
Current trends A frequent wayside
plant apparently favoured by modern
forestry practices, and probably
increasing.

249 *Stellaria alsine*   **Bog Stitchwort**

Established strategy Between C–S–R
and C–R.
Gregariousness Intermediate or rarely
stand-forming.
Flowers White, hermaphrodite, insect-
or self-pollinated.
Regenerative strategies V, B$_s$.
Seed 0.06 mg.
British distribution Widespread (100%
of VCs).
Commonest habitats Unshaded mire
18%, river banks 10%, running
aquatic/shaded mire 8%, enclosed
pastures/soil heaps 2%.
Spp. most similar in habitats *Juncus
effusus* 86%, *Phalaris arundinacea*
83%, *Juncus articulatus* 83%,
Epilobium obscurum 82%,
Cardamine amara 81%.
Full Autecological Account *CPE* page
550.

Synopsis *S.a.* is a prostrate
mat-forming herb of fertile wetlands
and is essentially an opportunist
colonizer of disturbed sites. In Japan,
at least, the species may flower within
five months and seed-set takes a
further three weeks. Thus, the species
often colonizes bare mud in advance of
potentially more-dominant species,
and may survive in trampled mire and
on river banks subject to erosion. *S.a.*
also displays a modest degree of shade
tolerance although, in shade, flowering
is often suppressed and the species
may form a sparse understorey
beneath taller vegetation. In tall
patchy vegetation such as stands of
Juncus effusus, ascending stems,
supported by the surrounding
vegetation, may protrude through gaps
in the canopy. *S.a.* also forms floating

rafts in small upland streams, often
intermixed with species such as *Holcus
mollis* and *Montia fontana*. The species
is most characteristic of mildly acidic
sites, and in the Sheffield region has
not been recorded from calcareous
mire. Plants observed in autumn have
a creeping habit and numerous
narrower leaves of lighter hue.
Vegetative regeneration by means of
prostrate rooted stems enables *S.a.* to
form extensive clones. Detached
shoots readily re-root and are, we
suspect, important in the colonization
of river banks and mire beside open
water. In unshaded sites seed
production is often prolific, and a large
persistent seed bank may be formed.
Annual and perennial ecotypes of this
in-breeding species have been
identified in Japan.
Current trends A species of fertile and
disturbed wetlands, which may be
increasing as a result of human
activities.

250 *Stellaria holostea*   **Greater Stitchwort**

Established strategy C–S–R.
Gregariousness Can form patches.

Flowers White, hermaphrodite, insect-
or self-pollinated.

Regenerative strategies V.
Seed 3.7 mg.
British distribution Common except in
   Scotland and Ireland (99% of VCs).
Commonest habitats Hedgerows 10%,
   coniferous plantations/acidic
   woodland/limestone wasteland 3%,
   scrub/verges 2%.
Spp. most similar in habitats *Tamus
   communis* 78%, *Rubus fruticosus*
   75%, *Glechoma hederacea* 63%,
   Pteridium aquilinum 60%, *Lonicera
   periclymenum* 60%.
Full Autecological Account *CPE* page
   552.

Synopsis A species about which we
know little and which deserves further
study. *S.h.* is a low-growing,
scrambling, winter-green herb, and is
typical of sites where the growth of
potential dominants is restricted by
moderate shade. As in many species of
habitats subject to summer shade, *S.h.*
shows much early season growth, and
flowers in spring. However, *S.h.*
maintains foliage throughout the year,
and the species is one of the few
herbaceous plants to achieve its highest
frequency of occurrence in hedgerows.
At upland sites *S.h.* is frequent in
unshaded sites, and the low tolerance
of heavy shade is confirmed
experimentally. Despite an association
with a wide range of soil types, *S.h.* is
largely absent from highly calcareous
soils and from the most freely-drained
sands. It is most abundant on moist,
mildly acidic, moderately infertile
soils, where it is frequently associated
with *Holcus mollis*. The leaves of *S.h.*
contain unusually low concentrations
of Fe and Mn. *S.h.* is rarely found in
heavily-grazed sites, and appears to be
ecologically similar to, but less
calcicolous and more shade-tolerant
than, *Galium cruciata*. Shoots are
often supported by the surrounding
herbage; when this dies back the shoot
drops to the soil surface and may form
adventitious roots in vegetation gaps.
In this way *S.h.* may form large clonal
patches. Detached shoots appear
capable of regeneration in the field,
and thus infrequent disturbance may
both serve to check the growth of
potential dominants and effect
vegetative spread. The 'patchy'
distribution of *S.h.* suggests that
dispersal by seed does not occur as
consistently as regeneration by
vegetative means. However, the
mechanism and importance of
regeneration by seed require further
investigation. In Lincolnshire *S.h.* is
most frequent in ancient woodland and
in the Sheffield region, too, *S.h.* does
not colonize recent artificial habitats.
Current trends Probably decreasing.

251 *Stellaria media*   Chickweed

Established strategy R.
Gregariousness Intermediate.
Flowers White, hermaphrodite, usually
   self-pollinated.

Regenerative strategies B_s, (V).
Seed 0.35 mg.
British distribution Widespread (100%
   of VCs).

Commonest habitats Arable 66%,
manure and sewage waste/soil heaps
33%, bricks and mortar 18%, river
banks 15%.

Spp. most similar in habitats
Polygonum persicaria 88%,
Chenopodium album 88%,
Polygonum aviculare 84%, *Capsella
bursa-pastoris* 82%, *Atriplex patula*
82%.

Full Autecological Account *CPE* page
554.

Synopsis *S.m.* is a short-lived annual
herb. Although primarily an arable
weed capable of severe effects on crop
yield, the species also occurs in a wide
range of other habitats, but always in
association with recently disturbed,
fertile soils. *S.m.* is recorded in 50
countries, and is classified as one of the
world's worst weeds. *S.m.* is attuned to
arable land through several
mechanisms. Seed is produced rapidly
(within five weeks) and abundantly (up
to 13 000 per plant). The species has a
high growth rate and shows a swift
response to fertilizer additions,
sometimes accumulating levels of
nitrate which are toxic to stock. Leaves
have exceptionally high levels of Na
and contain high P, Mg and Fe.
Despite its wide climatic range, *S.m.*
thrives best in cool moist
environments, and in the Sheffield
region is noticeably more-abundant
and luxuriant during spring and
autumn than in summer. *S.m.* is
moderately persistent in heavy shade
under crops. *S.m.* is palatable to stock
and slugs, and has been used as a salad
plant. *S.m.* is extremely variable, both
phenotypically and genotypically.
Summer and winter phenotypes with
different growth forms occur. *S.m.* also
shows variation in seed characters,
particularly with respect to dormancy
and persistence in the soil. Seeds are
released without a light requirement
for germination, but this has been
shown to be capable of development
after burial. Germination occurs
throughout the year, with peaks in
spring and autumn, and two, or even
three, generations of plants may be
produced in a year. *S.m.* forms a
persistent seed bank. Cuttings root
readily in disturbed sites, allowing a
limited amount of vegetative
regeneration. Seed is widely dispersed
in the faeces of birds and animals, and
on farm machinery or with crops. *S.
neglecta*, a taller, often short-lived
perennial of moist woodland, and *S.
pallida*, a smaller, very early flowering
winter annual mainly of sandy soils,
both occur in Britain and are perhaps
best regarded as ssp. of *S.m.*. Two
further ssp. are recorded from Europe.
Current trends Another 'follower of
humans', likely to remain common as
long as disturbed, fertile habitats are
created.

---

252 *Succisa pratensis*

Devil's-bit Scabious

Established strategy S.
Gregariousness Intermediate.
Flowers Usually purple,
gynodioecious, insect- or
self-pollinated.
Regenerative strategies S.

Seed 1.54 mg.
British distribution Widespread (100% of VCs).
Commonest habitats Limestone pastures 21%, limestone wasteland 9%, lead mine spoil 3%, shaded mire/paths 2%.
Spp. most similar in habitats *Danthonia decumbens* 88%, *Carex caryophyllea* 87%, *Stachys officinalis* 87%, *Carex panicea* 80%, *Viola hirta* 79%.
Full Autecological Account *CPE* page 556.

Synopsis *S.p.* is a slow-growing, winter-green, perennial, rosette-forming herb which exploits sites where the growth of potential dominants is restricted by a low level of soil fertility and often also by grazing. The species appears to be long-lived and often does not flower until its fourth year. The root system of *S.p.* is composed of spreading, rather superficial roots and the species does not occur in droughted habitats. Though strongly associated with continuously moist habitats, *S.p.* is seldom found in waterlogged sites other than in soligenous mire or on tussocks of other species. In unproductive grassland, wood margins and maritime and montane habitats, *S.p.* is characteristic of soils of intermediate pH. However, *S.p.* is also found on a wide range of other soil types, including calcareous mineral soils, but the species is infrequent on freely-drained sandy soils and on soils of low pH. In grazed habitats the leaves of *S.p.* often escape predation, and *S.p.* is tolerant of light trampling. Regeneration is mainly by seed which germinates in spring and, as far as is known, *S.p.* does not develop a persistent seed bank. The seed has no well-defined dispersal mechanism and is in fact poorly dispersed. *S.p.*, normally an out-breeding species, shows a modest degree of phenotypic and genotypic variation. Dwarf plants with short flowering stems from maritime habitats usually, but not always, increase their size on cultivation. In species-rich short turf, in which *S.p.* reaches maximum abundance, the species tends to be represented by phenotypes with low rosettes. In taller vegetation ascending leaves with longer petioles are observed.

Current trends *S.p.* mainly exploits infertile pasture; owing to its poor ability to colonize new artificial habitats it is largely restricted to semi-natural vegetation. As a result of drainage and high fertilizer input, *S.p.* is decreasing, dramatically so in lowland habitats where it is becoming increasingly restricted to linear habitats such as railway banks, ditch banks and woodland rides.

253 *Tamus communis*   **Black Bryony**

Established strategy Between C and C–R.

Gregariousness Only as isolated individuals.

Flowers Green, dioecious, insect-pollinated.
Regenerative strategies S.
Seed 18.33 mg.
British distribution Mainly southerly (46% of VCs).
Commonest habitats Hedgerows 15%, scrub 7%, limestone woodland 4%, paths/limestone wasteland 2%.
Spp. most similar in habitats *Stellaria holostea* 78%, *Glechoma hederacea* 74%, *Arum maculatum* 72%, *Geum urbanum* 71%, *Stachys sylvatica* 68%.
Full Autecological Account *CPE* page 558.

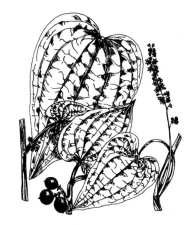

Synopsis *T.c.* has long twining stems and a large underground tuber which in extreme cases may have a fresh weight of 10–15 kg. *T.c.* is the only British member of the predominantly tropical Dioscoreaceae, the yam family. The species is virtually restricted to lowland areas, and reaches its N distributional limit in Britain. *T.c.* occupies two distinct habitats. In hedgerows and woodland margins large plants occur and seed-set is generally high, whereas in woodlands plants normally persist in a non-flowering, suppressed state, but are presumably released when gaps form in the tree canopy. The location of the tuber 50 mm or more below the soil surface, and the relatively deep root system, appear to restrict the capacity of *T.c.* to exploit waterlogged soils, and probably explain its virtual absence from shallow soils. Although the leaves and shoots may be used as a vegetable, the plant is little predated and when eaten raw may cause

fatalities to humans and stock. *T.c.* is normally dioecious and populations tend to contain more male than female plants. Effective regeneration is by means of seeds which germinate in spring. Hence, the survival of hedgerow and to a lesser extent woodland margin populations is essential for the success of *T.c.*. Large plants of *T.c.* may be 20 years old or more and plants do not flower until at least five years old. Seeds retained on the plant until late winter do not germinate in the subsequent spring and remain dormant for a further year. Seeds are enclosed within berries, and may be distributed by birds. However, *T.c.* tends to be a poor colonist of new sites and to be principally associated with those species-rich (older) hedgerows. No persistent seed bank has been detected. Occasionally, when tubers are broken, each half will form a new plant. Otherwise there is no vegetative regeneration.
Current trends Probably decreasing.

---

254 *Taraxacum agg.*   **Dandelion**

Established strategy Mostly between R and C–S–R.
Gregariousness Intermediate.
Flowers Florets yellow,

hermaphrodite, ligulate, mostly apomictic.
Regenerative strategies W.
Seed 0.64 mg.

British distribution Widespread (100% of VCs).

Commonest habitats Meadows 51%, verges 49%, bricks and mortar 38%, limestone quarries 37%, cinder heaps 35%.

Spp. most similar in habitats *Dactylis glomerata* 79%, *Poa pratensis* 67%, *Festuca rubra* 66%, *Plantago lanceolata* 64%, *Leucanthemum vulgare* 57%.

Full Autecological Account *CPE* page 560.

Synopsis *T.* agg. is a winter-green, rosette-forming polycarpic perennial with a long stout tap-root, and is found in a wide range of habitats. The aggregate is well known as a troublesome weed of gardens, waysides and pasture, but equally extends onto sand dunes, mountains and into wetland. This diversity is related to the existence of many ecotypes. Thus, *T.* agg., a taxonomically difficult grouping, has been divided into a number of Sections. Over 150 microspecies have been recognized in Britain, and there are possibly 1500 in Europe. The following generalizations may be made. Plants from Section *Taraxacum* are robust and are mainly found in fertile, disturbed, artificial habitats, and include the taxa which are familiar garden and pasture weeds. Those from Section *Erythrosperma* are associated with dry sandy or calcareous soils and probably have a bias towards S-facing slopes. They are generally much smaller in all their parts than taxa from Section *T.*. Taxa from Section *Spectabilia sensu lato* occur in moist, often less-fertile and less-disturbed sites. Most wetland records refer to this group, which tends to be more common on N-facing slopes. Other Sections are associated with montane habitats, with sand dunes or with mire. In moist habitats in Holland a reduction in the intensity of land management has been shown to result in the gradual replacement of Section

T. by Sections *S.* and *Palustria*. In drier habitats Section *E.* becomes more prominent. The tap-root often penetrates deeply and permits the exploitation of sites where mineral subsoil is overlain with debris such as building rubble or paving stones. In rocky or sandy habitats subject to summer drought the tap-root appears to be of critical importance in affording access to subsoil moisture. In pastures *T.* agg. is among the species preferred by stock; it is capable of rapid recovery after defoliation, and is also resilient under trampling damage. In part this resilience is due to the presence of contractile roots which pull the apical meristem 10–20 mm below the soil surface. Early growth of foliage and flowering, before appreciable growth of many grasses has occurred, may also be a key feature in the success of this species in pasture. Fruits of *T.* agg. are buoyant in air, released in large numbers during early summer (up to 2000 or more per plant or *c.* 5000 m^{-2} in pastures), and are capable of rapid germination over a wide range of temperatures. This combination enables *T.* agg. to be a highly effective colonizing species of sites with exposed soil. *T.* agg. is also effectively dispersed through human activities, and alien taxa appear to have been introduced from Continental Europe.

Only two native taxa are diploid and sexual; the remainder are apomictic. Thus, regeneration by seed is essentially an asexual process. No persistent seed bank is formed, but in the event of disturbance *T.* agg. may regenerate by means of fragments of tap-root. Large plants, particularly after close grazing, may also form multiple rosettes. Plants are phenotypically plastic, so much so that only those collected in early summer from typical habitats can be identified with any certainty. As implied already from the wide ecological range of *T.* agg., populations differ genotypically in the extent of their allocation of resources to vegetative growth and to seed and in a variety of characters related to growth and morphology, including relative growth rate. Microspecies from less-strongly fertilized and less-heavily grazed pastures tend to have a broad ecological range; those of fertile, intensively grazed habitats are more specialized and are probably among the most recently evolved taxa. Hybridization occurs between diploid sexual taxa and pollen-bearing apomictic species, and this has been important during episodes of speciation. In Europe an estimated 10% of taxa are diploid, 45% triploid, 28% tetraploid, 5% pentaploid and *c.* 11% aneuploid.

Current trends Overall the aggregate species is probably still increasing, particularly in Section *T.*, which includes the majority of populations exploiting fertile and disturbed lowland habitats. The status of other Sections is less clear but some which exploit fens and water-meadows, particularly Section *Palustria*, are clearly declining. Sexuality is not uncommon in Sect. *T.* and the evolution of new taxa capable of exploiting disturbed fertile habitats is doubtless proceeding.

255 *Teucrium scorodonia*   Wood Sage

Established strategy C–S–R.
Gregariousness Intermediate.
Flowers Green, hermaphrodite, protandrous, partially self-incompatible.
Regenerative strategies V, B_s.
Seed 0.87 mg.
British distribution Widespread (99% of VCs).
Commonest habitats Scree 31%, limestone pastures 16%, lead mine spoil 15%, limestone quarries 8%, cliffs/acidic quarries/verges/limestone wasteland 6%.
Spp. most similar in habitats *Galium sterneri* 82%, *Thymus praecox* 73%, *Scabiosa columbaria* 73%, *Carex flacca* 72%, *Helianthemum nummularium* 72%.
Full Autecological Account *CPE* page 562.

Synopsis *T.s.* is a long-lived, erect, somewhat woody herb forming extensive patches by means of rhizome growth. The species has a rather slow growth rate and, despite producing robust ascending shoots, tends not to form a dense canopy, and is excluded by species with taller or more-consolidated clonal populations. Consequently, *T.s.* is mainly restricted to open infertile habitats, and is particularly associated with scree and steeper slopes, which may be inherently unstable. *T.s.* is usually found on well-drained mineral soils, but is less frequent in areas with high summer temperatures and is sensitive to severe drought. The widespread occurrence of *T.s.* on scree appears to be in part a result of the continuously moist soil maintained by the 'mulch effect' of talus, even on S-facing slopes. The species is sensitive to grazing and trampling. *T.s.* is eaten by sheep, but not cows, and only survives in lightly grazed pasture. Physiological tolerance of shade has been recognized, but even woodland ecotypes of *T.s.* are only moderately shade-tolerant and the species is much more frequent in open scrub than dense woodland. The species is a European endemic with an Atlantic distribution. *T.s.* regenerates both vegetatively by rhizome proliferation and by seed. The seeds, which may form a persistent seed bank, have no well-defined dispersal mechanism, although they may be moved by ants. Regeneration by seed is infrequent in shaded sites, and fatalities on calcareous soils result mainly from summer drought and winter frost; on acidic soils fungal attack appears critical to seedling establishment. *T.s.* exploits a wide range of well-drained soils. There is ecotypic differentiation between populations exploiting acidic and those on calcareous soils, and between plants in shaded and unshaded habitats. *T.s.* is rather uniform morphologically except in the Mediterranean region, where other ssp. are recognized.

Current trends Restricted to infertile habitats. Probably underdispersed to suitable new sites, and decreasing.

256   *Thymus praecox*

Ssp. *arcticus*

Established strategy S.
Gregariousness Potentially carpet-forming.
Flowers Purple, gynodioecious but most plants hermaphrodite, insect-pollinated or selfed.
Regenerative strategies V, B_s.
Seed 0.11 mg.
British distribution Widespread (95% of VCs).
Commonest habitats Limestone pastures 27%, lead mine spoil/scree 26%, rock outcrops 14%, limestone quarries 3%.
Spp. most similar in habitats *Galium sterneri* 95%, *Campanula rotundifolia* 87%, *Helianthemum nummularium* 85%, *Sanguisorba minor* 85%, *Koeleria macrantha* 82%.

Wild Thyme

Full Autecological Account *CPE* page 564.

Synopsis *T.p.* is a long-lived, slow-growing, prostrate, mat-forming, evergreen undershrub. The species is rapidly submerged by taller and faster-growing herbs, and consequently is restricted to open, often rocky, habitats and short turf, extending from sea level (cliffs and sand dunes) to mountain sides. *T.p.* is most characteristic of sites which are droughted in summer. Although the shoot is xeromorphic, *T.p.* is essentially a drought-avoiding species and the deep primary root system, which may on occasion descend to 2 m, may reach sources of water unavailable to associated species. The seedling root may reach 40 mm in length before the first true leaves develop, a capacity which must be important in drought avoidance. The leaves of *T.p.* are generally avoided by grazing animals, and the relative abundance of *T.p.*, which is not shade-tolerant, is enhanced by grazing of taller turf components. *T.p.* is a strict calcicole over much of S, E and SE England, but extends into a wider range of habitats including acidic soils in the N and W. Seedlings establishing upon acidic soil, and presumably subject to some root stunting, are vulnerable to summer droughts. This may explain the capacity of *T.p.* to colonize more-acidic sites in areas of heavier rainfall and higher humidity, i.e. in the N and W. Nevertheless, even here *T.p.* is usually associated with elevated levels of exchangeable Ca, Mg or K and may, for example, grow from calcite veins in mountain rocks or in turf subject to salt spray. *T.p.* is a poor colonist, and is most characteristic of semi-natural habitats. *T.p.* regenerates both vegetatively and by seed. Seed is shed in autumn or may be retained on the plant over winter, and forms a persistent seed bank. At higher latitudes viable seed may be produced only during warmer, sunnier years, and here regeneration by vegetative means probably assumes greater importance. *T.p.* also produces extensive mats by long creeping branches bearing adventitious roots. These mats may break up with age. The production of long runnering stems, tolerant of shallow burial in the soil, is particularly important in the colonization of anthills which are a favoured habitat in S England and in exploiting sand dunes, where *T.p.* can survive at least 50 mm of sand deposition and contributes to the stabilization of dune blow-outs. A heavy-metal-tolerant race of *T.p.* in Derbyshire has been described. The species is morphologically variable in Europe where five sspp. have been recognized. European sspp. include diploids, but British plants are tetraploid.

Current trends *T.p.* is rapidly decreasing in many lowland areas. However, in upland regions where pressures of land-use are less intense, *T.p.* is probably declining more slowly, and a restricted capacity to colonize quarries and other artificial habitats is evident.

257 *Torilis japonica* Upright Hedge-parsley

Established strategy Between C–S–R and S–R.
Gregariousness Typically sparse.
Flowers Pale pink or purplish-white, hermaphrodite, insect-pollinated.
Regenerative strategies S.
Seed 1.98 mg.

British distribution Widespread except in N Scotland (98% of VCs).
Commonest habitats Rock outcrops 5%, soil heaps 4%, hedgerows/limestone quarries/limestone woodland 3%.
Spp. most similar in habitats *Trifolium*

dubium 61%, *Arabidopsis thaliana*
59%, *Saxifraga tridactylites* 58%,
Desmazeria rigida 53%, *Geranium
molle* 52%.

Full Autecological Account *CPE* page
566.

Synopsis *T.j.* is a late-flowering
umbellifer found in a range of habitats,
including areas on steep roadsides and
ditch banks where the growth of
perennials has been reduced by soil
creep, and at more-stable sites on
limestone rock outcrops, where
summer drought and low levels of
nutrients restrict productivity.
Although broadly similar in habitat
range to the diminutive winter annuals
(see 'Spp. most similar'), *T.j.* is more
robust and persists longer into
summer; this may be related to its
tendency to exploit pockets of deep
soil and its modest potential for
competing with perennial plants. The
species also occurs along woodland
margins and in hedgerows, where
populations show a capacity to persist
in tall but open vegetation dominated
by perennial species. *T.j.* has an erect
habit and is seldom recorded from cut
or grazed habitats. It has a
predominantly lowland distribution,
reaching only 415 m in the British Isles
and tends to be most abundant on
calcareous strata. The species is
dependent entirely upon the relatively
large seeds for regeneration. Detailed
studies of the germination behaviour of
T.j. have been undertaken in N
America. Seed is retained on the plant
after the death of the parent, and falls
to the ground in the late summer and
autumn. Some seed germinates to
produce plants which flower in the
following year, but seed which fails to
germinate in autumn, mainly because
it is retained longer on the plant, is
induced into dormancy by low winter
temperatures and does not germinate
until the following autumn. Thus, *T.j.*
has a transient seed bank in which the
maximum length of the dormancy
period appears to be 12 months.
However, in a field experiment in S
England it was found that germination
was delayed until spring and,
consistent with this, work in the
Sheffield region showed that the seeds
have a chilling requirement. Clearly,
these observations are at variance with
the N American studies and with field
observations from the Sheffield region
that young plants overwinter.
Additional work on the germination
behaviour of this species, collected
from various geographical locations
and vegetation types, is required. The
mericarps of *T.j.* have hooked bristles
and can be widely dispersed by animals
and humans. Dwarf plants resembling
T. arvensis, which fruit after harvest-
time, may be genetically distinct.
Otherwise, nothing is known of the
extent of ecotypic variation in this
wide-ranging diploid species.

Current trends A mobile and perhaps
increasing species.

Established strategy Between R and S–R.
Gregariousness Intermediate.
Flowers Flowers yellow, hermaphrodite, self-pollinated.
Regenerative strategies S, B$_s$.
Seed 0.32 mg.
British distribution Widespread (100% of VCs).
Commonest habitats Rock outcrops 8%, meadows 6%, limestone quarries 5%, soil heaps 4%, arable/cinder heaps/verges 2%.
Spp. most similar in habitats *Saxifraga tridactylites* 86%, *Arabidopsis thaliana* 82%, *Myosotis ramosissima* 75%, *Arenaria serpyllifolia* 72%, *Geranium molle* 71%.
Full Autecological Account *CPE* page 568.

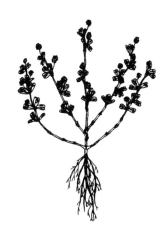

Synopsis *T.d.* is a creeping annual legume found in a wide range of relatively infertile grasslands and open habitats. Observations in N England suggest that *T.d.* is usually a winter annual, rather than a summer annual. Nitrogen-fixing root nodules are formed in conjunction with *Rhizobium trifolii*, but the capacity to fix nitrogen on sand and mine waste is inferior to that of most common legumes. At the ecological extremes of its distribution *T.d.* appears to exploit two very different types of habitat. On calcareous rock outcrops and dry sandy soils *T.d.* grows in winter-annual communities and has a restricted flowering period due to the onset of summer drought. Here *T.d.* is among the later-flowering winter annuals. It has a relatively deep tap-root which may afford some capacity for drought avoidance. In lawns, a habitat in which the species is not usually droughted, *T.d.* continues to set seed until autumn. Survival here appears to be related to the capacity of *T.d.* to adopt a low growth form. This habit is less effective against grazing, however, and

T.d. is largely absent from pasture, although it does occur occasionally in unproductive meadows. *T.d.* regenerates entirely by seed, and the presence of a buried seed bank has been reported. The quantity of *T.d.* present as a contaminant of agricultural seed stocks of *Trifolium repens* is carefully controlled but, bearing in mind the vast quantity of *T.r.* sown, this source is probably sufficient to have a significant effect on the distribution of *T.d.*. Only one cultivar of this low-yielding species has been grown on a small scale. Plant size and morphology depend to a great extent upon management and site fertility and, in the absence of severe drought, populations of *T.d.* on rock outcrops flower throughout summer. The extent to which phenotypic plasticity is complemented by ecotypic differentiation is not known. The wide-ranging distribution of *T.d.* overlaps with those of the related but more ecologically restricted *T. micranthrum* on lawns and *T. campestre* on outcrops, but there are strong breeding barriers between most species of *Trifolium*, and hybrids have not been reported.
Current trends *T.d.* is an effective

colonist of many habitats of intensively exploited landscapes. Probably

increasing, at least within lawns and other sown grasslands.

259 *Trifolium medium* — Zig-zag Clover

Established strategy Between S–C and C–S–R.
Gregariousness Potentially stand-forming.
Flowers Purple, hermaphrodite, self-incompatible, insect-pollinated.
Regenerative strategies V, ?B$_s$.
Seed 2.34 mg.
British distribution Widespread except in E Anglia (96% of VCs).
Commonest habitats Limestone wasteland 15%, coal mine waste/limestone pastures/limestone quarries 2%, enclosed pastures 1%.
Spp. most similar in habitats *Bromus erectus* 92%, *Brachypodium pinnatum* 86%, *Lathyrus montanus* 83%, *Centaurea nigra* 76%, *Avenula pubescens* 72%.
Full Autecological Account *CPE* page 570.

Synopsis *T.m.* is a long-lived legume particularly associated with grassland on heavy soils of intermediate fertility. Nitrogen-fixing root nodules are formed in conjunction with *Rhizobium trifolii*. *T.m.* is a robust species producing long straggling shoots which are capable of emergence through the tall canopies of derelict grasslands. The species is the only common legume consistently to achieve a large biomass in communities dominated by *Brachypodium pinnatum*. However, in some situations, particularly in stands of *Arrhenatherum elatius*, *T.m.* may be represented by a few suppressed individuals. The possibility that the balance between *T.m.* and its associated grasses experiences cyclical fluctuations which depend upon the changing nitrogen level of the soil (as is the case in *Lolius perenne* and *Trifolium repens*) deserves investigation. *T.m.* is tolerant of frost and is very winter-hardy. The species is also moderately tolerant of shade and is occasionally found at woodland margins where it is often subject to sublethal attack by mildew in autumn. On the magnesian limestone of South Yorkshire *T.m.* is particularly associated with soils of intermediate pH which are neither droughted nor waterlogged. *T.m.* forms extensive clonal patches by means of rhizomes. Breakdown of these rhizomes results in the isolation of daughter colonies. *T.m.* also regenerates by seeds, which, according to some, germinate in autumn. Other species of *Trifolium* that are common in Britain form a persistent seed bank and on this basis one is predicted for *T.m.*. *T.m.* is considered self-incompatible, and some clones produce little seed; this may partly explain the tendency of the species to be associated with older grassland habitats. *T.m.* varies in growth form according to habitat, and it has been suggested that it should be possible to select types suitable for grazing and others for hay production.

The degree of variation in *T.m.* in the British Isles is clearly less than that in C and S Europe, where *T.m.* is divided into four ssp.

Current trends *T.m.* has in the past colonized artificial habitats such as railway banks, roadsides, coal-mine spoil and cinders but shows little capacity to exploit habitats of more-recent origin. The species is decreasing, particularly in lowland areas. Selection for a higher degree of self-compatibility is expected.

260 *Trifolium pratense*

Includes vars. *pratense* and *sativum*

Established strategy C–S–R.
Gregariousness Sparse to intermediate.
Flowers Purple, hermaphrodite, insect-pollinated, virtually self-sterile.
Regenerative strategies S, B_s.
Seed 1.35 mg.
British distribution Widespread (100% of VCs).
Commonest habitats Meadows 32%, arable 13%, lead mine spoil 9%, cinder heaps 7%, manure and sewage waste 6%.
Spp. most similar in habitats *Phleum pratense* 79%, *Cerastium fontanum* 73%, *Bromus hordeaceus* 72%, *Ranunculus acris* 71%, *Rhinanthus minor* 65%.
Full Autecological Account *CPE* page 572.

Synopsis *T.p.* is an agriculturally important, tufted, winter-green, grassland legume with a pronounced summer peak in above-ground biomass. Nitrogen-fixing root nodules are formed in conjunction with *Rhizobium trifolii*. *T.p.* is strongly suppressed by nitrogenous fertilizer, whether applied as ammonium salts, sodium nitrate or farmyard manure. Records include var. *p.*, a native, long-lived perennial of semi-natural habitats and hay meadows of low fertility and var. *s*, a shorter-lived (often 2–3 years) more erect, robust taxon cultivated mainly for hay and silage in short-term leys. We suspect that var. *s.*, which is a frequent escapee from cultivation, is the more common, but the exact ecological distribution of the two varieties has not been determined.

Red Clover

T.p. is found in relatively moist but freely-drained soils, and var. *p.* appears to be particularly associated with slightly acidic soils (pH 5.0–6.0). Both varieties are intolerant of heavy grazing and trampling, var. *s.* perhaps more so because of its robust erect habit. *T.p.* is a poor competitor in fertile undisturbed habitats. Hence var. *p.* is characteristic of lightly grazed turf and old hay-meadows while var. *s.* is a commonly sown plant of fertile grassland. *T.p.*, which is of tufted habit, is dependent upon seed for regeneration and a persistent seed bank is recorded. Seeds may survive ingestion by cattle and horses. Of the cultivated strains tetraploids have a lower seed-setting ability. Estimation of the extent to which *T.p.* is an effective colonist of new sites is confounded by the fact that > 200 t of seed are sown annually. The degree of persistence of colonies formed from escaped cultivars is not known, nor have the ecological consequences of interbreeding between native and cultivated varieties been investigated. *T.p.* shows considerable genetic variation both within native European populations and within cultivars, which include early-, intermediate- and late-flowering types. Ten cultivars are recommended for cultivation. *T.p.* also shows potential for the reclamation of derelict land. Heavy-metal-tolerant populations have been recorded in Derbyshire.

Current trends Var. *p.* is presumably decreasing, whereas var. *s.* is held at an artificially high level through its use in agriculture. With the fragmentation of older grassland systems, we expect (a) that hybridization between native and cultivated populations will be increasingly likely and (b) that selection will favour genotypes which combine persistence with the capacity to exploit relatively fertile habitats.

261 *Trifolium repens*   White Clover, Dutch Clover

Established strategy Between C–S–R
 and C–R.
Gregariousness Intermediate.
Flowers White or pink, hermaphrodite;
 mainly outbreeding, insect
 pollinated.
Regenerative strategies (V), B_s.
Seed 0.56 mg.
British distribution Widespread (100%
 of VCs).
Commonest habitats Meadows 77%,
 verges 46%, limestone pastures
 38%, paths 32%, arable 27%.
Spp. most similar in habitats
 Cynosurus cristatus 85%, *Bellis
 perennis* 78%, *Phleum pratense*
 77%, *Lolium perenne* 76%,
 Cerastium fontanum 74%.
Full Autecological Account *CPE* page
 574.

Synopsis *T.r.* is a low-growing but far-creeping, stoloniferous legume of greatest abundance in moist, fertile habitats. In common with the majority of British legumes, *T.r.* is intolerant of shade. *T.r.* is rapidly suppressed in tall vegetation. In many sites nitrogen fixation by *T.r.* often creates soil conditions which encourage invasion and temporary dominance by grasses. This suppresses *T.r.* and eventually leads to reduced N status and conditions favouring renewed vigour of

T.r.. T.r. is intolerant of drought and severe frosts. *T.r.* has been important in fodder since the 17th century, and is by far the most important pasture legume in Britain (*c.* 900 t are sown annually) and 75 cultivars are listed for Europe. The value of *T.r.* to agriculture stems from the capacity to fix nitrogen in habitats where other major nutrients are not limiting. The morphology and phenology also complements that of *Lolium perenne*, with which it often grows. The creeping shoots allow colonization of gaps between tufts of *L.p.* and many other pasture grasses with a consolidated growth form. In keeping with the low nuclear DNA amount, *T.r.* grows predominantly during summer in contrast with the spring and autumn peaks for *L.p.*. In addition, *T.r.* is tolerant of heavy grazing, trampling and cutting. *T.r.* regenerates almost exclusively by rooted stolons in closed communities to form large diffuse clonal patches but colonizes new sites mainly by seed, which germinates mainly in spring and may form a potentially long-lived persistent seed bank. Seeds can survive ingestion by a range of animals and birds. Although the foliage of *T.r.* is highly palatable to stock, some genotypes contain cyanogenic glucosides, which afford varying degrees of protection against invertebrate herbivores and small mammals and, when consumed in large quantities, can disrupt the rumen of domestic animals. Cyanogenetic phenotypes are less common in cold climates, and at high altitudes show evidence of poorer flowering susceptibility to frost and drought. The wide ecological amplitude of *T.r.* in many temperate zones is correlated with genetic variation in morphology, growth rate, phenotypic plasticity in growth form, and with widespread sowing of agricultural cultivars of *T.r.* and the association with several strains of the nitrogen-fixing *Rhizobium trifolii*. A number of ssp. are recognized in Europe. Evidence has been obtained of local variation in *T.r.* coinciding with changes in the identity of the grass species with which *T.r.* is associated in the field. More-recent studies suggest that these differences are not genetic in origin, and it now seems likely that they may arise from changes in the phenotype induced by particular strains of *Rhizobium trifolii*, the distribution of which may be influenced by the rhizosphere associated with particular turf grasses.

Current trends Sowings for agriculture and more recently for amenity purposes maintain the abundance of *T.r.* at an artificially high level. The popularity of *T.r.* is unlikely to wane, since there are few native legumes with as great a capacity to persist in fertile habitats. A greater use of indigenous forms which are shorter, more persistent and drought-tolerant is predicted.

262 *Tripleurospermum inodorum*   Scentless Mayweed

Also known as *Matricaria perforata*

Established strategy R.
Gregariousness Intermediate.
Flowers Florets insect-pollinated, largely self-incompatible.
Regenerative strategies S, B_s.
Seed 0.29 mg.
British distribution Lowland (100% of VCs).
Commonest habitats Arable 23%, bricks and mortar 21%, soil heaps 16%, cinder heaps 13%, coal mine waste 10%.
Spp. most similar in habitats *Polygonum aviculare* 85%, *Capsella bursa-pastoris* 81%, *Matricaria matricarioides* 79%, *Artemisia*

vulgaris 79%, *Lamium purpureum* 78%.

Full Autecological Account *CPE* page 576.

Synopsis *T.i.* is a short-lived herb which exploits a range of disturbed and relatively fertile artificial habitats. *T.i.* is intolerant both of dense shade and of waterlogging, and the erect growth form renders the plant susceptible to grazing and frequent mowing. The persistence of *T.i.* on ballast beside railways may be related to its resistance to herbicides. Like many other common ruderals, the species exploits both 'new' open sites (e.g. soil heaps) and continuously disturbed habitats (e.g. arable). Success appears to be related to a high potential for seed production: in an open, fertile habitat an individual may produce more than 10 000 seeds, while stunted individuals with a single capitulum may form as few as 100 viable seeds. Seed is dispersed effectively by human and other agencies, including animals; *T.i.* is widely distributed during harvesting and as an impurity in grass seed and the achenes are eaten by birds and may survive ingestion by stock. Seed shows little innate dormancy, but a persistent seed bank is formed and confers the capacity for seed germination over much of the year (but particularly in spring and autumn). Unlike many weeds, however, *T.i.* is a long-day, relatively slow-maturing plant and flowering is mainly restricted to the period July to September. This inflexibility may partially explain why *T.i.* reaches maximum frequency in

sites with some habitat continuity, e.g. arable fields, and on those areas of spoil heaps where the rate of vegetation re-colonization is slow. Although normally an annual, a small percentage of the populations on demolition sites and cinder tips may be short-lived, polycarpic perennials. *T.i.* is usually strongly self-incompatible, and viable populations consist of several individuals; this may further explain the restriction to sites with some habitat continuity. The degree of ecotypic differentiation in this outbreeding species is not certain. There is no evidence of genetically differentiated winter- and summer-annual forms. However, the mean size of achenes varies from 1.5×0.7 mm in S and E England to 2.2×1.2 mm in Scotland; some populations from cereal fields may assume a prostrate habit in cultivation, and differences in resistance to herbicides have been reported. Native British populations of *T.i.*, like those from W Europe, appear diploid and the tetraploids, which are sometimes more vigorous but otherwise ecologically indistinguishable, occur only as a rare introduction from within their native range in E and N Europe. Maritime populations have recently been separated from *T.i.* as *T. maritimum*. These, too, are diploid and self-incompatible and introgression with *T.i.* is widespread near the coast. Hybrids with *Anthemis cotula* have been reported from Britain.

Current trends A common colonist of artificial habitats and perhaps still increasing.

263 *Trisetum flavescens*

Yellow Oat-grass

Established strategy C–S–R.
Gregariousness Sparse to intermediate.
Flowers Yellow-green, hermaphrodite, wind-pollinated, self-incompatible.
Regenerative strategies S, V.
Seed 0.18 mg.

British distribution Common, except in the N and W (94% of VCs).
Commonest habitats Meadows 37%, limestone quarries 29%, limestone wasteland 28%, rock outcrops 25%, limestone pastures 23%.

Spp. most similar in habitats *Plantago lanceolata* 85%, *Festuca rubra* 75%, *Lotus corniculatus* 71%, *Linum catharticum* 71%, *Hieracium pilosella* 68%.
Full Autecological Account *CPE* page 578.

Synopsis *T.f.* is a fairly tall, tufted grass of dry grassland and, to a lesser extent, rocky habitats, particularly on base-rich soils. *T.f.* resembles a 'small *Arrhenatherum elatius*' in habit and in producing a midsummer peak in above-ground biomass. In unmanaged grassland the species is usually restricted to sites where the growth of more-productive species is checked by, for example, periodic burning. *T.f.* also tends to be absent from severely droughted habitats. In common with other grasses producing erect nodal stems, *T.f.* appears to be susceptible to heavy grazing by stock. The leaves are palatable to cattle and the stems also are consumed by sheep. Towards the base of the plant long reflexed hairs are often present in abundance on the leaf sheaths; there is some evidence that such hairs restrict the activities of invertebrate herbivores. *T.f.* is susceptible to trampling. In Britain *T.f.* is associated with a range of soil types, moisture and fertility levels, and a variety of management regimes. *T.f.* appears to show only modest specialization towards any ecological factor or turf structure, and we conclude that the species is a 'congenital subordinate', i.e. never more than a minor component of grassland communities, irrespective of management or turf height, but capable of persisting through fluctuations of environment or management which are sufficient to bring about drastic changes in the abundance of the more-specialized potential dominants. This is consistent with descriptions of *T.f.* as 'a very insignificant member of all the associations in which it occurs, except occasionally on the limed sections of plots receiving farmyard manure'. Perhaps because *T.f.* is such a minor species, its ecological distribution has been little studied. Early seed-set and almost immediate germination has been documented in sites subject to summer drought, and this may be expected to confer an advantage in hayfields. No persistent seed bank has been reported. Populations tolerant of heavy metals have been detected in the Derbyshire flora. There is considerable taxonomic complexity and genetic diversity in Europe as a whole, and two additional subspecies have been recognized.

Current trends *T.f.* is no longer included in agricultural seed mixtures and is now restricted to less intensively managed grassland. *T.f.* is not among the most frequent colonists of artificial habitats, and is probably decreasing slightly.

Established strategy C–R.
Gregariousness Forms stands at low
frequencies.
Flowers Florets yellow, insect- or self-
pollinated.
Regenerative strategies (V), W.
Seed 0.25 mg.
British distribution Widespread (100%
of VCs).
Commonest habitats Limestone
quarries 45%, bricks and mortar
43%, coal mine waste 31%, cinder
heaps 28%, lead mine spoil 17%.
Spp. most similar in habitats *Senecio
squalidus* 70%, *Artemisia vulgaris*
69%, *Senecio viscosus* 68%, *Atriplex
prostrata* 68%, *Rumex crispus* 64%.
Full Autecological Account *CPE* page
580.

Synopsis *T.f.* is a rhizomatous, stand-
forming perennial which colonizes
bared ground. Extensive lateral spread
underground, coupled with the shade
created by the large, entire, radical
leaves contribute to its subsequent
capacity for dominance of relatively
short (< 200 mm) perennial
vegetation. In urban areas *T.f.* is most
frequent on spoil tips and demolition
sites. As a result of its low canopy
height, the persistence of *T.f.* varies
according to the productivity of the
site. In fertile habitats *T.f.* tends to be
replaced eventually by taller
dominants, while in less-fertile sites,
such as coal-mine spoil, *T.f.* may
persist for many years. Rhizomes are
produced at various depths in the soil,
and this may be important in allowing
T.f. to survive in such semi-natural
habitats as stream banks and boulder-
clay cliffs, where soil movement may
occur annually. However, because of
its robust stature and long-lived leaf
canopy, *T.f.* is vulnerable to the effects
of regular grazing, cutting and
trampling. The species is little affected
by waterlogging, but requires exposed
mineral soil for seedling establishment.

Early spring flowering from buds
initiated in autumn, followed by
vegetative growth in summer and die-
back over winter constitutes an
unusual phenological pattern within
the British flora and may be related to
the ability of *T.f.* to exploit sub-alpine
'snow patch' vegetation in Europe.
The levels of Na, Ca and Mg in the
leaves are unusually high, but the
concentrations of N and P are low. In
its ability to colonize disturbed
artificial habitats *T.f.* is most similar to
Chamerion angustifolium. On heavy
clay and on calcareous soils, *T.f.* is the
more frequent, despite the fact that
C.a. is a taller and more effective
potential dominant. Seed is short-lived
and under natural conditions over 50%
is non-viable within two months.
However, seeds are produced in
abundance during spring and early
summer, with as many as 1000 in each
tuft of flower heads. The small-plumed
achenes are extremely buoyant in air
and may be carried up to 4 km from
the parent plant. As a result, *T.f.* is a
frequent colonist of bare ground in
early summer by means of freshly-shed
seed. Germination is exceptionally
rapid, even on substrates with a low
water content and at low temperatures.
Subsequent establishment of the

seedling is also speedy in unshaded sites where plants may flower in their second year. Established colonies increase in size mainly by vegetative means, and large clonal patches may be formed through the expansion of short-lived rhizomes by up to 1 m year^{-1}. Rhizomes may be dispersed in soil by humans in the same way as those of *Reynoutria japonica*. However, the balance between vegetative and sexual reproduction varies according to the stage of colonization. Thus, in diffuse patches a greater proportion of resources are allocated to rhizomes,

while in crowded stands of *T.f.* more flowers are produced. The existence of ecotypes of this ecologically wide-ranging species is suspected but has not been demonstrated.

Current trends A species which has increased dramatically due to the artificial creation of disturbed habitats. Probably still increasing but, in view of the short-lived nature of its seed and the limited persistence of the adult plant in grazed sites and tall-herb communities, *T.f.* is potentially vulnerable in the event of a reduction in the level of disturbance within the landscape.

265 *Typha latifolia*   Cat's-tail, Great Reedmace

Established strategy C.
Gregariousness Stand-forming at low frequencies.
Flowers Flowers unisexual, wind-pollinated, usually self-pollinated.
Regenerative strategies V, W, B$_s$.
Seed 0.03 mg.
British distribution Widespread except in the N and W (94% of VCs).
Commonest habitats Still aquatic 7%, unshaded mire 6%, coal mine waste/shaded mire 2%.
Spp. most similar in habitats *Ranunculus peltatus* 96%, *Equisetum fluviatile* 96%, *Lemna minor* 92%, *Polygonum amphibium* 92%, *Eleocharis palustris* 92%.
Full Autecological Account *CPE* page 582.

Synopsis *T.l.* is a robust dominant of ungrazed and uncut eutrophic wetlands. Although common in mire, locally *T.l.* achieves maximum abundance as an emergent aquatic in shallow water and, like other species exploiting this habitat, has well-developed aerenchyma. Unusually among dominants, *T.l.* produces large, ascending basal or sub-basal leaves

and, despite its high productivity, the species has a rather open leaf canopy. The foliage of plants sampled in the Sheffield region contains relatively low levels of N, P and Fe, whereas the concentrations of Na and Mn are often unusually high. *T.l.* produces an extensive rhizome system, and offshoots are produced throughout the growing season. *T.l.* forms large clonal patches which may expand at up to 4 m year^{-1}. Detached rhizome pieces readily re-root. These may float in water until beached, and have the potential to form new colonies. Each inflorescence produces thousands of minute fruits, and these provide the main mechanism of colonization. They may be transported by wind under dry conditions and may float in water, but seeds are released and sink once the pericarp comes into contact with water. Seeds exhibit hard-coat dormancy, and may be induced to germinate by scarification. Germination is said to be promoted by light in combination with low oxygen concentrations or, at low light intensities, by fluctuating temperatures. A persistent seed bank is also recorded. Regeneration by seed appears not to occur within established colonies. *T.l.* is grown for ornament and may sometimes be a garden escapee. The female part of the inflorescence is receptive to pollen for four weeks and pollen, which is released in tetrads, is only effectively dispersed in high winds. Thus, despite its protogyny, *T.l.* is mainly in-breeding. Nevertheless, in N America a range of ecotypes has been identified. *T.l.* is widely distributed on nutrient-rich mineral soils and may occur in peaty water. Under calcareous or saline conditions, however, *T.l.* may be replaced by *Phragmites australis*. *T.l.* hybridizes with *T. angustifolia*. It has been suggested that in water more than 150 mm deep *T.l.* tends to give way to *T. angustifolia*, which is taller than *T.l.* and exhibits a greater allocation to sexual reproduction. Other workers suggest that *T.a.* replaces *T.l.* under saline conditions; both observations are consistent with features of the British distributions of the two species. The species has potential for use as a reed crop.

Current trends One of the commonest semi-aquatic dominants in Britain, and a very effective colonist even of landlocked sites; *T.l.* is one of the few wetland species which is clearly increasing.

---

266 *Ulex europaeus*

(Field records include seedlings)

Established strategy S–C.
Gregariousness Sparse except after fires.
Flowers Yellow, hermaphrodite, usually insect-pollinated.
Regenerative strategies B_s.
Seed 6.2 mg.
British distribution Widespread (98% of VCs).
Commonest habitats River banks/limestone wasteland 3%, cinder heaps/limestone pastures/acidic quarries/limestone quarries/acidic wasteland 2%.
Spp. most similar in habitats Insufficient data.
Full Autecological Account *CPE* page 584.

Furze, Gorse, Whin

Synopsis *U.e.* is a tall, colonizing shrub up to 2 m in height, with an Atlantic distribution. Most roots are superficial, but *U.e.* also has a deep tap-root. The species forms nitrogen-fixing root nodules and, unlike those of most herbaceous species, these are markedly perennial. *U.e.* is most characteristic of roadsides, railway banks, wasteland, derelict pasture and sea-cliffs on infertile soils, and has been planted in certain areas as hedging. The species is not persistent under the heavy shade of other species. As a seedling the plant forms trifoliate leaves. but these are replaced by woody spines or scales in the established plant. Both short days and low light intensities (but not high humidity) reduce the growth of the shoot and prolong the duration of the leafy juvenile phase. The species frequently forms impenetrable thickets with persistent spiny litter. Studies in New Zealand suggest that the amount of litter shed each year corresponds to about half the annual growth increment, and *U.e.* may frequently represent a fire hazard. The seed is relatively large, and although seeds are explosively discharged and may be dispersed by ants, the seedlings are mostly restricted to the vicinity of mature bushes. For this reason the field data presented for juveniles in this account are also relevant to the distribution and ecology of the adult bush. In young bushes the erect stems provide a dense canopy and often have no vegetation beneath them. After 15–20 years bushes become leggy and the branches more spreading, leaving a central, little-shaded area. *Betula* spp. frequently colonize and eventually replace degenerating gorse bushes. However, gorse may be maintained by controlled burning. The presence of *U.e.* often indicates a site of former disturbance, particularly by fire. The seed has a hard coat, which may afford it some protection from high temperatures, and seeds have been rendered germinable after heating to 88°C. After a fire it has been reported that about half of the seed bank is killed, but the remainder may subsequently germinate. Seed may persist for up to 28 years in the soil and areas may become carpeted with *U.e.* seedlings, and finally a gorse thicket, following a fire. Germination of seeds and survival of seedlings is lower under a gorse canopy than in the open. The species does occur on shallow calcareous soil, but is much more characteristic of mildly acidic soils within the pH range 4.0–6.0. The soil around the roots is frequently more acidic than the surrounding soil. *U.e.* is relatively sensitive to lime-chlorosis when grown on calcareous soil. *U.e.* was widely used for fodder and for other purposes, and is still an important source of food for cattle and ponies on marginal land. Young sprouting shoots may be severely predated by rabbits. The biological control of *U.e.* in New Zealand, using the seed-eating weevil *Apion ulicis*, was unsuccessful even though 90% of seed was destroyed. Intra- and inter-population variation in morphology is largely plastic, but shows a genetic component. On the infertile, highly acidic soils of heathland and moorland *U.e.* is replaced in the N and W by *U. gallii*, and in the SE mainly by *U. minor*. Hybrids between *U.e.* and *U.g.* occur in the SW, where the flowering period of the mainly spring-blooming *U.e.* and the autumnal *U.g.* overlap. A further ssp. is recorded from Europe.

Current trends Probably still increasing as a result of the dereliction of poor pasture and by colonizing motorway verges and the margins of woodland rides in plantations on acidic soils. An increase in grassland invasion has also been reported following myxomatosis. Has some potential in land reclamation as a result of its persistent growth under conditions of extreme infertility and its capacity to fix atmospheric nitrogen.

267 *Ulmus glabra*   Wych Elm

(Field records as seedlings)

Established strategy C.
Gregariousness Seedling populations
 locally dense.
Flowers Reddish, hermaphrodite,
 mainly wind-pollinated and
 self-incompatible.
Regenerative strategies W.
Seed 3.5 mg.
British distribution Common,
 expecially in the N and W.
Commonest habitats Lead mine spoil
 7%, limestone quarries 6%,
 scrub/verges 5%, shaded
 mire/broadleaved plantations 4%.
Spp. most similar in habitats *Fraxinus
 excelsior* (juv.) 68%, *Betula* sp.
 (juv.) 66%, *Melica uniflora* 64%,
 Bromus ramosus 64%, *Hedera helix*
 62%.
Full Autecological Account *CPE* page
 586.

Synopsis A species in need of further
study. *U.g.* is a tall, deciduous timber
tree with a furrowed bark, extending
to 40 m in height and forming ring-
porous wood with very large vessels in
spring. Bud break is rather earlier than
in other ring-porous trees. On
theoretical grounds this would render
vessels liable to cavitation during cold
springs. The effect of this unusual
feature on the distribution of *U.g.* is
uncertain, but it is interesting to note
that, unlike most other common trees,
U.g. does not form monocultures
within the Sheffield region. In addition
the species is vulnerable to drought in
East Anglia, and is most frequent in
the N and W of Britain. All species of
elm contribute only *c.* 11 000 ha to
broad-leaved woodland (< 2% of
total). Trees over 200 years old have
been reported in the Derbyshire Dales.
U.g. is tolerant of coppicing. The
species regularly produces an
abundance of winged fruits which are
widely wind-dispersed and capable of
germinating without delay. As a result
seedlings occur in a wider range of

habitats than the adult tree. In N
England *U.g.* occurs mainly as
scattered trees on moist, relatively
fertile soils, and is particularly
common in mixed woodland on
limestone. Less frequently the species
is recorded from scrub and even
occasionally as free-standing trees and
in hedges. Saplings are mainly found
under the shade of other trees, but
sites with extensive regeneration have
not been observed within the region.
The species is a frequent associate of
Fraxinus excelsior in secondary
woodland of the Derbyshire Dales and
seedlings regenerate within ashwoods
since they are much more tolerant of
shade than *F.e.*. As a result, the
proportion of elm in ashwoods tends to
increase and ash–elm woods develop.
U.g. is also a consistent associate of
Tilia spp. in ancient woodlands.
Observations indicate that predation of
seeds and mortality of seedlings is
extremely high. Furthermore, a high
proportion of the seeds shed,
particularly those released early, may
not be viable. No persistent seed bank
is formed. *U.g.* does not normally
sucker, and thus is more vulnerable to
Dutch elm disease than *U. procera*,
which may survive by means of root
suckers. *U.g.* produces a dense

canopy, and litter is short-lived. Ssp. *glabra*, with broad leaves, occurs mainly in the S, and ssp. *montana*, in the N. *U.g.* hybridizes with the two other native British species. However, there is little overlap between *U.g.* and the other native or naturalized species, *U.p.*. One authority does not subdivide *U.g.* and only recognizes one other species, *U. minor*, and hybrids. *U.m.* is restricted to lowland coppices, ornamental woodland and to hedgerows. The number of insects

associated with elms is moderately low for a deciduous tree, implying that the leaves are relatively palatable. Elm leaves were formerly fed to animals, but in the Middle Ages its forestry began to be superseded by that of *Acer pseudoplatanus*, which has wood which is easier to work.

Current trends Decreasing, through the destruction of older woodland and potentially much more vulnerable to the widespread and severe effects of Dutch elm disease than *U. procera*.

268 *Urtica dioica*   Stinging Nettle

Established strategy C.
Gregariousness Potentially stand-forming.
Flowers Green, normally dioecious, wind-pollinated.
Regenerative strategies V, B_s.
Seed 0.19 mg.
British distribution Widespread (100% of VCs).
Commonest habitats Hedgerows/soil heaps 53%, river banks 49%, manure and sewage waste 43%, shaded mire 33%.
Spp. most similar in habitats *Galium aparine* 78%, *Poa trivialis* 74%, *Stachys sylvatica* 73%, *Glechoma hederacea* 70%, *Calystegia sepium* 61%.
Full Autecological Account *CPE* page 588.

Synopsis *U.d.* is a dioecious, tall, rhizomatous herb with stinging hairs. In summer the dense, rapidly-ascending canopy of *U.d.* often precludes the growth of other herbaceous plants, and this, coupled with the impact of relatively persistent stem litter, often causes the species to form monospecific stands. However, because the shoots die back completely in winter, species such as the scrambling annual *Galium aparine*, the low-growing *Poa trivialis* and stem

litter-exploiting bryophytes, e.g. *Brachythecium rutabulum*, are able to benefit from the seasonal availability (between autumn and spring) of relatively high light intensity. *U.d.* is probably only native in fen carr; elsewhere it is a 'follower of humans', restricted to moist, fertile habitats by features such as its high potential growth rate and its high requirement for mineral nutrients. Leaves contain unusually high concentrations of N, Ca, Mg and Fe. Flowering is strongly inhibited by drought, but the species shows a modest capacity for drought-hardening and desiccation resistance. A marked increase in specific leaf area in response to shading is observed in plants established under moderately dense woodland canopies, although *U.d.* is intolerant of dense shade and exhibits poor flowering and a truncated shoot phenology in woodland habitats. *U.d.* is suppressed by repeated cutting, but occurs in pastures, where, perhaps because of the stinging hairs, it is rarely eaten by cattle. However, old leaves are palatable to invertebrates, and the young shoots have been used as a green vegetable. *U.d.* forms large patches by means of rhizome growth, which may increase at 450 mm year^{-1}. Rhizomes broken by digging, or other disturbances, readily re-root to form new colonies. Despite being dioecious and forming large clonal patches, *U.d.* also produces great quantities of seed and accumulates large and persistent banks of buried seeds which have the potential for long-term survival. Germination is stimulated by light and fluctuating temperatures, and occurs mainly on open and disturbed ground. Thus, many of the sites with a continuous vegetation cover in which *U.d.* persists as established clones are unsuitable for regeneration by seed. Seeds survive ingestion by a variety of animals and may be carried far. Seeds and rhizome pieces are also transported in soil and, for this reason, despite the lack of a well-defined dispersal mechanism, the species is highly mobile. *U.d.* shows considerable genetically-based variation, and includes populations, var. *subinermis*, virtually devoid of stinging hairs. Ssp. *gracilis*, which is found in N Europe and N America, is monoecious.

Current trends A very effective colonist of newly-disturbed sites, which, once established, is very persistent and invasive. Appears to be increasing in abundance.

269   *Urtica urens*   Small Nettle

Established strategy Between R and C–R.

Gregariousness Plants usually widely spaced.

Flowers Green, monoecious, wind-pollinated.

Regenerative strategies B$_s$.

Seed 0.50 mg.

British distribution Widespread, particularly in the E (99% of VCs).

Commonest habitats Arable 8%, bricks and mortar 6%.

Spp. most similar in habitats *Lamium purpureum* 81%, *Spergula arvensis* 79%, *Fallopia convolvulus* 77%, *Myosotis arvensis* 76%, *Anagallis arvensis* 74%.

Full Autecological Account *CPE* page 590.

Synopsis *U.u.* is an often tall, potentially robust, summer-annual herb of fertile, disturbed habitats, particularly on sandy soils. The species, sometimes regarded as doubtfully native, is a 'follower of humans' and is most common in arable areas in which broad-leaved crops are grown. In morphology and habitat range *U.u.* resembles *Chenopodium album*. *U.u.* also occurs in some semi-natural sandy habitats near the coast and is often more frequent in coastal than in inland areas. *U.u.* is usually absent from habitats that are grazed, cut or heavily shaded. The importance of stinging hairs is obscure, since *U.u.* does not normally occur in areas subject to heavy grazing pressure. Under favourable conditions *U.u.* has the potential to become a large plant producing abundant seed. In common with many arable weeds, the plant may flower when very small in less-suitable sites. This is consistent with experimental findings showing that reproductive effort is maintained in plants severely stunted by drought. The species regenerates entirely by seed, and effective seedling establishment is probably initiated by spring germination. To an unusual extent, freshly-shed seeds are strongly inhibited by high irradiance. On sandy soils this inhibition could provide a mechanism preventing germination in sites likely to be subject to desiccation during seedling establishment. Seeds persist in the soil, especially when deeply buried in uncultivated soil. Seeds may survive ingestion by animals and are probably transported in a manner similar to that of *U. dioica*. Despite the absence of well-defined dispersal mechanisms, both *U.u.* and *U.d.* appear to be highly mobile species, frequently occurring as casuals outside their main geographical ranges. **Current trends** *U.u.* was not considered an important arable weed in the early years of this century, but has probably increased since then. Now declining on arable land, but its general range may be expanding in response to the creation of an increasing diversity of fertile, disturbed, artificial habitats.

---

270 *Vaccinium myrtillus*

Bilberry, Blueberry, Whortleberry, Huckleberry

Established strategy S–C.
Gregariousness Stand-forming.
Flowers Pink, hermaphrodite, insect- or occasionally self-pollinated.
Regenerative strategies V, B$_s$.
Seed 0.26 mg.
British distribution Uplands (93% of VCs).
Commonest habitats Acidic pastures 37%, acidic quarries 20%, acidic wasteland 18%, limestone pastures 14%, acidic woodland 9%.
Spp. most similar in habitats *Galium saxatile* 94%, *Deschampsia flexuosa* 89%, *Empetrum nigrum* 85%, *Vaccinium vitis-idaea* 83%, *Carex pilulifera* 82%.
Full Autecological Account *CPE* page 592.

Synopsis *V.m.* is a strongly rhizomatous, long-lived, slow-growing, dwarf, deciduous shrub which, with the exception of very local sites in N-facing limestone grasslands, behaves as a strict calcifuge and often occurs on peaty soils. The leaves usually contain high concentrations of Mn and Al. *V.m.* is more tolerant of shade than

Calluna vulgaris, and in woodlands and scrub often forms a continuous layer consisting of large clonal patches. In these, as in other habitats, the angled shoots, although deciduous, provide a considerable photosynthetic area during winter and early spring. *V.m.* exhibits a N bias in its British distribution. This appears to be largely determined by the availability of siliceous substrata. The species is also especially common in woods of N aspect in the survey area and in S England appears to be largely restricted to humid or shaded sites. *V.m.* is relatively frost-sensitive, and in more-extreme climates may have a distribution which is dependent upon snow protection. *V.m.* is tolerant of sheep grazing, and in pastures a low canopy may be established by the development of phenotypes with a very high density of short erect shoots. Release from grazing usually leads to the formation of tall phenotypes with many erect robust shoots forming a dense, elevated leaf canopy in summer. The rhizome system is 150–200 mm below the soil surface, and consequently *V.m.* is much less vulnerable to burning than *C.v.*. The depth of the rhizome system may also contribute to the poor tolerance of waterlogged conditions; in ombrogenous mire *V.m.* is restricted to better-drained areas. *V.m.* regenerates mainly by means of rhizome growth to form over many years clonal patches up to 15 m in diameter. Investigations in the Sheffield region and elsewhere in Britain have failed to detect persistent seeds, but buried viable seeds have been reported in boreal forest areas in N Sweden. More research is needed to establish the general significance of seed persistence in the population biology of *V.m.*. Seedling establishment is slow and infrequent, and is probably restricted to areas of bare soil. Regeneration from seed is likely to be very uncommon at high altitudes, since few flowers are formed in habitats above 1000 m.

Current trends As a result of factors such as increased grazing pressure, *V.m.* may be increasing in upland areas at the expense of *C.v.*. However, in lowland habitats *V.m.* has decreased to a marked extent, a change which is at least partly attributable to habitat destruction.

271   *Vaccinium vitis-idaea*

Cowberry, Red Whortleberry

Established strategy Between S and S–C.

Gregariousness Stand-forming.

Flowers Pink, hermaphrodite, insect- or self-pollinated.

Regenerative strategies V, ?B_s.

Seed 0.26 mg.

British distribution Uplands only (61% of VCs).

Commonest habitats Acidic quarries 9%, acidic pastures 7%, acidic wasteland 4%, coniferous plantations/scrub 2%.

Spp. most similar in habitats *Carex pilulifera* 87%, *Vaccinium myrtillus* 83%, *Erica cinerea* 82%, *Deschampsia flexuosa* 81%, *Galium saxatile* 71%.

Full Autecological Account *CPE* page 594.

Synopsis *V.v-i.* is, like *V. myrtillus*, with which it commonly occurs, a long-lived, slow-growing, dwarf rhizomatous shrub of acidic peaty soils. However, *V.v-i.* differs from *V.m.* in a number of important respects. The leaves are evergreen; this may enable *V.v-i.* to maximize production per unit of limiting mineral nutrient in nutrient-deficient soils. *V.v-i.* also has very tough leaves and is little grazed. *V.v-i.* tends to be restricted to upland areas and seldom occurs below 200 m. The species is at the S edge of its British distribution within the region and is classified as an Arctic–Alpine species, rather than as a plant characteristic of the Continental Northern Element as is the case for *V.m.*. Germination of fresh seed requires high temperatures, a feature characteristic of many northern species (cf. *Eriophorum angustifolium*). *V.v-i.* is more drought-tolerant than *V.m.*, and within the Sheffield region comes closest to forming monocultures on well-drained S-facing slopes. *V.v-i.* is shorter than *V.m.* and may form an understorey in stands of *V.m.*. In many other aspects *V.v-i.* and *V.m.* are similar. Each forms extensive long-lived clones by means of underground rhizomes which, in each species, confer tolerance of burning and lead to susceptibility to waterlogging. At high altitudes each is particularly associated with sites protected in winter by snow cover. In both, regeneration by seed is infrequent, particularly at high altitudes, and seedling establishment is slow. *V.v-i.* also is tolerant of shade but, unlike *V.m.*, reputedly exhibits its maximum vegetative and reproductive performance in pine woods. The reasons for the greater abundance of *V.m.* in deciduous woodland and in the oceanic montane areas of N Wales require investigation. *V.m.* and *V.v-i.* hybridize within the unshaded habitats of the Sheffield region, and elsewhere. Another ssp. with an arctic distribution is recorded.

Current trends *V.v-i.* is now extinct within lowland Britain. In upland areas the species may have increased at the expense of *Calluna vulgaris* because of its greater tolerance of uncontrolled burning and low palatability in grazed sites.

272 *Valeriana officinalis* Valerian

Includes sspp. *collina* and *sambucifolia*

Established strategy C–S–R.
Gregariousness Intermediate.
Flowers Pink, hermaphrodite, insect-pollinated, self-incompatible.
Regenerative strategies V, S.
Seed 0.95 mg.
British distribution Widespread (99% of VCs).
Commonest habitats Limestone wasteland 11%, limestone quarries 5%, river banks/verges 3%, shaded mire/rock outcrops 2%.
Spp. most similar in habitats Insufficient data.
Full Autecological Account *CPE* page 596.

Synopsis *V.o.* is a mesophytic herb with a basal tuft consisting of a few ascending compound leaves, and produces a tall flowering shoot in which the lower leaves are the largest. The low shoot density and modest height of leaf canopy, coupled with the very restricted capacity for lateral vegetative spread, limit the capacity of *V.o.* for vegetation dominance. The shoot consists of a small number of succulent compound leaves; this renders *V.o.* vulnerable to grazing and, consequently, the species is infrequent in pastures. *V.o.* often sets seed in lightly shaded habitats, and we suspect that its survival in tall vegetation involves a modest degree of shade tolerance. It also seems likely that the persistence of *V.o.* depends upon the operation of factors, e.g. occasional grazing, mowing or fire, which limit the vigour of potential dominants. Occurrences over the range from moist to waterlogged base-rich habitats owe much to the existence of two cytotypes which have been afforded subspecific rank. *V.o.* shows considerable genotypic variation at each level of ploidy. All or most records from the Sheffield region appear to refer to ssp. *s.*, which is widely distributed in Britain and occurs in N, NC and EC Europe. Ssp. *s.*, an octoploid, occurs typically in wetter habitats within the range of ssp. *c.*, but in the N occupies the whole range of sites exploited by the species. In this ssp., plants regenerate vegetatively by both epigeal and hypogeal stolons which are produced in autumn, albeit very slowly. Regeneration by means of bulbils formed in the axils or the leaves, or by wind-dispersed seed, is infrequent. In common with many other wetland species, the seeds exhibit a high temperature requirement for germination. This suggests that germination will be a spring event, although further research is needed to define the conditions favourable to seedling establishment. Ssp. *c.*, a tetraploid, is in Britain distributed mainly S of a line between the Wash and the Bristol Channel, and occurs in W and C Europe. The seeds are smaller and the epigeal stolons (and sometimes also the hypogeal stolons) are absent. Stolons may be shorter, reducing still further the capacity for vegetative spread. The ssp. is also largely restricted to drier habitats, particularly those on calcareous soils. The limits of the two ssp. are confused by the existence of a wide range of intermediates, and in Europe a further diploid ssp., *officinalis*, is recorded.
Current trends Infrequent as a colonist of recent artificial habitats, and presumably decreasing.

273 *Veronica arvensis*

Wall Speedwell

Established strategy S–R.
Gregariousness Typically sparse.
Flowers Blue, hermaphrodite, probably often selfed.
Regenerative strategies B_s.
Seed 0.11 mg.
British distribution Widespread (100% of VCs).
Commonest habitats Limestone pastures/rock outcrops 8%, arable 6%, cinder heaps 4%, meadows/limestone quarries 3%.

Spp. most similar in habitats *Myosotis ramosissima* 75%, *Saxifraga tridactylites* 71%, *Aphanes arvensis* 70%, *Trifolium dubium* 66%, *Arabidopsis thaliana* 66%.
Full Autecological Account *CPE* page 598.

Synopsis *V.a.* is a small, facultatively winter- or summer-annual herb found in habitats where the growth of perennials is restricted by drought, nutrient deficiency, disturbance or by a combination of these factors. Like *Aphanes arvensis* and *Arenaria serpyllifolia*, the species extends from droughted, nutrient-deficient calcareous rock outcrops and dry sandy soils to fertile but highly disturbed arable fields, particularly those with a winter-sown cereal crop. In addition to exploiting sites which remain open for many years, the species also occupies transient areas of exposed soil in grassland and wasteland. *V.a.* is generally absent from grazed sites, and does not persist in waterlogged or woodland habitats. Germination requires relatively high levels of soil moisture. The species requires both vernalization and long days and is one of the later-flowering winter annuals. Consequently, the species tends to be absent from the most drought-prone microsites. A majority of seeds have an after-ripening requirement, and germination in summer is further inhibited by sensitivity to low moisture and high soil temperatures. Hence, although seeds are typically shed in summer, germination does not occur until autumn or the following spring. Germination is inhibited by darkness and by the presence of a leaf canopy, and the species forms a persistent seed bank. Under optimal conditions seed

production by large plants may reach 17 000. Seeds can survive ingestion by cattle. Despite the tendency of seeds to come to rest close to the parent, *V.a.* is capable of long-distance dispersal, often apparently through human agency. The minute, saucer-shaped seeds are buoyant in air currents and, as in the case of *Saxifraga tridactylites*, local dispersal by wind facilitates colonization of wall-tops. Stem pieces root readily but regeneration by vegetative means does not occur in the field. The extent of ecotypic differentiation in this ecologically wide-ranging species is unknown.
Current trends Uncertain, but not under any immediate threat.

274   *Veronica beccabunga*

Brooklime

Established strategy C–R.
Gregariousness Can form patches.
Flowers Blue, hermaphrodite, protogynous, insect-pollinated or often selfed.
Seed 0.34 mg.
British distribution Widespread (100% of VCs).
Commonest habitats Running aquatic 19%, shaded mire 6%, unshaded mire 5%, soil heaps 2%.
Spp. most similar in habitats
Nasturtium officinale (agg.) 97%,

Apium nodiflorum 93%, *Ranunculus peltatus* 91%, *Epilobium parviflorum* 91%, *Equisetum palustre* 87%.
Full Autecological Account *CPE* page 600.

Synopsis *V.b.* is a winter-green, patch-forming herb typically straddling the boundary between mire and aquatic habitats, particularly where the water is flowing. *V.b.* is restricted to fertile sites where disturbances such as winter-flooding, water currents, trampling or grazing restrict the growth of taller perennials. *V.b.* extends into calcareous sites, but is absent from acidic peats. In many respects the ecology of *V.o.* is very similar to that of *Nasturtium officinale* (agg.). Both form extensive clonal patches by means of prostrate rooted stems. In each, shoot pieces readily root, and regeneration along stream courses by means of detached fragments appears to be a frequent event. Both produce large quantities of small, readily germinable seeds, and in the case of *V.b.* a persistent seed bank has been demonstrated. A major difference between the two lies in their tolerance of submergence; whereas in *V.b.* a high proportion of the canopy consists of floating or emergent stems rooted to the bank, *N.o.* is strongly amphibious and exhibits considerable plasticity in growth form and leaf shape. A further contrast with *N.o.* arises from the fact that *V.b.* occurs more frequently in shaded habitats and persists to a greater extent than *N.o.* in taller stream-side vegetation. The seeds of

V.b. are mucilaginous and adhesive and the species is an effective colonist, even in land-locked sites.

Current trends *V.b.* may have become more abundant beside upland streams in response to increased eutrophication. In lowland areas, where changes to wetland systems have been more marked, the status of *V.b.* is less certain and the species may be decreasing.

---

275 *Veronica chamaedrys*

Ssp. *chamaedrys*

Established strategy Between C–S–R and S.
Gregariousness Intermediate.
Flowers Blue, hermaphrodite, homogamous, insect-pollinated.
Regenerative strategies V.
Seed 0.18 mg.
British distribution Widespread (100% of VCs).
Commonest habitats Limestone pastures 25%, scree 17%, meadows/limestone quarries 11%, verges 9%.
Spp. most similar in habitats Insufficient data.
Full Autecological Account *CPE* page 602.

Germander Speedwell

Synopsis *V.c.* is a creeping, snallow-rooted, tetraploid perennial herb of moist, base-rich, usually rather infertile soils. *V.c.* is locally frequent in broken short turf in lightly grazed unproductive pastures, where it tends to colonize gaps in the vegetation by means of the stolons, which have the capacity to root along their entire length. *V.c.* is particularly common on stabilized scree where the opportunities for rooting are localized and leaves are often subtended over bare rock at some distance from the roots. A consistent feature of all sites colonized by *V.c.* is that they are liable to some disturbance, and hence present gaps which *V.c.* may exploit by vegetative growth. Suitable sites also include mown hedgebanks and roadsides. In fertile pasture *V.c.* is less common, and gap colonization is effected by larger, faster-growing species such as *Ranunculus repens* and *Trifolium repens*. *V.c.* is occasionally observed in a non-flowering state in open woodland and a degree of shade tolerance is further suggested by the persistence of *V.c.* as an understorey in some types of taller grassland. The plant contains appreciable quantities of mannitol. *V.c.* flowers early, and rapid shoot growth appears restricted to autumn and spring; the significance of this phenology requires further study. Apart from the capacity to spread by means of stolons, *V.c.* may also regenerate vegetatively from shoots detached as a result of disturbance. This may be an important mechanism of spread in trampled pastures and on roadsides. As in many other perennial species of less-productive habitats, regeneration by seed appears to be comparatively rare. It is not known whether *V.c.* forms a persistent seed bank (although all other species of *Veronica* so far examined do so). Neither is it known when the seed normally germinates. The small flattened seeds survive ingestion by cattle. Some morphological plasticity is evident in field populations; plants growing in short turf are appressed closely to the ground, and stolons are rooted along most of their length. In taller vegetation, however, *V.c.* adopts a more ascending growth form. Another ssp. with $2n = 16\ (2x)$ also occurs elsewhere in Europe.

Current trends *V.c.* is not a threatened species but it may be expected to fall in abundance as the quantity of unproductive pasture declines further.

276 *Veronica montana*   **Wood Speedwell**

Established strategy Between S and C–S–R.

Gregariousness Intermediate to patch-forming.

Flowers Lilac-blue, hermaphrodite, insect-pollinated.

Regenerative strategies V, ?B$_s$.

Seed 0.34 mg.

British distribution Common except in N Scotland and Ireland (85% of VCs).

Commonest habitats Shaded mire 6%, river banks/acidic woodland/limestone woodland 3%, scrub 1%.

Spp. most similar in habitats *Circaea*

lutetiana 91%, *Filipendula ulmaria* 83%, *Festuca gigantea* 81%, *Chrysosplenium oppositifolium* 77%, *Ranunculus ficaria* 77%.

Full Autecological Account *CPE* page 604.

Synopsis *V.m.* is a sprawling, winter-green herb which exploits moist, often clayey soils in sites where the vigour of dominant species is suppressed by shade and, often, by a degree of disturbance. Mineral nutrient stress may also be an important factor in some habitats. *V.m.* is shade-tolerant and has been observed to flower over a wide range of irradiance. In competition experiments conducted in garden plots with and without summer shade, *V.m.* was comparatively resistant to shade and better able to maintain its status as a subordinate component in vegetation where shade had reduced the vigour of potential dominants such as *Holcus mollis*. The species is most floriferous in lightly-shaded sites, and in upland areas is occasionally observed in unshaded habitats. *V.m.* is shallow-rooted and it is uncertain to what extent the N-facing bias in its distribution is related to a high level of shade tolerance or to a requirement for unusually moist conditions. Although extending onto calcareous strata within the region, *V.m.* is most abundant on mildly acidic soils. Further, the restriction of *V.m.* to sites below 320 m in altitude suggests that the distribution of *V.m.* is strongly influenced by climatic factors. *V.m.* regenerates by means of prostrate rooted stems to form extensive clonal patches. It is particularly frequent on stream banks and woodland rides, and appears to spread considerable distances along these linear habitats by means of regenerating shoot fragments. Seed-set is often low and seeds, which have a small embryo relative to their internal volume, have a chilling requirement for germination. A persistent seed bank has been reported but so far this has not been confirmed by our own studies. The seed appears to be poorly dispersed and, with the exception of stream-sides and woodland rides, the habitats to which *V.m.* appears to be restricted have been long-established. In Lincolnshire, *V.m.* tends to be associated with ancient woodlands. **Current trends** Perhaps decreasing as a result of woodland destruction but on heavy soils appears to be stable along woodland rides in broad-leaved plantations.

277 *Veronica persica*

Large Field Speedwell

Established strategy R.
Gregariousness Sparse to intermediate.
Flowers Blue, hermaphrodite, visited by various insects, but often selfed.
Regenerative strategies B$_s$, V.
Seed 0.52 mg.
British distribution Widespread (100% of VCs).
Commonest habitats Arable 38%, soil heaps 11%, bricks and mortar 3%, paths/verges 2%.
Spp. most similar in habitats *Spergula arvensis* 99%, *Fallopia convolvulus* 97%, *Myosotis arvensis* 95%, *Polygonum persicaria* 92%, *Papaver rhoeas* 89%.
Full Autecological Account *CPE* page 606.

Synopsis *V.p.* is a fast-growing, annual herb of disturbed but fertile soils, with indeterminate procumbent stems. It occurs on fallow arable land or beneath the crop canopy (usually of cereals). The species is strongly suppressed in shade. Flowers and fruits are formed sequentially in the axils of the leaves. The species is a native of SW Asia, and was first recorded from Europe in *c.* 1800 and from the British Isles in 1825. Subsequently the species has spread rapidly, perhaps mainly as an impurity in agricultural seeds, in fodder and in manure. *V.p.* is now common throughout the British Isles. It has a high potential seed production, and an average plant may produce > 6000 seeds. The species also forms a persistent seed bank, and the seeds, which show little innate dormancy, are capable of germination during every month of the year. Further, two generations may be produced in a single season. Although capable of exploiting both transient and semi-permanent disturbed habitats, the species is mainly recorded as an arable and garden weed. Its tendency to be restricted to almost continuously disturbed sites may be related to its inferior colonizing potential compared with many other ruderals. Though the seed output per plant is comparable, the seeds are dispersed from capsules held in close contact with the soil, with the result that many appear to be directly incorporated into a buried seed bank. The large seeds are specialized for dispersal by ants, but transport by human activity is more important in the agricultural and horticultural systems exploited by most contemporary populations. Shoot fragments regenerate if placed on moist garden soil or when cultivated in the laboratory, and vegetative regeneration is potentially a significant process in frequently disturbed situations. An unusual degree of herbicide resistance has been recorded

for certain populations. Two possibly native species, *V. agrestis* and *V. polita*, also exploit arable land. Both are less robust than *V.p.*, and the rate of dry-matter production of *V.a.* is less than that of *V.p.*. Within the Sheffield area the altitudinal limit of *V.a.* is higher. In the lowland, but not the upland, part of the region this species is also restricted to non-calcareous strata. *V.p.* is more strictly a lowland plant, and is most frequent on calcareous clays. The two species appear to overlap ecologically and often occur in association with *V.p.*. It is not clear which features currently limit their effectiveness as arable weeds.

Current trends As a result of the application of herbicides, and other methods of weed control, *V.p.* may be decreasing. However, the species is likely to remain a common arable and garden weed.

278 *Vicia cracca*   Tufted Vetch

Established strategy Between C and C–S–R.
Gregariousness May form small patches.
Flowers Blue-purple, hermaphrodite, insect-pollinated.
Regenerative strategies S, V.
Seed 14.29 mg.
British distribution Widespread (100% of VCs).

Commonest habitats Limestone
wasteland 4%, bricks and
mortar/hedgerows/river
banks/scree/verges/acidic wasteland
3%.
Spp. most similar in habitats
Insufficient data.
Full Autecological Account *CPE* page
608.

Synopsis A species much in need of
further study (see *V. sepium*). *V.c.* is a
large scrambling legume supported on
the surrounding vegetation by tendrils.
Nitrogen-fixing root nodules are
formed in conjunction with *Rhizobium
leguminosarum*. *V.c.*, which is found
on relatively fertile, moist soils, is
capable of forming small clonal patches
and produces a moderately dense
canopy. However, its capacity for
dominance is restricted by its
dependence upon other species for
support. Studies of the dynamic
interactions between *V.c.* and its
supporting species are required to
understand the factors allowing *V.c.* to
attain local dominance. Robust plants
(to 2 m) are often found at the margins
of hedgerows. *V.c.* has the capacity to
persist in a non-flowering condition in
shaded sites, and in tall stands, such as
those of *Phragmites australis*, *V.c.*
frequently occurs as stunted
individuals. The species is virtually
absent from pasture, and its sensitivity
to grazing is presumably related to the
erect growth form. However, *V.c.*
persists in meadows, including those
which are grazed later in the year. *V.c.*
has a very limited capacity for

vegetative spread below ground, and
plants may produce 1–25 stems from a
short rhizome. *V.c.* regenerates
primarily by means of the large seeds,
many of which are said to germinate in
autumn, soon after shedding. There is
no information concerning the
persistence of ungerminated seeds in
the field. Much of the considerable
variability with which *V.c.* is associated
represents phenotypic plasticity.
Within Europe *V.c.* is included within
a critical group of four species. *V.c.*
occurs both as a diploid and a
tetraploid in Europe. The diploid has a
narrow geographical and ecological
range, whereas the tetraploid, which is
presumably of more-recent origin, is
widespread and is probably the only
level of ploidy occurring in N Europe,
including the British Isles.
Current trends Not an efficient colonist
of new artificial habitats, and probably
decreasing.

279 *Vicia sepium*   **Bush Vetch**

Established strategy Between C–S–R
and C.
Gregariousness Often forms small
patches.
Flowers Purple, hermaphrodite, insect-
pollinated.
Regenerative strategies V, ?B$_s$.

Seed 26.0 mg.
British distribution Widespread (100%
of VCs).
Commonest habitats Soil heaps 6%,
verges/limestone wasteland 5%, coal
mine waste/hedgerows/limestone
woodland 3%.

Synopsis Many aspects of the ecology of *V.s.* require further study. *V.s.* is a scrambling legume with tendrils. The species forms nitrogen-fixing root nodules in association with *Rhizobium leguminosarum*. In lowland Britain *V.s.* is particularly characteristic of hedgerows and woodland margins, while in upland areas the species is frequent on unshaded road verges and ungrazed dale-sides. Thus, *V.s.* appears to be a 'semi-shade' plant, although this may be a reflection of moisture demand rather than shade tolerance. *V.s.* rarely occurs in tall, unmanaged vegetation, tending instead to grow where the vigour of potential dominants is checked either by ● disturbance, such as that imposed by the infrequent cutting of roadsides, or by lower productivity, as in an unfertilized dale-side grassland. *V.s.* forms relatively long-lived patches but, as with most other native legumes, it is not known whether this local monopoly of the leaf canopy persists over long periods or whether the fixation of nitrogen alters the equilibrium between *V.s.* and its competitors. The possibility that the balance between *V.s.* and associated grasses experiences cyclical fluctuations dependent upon the changing nitrogen level of the soil (as is the case for *Lolium perenne* and *Trifolium repens*) requires investigation. *V.s.* has a limited capacity for vegetative spread, and can produce flowers and seeds over a major part of the summer. However, it is not clear whether regeneration from seed or by vegetative means is the more important and no seed bank studies seem to have been undertaken. As with most native legumes, there is poor understanding of the mechanisms and significance of hard-coat dormancy under British climatic conditions. The large seeds of *V.s.* appear to be ineffectively dispersed, and *V.s.* is seldom a colonist of new artificial habitats.

Current trends Probably decreasing.

280 *Viola hirta*   Hairy Violet

Established strategy S.
Gregariousness Intermediate.
Flowers Blue-violet, hermaphrodite, insect-pollinated.
Regenerative strategies V, ?S.
Seed 2.81 mg.
British distribution Mainly S (54% of VCs).
Commonest habitats Limestone pastures 16%, limestone wasteland 12%, scrub/scree/limestone woodland 3%.

Synopsis *V.h.* is a winter-green herb of short turf on calcareous, relatively infertile soils and is most frequent in S England. Despite a marked bias towards S-facing slopes *V.h.* is restricted to less-droughted sites; this may be related in part to the fact that the root system is relatively shallow. These various features lead us to suspect that the distribution of *V.h.* is determined to a considerable extent by climatic factors. *V.h.* has some tolerance of shade, since it occurs infrequently in open woodland and scrub and the leaves also often experience some shading within the grassland canopy. However, *V.h.* does not normally persist in tall grassland. *V.h.* is one of the first species to flower in calcareous turf, but the species presents its maximum leaf canopy during summer, and spring- and summer-produced leaves are recognizably different in that the latter are considerably larger. The floral and regenerative biology of *V.h.* is similar to that of *V. riviniana*. However, since the seeds are passively discharged *V.h.* is dependent upon ants for dispersal. At least within the Sheffield region, *V.h.* has extremely limited mobility. It has been suggested that ants' nests, to which seeds are taken, represent a favourable environment for seedling establishment. *V.h.* also regenerates vegetatively through the production of a branched rhizome. Regeneration by

seed involves high fatalities and vegetative spread is slow. More research is needed to establish the relative importance of these two mechanisms of regeneration in natural populations. A small, late-flowering variant was formerly given subspecific rank, otherwise the extent of ecotypic variation is uncertain. Despite its comparatively narrow habitat range, the distribution of *V.h.* overlaps that of *V.r.* to a limited extent. In shaded habitats local contact also takes place with *V. odorata* (with which it hybridizes) and with *V. reichenbachiana*.

Current trends The species is an indicator of long-established semi-natural turf, and is decreasing.

281 *Viola riviniana*   Common Violet

Established strategy S.
Gregariousness Intermediate.
Flowers Violet, hermaphrodite, insect-pollinated.
Regenerative strategies V, S.
Seed 1.01 mg.
British distribution Widespread (100% of VCs).
Commonest habitats Limestone pastures 42%, scree 39%, limestone wasteland 19%, limestone woodland 16%, lead mine spoil 11%.
Spp. most similar in habitats *Carex flacca* 86%, *Helianthemum nummularium* 81%, *Potentilla sterilis* 80%, *Avenula pratensis* 80%, *Polygala vulgaris* 80%.
Full Autecological Account *CPE* page 614.

Synopsis *V.r.* is a long-lived, slow-growing, small, winter-green herb mainly restricted to infertile habitats. Exceptionally among British herbs, *V.r.* exploits open habitats, particularly limestone screes and woodland herb layers. *V.r.* is equally common on highly calcareous and on mildly acidic soils, and the leaves tend to contain high concentrations of Mg and low levels of P. The species is tolerant of burning, and in pastures tends to be little grazed. The niche of *V.r.* in grassland habitats, where it is generally a minor component, is poorly understood, but foliage tends to occupy lower layers of the leaf canopy and the species appears to be shade-tolerant. In woodland *V.r.* is usually scarce in areas with a high density of vernal species, but may be a relatively conspicuous component on shallow skeletal soils. In transplant experiments *V.r.* was found to be capable of surviving moderate depositions of persistent tree leaf litter, mainly by marked extension growth of the petioles. *V.r.* produces both insect-pollinated and, later, cleistogamous flowers. A majority of the seeds produced appears to originate from the latter. The seeds, which are heavily predated, are shed by explosive dehiscence and are subsequently dispersed by ants. Seeds have a chilling requirement which, in calcareous grassland, may result in the appearance of large, even-aged populations of seedlings in the spring. Vegetative regeneration through adventitious sprouting from the roots is also possible in some plants. Despite the fact that *V.r.* regenerates mainly by vegetative means and from cleistogamous flowers, the species is

variable in chromosomal complement. Ecotypic differences have been described, although their relationship to cytology is obscure. Populations occurring in woodlands are genetically and physiologically different from those established in unshaded habitats. In woodland habitats on heavy, usually base-rich, soils, *V.r.* often occurs with *V. reichenbachiana*. However, the latter flowers slightly earlier, produces fewer and heavier seeds and is, we suspect, even more shade-tolerant than *V.r.*. The two species occasionally hybridize. In dry, often sandy, habitats on acidic soils *V.r.* shows a tendency to be replaced by the later-flowering *V. canina* ssp. *montana*, with which it also hybridizes.

Current trends Seed dispersal by ants is ineffective for long-distance dispersal to new artificial sites. Particularly in lowland areas, the species is probably decreasing, the more so in grassland than in woodland.

4 Tables of attributes

Introduction

This chapter sets out in tabular form a collection of standardized autecological information for the more common vascular species of the British flora.

In addition to providing summaries for the 281 species which are the subjects of accounts in Chapter 3, the tables include data on a further 221 species, which each satisfy all of three criteria: (a) they are recorded in at least 66% of Watsonian vice-counties of the British Isles; (b) they are either native or well-naturalized alien throughout the majority of their British range; and (c) they have an extensive distribution in inland Britain. Ecological information for these additional species (and for *Carlina vulgaris*, *Phragmites australis* and *Reynoutria japonica*, which were recorded from fewer than ten quadrats in the UCPE Survey II) is derived from unpublished UCPE data. Only one of the species tabulated, *Sedum anglicum* Hudson, was not recorded during fieldwork in the Sheffield area.

The quantity and quality of information available varies considerably according to species. Certain species are variable in certain biological characteristics, and field distributions with respect to habitat features such as slope and aspect may be too strongly affected by geographical location to allow reliable generalizations at a national scale. Despite these problems, it has been possible to present data for 22 characteristics which, apart from caveats included in the explanatory notes, can be applied with some confidence to the species observed in the British Isles; these relate to ecology (Table 4.1), attributes of the established phase (Table 4.2) and attributes of the regenerative phase (Table 4.3). The significance and use of each of the 22 characteristics is discussed in Chapter 2 and in *CPE*.

Explanation of tables

Nomenclature

Nomenclature follows that of *CTW* (see Chapter 2, 'Nomenclature') except for the following species groupings: *Betula* spp. (includes *B. pendula*, *B. pubescens* and their hybrid), *Festuca ovina* (also includes *F. tenuifolia*), *Festuca rubra* (also includes *F. nigrescens*), *Lolium perenne* (refers only to ssp. *perenne*; the monocarpic, and introduced, ssp. *multiflorum* (Lam.) Husnot is a non-persistent escapee from cultivation),

Medicago sativa (excludes ssp. *falcata* (L.) Arcangeli which is native in East Anglia), *Nasturtium officinale* agg. (*N. microphyllum, N. officinale* and their hybrid), *Poa pratensis* agg. (*P. angustifolia, P. pratensis sensu stricto* and *P. subcaerulea*), *Quercus* agg. (*Q. petraea, Q. robur* and their hybrid), *Rosa* spp. (*R. canina* group, *R. rubiginosa* group and *R. tomentosa* group), *Salix cinerea* agg. (*S. aurita* L., *S. caprea, S. cinerea* and hybrids), *Salix fragilis* agg. (*S. alba* L., *S. fragilis* L. and related hybrids) and *Taraxacum* agg. (Section *Erythrosperma*, Section *Obliqua*, Section *Palustria*, Section *Spectabilia* and Section *Taraxacum*).

Two columns of symbols precede the scientific name. In the first, species which are the subject of Accounts in Chapter 3 are indicated by a '+'; in the second, alien species are denoted by an asterisk.

Only in Table 4.1 are the authorities for the scientific names cited. In the case of seven of the species, the combined entry for species binomial and authority is too long for the space available in the table. The following are the full citations of the authorities omitted in such cases:

[1] P. Gaertner, B. Meyer & Scherb.
[2] (L.) Beauv. ex J. & C. Presl.
[3] P. Gaertner, B. Meyer & Scherb.
[4] (L.) Chouard ex Rothm.
[5] (L.) Ehrend & Polatschek
[6] (L.) Tausche ex L. H. Bailey
[7] (L.) P. Gaertner, B. Meyer & Scherb.

Ordering of species

Species lists are divided into three groups: (a) herbs, and woody species up to 1.5 m in height, (b) woody species exceeding 1.5 m in height, and (c) pteridophytes. Within each group species are arranged in alphabetical order.

Ecological attributes (Table 4.1)

For most species only one set of data is given. However, in the case of individual trees and shrubs ecological attributes are, wherever possible, presented separately for seedlings and small saplings, 'juveniles', and for the more mature individuals which contribute to the canopy.

(1) Habitat range

General policy The habitat key (Fig. 1.3) has been used as a basis for identifying the habitat range of each species. First, an estimate of the abundance in each of the seven primary habitat groups (wetland, skeletal, arable, pasture, spoil, wasteland and woodland) is provided.

Second, the terminal habitat in which the species is most frequent is indicated.

(a) Abundance in the seven primary habitats
The percentage frequency of the species in each of these major groups is compared with the percentage frequency of the species in the overall survey. The symbols used are as follows: ++, very common and characteristic of the particular habitat (percentage frequency > 4 times the overall value); +, common within habitat (> 2–4 times the overall value); ., widespread in the habitat (> 0.5–2 times the overall value); –, infrequent and uncharacteristic of the habitat (0.25–0.5 times the overall value); – –, largely absent from habitat or confined to an uncommon variant (< 0.25 of the overall value).

In 'Woodland' in Survey II and 'Wasteland' in Survey III these primary habitat groups each constitute over 25% of all records, rendering the '++' rating an impossibility. It was decided that all species in which at least 80% of records fall within a single primary habitat should also be classified as '++'.

For species for which there are insufficient field records (*Hypericum androsaemum* L., *Salix repens* L. and *Sedum anglicum* Hudson) only the identity of the most characteristic major habitat is indicated.

(b) Commonest terminal habitat
The commonest terminal habitat for the subject species is identified as follows: AQUATp (lakes, canals, ponds and ditches), AQUATr (rivers and streams), ARABLE (arable), BRICK (bricks and mortar rubble), CANALB (canal banks, +), CINDER (cinder tips and cindery railway tracks), CLIFF (cliffs), COAL (coal-mine spoil), HEDGE (hedgerows), LEAD (lead-mine spoil), MANURE (manure and sewage spoil), MEADOW (meadows), MIREs (shaded mire), MIREu (unshaded mire), OUTCRP (rock outcrop), PASTa (pasture on acidic strata), PASTe (enclosed pasture), PASTl (pasture on limestone strata), PATH (paths), PLANTb (broadleaved plantations), PLANTc (coniferous plantations), PONDBK (banks of lakes, canals, ponds and ditches, +), QRYa (quarry spoil on acidic strata), QYRl (limestone quarry spoil), RD/RLY (road verges/railway banks, +), RDVRGE (road verge), RIVBNK (river and stream banks), RLYBNK (railway banks, +), SCREE (limestone scree), SCRUB (scrub), SNDPIT (sand and gravel pits, +), SOIL (soil heaps), WALL (walls), WASTEa (wasteland on acidic strata), WASTEd (dry sandy wasteland droughted during summer, +), WASTEl (wasteland on calcareous strata), WOODa (woodland on acidic strata), WOODl (woodland on limestone strata).

Habitats suffixed '+' are minor habitats which, because of their infrequent occurrence, were excluded from the habitat classification of

Survey II (Fig. 1.3).

These data are particularly useful for categorizing species of narrow ecological range (e.g. *Chaenorhinum minus*, associated with cinders, and *Lemna minor*, associated with lakes, canals, ponds and ditches). However, three grassland species (*Euphrasia officinalis*, *Leontodon autumnalis* and *Rumex acetosa*), and seedlings of *Ulmus glabra*, were most commonly recorded from lead-mine spoil even though they are characteristic of a range of base-rich habitats. In order to prevent the impression that these species are 'metallophytes' and largely restricted to soils contaminated with heavy metals, the habitat in which these species are *second most common* is substituted. Rather anomalous data were obtained for two wetland species which are persistent after drainage. *Eriophorum vaginatum* was recorded most commonly in pasture on limestone strata and *Lotus uliginosus* most commonly in wasteland on limestone. In both species the second most common habitat (unshaded mire) has been promoted to first place. Four species which do not normally reach maturity in their most common habitat (*Athyrium felix-femina*, cliffs and walls, *Dryopteris filix-mas*, walls, *Equisetum palustre*, rivers and streams, and *Ulex europaeus*, river banks) have been similarly treated. For each of these ten species the habitat code is preceded by an asterisk to warn that the information has been modified in this manner.

These examples illustrate that data concerning the most common terminal habitat could, if utilized in isolation, lead to misinterpretation. Their use in conjunction with the distribution of the species in the seven major habitats is advised.

N.B. The range and diversity of major habitats differs according to region. For example, in areas where (unlike that sampled here) there is little limestone, species such as *Carex panicea* would be regarded as a wetland rather than as a grassland species.

(2) Soil pH

The data take the form of a numeral indicating the modal pH class for the species, followed by a letter indicating the number of pH classes in which the frequency of the species exceeds 50% of that in its modal class. Thus, '7a' would indicate that the species is most frequent within the range 7.0–7.9 and has a range of only 1 pH unit. In contrast, '3d' would indicate a mode within the interval pH 3.0–3.9 and a range of 4 pH units. Where information on soil surface pH is scarce, the estimate of modal class is preceded by a question mark.

A few species exploit both strongly acidic and highly calcareous soils and thus possess a bimodal pH distribution. They are identified by a 'W' (for wide-ranging) followed by a numeral indicating the pH class in which the species has been most commonly recorded during survey work.

N.B. pH range varies according to region. Some species which are narrowly restricted to soils of high pH in SE England extend onto more acidic soils in the N and W.

(3) Floristic diversity

The mean number of species m^{-2} with which each species is associated has been calculated by reference to survey data from the Sheffield region. The classes recognized are as follows: 1, 10.0 species, or fewer; 2, 10.1–14.0; 3, 14.1–18.0; 4, 18.1–22.0; 5, greater than 22.0.

(4) Distribution in N Europe

The latitudinal and longitudinal range of species in N Europe has been assessed by reference to the standard sources cited in *CPE*. The area excludes most ·of Finland and Russia. The symbols utilized are as follows. *Restriction with respect to latitude*: S, largely restricted to southern areas and absent from parts of both northern Britain and Scandinavia (e.g. *Anagallis arvensis* and *Daucus carota*); s, similar to S but distributed throughout either northern Britain or Scandinavia; n, species largely restricted to northern areas (e.g. *Empetrum nigrum*). *Restriction with respect to longitude*: W, largely restricted to NW Europe with distribution centred on the Atlantic seaboard (e.g. *Conopodium majus* and *Hyacinthoides non-scripta*); w, distribution similar to W but extending to a considerable extent into central regions (e.g. *Chrysosplenium oppositifolium*, *Digitalis purpurea*); e, more widespread in NE than NW Europe (e.g. *Vaccinium vitis-idaea*).

Species without well-marked geographical restriction in N Europe are marked '−'. *Minuartia verna*, which is largely restricted to montane areas and metalliferous sites, shows a disjunct distribution (indicated as D) and aggregate species, e.g. *Alchemilla vulgaris* agg., whose segregates show contrasted distribution patterns, are denoted 'X'. Some species (e.g. *Apium nodiflorum*, SW) show restriction with respect both to latitude and to longitude. Cases where insufficient data exist for accurate mapping are indicated '?'.

Data are abstracted from a variety of sources and the classification presented is only approximate.

(5) Present status

The extent to which the abundance of species is changing in response to modern methods of land-use has been estimated. The abbreviations used are as follows: +, species increasing; −, species decreasing; ?, present and future status uncertain.

Attributes of the established phase (Table 4.2)

(6) Life-history

The life-cycles classified are as follows: As, summer annual; Aw, winter annual; B, usually biennial; M, monocarpic perennial (of duration 2 years or, usually, more); P, polycarpic perennial.

Some variable species were classified into more than one group; in such cases the most common form of life-cycle is listed first. In the case of *Plantago coronopus* L., the symbol M+ indicates that the species is typically monocarpic, but is also capable of being either a winter annual or a polycarpic perennial. Only a minority of plants of *Lemna minor* and *L. trisulea* L. overwinter and these species are denoted as (P).

(7) Established strategy

An assessment of the established strategy of the subject species. Abbreviations are as follows: *primary strategies* C, competitor; R, ruderal; S, stress-tolerator; *secondary strategies* CR, competitive-ruderal; SC, stress-tolerant competitor; SR, stress-tolerant ruderal; CSR, C–S–R strategist. Strategy types intermediate between these seven are also recognized, e.g. CR/CSR.

(8) Life-form

The life-form according to Raunkiaer's classical system, supplemented by field observations. The following classes have been separated: Ph, phanerophyte (woody plant with buds more than 250 mm above soil surface); Ch, chamaephyte (herbaceous, or woody, plant with buds not in contact with but less than 250 mm above the soil surface); H, hemicryptophyte (herb with buds at soil level); G, geophyte (herb with buds below the soil surface); Hel, helophyte (marsh plant); Hyd, hydrophyte (aquatic plant); Th, therophyte (plant passing the unfavourable season as seeds); Wet, wide-ranging wetland species (facultatively a helophyte or a hydrophyte).

Monocarpic and biennial species are classified according to their vegetative overwintering condition, though they could equally well have been classified as therophytes. *Helianthemum nummularium* and *Thymus praecox* are denoted Chw (woody chamaephytes). All other woody species are, at least partially, phanerophytes.

(9) Canopy structure

The following classes are recognized: R, rosette (leaves confined to a basal rosette, or to a prostrate stem); S, semi-rosette (stems leafy but with the largest leaves towards their base); L, leafy (no basal rosette, leaves of approximately equal size all the way up the stem); −, leaves small, reduced to scales or spines, with the stem as the main photosynthetic organ.

Monocarpic species (e.g. *Cirsium vulgare*), which spend their juvenile phase as a rosette and subsequently develop a leafy stem, have been classified as semi-rosette species. *Hieracium* subgenus *Hieracium* includes both semi-rosette and leafy taxa and is denoted as V (for various).

A number of additional categories have been used for non-emergent aquatic species to identify the characteristic positions of leaves or, in the case of *Lemna* spp., the photosynthetic thallus relative to the water surface: F, floating (some or all of the leaves floating on the water surface); U, underwater (all leaves submerged).

In three instances, *Callitriche stagnalis*, *Polygonum amphibium* and *Sparganium emersum* Rehmann, information is included on both terrestrial and aquatic leaf forms.

(10)   Canopy height
The maximum height of the leaf canopy has been assessed from field observations, and the following broad classes have been constructed: 1, foliage less than 100 mm in height; 2, 101–299 mm; 3, 300–599 mm; 4, 600–999 mm; 5, 1.0–3.0 m; 6, 3.1–4.0 m; 7, 4.1–15.0 m; 8, greater than 15.0 m; V, various (*Hieracium* subgenus *Hieracium*); −, submerged or floating aquatics.

(11)   Lateral spread
Based upon field observations, the following classes have been recognized: 1, therophytes (i.e. lateral spread of exceedingly limited extent and duration); 2, perennials with compact unbranched rhizomes or forming small tussocks (less than 100 mm in diameter); 3, perennials with rhizomatous systems or tussocks attaining 100–250 mm; 4, perennials attaining diameter of 251–1000 mm; 5, perennials attaining a diameter of more than 1000 mm; V, various (*Hieracium* subgenus *Hieracium*).

N.B. Patch size relates to the dimensions of connected branching systems of stems (excluding the leaf canopy). Ancient, fragmented, clonal colonies are often considerably larger but estimates of their size are unavailable.

(12) Mycorrhizas
The frequency and nature of mycorrhizal infection have been summarized for each species as follows: *Normally mycorrhizal* (75% or more records report infection): EC, with ectomycorrhizas; ER, with ericoid mycorrhizas; OR, with orchid mycorrhizas; VA, with vesicular-arbuscular mycorrhizas. *Intermediate*: +, 26-74% of records report infection with VA mycorrhizas. *Non-mycorrhizal*: −, 25% or less of records report mycorrhizas; HP, non-mycorrhizal hemiparasite. *No information*: ?.

This classification of species is only approximate. Although data have been collated from a wide variety of sources, for most species few records are available.

(13) Leaf phenology
The phenology of the leaf canopy, as observed in the Sheffield region. The following classes have been recognized. *Canopy seasonal, (S)*: Sa, aestival (duration of canopy spring to autumn); Sh, hibernal (mainly autumn to early summer); Sv, vernal (winter to spring). *Canopy evergreen (E)*: Ea, always evergreen; Ep, partially evergreen (species evergreen in some habitats and not in others, or species evergreen only in mild winters, or leaves slowly but incompletely senescing over winter, or overwintering with small leafy shoots, formed in autumn).

In *Hieracium* subgenus *Hieracium*, denoted V (for various), some taxa are winter-green and others are not.

N.B. The winter-greenness of leaves is very much determined by climate and the leaves of some species may be more persistent in S England and less persistent in Scotland than stated here.

(14) Flowering time and duration
Here the time of first flowering and its duration are presented. The month of first flowering is abbreviated to its first three letters and is immediately followed by the span, in months, of the flowering period, e.g. 'Jun3' refers to a species flowering from June to August.

N.B. Like leaf phenology, flowering time shows variation according to region.

(15) Polyploidy
Using the sources of information on chromosome number, the ploidy of species relative to the base number of the genus has been assessed. Genera with a base number of 13 or less are considered to be predominantly diploid and those with a base number greater than 13 to be mainly polyploid. This arbitrary separation, though not ideal, allows identification of cytological groups which, at least within the Sheffield flora, appear to differ with respect to ecological characteristics.

The following abbreviations have been used: D, diploid; P, polyploid with extant diploid relatives (from a genus where the base number is less than or equal to 13); R, polyploid relic (diploid for genus, genus with a polyploid base number of more than 13, ancestral diploids probably extinct); rP, secondary polyploids (species which have originated through two distinct cycles of polyploidy, the first leading to the creation of a genus of species type R, the second involving further polyploidy of type R polyploids).

Where the chromosome count has been obtained using British

material, the abbreviation is following by an asterisk. No estimates are presented for the Cyperaceae, which have a diffuse centromere and form aneuploid series.

Attributes of the regenerative phase (Table 4.3)

(16)   Regenerative strategies

These estimates are based on laboratory and field studies supplemented by data from the literature. They are presented using the format adopted for the Accounts. The regenerative strategies are abbreviated as follows: S, seasonal regeneration by seed; Sv, seasonal regeneration by vegetative means (offsets soon independent of parent); V, lateral vegetative spread (offsets remaining attached to the parent for a long period, usually for more than one growing season); (V), denotes instances where the period of attachment is intermediate between those of Sv and V, although in most instances we have insufficient demographic data to distinguish fully between Sv and V; W, regeneration involving numerous widely-dispersed seeds or spores; Bs, a persistent bank of buried seeds or spores; ?, strategies of regeneration by seed are uncertain.

N.B. Readers using this information are strongly advised to refer to the rationale concerning the identification of regenerative strategies given in Chapter 2 before attempting to interpret the data presented under this heading.

Data on seed banks have been revised since the publication of *CPE* and now use the protocols and unpublished database of K. Thompson, J. Bakker and R. Bekker. The classes of seed bank may be summarised as follows:

1 transient: seed rarely persisting for more than one year;
2 short-term persistent: seed persisting for more than one year but usually less than five;
3 long-term persistent: seeds persisting for at least five years, and often much longer;
– no seed produced.

Only species with seed bank type 3 are classified as possessing a persistent seed bank (Bs). A number preceded by ? denotes that the classification is provisional, while a ? alone means that data are absent or contradictory. Note that although the seed bank classification presented here is based on a thorough survey of the available literature, many classifications are based on relatively few data, often of rather poor quality, and the information should therefore be interpreted with care.

(17) Agency of dispersal

The extent of dispersal and the nature of the dispersal mechanism are clearly critical in determining the distribution of species, particularly in a disturbed and changing landscape such as that occupying most of Lowland Britain. Unfortunately, although this topic is too important to ignore, most of the information available on dispersal is anecdotal. It is hoped that the inadequacy of the data presented here will stimulate the production of a more complete appraisal of the mobility of species and of their most important vectors of transport. Until then, readers are advised to utilize these data with caution and in conjunction with the assessment of the extent to which the species is currently increasing or decreasing (attribute 5). What is clear to date is that many of the species, which are becoming increasingly abundant in the British landscape, lack obvious specialization for long-distance dispersal.

Types of dispersal for which the seeds of each species appear to be specialized are assessed by reference to standard sources and from observations of the morphology of seeds and the habitats in which they are shed. The following abbreviations are used: ANIM, dispersal by means of animals; *dispersal a direct consequence of food gathering*: ANIMi, an ingested berry; ANIMn, a nut or related type of hard-coated dispersule; ANIMe, seed with an elaisome (dispersed short distances by ants); *dispersal adhesive*: ANIMb, an adhesive burr; ANIMa, dispersule with an awn, or with spiny calyx teeth; ANIMm, dispersule adhesive through the secretion of mucilage; AQUAT, dispersal by water, either by means of buoyant seeds or of floating seedlings; WIND, wind-dispersed (WINDm, dispersule minute – orchid seeds and fern spores; WINDp, dispersule plumed or wrapped in woolly hairs; WINDc, seeds small and shed from a capsule, held above the level of the surrounding vegetation; WINDw, seeds winged or strongly flattened); UNSP, unspecialized, with morphological features facilitating dispersal absent or undetected (includes UNSPag, unspecialized but dispersed widely as a result of agricultural practices). In some instances the symbols 'c' for capsule and 'w' for winged or flattened seeds are used in conjunction with UNSP. This occurs where seeds of a morphological character which favours wind dispersal are also associated with a growth habit which renders long-distance dispersal by wind unlikely, e.g. in the case of *Veronica montana*, UNSPcw, where fruits are borne close to the ground in sheltered woodland sites. In such cases, and also where there is explosive discharge of seeds, the main significance of wind-dispersal is likely to lie in its movement of progeny away only from the immediate vicinity of the parent.

Question marks are used in some instances to indicate uncertainty, and an asterisk indicates that dispersal of vegetative shoots or root fragments may be as, or more, important for the spread of the species than dispersal of seeds. Species which do not normally produce seed are identified by '−'. In some cases more than one mechanism is cited, and the following abbreviations have been adopted: AN/AQ, by means of animals or water; AQ/WI, by means of water or wind.

We suspect that in some instances hard-coat dormancy, a character included in '(21) Germination requirements', allows seeds to survive ingestion by animals.

(18)   Dispersule and germinule form
The following classes are identified: Fr, dispersule and germinule a fruit (or part of a fruit, e.g. nutlet or mericarp); Sd, dispersule and germinule a seed; Sp, dispersule and germinule a spore; F/S, dispersule a fruit, germinule a seed (as in berries and other fleshy fruits); f/S, germinule a seed, dispersed either within a fruit or as a seed (e.g. *Atriplex patula*); Bul, dispersule vegetative with flowers replaced by bulbils; X, seeds, or bulbils, not produced.

(19)   Dispersule weight
The weight of seed, achene or other indehiscent germinule collected from the Sheffield region and dried at room temperature (excepting the minute seeds of orchids). The following classes are identified: S, too small to be measured easily (orchid seeds and fern spores); 1, weight less than or equal to 0.20 mg; 2, 0.21–0.50 mg; 3, 0.51–1.00 mg; 4, 1.01–2.00 mg; 5, 2.01–10.00 mg; 6, greater than 10 mg; X, no seed produced; −, no information available.

(20)   Dispersule shape
The following classes have been separated: 1, length/breadth ratio less than 1.5; 2, 1.5–2.5; 3, greater than 2.5; −, seed normally absent.

(21)   Germination requirements
The treatments required to achieve a high percentage of germination have been determined by UCPE for many common species. Supplementary data from the literature are also included. Treatments are abbreviated as follows: Chill, chilling; Dry, dry storage at room temperature; Fluct, fluctuating temperatures; Freeze, alternate freezing and thawing treatments; Heat, heat treatment; Orchid, fungal symbiont (usually a *Rhizoctonia* spp.) necessary for establishment in the field; Scar, scarification (many species with hard-coat dormancy also respond to chilling); Warm, warm moist incubation; Wash, water-washing to remove inhibitor in seed coat.

Species capable of immediate germination are identified as '−' and those which rarely or never produce seed are indicated as 'X'. Where germination requirement differs between collections, e.g. *Arctium minus*, or where seeds are polymorphic in their germination characteristics even on the same plant, e.g. *Chenopodium album*, the various treatments resulting in germination are separated by '/'. Where several alternative dormancy-breaking mechanisms are effective for a species they are separated by a comma. For a few species two treatments in combination are required, e.g. *Hyacinthoides non-scripta* (warm moist incubation following by chilling, Warm+Chill). 'Unclassified' species lack the capacity for immediate germination, but the dormancy-breaking mechanism(s) have not been identified. This grouping is probably dominated by species with an, as yet, unrecognized chilling requirement. Where parentheses are present the final germination percentage was low (less than 50%) and the time to 50% germination high (more than 20 days). The use of '?' indicates that confirmation of the dormancy-breaking mechanism is required.

(22) Family
The family name is abbreviated to its first three letters except where this leads to ambiguity (as in the cases of Polygalaceae and Polygonaceae). The abbreviations used are as follows:

Ace, Aceraceae; Ali, Alismataceae; Api, Aspidiaceae; Apl, Aspleniaceae; Aqu, Aquifoliaceae; Ara, Araceae; Bet, Betulaceae; Ble, Blechnaceae; Bor, Boraginaceae; Cal, Callitrichaceae; Cam, Campanulaceae; Can, Cannabaceae; Cap, Caprifoliaceae; Car, Caryophyllaceae; Cel, Celastraceae; Che, Chenopodiaceae; Cis, Cistaceae; Com, Compositae; Con, Convolvulaceae; Cor, Corylaceae; Cra, Crassulaceae; Cru, Cruciferae; Cyp, Cyperaceae; Dio, Dioscoreaceae; Dip, Dipsacaceae; Emp, Empetraceae; Equ, Equisetaceae; Eri, Ericaceae; Eup, Euphorbiaceae; Fag, Fagaceae; Fum, Fumariaceae; Gen, Gentianaceae; Ger, Geraniaceae; Gra, Gramineae; Gro, Grossulariaceae; Hal, Haloragaceae; Hip, Hippuridaceae; Hyd, Hydrocharitaceae; Hyo, Hypolepidaceae; Hyp, Hypericaceae; Iri, Iridaceae; Jce, Juncaceae; Jcg, Juncaginaceae; Lab, Labiatae; Leg, Leguminosae; Lem, Lemnaceae; Lil, Liliaceae; Lin, Linaceae; Lyt, Lythraceae; Mal, Malvaceae; Men, Menyanthaceae; Nym, Nymphaeaceae; Ole, Oleaceae; Ona, Onagraceae; Orc, Orchidaceae; Oxa, Oxalidaceae; Pap, Papaveraceae; Pga, Polygalaceae; Pgo, Polygonaceae; Pla, Plantaginaceae; Por, Portulacaceae; Pot, Potamogetonaceae; Ppo, Polypodiaceae; Pri, Primulaceae; Ran, Ranunculaceae; Res, Resedaceae; Ros, Rosaceae; Rub, Rubiaceae; Sal, Salicaceae; Sax, Saxifragaceae; Scr, Scrophulariaceae; Sol, Solanaceae; Spa, Sparganiaceae; Typ, Typhaceae; Ulm, Ulmaceae; Umb, Umbelliferae; Urt, Urticaceae; Val, Valerianaceae; Vio, Violaceae.

Ie 4.1 Ecological attributes of species (symbols and notes are explained in the :eding text).

Introduced		Abundance in Wetland	Abundance in Skeletal	Abundance in Arable	Abundance in Pasture	Abundance in Spoil	Abundance in Wasteland	Abundance in Woodland	Commonest terminal habitat	Soil pH	Floristic diversity	Distribution in N Europe	Present status
	(a) Herbs and woody species up to 1.5 m tall												
	Achillea millefolium L.	--	*	*	+	*	+	--	PASTe	5c	4	-	?
	Achillea ptarmica L.	+	--	--	+	--	*	--	MIREu	5c	4	-	-
*	Aegopodium podagraria L.	--	--	*	-	-	*	+	SOIL	6b	2	s	+
	Aethusa cynapium L.	--	--	++	-	*	*	*	ARABLE	7b	4	S	?
	Agrimonia eupatoria L.	--	--	--	*	--	++	-	RDVRGE	6b	4	S	-
	Agrostis canina L.	++	*	--	--	--	*	--	MIREu	4c	2	-	-
	Agrostis capillaris L.	-	*	-	+	*	*	--	PASTe	4c	2	-	-
	Agrostis gigantea Roth	--	--	++	*	+	*	--	ARABLE	6c	3	s	+
	Agrostis stolonifera L.	*	-	+	*	*	*	--	ARABLE	7c	2	-	+
	Agrostis vinealis Schreber	--	--	--	+	-	++	--	WASTEl	4b	2	?	-
	Aira caryophyllea L.	--	*	--	-	+	+	--	CINDER	5c	3	sw	?
	Aira praecox L.	--	*	--	*	+	*	--	QRYa	5b	2	sw	?
	Ajuga reptans L.	*	-	-	+	-	-	+	SCRUB	6b	4	s	-
	Alchemilla vulgaris agg.	--	-	--	+	*	*	*	PASTl	6b	5	X	-
	Alisma plantago-aquatica L.	++	--	--	--	--	--	--	MIREu	6a	1	s	?
	Alliaria petiolata (Bieb.) Cavara & Grande	--	*	-	-	*	++	*	RIVBNK	6b	3	S	+
	Allium ursinum L.	-	--	--	--	--	--	++	WOODl	6b	1	s	-
	Allium vineale L.	--	++	--	-	*	+	--	OUTCRP	7b	3	S	-
	Alopecurus geniculatus L.	++	--	*	*	*	--	--	MIREu	6b	2	-	?
	Alopecurus pratensis L.	--	--	-	+	*	-	-	MEADOW	5b	3	-	-
	Anacamptis pyramidalis (L.) L.C.M. Richard	--	*	--	+	++	*	--	QRYl	7a	5	S	-
	Anagallis arvensis L.	--	-	++	-	*	--	--	ARABLE	7c	4	S	-
	Anchusa arvensis (L.) Bieb.	--	--	++	-	*	*	--	ARABLE	6c	4	S	?
	Anemone nemorosa L.	--	--	--	*	--	--	++	WOODl	6d	2	-	-
	Angelica sylvestris L.	*	--	--	-	*	+	*	RIVBNK	5d	3	-	?
	Anthoxanthum odoratum L.	*	*	--	+	*	*	-	PASTl	5b	3	-	-
	Anthriscus sylvestris (L.) Hoffm.	--	*	-	*	*	*	*	RDVRGE	7c	2	-	?
	Anthyllis vulneraria L.	--	+	-	+	+	*	--	OUTCRP	7c	4	-	-
	Aphanes arvensis L.	--	*	++	*	*	*	--	ARABLE	6c	4	S	?
	Apium nodiflorum (L.) Lag.	++	--	--	-	--	--	--	AQUATr	7b	1	SW	?
	Arabidopsis thaliana (L.) Heynh.	--	+	--	*	*	*	--	OUTCRP	6b	4	-	+
	Arabis hirsuta (L.) Scop.	--	++	--	*	*	*	--	SCREE	7b	4	-	-
	Arctium lappa L.	*	--	--	-	--	++	*	RIVBNK	6b	2	S	+
	Arctium minus agg.	--	--	*	-	+	+	*	SOIL	7a	3	s	?
	Arenaria serpyllifolia L.	--	+	--	-	+	*	--	OUTCRP	7a	4	-	?
*	Armoracia rusticana [1]	--	--	*	-	+	+	--	RDVRGE	6c	2	S	?
	Arrhenatherum elatius [2]	--	+	--	*	*	*	*	SCREE	7c	2	s	+
	Artemisia absinthium L.	--	*	--	-	+	*	--	BRICK	7a	3	s	+
	Artemisia vulgaris L.	--	-	+	-	+	*	--	SOIL	7c	3	-	+
	Arum maculatum L.	--	--	--	-	--	--	++	WOODl	6c	1	Sw	?
	Atriplex patula L.	--	--	++	-	+	-	--	MANURE	7c	3	s	+
	Atriplex prostrata Boucher ex DC.	*	--	--	-	++	*	--	BRICK	7c	2	S	+
	Avenula pratensis (L.) Dumort.	--	+	--	+	-	*	--	PASTl	5c	5	s	-
	Avenula pubescens (Hudson) Dumort.	--	*	--	+	-	+	--	WASTEl	5c	5	s	-
	Barbarea vulgaris R.Br.	--	--	--	-	*	++	--	RIVBNK	7b	4	s	?
	Bellis perennis L.	-	*	*	+	*	*	--	MEADOW	7c	4	s	+
	Berula erecta (Hudson) Coville	++	--	--	--	--	--	--	AQUATp	6b	1	S	-
	Brachypodium pinnatum (L.) Beauv.	--	--	--	-	*	++	--	WASTEl	7c	4	S	?
	Brachypodium sylvaticum (Hudson) Beauv.	--	*	--	*	-	--	+	WOODl	7c	2	s	-
*	Brassica rapa L.	-	-	+	-	*	+	--	RIVBNK	7c	3	s	+
	Briza media L.	--	*	--	+	*	+	--	PASTl	7c	5	S	-

Autecological account	Introduced	Species	Abundance in Wetland	Abundance in Skeletal	Abundance in Arable	Abundance in Pasture	Abundance in Spoil	Abundance in Wasteland	Abundance in Woodland	Commonest terminal habitat	Soil pH	Floristic diversity	Distribution in N Europe
+		Bromus erectus Hudson	--	--	--	*	*	++	--	WASTEl	7c	4	S
+		Bromus hordeaceus L.	--	*	*	+	*	-	--	MEADOW	5b	4	s
+		Bromus ramosus Hudson	-	*	--	-	-	*	+	WOODl	7c	2	Sw
+		Bromus sterilis L.	--	+	+	*	*	*	*	HEDGE	7b	2	s
+		Callitriche stagnalis Scop.	++	--	-	-	--	--	--	AQUATr	6c	1	-
+		Calluna vulgaris (L.) Hull	-	*	--	+	*	*	--	QRYa	3a	1	s
+		Caltha palustris L.	++	-	-	-	-	-	--	MIREs	6a	2	-
+		Calystegia sepium (L.) R.Br.	*	--	-	-	+	+	--	RIVBNK	7b	2	S
+		Campanula rotundifolia L.	--	+	--	+	*	*	--	PASTl	W7	4	-
+		Capsella bursa-pastoris (L.) Medicus	--	--	++	-	*	*	--	ARABLE	7c	4	-
+		Cardamine amara L.	++	--	--	-	-	*	*	MIREs	6a	2	S
+		Cardamine flexuosa With.	+	*	--	-	--	*	*	MIREs	6b	2	s
+		Cardamine hirsuta L.	--	*	--	*	+	*	--	QRYl	7a	3	s
+		Cardamine pratensis L.	++	--	--	+	--	*	--	MIREu	5b	3	-
	*	Cardaria draba (L.) Desv.	--	--	*	-	*	++	--	RLYBNK	7b	2	S
		Carduus acanthoides L.	--	-	-	-	*	+	+	RIVBNK	7b	3	s
+		Carex acutiformis Ehrh.	++	--	--	-	--	*	--	RIVBNK	7c	1	S
		Carex binervis Sm.	*	--	--	+	--	*	-	PASTa	3b	2	sW
+		Carex caryophyllea Latour.	--	*	--	+	*	+	--	PASTl	5d	5	s
		Carex demissa Hornem.	++	*	--	-	--	--	--	MIREu	5c	4	s
		Carex echinata Murray	++	--	--	-	--	--	--	MIREu	5b	4	-
+		Carex flacca Schreber	--	*	--	+	*	+	-	SCREE	7c	5	s
		Carex hirta L.	+	--	--	+	-	*	--	MIREu	6d	4	S
		Carex hostiana DC.	++	--	--	+	--	--	--	MIREu	5b	5	s
+		Carex nigra (L.) Reichard	+	--	*	+	--	*	--	MIREu	3c	2	-
		Carex otrubae Podp.	++	--	--	-	--	++	--	MIREu	6c	3	S
		Carex ovalis Good.	+	--	--	+	--	*	--	MIREu	5b	4	s
		Carex pallescens L.	*	--	--	+	--	-	-	PASTa	5a	5	-
+		Carex panicea L.	+	--	--	+	--	++	--	PASTl	5b	5	-
		Carex pendula Hudson	*	--	--	-	--	--	++	MIREs	6c	2	SW
+		Carex pilulifera L.	--	*	--	+	*	*	*	PASTa	3a	1	s
+		Carex remota L.	++	-	--	-	--	--	+	MIREs	5b	2	S
		Carex sylvatica Hudson	--	--	--	*	--	*	++	SCRUB	6d	3	s
+		Carlina vulgaris L.	--	++	--	-	+	*	--	QRYl	7a	4	S
+		Centaurea nigra L.	--	*	--	*	*	+	--	WASTEl	7c	4	sW
+		Centaurea scabiosa L.	--	+	--	-	*	+	--	WASTEl	7b	3	S
+		Centaurium erythraea Rafn.	--	*	--	-	*	+	--	WASTEl	7a	4	S
	*	Centranthus ruber (L.) DC.	--	++	--	-	+	*	--	CLIFF	7a	1	SW
+		Cerastium fontanum Baumg.	--	*	*	+	*	*	--	MEADOW	5c	4	-
		Cerastium glomeratum Thuill.	--	*	+	-	+	*	--	OUTCRP	7c	4	s
		Cerastium semidecandrum L.	--	++	--	*	*	*	--	OUTCRP	7b	4	S
	*	Cerastium tomentosum L.	--	++	--	-	*	*	--	OUTCRP	7b	2	?
+		Chaenorhinum minus (L.) Lange	--	--	--	-	++	--	--	CINDER	6b	3	S
		Chaerophyllum temulentum L.	--	++	--	-	-	-	++	HEDGE	6b	2	S
+		Chamerion angustifolium (L.) J. Holub	--	*	--	-	+	*	*	CINDER	4d	2	-
	*	Cheiranthus cheiri L.	--	++	--	-	--	--	--	CLIFF	6b	1	SW
+		Chelidonium majus L.	--	*	--	-	+	+	--	RDVRGE	6b	2	S
		Chenopodium album L.	--	--	++	-	*	*	--	ARABLE	6b	3	-
	*	Chenopodium bonus-henricus L.	--	--	--	+	*	++	--	RDVRGE	6b	2	Se
+		Chenopodium rubrum L.	*	*	*	-	++	--	--	MANURE	6b	2	S
	*	Chrysanthemum segetum L.	--	--	++	-	*	--	--	ARABLE	6c	3	s
+		Chrysosplenium oppositifolium L.	+	*	--	-	--	*	*	MIREs	6b	2	sw

	Abundance in Wetland	Abundance in Skeletal	Abundance in Arable	Abundance in Pasture	Abundance in Spoil	Abundance in Wasteland	Abundance in Woodland	Commonest terminal habitat	Soil pH	Floristic diversity	Distribution in N Europe	Present status
Circaea lutetiana L.	*	--	--	-	--	--	+	MIREs	6b	2	S	+
Cirsium arvense (L.) Scop.	--	-	+	*	+	*	--	COAL	5c	3	-	+
Cirsium palustre (L.) Scop.	+	-	--	-	+	-	*	-- PASTl	5c	4	-	?
Cirsium vulgare (Savi) Ten.	--	*	*	*	-	+	*	PASTl	5c	4	s	+
Clinopodium vulgare L.	--	+	--	-	+	*	-	QRYl	7a	4	S	?
Conium maculatum L.	--	*	*	-	*	+	*	RIVBNK	6c	3	S	+
Conopodium majus (Gouan) Loret	--	--	--	+	--	*	*	PASTl	6c	3	sW	-
Convolvulus arvensis L.	--	-	++	-	*	*	-	ARABLE	7c	3	S	?
Coronopus squamatus (Forskal) Ascherson	--	--	+	+	*	*	--	PATH	7b	2	S	-
Corydalis claviculata (L.) DC.	-	--	--	-	--	*	++	WOODa	3b	1	sW	?
Corydalis lutea (L.) DC.	--	++	--	-	-	--	--	WALL	7a	1	?	?
Crepis capillaris (L.) Wallr.	--	*	-	*	+	*	-	QRYl	7b	4	s	+
Cymbalaria muralis [3]	--	++	--	-	*	--	-	WALL	7a	1	S	?
Cynosurus cristatus L.	-	-	-	+	-	*	--	MEADOW	6c	4	s	-
Dactylis glomerata L.	--	*	*	*	*	*	-	MEADOW	7c	3	-	?
Dactylorhiza fuchsii (Druce) Soo	+	-	--	+	*	+	-	PASTl	7c	5	s	?
Dactylorhiza incarnata (L.) Soo	++	--	--	+	--	--	--	MIREu	6a	3	-	-
Dactylorhiza maculata (L.) Soo	++	--	--	+	--	--	--	MIREu	4b	3	-	-
Danthonia decumbens (L.) DC.	--	-	-	+	--	+	--	PASTl	W4	4	s	-
Daucus carota L.	--	*	-	-	*	+	--	WASTEl	7a	5	S	-
Deschampsia cespitosa (L.) Beauv.	*	-	*	*	*	*	*	WOODl	5c	2	-	?
Deschampsia flexuosa (L.) Trin.	--	*	--	+	*	*	*	PASTa	3a	1	-	-
Desmazeria rigida (L.) Tutin	--	+	--	-	++	--	--	QRYl	7a	4	Sw	+
Digitalis purpurea L.	--	*	*	-	*	-	+	PLANTb	4c	3	sw	-
Eleocharis palustris (L.) Roemer & Schultes	++	--	--	-	--	--	--	MIREu	6b	1	-	-
Elodea canadensis Michx	++	--	--	-	-	--	--	AQUATp	7c	1	S	-
Elymus caninus (L.) L.	--	+	--	-	-	--	+	WOODl	6b	2	-	-
Elymus repens (L.) Gould	--	--	++	*	*	*	-	ARABLE	7c	2	-	+
Empetrum nigrum ssp. nigrum L.	-	*	--	+	-	+	--	PASTa	3a	1	n	?
Epilobium ciliatum Rafin.	*	*	--	-	*	+	*	CINDER	6c	3	S	+
Epilobium hirsutum L.	+	*	--	-	*	*	--	RIVBNK	7b	2	S	+
Epilobium montanum L.	--	+	--	-	*	*	*	QRYl	7c	3	-	?
Epilobium obscurum Schreber	++	*	--	-	--	--	--	MIREu	6a	3	S	?
Epilobium palustre L.	++	--	--	-	--	--	--	MIREu	5a	3	-	-
Epilobium parviflorum Schreber	++	-	--	-	*	--	--	QRYl	6c	3	S	+
Epipactis helleborine (L.) Crantz	--	--	--	-	*	--	++	SCRUB	7c	2	S	?
Erica cinerea L.	--	*	--	+	+	*	--	QRYa	3b	1	sW	-
Erica tetralix L.	++	--	--	*	--	-	--	MIREu	3b	1	sw	-
Erigeron acer L.	--	+	--	-	++	--	--	QRYl	7a	4	se	?
Eriophorum angustifolium Honckeny	++	*	--	*	--	--	--	MIREu	3b	1	-	?
Eriophorum vaginatum L.	++	--	--	-	--	--	--	*MIREu	3a	1	-	?
Erodium cicutarium (L.) L'Her.	--	+	*	+	*	+	--	OUTCRP	6c	4	S	-
Erophila verna (L.) Chevall.	--	++	-	-	*	*	--	OUTCRP	7b	4	s	-
Eupatorium cannabinum L.	+	*	--	-	*	-	--	MIREs	7b	3	S	?
Euphorbia helioscopia L.	--	--	++	-	*	*	-	ARABLE	6c	5	S	?
Euphorbia peplus L.	--	--	++	-	+	*	-	SOIL	6b	3	S	?
Euphrasia officinalis L., sensu lato	--	*	--	+	+	*	--	*PASTl	7b	4	X	-
Fallopia convolvulus (L.) A. Love	--	--	++	-	*	--	--	ARABLE	5c	3	-	-
Festuca arundinacea Schreber	*	*	--	*	*	*	-	WASTEl	7b	4	S	?
Festuca gigantea (L.) Vill.	*	-	--	-	--	*	+	RIVBNK	6b	2	S	?
Festuca ovina agg.	--	*	--	+	*	*	--	PASTl	W3	2	-	-
Festuca pratensis Hudson	-	--	--	+	--	+	--	MEADOW	5b	4	s	-

		Autecological account	Abundance in Wetland	Abundance in Skeletal	Abundance in Arable	Abundance in Pasture	Abundance in Spoil	Abundance in Wasteland	Abundance in Woodland	Commonest terminal habitat	Soil pH	Floristic diversity	Distribution in N Europe
+		Festuca rubra L.	--	*	.	*	*	+	--	RDVRGE	7c	3	-
+		Filipendula ulmaria (L.) Maxim.	+	--	--	-	--	*	*	MIREs	5c	2	-
+		Fragaria vesca L.	--	*	--	*	*	*	*	SCREE	7a	4	-
		Fumaria muralis Sonder ex Koch	--	--	++	-	--	*	--	ARABLE	6b	5	SW
		Fumaria officinalis L.	--	-	++	-	+	*	--	ARABLE	6b	3	s
+		Galeopsis tetrahit L., sensu lato	--	-	++	-	-	*	*	ARABLE	6b	2	s
+		Galium aparine L.	-	-	+	-	-	*	*	HEDGE	7c	2	s
+		Galium cruciata (L.) Scop.	-	-	--	*	*	+	-	WASTEl	7c	3	S
		Galium odoratum (L.) Scop.	-	*	--	-	--	-	++	SCRUB	6c	2	s
+		Galium palustre group	++	--	--	-	--	--	--	MIREu	5b	3	-
+		Galium saxatile L.	-	--	--	+	--	*	*	PASTa	3b	1	sw
+		Galium sterneri Ehrend.	--	+	--	+	*	-	--	SCREE	7c	4	sw
		Galium uliginosum L.	++	--	--	-	--	--	--	MIREu	6b	4	s
+		Galium verum L.	--	*	--	+	--	+	--	PASTl	5d	5	-
		Gentianella amarella (L.) Borner	-	*	--	+	+	-	--	QRYl	7a	5	n
		Geranium columbinum L.	--	++	--	-	-	*	--	OUTCRP	6b	5	S
		Geranium dissectum L.	--	+	*	-	*	+	--	RDVRGE	7c	4	S
		Geranium lucidum L.	--	++	--	-	*	-	-	OUTCRP	7b	3	Sw
+		Geranium molle L.	--	+	+	*	*	*	--	OUTCRP	7b	5	s
		Geranium pyrenaicum Burm. fil.	--	--	--	-	+	+	--	RDVRGE	6b	2	S
+		Geranium robertianum L.	-	+	--	-	*	*	*	SCREE	7b	2	s
		Geum rivale L.	*	*	--	+	--	*	+	SCRUB	6c	4	e
+		Geum urbanum L.	--	*	--	-	--	--	++	WOODl	5c	2	s
+		Glechoma hederacea L.	*	-	--	-	*	*	+	HEDGE	6c	2	S
		Glyceria declinata Breb.	++	--	--	-	--	--	--	MIREu	6c	2	sw
+		Glyceria fluitans (L.) R.Br.	++	--	--	*	--	-	--	AQUATp	5b	1	-
+		Glyceria maxima (Hartman) Holmberg	++	--	--	-	--	*	--	MIREu	7c	1	S
		Glyceria plicata Fries	++	--	--	-	--	--	--	MIREu	7b	2	S
		Gnaphalium uliginosum L.	++	--	+	-	*	--	--	MIREu	5c	3	s
		Gymnadenia conopsea (L.) R.Br.	*	*	*	+	+	*	--	QRYl	7a	5	-
+		Hedera helix L.	--	*	--	-	-	--	+	WOODl	7d	1	s
+		Helianthemum nummularium (L.) Miller	--	+	--	+	*	*	--	PASTl	7c	5	S
+		Heracleum sphondylium L.	--	-	*	*	*	+	*	WASTEl	5c	3	-
	*	Hesperis matronalis L.	--	--	--	-	-	*	++	RIVBNK	7b	2	?
+		Hieracium pilosella L.	--	*	--	*	+	*	--	PASTl	7c	4	-
+		Hieracium subgenus Hieracium L.	--	*	--	-	+	*	--	QRYl	W7	3	-
		Hippuris vulgaris L.	++	--	--	-	--	--	--	AQUATp	5b	1	-
+		Holcus lanatus L.	*	*	*	+	+	*	-	PASTe	5c	3	s
+		Holcus mollis L.	*	-	--	*	-	*	*	WOODa	4b	1	s
+		Humulus lupulus L.	-	-	--	-	--	*	++	HEDGE	6a	1	S
+		Hyacinthoides non-scripta [4]	--	--	--	-	--	--	++	PLANTb	4b	1	sW
+		Hydrocotyle vulgaris L.	++	*	--	-	--	--	--	MIREu	5a	1	s
		Hypericum androsaemum L.							++		-	-	SW
+		Hypericum hirsutum L.	--	*	--	*	-	-	*	SCREE	7b	3	S
		Hypericum humifusum L.	--	--	--	*	-	+	--	PATH	5c	3	S
		Hypericum maculatum Crantz	--	--	--	*	+	+	--	WASTEa	5c	3	S
+		Hypericum perforatum L.	--	*	*	*	*	+	--	WASTEl	7c	4	S
		Hypericum pulchrum L.	--	+	--	+	-	*	*	PASTa	5b	4	sw
		Hypericum tetrapterum Fries	++	*	--	-	*	-	--	MIREu	6c	4	S
+		Hypochaeris radicata L.	--	*	-	*	+	+	--	QRYa	5b	3	s
+	*	Impatiens glandulifera Royle	*	--	--	-	--	*	*	RIVBNK	6a	2	S
+		Inula conyza DC.	--	*	--	*	+	+	--	QRYl	7a	4	S

Introduced		Abundance in Wetland	Abundance in Skeletal	Abundance in Arable	Abundance in Pasture	Abundance in Spoil	Abundance in Wasteland	Abundance in Woodland	Commonest terminal habitat	Soil pH	Floristic diversity	Distribution in N Europe	Present status
	Iris pseudacorus L.	++	--	--	-	--	--	--	MIREs	5c	1	s	-
	Isolepis setacea (L.) R.Br.	++	*	--	-	--	*	-	MIREu	6c	4	sw	?
	Juncus acutiflorus Ehrh. ex Hoffm.	++	--	--	-	--	--	--	MIREu	5c	3	sw	-
	Juncus articulatus L.	++	--	--	*	--	*	--	MIREu	6b	3	-	-
	Juncus bufonius L.	+	*	++	-	--	*	--	MIREu	6b	3	-	+
	Juncus bulbosus L.	++	--	--	-	--	-	--	AQUATp	5a	2	-	?
	Juncus conglomeratus L.	+	--	--	*	--	*	--	MIREu	5c	4	-	?
	Juncus effusus L.	++	--	--	*	--	*	--	MIREu	5d	2	s	+
	Juncus inflexus L.	++	--	--	*	-	-	--	MIREu	6c	3	S	?
	Juncus squarrosus L.	*	*	--	+	--	+	--	PATH	3b	1	sw	?
	Knautia arvensis (L.) Coulter	--	--	--	+	*	+	--	WASTEl	7c	4	s	-
	Koeleria macrantha (Ledeb.) Schultes	--	+	--	+	*	*	--	PASTl	7d	5	s	-
	Lamiastrum galeobdolon [5]	-	--	--	-	--	--	++	WOODl	4d	1	S	-
	Lamium album L.	--	*	--	-	+	*	--	SOIL	7a	2	S	+
	Lamium purpureum L.	--	--	++	-	+	--	--	ARABLE	7b	4	s	+
	Lapsana communis L.	--	*	++	-	*	--	*	ARABLE	6c	3	s	?
	Lathyrus montanus Bernh.	--	--	--	+	--	++	--	WASTEl	4a	4	s	-
	Lathyrus pratensis L.	--	--	--	+	*	+	--	MEADOW	5c	4	-	-
	Lemna minor L.	++	-	--	-	--	--	--	AQUATp	7c	1	S	+
	Lemna trisulca L.	++	--	--	-	--	--	--	AQUATp	6b	1	S	-
	Leontodon autumnalis L.	--	*	*	*	+	*	--	*BRICK	6c	3	-	+
	Leontodon hispidus L.	--	+	--	*	+	*	--	QRYl	7a	4	S	?
	Leontodon taraxacoides (Vill.) Merat	-	*	--	+	-	*	--	RDVRGE	7c	2	S	?
	Lepidium campestre (L.) R.Br.	--	--	--	*	*	+	--	RLYBNK	6c	2	S	-
	Lepidium heterophyllum Bentham	--	--	--	-	++	*	--	CINDER	6c	2	sW	+
	Leucanthemum vulgare Lam.	--	*	*	-	+	*	--	QRYl	7a	4	-	-
	Linaria vulgaris Miller	--	-	--	-	++	--	--	CINDER	6c	2	-	+
	Linum catharticum L.	--	+	--	*	+	*	--	QRYl	7a	4	s	?
	Listera ovata (L.) R.Br.	--	--	--	*	*	+	+	SCRUB	6c	4	s	-
	Lolium perenne ssp perenne L.	--	*	+	+	*	*	--	MEADOW	6c	3	s	+
	Lonicera periclymenum L.	--	--	--	-	-	-	++	WOODl	4a	1	ws	?
	Lotus corniculatus L.	--	*	--	+	+	+	--	PASTl	7c	4	-	-
	Lotus uliginosus Schkuhr	+	--	--	*	--	+	--	*MIREu	5b	4	S	-
	Luzula campestris (L.) DC.	--	--	--	+	+	+	--	PASTl	5c	3	s	?
	Luzula multiflora (Retz.) Lej.	*	--	--	+	*	*	--	QRYa	4d	3	-	-
	Luzula pilosa (L.) Willd.	--	*	--	-	--	--	++	PLANTc	3b	1	-	?
	Luzula sylvatica (Hudson) Gaudin	--	+	--	-	--	--	++	WOODa	4c	2	w	-
	Lychnis flos-cuculi L.	++	--	--	*	--	-	--	MIREu	6c	4	-	-
	Lycopus europaeus L.	++	--	--	-	--	-	--	PONDBNK	6c	2	S	-
	Lysimachia nemorum L.	+	--	--	*	--	-	+	MIREs	5d	3	s	-
	Lysimachia nummularia L.	++	--	--	+	--	-	*	MIREu	6c	3	S	-
	Lysimachia vulgaris L.	++	--	--	*	-	-	--	MIREs	5c	1	S	-
	Lythrum portula (L.) D.A. Webb	++	--	--	-	--	-	--	MIREu	5c	2	S	+
	Lythrum salicaria L.	++	--	--	-	--	-	--	MIREu	6c	2	S	-
	Malva moschata L.	--	+	--	*	*	+	--	OUTCRP	7c	3	Sw	-
	Malva neglecta Wallr.	--	*	--	-	*	+	--	RDVRGE	6b	2	S	+
	Malva sylvestris L.	--	+	*	-	+	+	--	OUTCRP	6b	3	S	+
	Matricaria matricarioides (Less.) Porter	--	--	++	-	+	*	--	ARABLE	7c	3	-	+
	Medicago lupulina L.	--	+	+	*	*	+	--	OUTCRP	7a	4	S	?
	Medicago sativa L.	--	-	--	-	*	++	--	RDVRGE	6b	2	S	+
	Melampyrum pratense L.	--	--	--	+	*	--	++	WOODa	3b	1	-	-
	Melica uniflora Retz.	--	*	--	-	--	--	++	WOODl	7b	1	S	-

Autecological account	Introduced	Species	Abundance in Wetland	Abundance in Skeletal	Abundance in Arable	Abundance in Pasture	Abundance in Spoil	Abundance in Wasteland	Abundance in Woodland	Commonest terminal habitat	Soil pH	Floristic diversity	Distribution in N Europe
+		Mentha aquatica L.	++	--	--	-	--	--	--	MIREs	6b	2	s
		Mentha arvensis L.	+	*	+	-	--	-	*	MIREu	6c	3	s
	*	Mentha spicata L.	--	-	*	+	+	+	--	SOIL	6c	2	s
		Mentha x verticillata L.	++	--	--	*	--	--	--	MIREu	6c	3	?
		Menyanthes trifoliata L.	++	--	--	-	--	--	--	MIREu	5c	2	-
+		Mercurialis perennis L.	--	--	--	-	--	--	++	WOODl	7c	2	s
+		Milium effusum L.	--	--	--	-	--	--	++	SCRUB	4c	1	s
	*	Mimulus guttatus DC.	++	--	--	-	--	--	--	RIVBNK	6b	2	se
+		Minuartia verna (L.) Hiern	--	--	--	-	++	--	--	LEAD	6a	2	D
+		Moehringia trinervia (L.) Clairv.	--	*	--	-	--	--	++	WOODl	4a	1	s
+		Molinia caerulea (L.) Moench	*	-	--	+	--	*	--	PASTa	W3	1	-
		Montia fontana L.	++	--	--	-	--	--	--	MIREu	5b	2	-
+		Mycelis muralis (L.) Dumort.	--	++	--	-	-	--	*	OUTCRP	7b	3	S
+		Myosotis arvensis (L.) Hill	--	--	++	*	-	--	--	ARABLE	6b	4	-
		Myosotis discolor Pers.	-	-	-	+	-	+	--	WASTEd	5c	4	sw
		Myosotis laxa ssp caespitosa Lehm.	++	--	--	-	--	--	--	MIREu	5c	3	w
+		Myosotis ramosissima Rochel	--	+	--	*	+	*	--	OUTCRP	7b	5	S
+		Myosotis scorpioides L.	++	--	--	-	--	--	--	MIREs	6c	2	s
		Myosotis secunda A. Murray	++	--	--	-	--	--	--	AQUATr	5b	3	sW
		Myriophyllum spicatum L.	++	--	--	-	--	--	--	AQUATp	6c	1	-
+	*	Myrrhis odorata (L.) Scop.	--	--	*	-	--	+	+	RIVBNK	6a	2	se
+		Nardus stricta L.	*	--	--	+	-	*	--	PASTa	3a	1	-
+		Nasturtium officinale agg.	++	--	--	-	--	--	--	AQUATr	7b	1	s
		Nuphar lutea (L.) Sm.	++	--	--	-	--	--	--	AQUATp	6a	1	s
		Odontites verna (Bellardi) Dumort.	--	--	+	*	-	+	--	RDVRGE	7c	4	sw
+		Oenanthe crocata L.	++	--	--	-	--	--	--	MIREu	6c	1	SW
		Oenanthe fistulosa L.	++	--	--	-	--	--	--	AQUATp	6a	1	Sw
		Ononis repens L.	--	+	--	*	+	+	--	OUTCRP	7b	4	Sw
		Orchis mascula (L.) L.	--	--	--	+	-	*	++	PASTl	6c	4	s
		Orchis morio L.	--	--	--	+	++	--	--	PASTl	6a	5	S
+		Origanum vulgare L.	--	*	--	*	+	+	--	QRYl	7b	4	S
+		Oxalis acetosella L.	-	*	-	-	--	--	++	SCREE	4b	1	-
		Papaver argemone L.	--	*	++	-	*	*	*	ARABLE	6c	4	S
		Papaver dubium L.	--	--	*	-	+	+	--	RLYBNK	6c	3	s
+		Papaver rhoeas L.	--	--	++	-	*	*	--	ARABLE	7b	4	S
		Parietaria judaica L.	--	++	--	-	-	-	*	WALL	6b	1	SW
		Pedicularis sylvatica L.	*	--	--	+	--	--	--	PASTa	4b	5	sw
	*	Pentaglottis sempervirens [6]	--	--	--	-	*	*	++	RDVRGE	6b	1	SW
	*	Petasites fragrans (Vill.) C. Presl	--	--	--	-	*	+	+	RDVRGE	6b	1	SW
+		Petasites hybridus [7]	*	*	*	-	-	*	*	RIVBNK	6b	2	S
+		Phalaris arundinacea L.	++	*	--	-	--	*	--	RIVBNK	7c	1	-
+		Phleum pratense L.	--	--	++	+	*	*	--	MEADOW	5b	3	-
+		Phragmites australis (Cav.) Trin ex Steudel	++	--	--	-	--	--	--	MIREu	6c	1	-
+		Pimpinella saxifraga L.	--	*	--	+	*	+	--	PASTl	7c	5	s
+		Plantago coronopus L.	--	*	--	-	--	++	--	PATH	6b	4	Sw
+		Plantago lanceolata L.	--	*	*	+	*	*	--	MEADOW	7c	4	s
+		Plantago major L.	--	*	+	*	*	+	--	PATH	7b	3	-
		Platanthera chlorantha (Custer) Reichenb.	--	--	--	+	*	*	*	PASTl	5a	5	s
+		Poa annua L.	-	*	++	*	*	*	--	PATH	7c	3	-
		Poa nemoralis L.	--	*	--	-	--	--	++	OUTCRP	6d	1	s
+		Poa pratensis agg.	--	*	-	*	*	+	--	RDVRGE	7c	3	-
+		Poa trivialis L.	*	*	*	*	*	*	*	MEADOW	7c	2	-

Introduced	Species	Abundance in Wetland	Abundance in Skeletal	Abundance in Arable	Abundance in Pasture	Abundance in Spoil	Abundance in Wasteland	Abundance in Woodland	Commonest terminal habitat	Soil pH	Floristic diversity	Distribution in N Europe	Present status
	Polygala serpyllifolia J.A.C. Hose	--	-	--	+	*	*	--	PASTa	4b	3	sw	-
	Polygala vulgaris L.	--	*	--	+	*	+	--	PASTl	5c	5	s	-
	Polygonum amphibium L.	++	--	--	-	--	--	--	MIREu	5c	1	-	-
	Polygonum aviculare group	--	--	++	-	*	*	--	ARABLE	6c	3	-	+
	Polygonum hydropiper L.	++	--	*	-	--	--	--	MIREu	5c	3	s	+
	Polygonum lapathifolium L.	-	--	++	-	*	-	--	ARABLE	6b	3	s	?
	Polygonum persicaria L.	--	--	++	-	*	-	--	ARABLE	5b	3	s	-
	Potamogeton crispus L.	++	--	--	-	--	--	--	AQUATp	7a	1	S	-
	Potamogeton natans L.	++	--	--	-	--	--	--	AQUATp	6b	1	-	-
	Potentilla anglica Laicharding	+	--	--	-	--	+	--	RDVRGE	5c	4	S	?
	Potentilla anserina L.	+	--	*	+	-	*	--	MIREu	6c	3	-	?
	Potentilla erecta (L.) Rauschel	*	--	--	+	--	+	--	PASTl	4b	3	-	-
	Potentilla palustris (L.) Scop.	++	--	--	-	--	--	-	MIREu	5b	2	e	-
	Potentilla reptans L.	-	--	--	*	*	+	-	RDVRGE	6c	3	S	+
	Potentilla sterilis (L.) Garcke	--	*	--	+	*	*	*	PASTl	5c	5	sw	-
	Primula veris L.	--	*	--	+	-	+	--	PASTl	6b	5	s	-
	Primula vulgaris Hudson	-	*	--	*	*	-	++	SCRUB	6d	3	Sw	-
	Prunella vulgaris L.	*	*	*	+	*	*	--	MEADOW	5a	5	-	+
	Pulicaria dysenterica (L.) Bernh.	++	--	-	*	-	*	--	MIREu	6c	3	S	?
	Ranunculus acris L.	*	--	*	+	*	*	--	MEADOW	6c	4	-	-
	Ranunculus auricomus L.	--	--	--	+	-	+	++	WOODl	6c	3	s	-
	Ranunculus bulbosus L.	--	-	--	+	*	*	--	MEADOW	5c	5	s	-
	Ranunculus ficaria L.	*	--	--	*	--	*	+	RIVBNK	5c	2	s	?
	Ranunculus flammula L.	++	--	--	-	--	*	--	MIREs	5b	3	s	-
	Ranunculus peltatus Schrank	++	--	--	-	--	--	--	AQUATp	5a	1	s	-
	Ranunculus penicillatus (Dumort.) Bab.	++	*	--	-	--	--	--	AQUATr	7a	1	S	-
	Ranunculus repens L.	*	--	+	+	*	*	--	MEADOW	6c	3	-	+
	Ranunculus sceleratus L.	++	--	--	-	*	-	--	MIREu	6b	2	S	+
	Raphanus raphanistrum L.	--	--	++	-	*	*	--	ARABLE	5c	4	-	-
•	Reseda luteola L.	--	-	-	-	++	*	--	COAL	6b	3	S	+
	Reynoutria japonica Houtt.	--	--	--	-	+	+	*	RDVRGE	7b	1	s	+
	Rhinanthus minor L., sensu lato	--	*	--	+	--	--	--	MEADOW	5b	5	-	-
	Ribes uva-crispa L.	--	--	--	-	--	--	++	WOODl	7d	1	s	?
	Rorippa palustris (L.) Besser	++	*	*	-	*	--	--	MIREu	6b	2	s	+
	Rubus caesius L.	--	*	--	-	*	*	+	SCRUB	7b	2	S	-
	Rubus fruticosus L., sensu lato	--	-	-	-	*	*	+	HEDGE	4b	1	s	?
	Rubus idaeus L.	--	*	--	-	-	*	+	PLANTb	4d	2	-	?
	Rumex acetosa L.	*	-	*	+	*	*	--	*MEADOW	5c	3	-	?
	Rumex acetosella agg.	--	*	*	*	*	+	-	COAL	4c	1	-	+
	Rumex conglomeratus Murray	++	--	--	*	--	-	--	MIREu	6c	3	s	+
	Rumex crispus L.	*	--	*	-	+	*	--	CINDER	6b	3	s	+
	Rumex hydrolapathum Hudson	++	--	--	-	--	--	--	CANALB	6c	1	S	-
	Rumex obtusifolius L.	*	-	++	-	+	*	--	SOIL	7c	3	s	+
	Rumex sanguineus L.	*	--	--	*	--	*	+	SCRUB	6c	3	S	+
	Sagina apetala Ard.	--	*	--	-	+	*	--	CINDER	6c	4	s	+
	Sagina nodosa (L.) Fenzl	++	*	--	-	*	--	--	MIREu	7b	4	-	-
	Sagina procumbens L.	*	*	*	-	+	*	--	CINDER	5c	3	-	+
	Salix repens L.	++								W	3	s	-
	Sanguisorba minor Scop.	--	*	--	+	*	*	--	PASTl	5c	5	S	-
	Sanicula europaea L.	--	--	--	-	--	--	++	WOODl	6b	2	s	-
•	Saponaria officinalis L.	--	--	--	-	+	++	*	WASTEa	6c	1	S	?
	Saxifraga tridactylites L.	--	++	--	-	+	--	--	OUTCRP	7b	4	S	-

Autecological account	Introduced		Abundance in Wetland	Abundance in Skeletal	Abundance in Arable	Abundance in Pasture	Abundance in Spoil	Abundance in Wasteland	Abundance in Woodland	Commonest terminal habitat	Soil pH	Floristic diversity	Distribution in N Europe
+		Scabiosa columbaria L.	--	+	--	+	*	*	--	PASTl	7b	4	Sw
		Schoenoplectus lacustris (L.) Palla	++	--	--	-	--	--	--	AQUATp	6b	1	s
		Scleranthus annuus L.	--	--	++	*	+	+	--	SNDPIT	5c	4	S
+		Scrophularia auriculata L.	+	-	--	-	*	*	*	RIVBNK	6c	3	SW
		Scrophularia nodosa L.	-	--	--	-	+	*	+	SCRUB	6c	3	s
		Scutellaria galericulata L.	++	*	--	-	--	--	--	MIREs	5c	2	-
+		Sedum acre L.	--	++	--	*	*	--	--	OUTCRP	7a	3	-
		Sedum anglicum Hudson		++							?5	-	sW
		Sedum telephium L.	--	++	--	-	--	*	--	OUTCRP	6b	3	s
		Senecio aquaticus Hill	++	--	--	*	--	*	--	MIREu	5c	4	sw
+		Senecio jacobaea L.	--	+	-	*	+	*	--	QRYl	7a	4	s
+	*	Senecio squalidus L.	--	*	--	-	++	-	--	BRICK	7a	2	S
		Senecio sylvaticus L.	--	--	-	-	*	++	--	WASTEd	5d	3	s
+		Senecio viscosus L.	--	-	--	-	++	--	--	CINDER	7c	2	S
+		Senecio vulgaris L.	--	-	++	-	+	-	--	BRICK	7a	3	-
		Sherardia arvensis L.	--	++	+	*	--	*	--	OUTCRP	7b	5	S
		Silene alba (Miller) E.H.L. Krause	--	*	+	*	*	+	--	RDVRGE	7c	3	-
+		Silene dioica (L.) Clairv.	*	-	-	-	--	*	+	RIVBNK	4e	2	-
		Silene vulgaris (Moench) Garcke	--	*	-	-	+	+	--	CINDER	6b	2	-
+		Sinapis arvensis L.	--	--	++	-	*	-	--	ARABLE	7c	4	-
	*	Sisymbrium altissimum L.	--	--	--	-	++	*	--	SOIL	7b	2	S
		Sisymbrium officinale (L.) Scop.	--	--	+	-	+	+	--	RDVRGE	6b	3	S
+		Solanum dulcamara L.	++	--	--	-	-	*	--	MIREs	7c	1	S
+		Solidago virgaurea L.	--	*	--	-	+	*	--	QRYl	W7	3	-
		Sonchus arvensis L.	*	--	+	-	*	+	--	RDVRGE	7c	3	s
+		Sonchus asper (L.) Hill	--	-	*	-	++	*	--	QRYl	7c	4	s
+		Sonchus oleraceus L.	--	*	+	-	+	*	--	BRICK	7b	3	s
		Sparganium emersum Rehmann	++	--	--	-	--	--	--	AQUATp	6b	1	-
+		Sparganium erectum L.	++	--	--	-	--	--	--	AQUATp	6b	1	S
+		Spergula arvensis L.	--	--	++	-	*	--	--	ARABLE	5b	3	-
		Spergularia rubra (L.) J. & C. Presl.	*	--	*	-	+	*	--	CINDER	5c	3	s
		Stachys arvensis (L.) L.	--	--	++	-	*	*	--	ARABLE	6c	4	S
+		Stachys officinalis (L.) Trevisan	--	--	--	+	--	++	--	PASTl	5a	5	S
		Stachys palustris L.	+	--	+	*	--	*	-	PONDBK	6c	2	s
+		Stachys sylvatica L.	-	*	*	-	*	*	*	HEDGE	6b	2	s
+		Stellaria alsine Grimm	++	--	--	*	--	*	--	MIREu	6b	3	s
		Stellaria graminea L.	*	--	--	+	-	+	--	WASTEa	5c	4	-
+		Stellaria holostea L.	--	--	--	-	--	*	++	HEDGE	3d	2	s
+		Stellaria media (L.) Vill.	--	-	++	-	*	*	-	ARABLE	6c	3	-
+		Succisa pratensis Moench	*	--	--	+	--	+	--	PASTl	5b	5	s
		Symphytum officinale L.	*	--	--	-	-	*	+	RIVBNK	6b	2	S
	*	Symphytum x uplandicum Nyman	--	--	--	*	*	+	*	RDVRGE	6b	1	?
+		Tamus communis L.	--	--	--	-	--	*	++	HEDGE	7a	2	SW
	*	Tanacetum parthenium (L.) Schultz Bip.	--	+	-	-	+	*	--	WALL	6c	3	?
		Tanacetum vulgare L.	--	--	--	-	++	+	--	CINDER	6b	2	-
+		Taraxacum agg.	--	*	*	*	*	*	--	MEADOW	7b	3	-
		Teucrium scorodonia L.	--	+	--	*	*	*		SCREE	W4	2	sw
		Thlaspi arvense L.	--	--	++	-	*	+	--	ARABLE	6b	3	s
+		Thymus praecox Opiz	--	+	--	+	*	--	--	PASTl	7c	5	w
+		Torilis japonica (Houtt.) DC.	--	*	--	*	+	*	*	OUTCRP	7b	5	S
		Tragopogon pratensis L.	--	*	--	*	*	+	--	RDVRGE	7b	3	S
		Trifolium arvense L.	--	-	--	-	+	+	--	SNDPIT	6b	3	S

Introduced		Abundance in Wetland	Abundance in Skeletal	Abundance in Arable	Abundance in Pasture	Abundance in Spoil	Abundance in Wasteland	Abundance in Woodland	Commonest terminal habitat	Soil pH	Floristic diversity	Distribution in N Europe	Present status
	Trifolium campestre Schreber	--	+	--	-	*	+	--	OUTCRP	6c	4	s	-
	Trifolium dubium Sibth.	--	+	+	*	+	*	--	OUTCRP	6c	5	s	+
*	Trifolium hybridum L.	*	--	--	*	+	+	--	RDVRGE	7c	3	e	?
	Trifolium medium L.	--	--	--	*	*	++	--	WASTEl	5a	4	s	-
	Trifolium pratense L.	--	--	++	+	*	*	--	MEADOW	5c	4	-	?
	Trifolium repens L.	-	--	+	+	*	*	--	PASTe	5c	3	-	+
	Triglochin palustris L.	++	--	--	-	--	--	--	MIREu	6c	4	e	-
	Tripleurospermum inodorum Schultz Bip.	--	--	++	-	+	-	--	ARABLE	6c	3	-	+
	Trisetum flavescens (L.) Beauv.	--	*	--	+	*	+	--	MEADOW	7c	4	S	-
	Tussilago farfara L.	-	*	*	-	++	*	--	QRYl	7b	3	-	+
	Typha latifolia L.	++	--	--	-	--	--	--	AQUATr	6a	1	S	+
	Umbilicus rupestris (Salisb.) Dandy	--	++	--	-	--	--	--	WALL	5c	1	SW	-
	Urtica dioica L.	*	*	*	-	*	*	*	SOIL	6c	2	-	+
	Urtica urens L.	--	--	++	-	*	--	--	ARABLE	6b	3	-	?
	Vaccinium myrtillus L.	--	-	--	+	-	*	*	PASTa	3a	1	e	?
	Vaccinium vitis-idaea L.	--	--	--	+	*	*	*	QRYa	3a	1	e	?
	Valeriana officinalis L.	-	*	--	-	*	++	-	WASTEl	5d	3	s	-
	Valerianella locusta (L.) Betcke	--	+	--	*	*	*	--	OUTCRP	7b	4	S	-
	Verbascum thapsus L.	--	*	--	-	++	*	--	CINDER	7b	3	S	?
	Veronica agrestis L.	--	--	++	-	*	*	--	ARABLE	6b	3	s	-
	Veronica anagallis-aquatica L.	++	--	--	-	--	--	--	MIREu	6b	1	S	?
	Veronica arvensis L.	--	+	++	+	*	--	--	PASTl	6b	4	s	?
	Veronica beccabunga L.	++	--	--	-	--	--	--	AQUATr	7b	1	s	?
	Veronica chamaedrys L.	--	*	--	+	*	*	--	PASTl	7c	4	-	-
*	Veronica filiformis Sm.	--	--	--	+	--	*	--	RDVRGE	7c	3	Sw	+
	Veronica hederifolia L.	--	*	++	*	--	-	+	WOODl	6c	2	S	+
	Veronica montana L.	*	--	--	-	--	--	+	MIREs	5c	2	Sw	-
	Veronica officinalis L.	--	*	--	+	*	*	*	QRYa	5d	5	-	-
*	Veronica persica Poiret	--	--	++	-	*	--	--	ARABLE	6b	3	sw	-
	Veronica polita Fries	--	--	++	-	*	--	--	ARABLE	7b	3	Sw	-
	Veronica scutellata L.	++	--	--	-	--	--	--	MIREa	5b	3	-	-
	Veronica serpyllifolia L.	*	--	*	+	--	+	*	RDVRGE	6c	4	-	+
	Vicia cracca L.	--	-	*	*	*	++	--	WASTEl	6b	2	-	-
	Vicia hirsuta (L.) S.F. Gray	--	*	-	*	*	++	--	WASTEd	6c	3	S	?
	Vicia sativa ssp nigra L.	--	-	*	*	*	++	--	WASTEd	6d	3	-	?
	Vicia sepium L.	--	--	--	*	*	+	*	SOIL	5c	3	-	-
	Viola arvensis Murray	--	--	++	-	*	*	--	ARABLE	5c	3	s	-
	Viola hirta L.	--	--	--	+	--	+	*	PASTl	7c	5	S	-
	Viola odorata L.	--	--	--	*	--	-	++	SCRUB	7b	2	S	-
	Viola palustris L.	++	--	--	-	--	--	--	MIREu	4c	2	ne	-
	Viola riviniana Reichenb.	--	*	--	+	-	*	*	PASTl	5d	3	-	-
	Viola tricolor L.	--	--	++	-	+	*	--	ARABLE	5c	3	-	-
	Vulpia bromoides (L.) S.F. Gray	--	*	--	-	+	+	--	WASTEd	6d	3	sw	?

(b) Woody species more than 1.5 m tall

Introduced		Abundance in Wetland	Abundance in Skeletal	Abundance in Arable	Abundance in Pasture	Abundance in Spoil	Abundance in Wasteland	Abundance in Woodland	Commonest terminal habitat	Soil pH	Floristic diversity	Distribution in N Europe	Present status
	Acer campestre L. (canopy)	--	--	--	-	--	--	++	SCRUB	7c	1	S	-
*	Acer pseudoplatanus L. (canopy)	*	-	--	-	--	--	++	WOODl	7e	1	S	+
	ditto (juvenile)	-	-	--	-	-	*	++	PLANTb	4b	1		
	Alnus glutinosa (L.) Gaertner (canopy)	+	--	--	-	--	--	+	MIREs	6d	1	s	-
	ditto (juvenile)	+	+	--	-	--	*	-	RIVBNK	5c	3		
	Betula spp. (canopy)	*	-	--	-	--	--	++	SCRUB	3b	1	-	+
	ditto (juvenile)	-	*	-	-	*	-	*	SCRUB	4c	2		

Autecological account	Introduced		Abundance in Wetland	Abundance in Skeletal	Abundance in Arable	Abundance in Pasture	Abundance in Spoil	Abundance in Wasteland	Abundance in Woodland	Commonest terminal habitat	Soil pH	Floristic diversity	Distribution in N Europe
		Corylus avellana L. (canopy)	*	--	--	-	--	--	++	SCRUB	7c	1	s
	+	Crataegus monogyna Jacq. (canopy)	-	-	--	-	--	--	++	HEDGE	7c	1	s
		ditto (juvenile)	--	*	--	*	*	*	*	PASTl	7d	3	
		Cytisus scoparius (L.) Link (canopy / juv)	--	--	--	-	+	+	--	RD/RLY	4c	1	S
		Euonymus europaeus L. (canopy)	--	--	--	-	--	*	++	SCRUB	??	1	S
	+	Fagus sylvatica L. (canopy)	--	--	--	-	--	--	++	PLANTb	3c	1	S
		ditto (juvenile)	--	--	--	-	--	--	++	PLANTb	3b	1	
	+	Fraxinus excelsior L. (canopy)	-	-	--	-	--	--	++	WOODl	7c	1	s
		ditto (juvenile)	*	*	--	*	*	*	+	WOODl	7d	2	
		Ilex aquifolium L. (canopy)	--	--	--	-	--	--	++	WOODa	3d	1	sw
		Ligustrum vulgare L. (canopy)	--	*	--	-	--	--	++	SCRUB	7b	1	Sw
		Malus sylvestris Miller (canopy)	--	--	--	-	--	--	++	HEDGE	5c	2	S
		Populus tremula L. (canopy)	--	--	--	-	--	--	++	SCRUB	?6	2	-
		Prunus avium (L.) L. (canopy)	--	--	--	-	-	--	++	WOODa	6d	2	S
		Prunus spinosa L. (canopy)	--	--	--	-	--	--	++	SCRUB	6c	2	s
	+	Quercus agg. (canopy)	-	-	--	-	--	--	++	WOODa	3b	1	s
		ditto (juvenile)	--	*	--	-	-	-	+	PLANTb	4b	1	
*	*	Rhododendron ponticum L. (canopy)	*	*	*	*	++	*	++	SCRUB	3a	1	?
		Rosa spp. (canopy)	--	-	--	-	--	--	++	HEDGE	7b	-	s
		ditto (juvenile)	--	*	*	*	*	*	*	OUTCRP	7c	4	
	+	Salix cinerea agg. (canopy)	*	--	--	-	--	--	++	SCRUB	5a	2	X
		ditto (juvenile)	*	*	--	-	+	--	--	CINDER	6b	3	
	*	Salix fragilis agg. (canopy)	++	--	--	-	--	--	*	MIREs	5b	1	S
		Salix purpurea L. (canopy)	+	--	--	-	--	--	*	MIREs	?6	-	s
	*	Salix viminalis L. (canopy)	++	--	--	-	--	--	*	MIREs	6b	-	S
	+	Sambucus nigra L. (canopy)	--	--	--	-	--	--	++	PLANTb	5c	1	s
		ditto (juvenile)	--	*	*	-	*	*	*	BRICK	5d	2	
	+	Sorbus aucuparia L. (canopy)	--	--	--	-	--	--	+	WOODa	3a	1	-
		ditto (juvenile)	--	--	--	-	--	-	++	WOODa	4b	1	
	+	Ulex europaeus L. (canopy / juvenile)	--	--	--	*	*	+	--	*WASTEl	5b	2	sw
	+	Ulmus glabra Hudson (canopy)	--	--	--	-	--	--	++	WOODl	6c	1	e
		ditto (juvenile)	*	*	*	-	*	-	*	*QRYl	6b	3	
		Ulmus procera Salisb. (canopy)	--	--	--	-	--	--	++	WOODa	6c	1	S
		Viburnum opulus L. (canopy)	*	--	--	-	--	--	++	WOODl	6d	1	s

(c) Pteridophytes

Autecological account	Introduced		Abundance in Wetland	Abundance in Skeletal	Abundance in Arable	Abundance in Pasture	Abundance in Spoil	Abundance in Wasteland	Abundance in Woodland	Commonest terminal habitat	Soil pH	Floristic diversity	Distribution in N Europe
		Asplenium adiantum-nigrum L.	--	++	--	-	--	--	--	WALL	7c	1	sw
+		Asplenium ruta-muraria L.	--	++	--	-	-	--	--	WALL	7a	1	s
+		Asplenium trichomanes L.	--	++	--	-	*	--	--	CLIFF	7a	2	-
+		Athyrium filix-femina (L.) Roth	*	++	--	-	--	--	*	*MIREs	4d	1	-
		Blechnum spicant (L.) Roth	--	++	--	+	--	*	*	OUTCRP	3b	1	-
+		Cystopteris fragilis (L.) Bernh.	--	++	--	-	--	--	--	WALL	7a	1	-
		Dryopteris affinis (Lowe) Fraser-Jenkins	--	++	--	-	*	-	+	OUTCRP	3c	1	sw
+		Dryopteris dilatata (Hoffm.) A. Gray	-	*	--	-	--	--	--	PLANTc	3b	1	s
+		Dryopteris filix-mas (L.) Schott	--	+	--	-	--	--	*	*WOODl	4e	1	-
+		Equisetum arvense L.	*	--	*	*	+	*	--	RIVBNK	6c	2	-
+		Equisetum fluviatile L.	++	--	--	-	--	--	--	AQUATp	6b	1	-
+		Equisetum palustre L.	++	--	-	-	--	--	--	*MIREu	6b	2	-
		Phyllitis scolopendrium (L.) Newman	--	++	--	-	--	--	--	CLIFF	7b	1	sw
		Polypodium vulgare group	--	++	--	-	--	--	--	CLIFF	5c	1	-
+		Pteridium aquilinum (L.) Kuhn	--	--	--	*	-	*	+	PLANTc	3b	1	-

4.2 Attributes of the established phase of species (symbols and notes are ~~ined~~ defined in the preceding text).

	Life history	Established strategy	Life form	Canopy structure	Canopy height	Lateral spread	Mycorrhizas	Leaf phenology	Flowering time and duration	Polyploidy
Herbs and woody species up to 1.5 m tall										
Achillea millefolium	P	CR/CSR	Ch	S	2	5	VA	Ea	Jun-03	P*
Achillea ptarmica	P	CR/CSR	H	L	3	3	-	Sa	Jul-02	D*
Aegopodium podagraria	P	CR/CSR	H	S	2	5	VA	Sa	May-03	DP
Aethusa cynapium	As	R	Th	L	5	1	VA	Sa	Jul-02	D
Agrimonia eupatoria	P	CSR	H	S	3	2	VA	Sa	Jun-03	R*
Agrostis canina	P	CSR	H	L	1	5	VA	Ea	Jun-02	D*
Agrostis capillaris	P	CSR	H	L	2	4	VA	Ea	Jun-03	P*
Agrostis gigantea	P	CR	H	L	2	5	VA	Ea	Jun-03	P*
Agrostis stolonifera	P	CR	H	L	2	5	+	Ea	Jul-02	P*
Agrostis vinealis	P	CSR	H	L	2	2	VA	Ea	Jun-03	P*
Aira caryophyllea	Aw	SR	Th	S	1	1	+	Sh	May-01	DP
Aira praecox	Aw	SR	Th	S	1	1	+	Sh	Apr-02	D
Ajuga reptans	P	CSR	H	S	2	4	VA	Ea	May-03	P*
Alchemilla vulgaris agg.	P	S/CSR	H	S	3	2	VA	Ea	Jun-04	P*
Alisma plantago-aquatica	P	R/CR	Wet	B	3	2	-	Sa	Jun-03	D*
Alliaria petiolata	A/M	CR	H	S	5	1	-	Sh	Apr-03	R*
Allium ursinum	P	SR/CSR	G	B	2	2	VA	Sv	Apr-03	D*
Allium vineale	P	S/CSR	G	L	4	2	VA	Sh	Jun-02	DP
Alopecurus geniculatus	P	CR	H	L	1	5	VA	Ea	Jun-02	P
Alopecurus pratensis	P	C/CSR	H	L	2	3	VA	Ea	Apr-03	P
Anacamptis pyramidalis	P	S/SR	G	S	2	2	OR	Sa	Jun-03	R
Anagallis arvensis	Asw	R/SR	Th/Ch	L	2	1	VA	Ep	Jun-03	P
Anchusa arvensis	A	R/CR	Th/H	S	3	1	?	Ep	Jun-04	P
Anemone nemorosa	P	SR/CSR	G	S	3	3	VA	Sv	Mar-03	P*
Angelica sylvestris	M/P	C/CR	H	LA	3	2	VA	Sa	Jul-03	D
Anthoxanthum odoratum	P	SR/CSR	H	L	2	2	VA	Ea	Apr-03	DP*
Anthriscus sylvestris	P	CR	H	LA	3	2	-	Sh	Apr-03	D*
Anthyllis vulneraria	P/M	S/SR	H	S	2	2	VA	Ea	Jun-04	D*
Aphanes arvensis	Aws	R/SR	Th	L	1	1	VA	Ap	Apr-07	P
Apium nodiflorum	P	CR	Hel	L	3	5	-	Ea	Jul-02	D*
Arabidopsis thaliana	Aws	SR	Th	S	1	1	+	Sh	Apr-02	D
Arabis hirsuta	P/M	S/SR	H	S	1	2	-	Ea	Jun-03	P*
Arctium lappa	M	CR	H	S	5	1	+	Sa	Jul-03	R*
Arctium minus agg.	M	CR	H	S	5	1	VA	Sa	Jul-03	R
Arenaria serpyllifolia	Aws	SR	Th	L	2	1	-	Sh	May-04	P*
Armoracia rusticana	P	C/CR	H	S	3	3	+	Sa	May-02	R*
Arrhenatherum elatius	P	C/CSR	H	L	5	4	+	Ep	Jun-02	DP*
Artemisia absinthium	P	CR/CSR	Ch	L	4	2	-	Ea	Jul-02	D*
Artemisia vulgaris	P	C/CR	H	L	5	3	VA	Ea	Jul-03	D*
Arum maculatum	P	SR/CSR	G	B	3	2	VA	Sv	Apr-02	rP*
Atriplex patula	As	R	Th	S	4	1	VA	Sa	Jun-05	P*
Atriplex prostrata	As	R	Th	S	4	1	-	Sa	Jul-03	D*
Avenula pratensis	P	S/SC	H	S	2	3	VA	Ea	Jun-01	P*
Avenula pubescens	P	S/CSR	H	L	2	2	VA	Ea	Jun-02	D*
Barbarea vulgaris	P/M	R/CR	H	S	3	2	-	Ea	May-04	D
Bellis perennis	P	R/CSR	H	B	1	3	VA	Ea	Mar-12	D*
Berula erecta	P	CR	Hel	L	4	5	-	Ea	Jul-03	D
Brachypodium pinnatum	P	SC	H	S	2	5	VA	Ea	Jul-01	DP*
Brachypodium sylvaticum	P	S/SC	H	S	3	3	VA	Ea	Jul-02	D*P
Brassica rapa	?B	CR	Th/Ch	S	3	1	-	Ea	May-04	D
Briza media	P	S	H	S	1	3	VA	Ea	Jun-02	D*P*

378   TABLES OF ATTRIBUTES

Autecological account	Introduced		Life history	Established strategy	Life form	Canopy structure	Canopy height	Lateral spread	Mycorrhizas	Leaf phenology	Flowering time and duration	Polyploidy
+		Bromus erectus	P	SC/CSR	H	S	2	3	VA	Ea	Jun-02	P
+		Bromus hordeaceus	Aw	R/CR	Th	L	3	1	+	Sh	May-03	P*
+		Bromus ramosus	P	CSR	H	L	3	2	?	Ea	Jul-02	DP
+		Bromus sterilis	Aws	R/CR	Th	S	3	1	-	Sh	May-03	D*
+		Callitriche stagnalis	P	R/CR	Wet	LF	1	4	?	Ea	May-05	D*
+		Calluna vulgaris	P	SC	Ch/Ph	L	4	4	ER	Ea	Aug-02	D
+		Caltha palustris	P	S/CSR	Hel	S	2	2	VA	Sa	Apr-03	DP
+		Calystegia sepium	P	C/CR	G	L	5	5	VA	Sa	Jun-04	D*
+		Campanula rotundifolia	P	S	H	S	2	3	VA	Ea	Jul-03	P*
+		Capsella bursa-pastoris	Asw	R	Th	S	1	1	-	Ep	Mar-12	P*
+		Cardamine amara	P	CR	H	L	3	5	-	Ea	May-02	D*
+		Cardamine flexuosa	A-P	R/SR	H	S	2	1	-	Ea	Apr-06	P*
+		Cardamine hirsuta	Aws	SR	Th	S	2	1	-	Sh	Apr-05	D*
+		Cardamine pratensis	P	R/CSR	H	S	2	2	-	Ea	Apr-03	DP
	*	Cardaria draba	P	CR	H/G	L	3	5	?	Sa	May-02	P*
		Carduus acanthoides	M	CR	H	S	5	1	VA	Ea	Jun-03	D*
+		Carex acutiformis	P	C/SC	Hel	S	5	5	-	Ep	Jun-02	?
		Carex binervis	P	S	H	S	2	3	?	Ea	Jun-01	?
+		Carex caryophyllea	P	S	H	S	1	3	-	Ea	Apr-02	?
		Carex demissa	P	S	H	S	1	2	-	Ea	Jun-01	?
		Carex echinata	P	S	Hel/H	S	1	2	-	Ea	May-02	?
+		Carex flacca	P	S	H	S	2	4	+	Ea	May-02	?
		Carex hirta	P	C/CSR	H	S	3	5	-	Sa	May-02	?
		Carex hostiana	P	S	H	S	2	2	-	Ea	Jun-01	?
+		Carex nigra	P	S/SC	H	S	3	4	+	Ep	May-03	?
		Carex otrubae	P	CR/CSR	H	S	4	3	-	Ea	Jun-02	?
		Carex ovalis	P	S/CSR	H	S	3	2	?	Ea	Jun-01	?
		Carex pallescens	P	S	H	S	2	2	-	Ea	May-02	?
+		Carex panicea	P	S	H	S	2	4	+	Ea	May-02	?
		Carex pendula	P	S/SC	H	S	4	4	-	Ea	May-02	?
+		Carex pilulifera	P	S	H	S	1	2	-	Ea	May-02	?
		Carex remota	P	CSR	H	S	3	3	-	Ea	Jun-01	?
		Carex sylvatica	P	S	H	S	1	2	-	Ea	May-03	?
+		Carlina vulgaris	M	SR	H	S	3	1	VA	Ep	Jul-04	D*
+		Centaurea nigra	P	CSR	H	S	3	2	VA	Sa	Jun-04	D*P
+		Centaurea scabiosa	P	S/CSR	H	S	2	2	VA	Sa	Jul-03	D*P
+		Centaurium erythraea	Aw	SR	Th	S	1	1	VA	Ea	Jun-05	DP
	*	Centranthus ruber	P	CSR	H	L	3	2	-	Ea	Jun-03	P
+		Cerastium fontanum	P/A	R/CSR	Ch/Th	L	1	2	-	Ea	Apr-06	P*
		Cerastium glomeratum	Aws	R/SR	Th	L	2	1	?	Sh	Apr-06	P*
		Cerastium semidecandrum	Aw	SR	Th	S	2	1	-	Sh	Apr-02	P*
	*	Cerastium tomentosum	P	CSR	Ch	L	2	4	?	Ea	May-04	P*
+		Chaenorhinum minus	As	R/SR	Th	L	2	1	-	Sa	May-06	D
		Chaerophyllum temulentum	M	R/CSR	H	S	4	1	+	Sh	Jun-02	D
+		Chamerion angustifolium	P	C	G	L	4	5	VA	Sa	Jul-03	R*
	*	Cheiranthus cheiri	P	S/CSR	Ch	L	2	2	?	Ea	Apr-03	D
		Chelidonium majus	P	CR/CSR	H	S	4	2	-	Ea	May-04	D*
+		Chenopodium album	As	R/CR	Th	L	5	1	-	Sa	Jul-04	P*
	*	Chenopodium bonus-henricus	P	CSR	H	S	3	2	-	Sa	May-03	P*
+		Chenopodium rubrum	As	R/CR	Th	L	4	1	-	Sa	Jul-03	P*
	*	Chrysanthemum segetum	As	R	Th	L	2	1	VA	Sa	Jun-03	D*
+		Chrysosplenium oppositifolium	P	CSR	Ch	L	2	4	-	Ea	Apr-04	P*

	Life history	Established strategy	Life form	Canopy structure	Canopy height	Lateral spread	Mycorrhizas	Leaf phenology	Flowering time and duration	Polyploidy
Circaea lutetiana	P	CR	G	L	2	5	VA	Sa	Jun-03	D*
Cirsium arvense	P	C	G	L	4	5	VA	Sa	Jul-03	R*
Cirsium palustre	M	CR	H	S	5	1	+	Ea	Jul-03	R*
Cirsium vulgare	M	CR	H	S	5	1	+	Ea	Jul-04	rP*
Clinopodium vulgare	P	S/CSR	H	L	4	2	+	Ep	Jul-03	D*
Conium maculatum	?B	CR	H	S	5	1	VA	Ea	Jun-02	D
Conopodium majus	P	SR	G	S	1	2	VA	Sv	May-02	D
Convolvulus arvensis	P	CR	G	L	4	5	VA	Sa	Jun-04	P*
Coronopus squamatus	A/M	R/CSR	Th	B	1	1	?	?	Jun-04	R
Corydalis claviculata	As	SR	Th	L	4	1	?	Sa	Jun-04	P
Corydalis lutea	P	CSR	H	L	2	2	-	Ea	May-04	P
Crepis capillaris	Aw	R/SR	Th	S	2	1	VA	Sh	Jun-02	D*
Cymbalaria muralis	P	CSR	Ch	B	1	4	VA	Ea	May-05	D
Cynosurus cristatus	P	CSR	H	S	1	2	VA	Ea	Jun-03	D*
Dactylis glomerata	P	C/CSR	H	S	3	3	VA	Ea	May-03	DP*
Dactylorhiza fuchsii	P	S/CSR	G	L	3	2	OR	Sa	Jun-02	R*
Dactylorhiza incarnata	P	S/CSR	G	L	3	2	OR	Sa	May-03	R*
Dactylorhiza maculata	P	S/CSR	G	L	3	2	OR	Sa	Jun-02	R*rP*
Danthonia decumbens	P	S	H	S	2	2	VA	Ea	Jul-01	P*
Daucus carota	M	SR/CSR	H	S	2	1	VA	Ea	Jun-02	D*
Deschampsia cespitosa	P	SC/CSR	H	S	3	4	VA	Ea	Jun-03	P*
Deschampsia flexuosa	P	S/SC	H	S	2	4	VA	Ea	Jun-02	P*
Desmazeria rigida	Aws	SR	Th	S	1	1	VA	Sh	May-02	D*
Digitalis purpurea	P/M	CR/CSR	H	S	4	2	+	Ea	Jun-04	R
Eleocharis palustris	P	SC/CSR	Hel	-	3	5	-	Sa	May-03	?
Elodea canadensis	P	CR	Hyd	U	-	5	-	Ep	Jun-04	D*P
Elymus caninus	P	CSR	H	L	4	2	VA	Ea	Jul-01	P
Elymus repens	P	C/CR	H	L	3	5	VA	Ea	Jun-04	P*
Empetrum nigrum ssp. nigrum	P	SC	Ch/Ph	L	2	5	ER	Ea	Apr-02	D*
Epilobium ciliatum	P	CSR	H/Ch	L	4	2	VA	Ea	Jul-02	R*
Epilobium hirsutum	P	C	H	L	5	5	VA	Ep	Jul-02	R*
Epilobium montanum	P	CSR	H/Ch	L	3	2	VA	Ea	Jun-03	R*
Epilobium obscurum	P	CSR	H/Ch	L	3	2	VA	Ea	Jul-02	R*
Epilobium palustre	P	S/CSR	H	L	3	2	VA	Sa	Jul-02	R*
Epilobium parviflorum	P	CSR	H/Ch	L	3	2	VA	Ea	Jul-02	R*
Epipactis helleborine	P	S	G	L	3	2	OR	Sa	Jul-04	P
Erica cinerea	P	S/SC	Ch/Ph	L	3	3	ER	Ea	Jul-02	D
Erica tetralix	P	S/SC	Ch/Ph	L	3	3	ER	Ea	Jul-03	D
Erigeron acer	M/P	SR	Th/H	S	1	1	-	Ep	Jul-02	D*
Eriophorum angustifolium	P	S/SC	Hel	S	3	5	-	Ep	May-02	?
Eriophorum vaginatum	P	SC	Hel	L	4	3	+	Ep	Apr-02	?
Erodium cicutarium	Aws	SR	Th	S	2	1	+	Ep	Jun-04	D*P*
Erophila verna	Aw	SR	Th	B	1	1	-	Ea	Mar-04	D*P*
Eupatorium cannabinum	P	C/CSR	Hel/H	L	5	3	VA	Sa	Jul-03	D*
Euphorbia helioscopia	As	R	Th	L	3	1	+	Sa	May-06	P
Euphorbia peplus	As	R	Th	L	2	1	+	Sa	Apr-08	D
Euphrasia officinalis s.l.	As	SR	Th	L	2	1	Hp	Sa	Jun-04	D*P*
Fallopia convolvulus	As	R/CR	Th	L	3	1	-	Sa	Jul-04	P
Festuca arundinacea	p	CSR	H	S	3	4	VA	Ea	Jun-03	P*
Festuca gigantea	P	CSR	H	S	3	2	-	Ea	Jun-02	P
Festuca ovina agg.	P	S	H	S	2	3	VA	Ea	May-03	D*P*
Festuca pratensis	P	CSR	H/Ch	S	2	2	VA	Ea	Jun-01	D*P

Autecological account	Introduced		Life history	Established strategy	Life form	Canopy structure	Canopy height	Lateral spread	Mycorrhizas	Leaf phenology	Flowering time and duration	Polyploidy
+		Festuca rubra	P	CSR	H	S	2	4	VA	Ea	May-03	P*
+		Filipendula ulmaria	P	C/SC	H	S	5	4	+	Sa	Jun-03	D*
+		Fragaria vesca	P	CSR	H	B	2	4	VA	Ea	Apr-03	D*
		Fumaria muralis	As	R/CR	Th	L	3	1	?	Sa	May-06	P*
		Fumaria officinalis	As	R	Th	L	3	1	-	Sa	May-06	P*
+		Galeopsis tetrahit s.l.	As	R/CR	Th	L	4	1	-	Sa	Jul-03	P*
+		Galium aparine	Aws	CR	Th	L	5	1	+	Sh	Jun-03	P
+		Galium cruciata	P	CSR	H	L	3	3	VA	Ea	May-02	D
		Galium odoratum	P	SC/CSR	G/H	L	3	4	+	Ea	May-02	P
+		Galium palustre gp	P	CR/CSR	H	L	4	4	+	Ep	Jun-02	D*
+		Galium saxatile	P	S	H	L	1	4	+	Ea	Jun-03	DP
+		Galium sterneri	P	S	H	L	1	3	?	Ea	Jun-02	D*
		Galium uliginosum	P	S/CSR	Hel	L	3	4	-	Ep	Jul-02	D*
+		Galium verum	P	SC/CSR	H	L	2	4	VA	Ea	Jul-02	DP
		Gentianella amarella	M	SR	H	S	2	1	VA	Sa	Aug-03	R
		Geranium columbinum	Aw	SR	Th	S	3	1	?	Sh	Jun-02	D*
		Geranium dissectum	Aws	R/SR	Th	L	3	1	VA	Sh	May-04	D*
		Geranium lucidum	Aw	R/SR	Th	S	3	1	+	Sh	May-04	D*
+		Geranium molle	Aws	R/SR	Th	S	2	1	+	Sh	Apr-06	D*
		Geranium pyrenaicum	P	CSR	H	S	3	2	?	Ea	Jun-03	D
+		Geranium robertianum	B-P	R/CSR	Th/H	S	3	1	+	Ea	May-05	R*r
		Geum rivale	P	S/CSR	H	S	3	3	VA	Ea	May-05	P*
+		Geum urbanum	P	S/CSR	H	S	3	2	VA	Ea	Jun-03	P
+		Glechoma hederacea	P	CSR	H	S	2	5	+	Ea	Mar-03	DP*
		Glyceria declinata	P	CR	Hel	L	2	4	?	Ea	Jun-04	D*
+		Glyceria fluitans	P	CR	Wet	L	4	4	-	Ea	May-04	P*
+		Glyceria maxima	P	C	Wet	L	4	5	-	Ep	Jul-02	P*
		Glyceria plicata Fries	P	CR	Wet	L	4	4	+	Ea	May-02	P*
		Gnaphalium uliginosum	As	R	Th	L	2	1	VA	Sa	Jul-02	D
		Gymnadenia conopsea	P	S/SR	G	S	2	2	OR	Sa	Jun-03	R*r
+		Hedera helix	P	SC	Ch/Ph	L	8	5	VA	Ea	Sep-03	R*r
+		Helianthemum nummularium	P	S	Chw	L	2	3	EC	Ea	Jun-02	D*
+		Heracleum sphondylium	P/M	CR	H	LA	3	2	+	Sa	Jun-04	D*
	*	Hesperis matronalis	P/M	CR/CSR	H	L	3	2	?	Ea	May-03	DP
+		Hieracium pilosella	P	S/CSR	H	B	1	4	VA	Ea	May-02	DP*
+		Hieracium subgen. Hieracium	P	S/CSR	H/G	V	V	V	VA	V	May-06	D*P
		Hippuris vulgaris	P	CSR	Hyd	U	-	5	?	Ea	Jun-02	R
+		Holcus lanatus	P	CSR	H	L	3	3	VA	Ep	Jun-02	D*
+		Holcus mollis	P	C/CSR	H	L	2	5	VA	Ea	Jun-02	DP*
		Humulus lupulus	P	C	H	L	6	5	+	Sa	Jul-02	D
		Hyacinthoides non-scripta	P	SR	G	B	2	2	VA	Sv	Apr-03	D*
+		Hydrocotyle vulgaris	P	CSR	H	B	2	5	+	Sa	Jun-03	P
		Hypericum androsaemum	P	S/CSR	Ph	L	4	2	?	Ep	Jun-03	P*
+		Hypericum hirsutum	P	SR/CSR	H	L	4	2	-	Ep	Jul-02	D
		Hypericum humifusum	P	S/CSR	Ch	L	1	3	-		Jun-04	D*
		Hypericum maculatum	P	CR/CSR	H	L	3	3	VA	Ep	Jun-03	D*P
+		Hypericum perforatum	P	CR/CSR	H	L	3	4	VA	Ep	Jun-04	P*
		Hypericum pulchrum	P	S	H	L	3	2	VA	Ea	Jun-03	D*
		Hypericum tetrapterum	P	CSR	H	L	3	2	+	Sa	Jun-04	D*
+		Hypochaeris radicata	P	CSR	H	B	1	2	VA	Ep	Jun-04	D*
+	*	Impatiens glandulifera	As	CR	Th	L	5	1	+	Sa	Jul-04	D*
+		Inula conyza	M/P	S/SR	H	S	3	1	VA	Ea	Jul-03	P*

	Life history	Established strategy	Life form	Canopy structure	Canopy height	Lateral spread	Mycorrhizas	Leaf phenology	Flowering time and duration	Polyploidy
Iris pseudacorus	P	C/SC	G	LA	5	5	-	Ea	May-03	P
Isolepis setacea	P/A	CSR	H	B	2	2	?	Ea	May-03	?
Juncus acutiflorus	P	SC	Hel/H	S	4	5	-	Sa	Jul-03	R*
Juncus articulatus	P	CSR	Hel/H	S	3	4	VA	Ea	Jun-04	rP*
Juncus bufonius	As	R/SR	Th	S	1	1	+	Sa	May-05	rP*
Juncus bulbosus	P	SR/CSR	Wet/H	S	1	5	-	Ea	Jun-04	R
Juncus conglomeratus	P	C/SC	H	-	4	3	?	Ea	May-03	R
Juncus effusus	P	C/SC	H	-	5	4	-	Ea	Jun-03	R*
Juncus inflexus	P	SC	H	-	5	4	VA	Ea	Jun-03	R*
Juncus squarrosus	P	S/SC	H	B	1	3	+	Ea	Jun-02	R
Knautia arvensis	P	CSR	H	S	3	2	VA	?S	Jul-03	DP
Koeleria macrantha	P	S	H	L	1	2	VA	Ea	Jun-02	DP
Lamiastrum galeobdolon	P	S/SC	Ch	L	2	4	+	Ea	May-02	D*P*
Lamium album	P	CR	H	L	3	3	VA	Ea	May-08	D*
Lamium purpureum	Aws	R	Th	S	3	1	VA	Sh	Mar-08	D*
Lapsana communis	Aws	R/CR	Th	S	3	1	VA	Ep	Jul-03	D*
Lathyrus montanus	P	S/CSR	H	L	3	3	VA	Sa	Apr-04	D*
Lathyrus pratensis	P	CSR	H	L	4	3	VA	Sa	May-04	D*P*
Lemna minor	(P)	CR	Hyd	F	-	2	-	Ep	Jun-02	DP
Lemna trisulca	(P)	SR	Hyd	U	-	2	-	Ep	May-03	P*
Leontodon autumnalis	P	R/CSR	H	B	1	2	VA	Ea	Jun-05	D*P
Leontodon hispidus	P	S	H	B	2	3	VA	Sa	Jun-04	D*
Leontodon taraxacoides	P/M	SR/CSR	H	B	2	2	?	Ea	Jun-04	D*
Lepidium campestre	A/M	R/CSR	Th	L	3	1	?	Sh	May-04	D
Lepidium heterophyllum	P	CSR	H	L	2	3	?	Ea	May-04	D
Leucanthemum vulgare	P	C/CSR	H	L	3	2	VA	Ea	Jun-03	D*P*
Linaria vulgaris	P	CR	H	L	3	4	VA	Ep	Jul-04	D
Linum catharticum	B/A	SR	Ch/Th	L	1	1	+	Ea	Jun-04	D
Listera ovata	P	S/CSR	G	L	2	3	OR	Sv	Jun-02	R
Lolium perenne ssp perenne	P	CR/CSR	H	S	2	3	VA	Ea	May-04	D*
Lonicera periclymenum	P	SC	Ph	L	6	5	VA	Ep	Jun-04	D*P*
Lotus corniculatus	P	S/CSR	H	L	2	2	VA	Sa	Jun-04	P*
Lotus uliginosus	P	C/CSR	H	L	3	4	VA	Sa	Jun-03	D*
Luzula campestris	P	S/CSR	H	S	1	3	+	Ea	Mar-04	D*
Luzula multiflora	P	S	H	S	1	3	?	Ea	May-02	P*
Luzula pilosa	P	S	H	S	1	3	-	Ea	Apr-03	P*
Luzula sylvatica	P	SC	H	S	2	4	-	Ea	May-02	DP*
Lychnis flos-cuculi	P	CSR	H	S	2	3	+	Ea	May-02	D*
Lycopus europaeus	P	CR	Hel/H	L	4	3	+	Sa	Jun-04	D*
Lysimachia nemorum	P	CSR	Ch	L	1	4	VA	Ea	May-05	D
Lysimachia nummularia	P	CSR	Ch	L	1	4	VA	Ea	Jun-03	P
Lysimachia vulgaris	P	C/SC	Hel/H	L	5	3	VA	Sa	Jul-02	P
Lythrum portula	As	R/SR	Th	L	1	1	?	Sa	Jun-05	D
Lythrum salicaria	P	C/CSR	Hel/H	L	4	2	+	Sa	Jun-03	P
Malva moschata	P	CSR	H	L	4	2	?	Ea	Jul-02	R
Malva neglecta	P-A	CR	Th/H	L	2	4	?	?E	Jun-04	R
Malva sylvestris	P	CR	H	L	4	3	VA	Ea	Jun-04	R
* Matricaria matricarioides	Asw	R	Th	S	2	1	VA	Sa	Jun-02	D*
Medicago lupulina	A/P	R/SR	Th/H	L	2	1	VA	Ea	May-04	D*
* Medicago sativa	P	C/CSR	H	L	4	2	VA	Ea	Aug-02	D*P*
Melampyrum pratense	As	SR	Th	L	3	1	Hp	Sa	May-06	D
Melica uniflora	P	S/SC	H	L	4	5	-	Sa	May-02	D

Autecological account	Introduced		Life history	Established strategy	Life form	Canopy structure	Canopy height	Lateral spread	Mycorrhizas	Leaf phenology	Flowering time and duration	Polyploidy
+		Mentha aquatica	P	C/CR	Hel/H	L	3	5	VA	Ea	Jul-03	P*
		Mentha arvensis	P	CR	H	L	3	4	VA	Ea	May-06	P*
	*	Mentha spicata	P	C/CR	H	L	3	4	?	Ea	Aug-02	P*
		Mentha x verticillata	P	CR	Hel/H	L	4	4	?	Ea	Jul-04	P*
		Menyanthes trifoliata	P	S/SC	Hel	B	2	4	-	Sa	Jul-02	R
+		Mercurialis perennis	P	SC	H	L	3	5	+	Ep	Feb-04	P
+		Milium effusum	P	S/CSR	H	L	3	2	+	Ea	Jun-01	P*
	*	Mimulus guttatus	P	CR	Hel/H	S	3	3	?	Ea	Jul-03	P*
+		Minuartia verna	P	S	Ch	L	1	2	-	Ea	May-05	D*P
+		Moehringia trinervia	A/P	SR	Th/Ch	L	2	1	-	Sa	May-02	D*
+		Molinia caerulea	P	SC	H	S	4	4	VA	Sa	Jun-03	P*
		Montia fontana	P/A	R/SR	Th/Wet	L	1	3	?	Ea	May-06	D*
+		Mycelis muralis	P	CSR	H	S	3	2	+	Ea	Jul-03	D*
+		Myosotis arvensis	Aw	R/SR	Th	S	2	1	-	Sh	Apr-06	P
		Myosotis discolor	Aw	SR	Th	S	2	1	VA	Sh	May-05	P
		Myosotis laxa ssp caespitosa	As	R/CR	Hel/H	S	2	1	VA	Sa	May-04	D*
+		Myosotis ramosissima	Aw	SR	Th	S	1	1	+	Sh	Apr-03	P
+		Myosotis scorpioides	P	CR	H/Hel	S	2	3	+	Ea	May-05	P
		Myosotis secunda	P	CR	Hel/H	S	2	3	?	Ea	May-04	P*
		Myriophyllum spicatum	P	CSR	Hyd	U	-	5	?	Ea	Jun-02	P
+	*	Myrrhis odorata	P	C/CSR	H	LA	5	3	VA	Sa	May-02	D
+		Nardus stricta	P	S	H	S	3	2	VA	Ea	Jun-03	D*
+		Nasturtium officinale agg.	P	CR	Hel	L	3	5	-	Ea	May-03	R*rP*
		Nuphar lutea	P	C/CSR	Hyd	F	-	5	?	Sa	Jun-03	R
		Odontites verna	As	R	T	L	3	1	Hp	Sa	Jun-03	DP
		Oenanthe crocata	P	CR/CSR	G	S	4	2	-	Ea	Jun-02	D*
		Oenanthe fistulosa	P	CSR	Hel/H	S	3	3	?	Ea	Jul-03	D*
		Ononis repens	P	S/CSR	H/Ch	L	3	4	VA	Sa	Jun-04	P*
		Orchis mascula	P	S/SR	G	S	2	2	OR	Sv	Apr-03	R
		Orchis morio	P	S/SR	G	S	2	2	OR	Sh	May-02	R
+		Origanum vulgare	P	SC/CSR	Ch/H	L	4	3	VA	Ep	Jul-03	R*
+		Oxalis acetosella	P	S/CSR	H/Ch	B	1	2	VA	Ea	Apr-02	D*
		Papaver argemone	As	R	Th	S	2	1	?	Sa	Jun-02	P*
		Papaver dubium	As	R	Th	S	2	1	?	Sa	Jun-02	P*
+		Papaver rhoeas	Asw	R	Th	S	2	1	-	Sa	Jun-03	D*
		Parietaria judaica	P	CSR	H	B	4	3	+	Ea	Jun-05	D
		Pedicularis sylvatica	M	SR	H	S	2	1	Hp	Sa	Apr-04	D
	*	Pentaglottis sempervirens	P	CSR	H	S	4	3	?	Ea	May-02	D*
	*	Petasites fragrans	P	C/CR	G	LA	2	4	?	Ep	Jan-03	R*
+		Petasites hybridus	P	C	G	LA	5	4	+	Sa	Mar-03	R*
+		Phalaris arundinacea	P	C	Hel	L	5	5	VA	Sa	Jun-02	DP*
+		Phleum pratense	P	CSR	H	L	3	3	VA	Ea	Jun-02	DP
+		Phragmites australis	P	C	Wet	L	5	4	VA	Sa	Aug-02	P*
+		Pimpinella saxifraga	P	S/SR	H	S	1	2	VA	Ea	Jul-02	DP
		Plantago coronopus	M+	SR/CSR	H	B	1	1	VA	Ea	May-03	D*P
+		Plantago lanceolata	P	CSR	H	B	2	2	VA	Ea	Apr-05	D*
+		Plantago major	P	R/CSR	H	B	2	2	VA	Ep	Jun-04	D*
		Platanthera chlorantha	P	S/SR	G	L	2	2	OR	Sv	May-03	R
+		Poa annua	A/P	R	Th/H	S	2	1	VA	Ep	Mar-12	P*
		Poa nemoralis	P	S/CSR	H	S	4	2	+	Ea	Jun-03	P*
+		Poa pratensis agg.	P	CSR	H	S	2	3	+	Ea	May-03	P*
+		Poa trivialis	P	CR/CSR	H/Ch	S	1	2	+	Ea	Jun-01	D*P

EXPLANATION OF TABLES 383

	Life history	Established strategy	Life form	Canopy structure	Canopy height	Lateral spread	Mycorrhizas	Leaf phenology	Flowering time and duration	Polyploidy
Polygala serpyllifolia	P	S	Ch	L	1	2	VA	Ea	May-04	R*
Polygala vulgaris	P	S	Ch	L	1	2	VA	Ea	May-03	rP*
Polygonum amphibium	P	CR	Wet	LF	4	5	-	Sa	Jul-03	P
Polygonum aviculare grp	As	R	Th	L	4	1	+	Sa	Jul-05	P*
Polygonum hydropiper	As	R	Th	L	4	1	-	Sa	Jul-03	D*
Polygonum lapathifolium	As	R/CR	Th	L	4	1	-	Sa	Jun-05	D*
Polygonum persicaria	As	R	Th	L	4	1	+	Sa	Jun-05	P*
Potamogeton crispus	P	CR	Hyd	U	-	4	-	Sh	May-03	P
Potamogeton natans	P	C/SC	Hyd	F	-	5	-	Sa	Jun-04	P
Potentilla anglica	P	CSR	H	L	1	5	VA	Ep	Jun-04	P*
Potentilla anserina	P	CR/CSR	H	L	2	5	VA	Sa	Jun-03	P*
Potentilla erecta	P	S/CSR	H	S	2	3	+	Sa	Jun-04	P*
Potentilla palustris	P	S/SC	Hel	L	3	5	+	Sa	May-03	P
Potentilla reptans	P	CR/CSR	H	L	2	5	VA	Ep	Jun-04	P*
Potentilla sterilis	P	S	H	S	1	3	VA	Ea	Feb-04	P*
Primula veris	P	S/CSR	H	B	2	2	VA	Ea	Apr-02	D
Primula vulgaris	P	S/CSR	H	B	2	2	VA	Ea	Dec-06	D*
Prunella vulgaris	P	CSR	H	S	1	3	VA	Ea	Jun-04	R*
Pulicaria dysenterica	P	SC	H	L	3	4	VA	Sa	Aug-02	D*
Ranunculus acris	P	CSR	H	S	2	2	VA	Ea	May-03	D*
Ranunculus auricomus	P	SR/CSR	H/G	S	2	2	VA	Sv	Apr-02	P
Ranunculus bulbosus	P	SR	H	S	1	2	VA	Sh	May-02	D*
Ranunculus ficaria	P	R/SR	G	S	2	2	+	Sv	Mar-03	D*P*
Ranunculus flammula	P	CR/CSR	Hel	S	2	4	+	Ea	May-04	P*
Ranunculus peltatus	P/A	R/CSR	Hyd	F	-	1	?	?S	May-04	DP*
Ranunculus penicillatus	P	?C	Hyd	U	-	5	?	Ea	May-04	P*
Ranunculus repens	P	CR	H	S	2	5	VA	Ea	May-02	P*
Ranunculus sceleratus	Asw	R	Th/Hel	S	3	1	+	Sa	May-05	P
Raphanus raphanistrum	As	R	Th	S	3	1	-	Sa	May-05	D*
Reseda luteola	B-P	R/CSR	H	S	4	1	-	Ea	Jun-03	D
Reynoutria japonica	P	C	G	L	5	5	?	Sa	Aug-03	R*rP*
Rhinanthus minor s.l.	As	R/SR	Th	L	3	1	Hp	Sa	May-04	D*
Ribes uva-crispa	P	SC	Ph	L	4	5	VA	Sa	Jun-03	D
Rorippa palustris	As	R	Th	S	3	1	-	Sa	Jun-04	P*
Rubus caesius	P	SC	Ph	L	4	5	VA	Sa	Jun-04	P*
Rubus fruticosus s.l.	P	SC	Ph	L	6	5	VA	Ep	Jun-04	D*P*
Rubus idaeus	P	SC	Ph	L	5	5	VA	Sa	May-03	DP
Rumex acetosa	P	CSR	H	S	2	2	-	Ea	May-02	D*
Rumex acetosella agg.	P	SR/CSR	H/G	S	1	4	-	Ea	May-03	P*
Rumex conglomeratus	P/M	CR	H	S	4	2	-	Ea	May-02	D
Rumex crispus	P/A	R/CR	H	S	3	2	+	Ea	May-06	P*
Rumex hydrolapathum	P	C/CR	Wet	LA	5	2	?	Ep	Jul-03	P
Rumex obtusifolius	P	CR	H	LA	2	2	-	Ea	Jun-05	P
Rumex sanguineus	P	CSR	H	S	4	2	?	Ea	Jun-03	D*
Sagina apetala	Aw	SR	Th	S	1	1	-	Sh	May-04	D*
Sagina nodosa	P	S/SR	H/Ch	L	1	2	?	Ea	Jul-03	DP*
Sagina procumbens	P/A	R/CSR	H	S	1	2	-	Ea	May-05	D*
Salix repens	P	SC	Ph/Ch	L	5	5	EC	Sa	Apr-02	R*
Sanguisorba minor	P	S	H	S	2	2	VA	Ea	Jun-02	R*
Sanicula europaea	P	S	H	B	2	2	VA	Ea	May-05	D
Saponaria officinalis	P	CR	H	L	4	4	-	Ea	Jul-03	R*
Saxifraga tridactylites	Aw	SR	Th	S	1	1	VA	Sh	Apr-03	D

			Life history	Established strategy	Life form	Canopy structure	Canopy height	Lateral spread	Mycorrhizas	Leaf phenology	Flowering time and duration	Polyploidy
+		Scabiosa columbaria	M/P	S/SR	H	S	1	2	VA	Ea	Jul-02	D*
		Schoenoplectus lacustris	P	SC	Hyd	-	5	4	?	Ea	Jun-02	?
		Scleranthus annuus	As	SR	Th	L	2	1	-	Sa	Jun-03	P*
		Scrophularia auriculata	P	CR	H	L	3	2	VA	Sa	Jun-04	P*
		Scrophularia nodosa	P	CR	H	L	3	2	+	Sa	Jun-04	P
		Scutellaria galericulata	P	CR/CSR	H	L	3	3	VA	Sa	Jun-04	P*
+		Sedum acre	P	S	Ch	L	1	3	-	Ea	Jun-02	P*
		Sedum anglicum	P	S	Ch	L	1	3	?	Ea	Jun-03	DF
		Sedum telephium	P	S	H	L	3	3	-	Sa	Jul-03	DF
		Senecio aquaticus	M/P	R/CR	H	S	3	1	?	Ea	Jul-02	P*
+		Senecio jacobaea	M/P	R/CR	H	S	4	1	VA	Ea	Jun-05	P*
+	*	Senecio squalidus	P/A	R/CSR	H/Th	L	2	2	VA	Ea	Jun-07	D*
		Senecio sylvaticus	As	R	Th	L	4	1	+	Sa	Jul-03	P*
+		Senecio viscosus	As	R	Th	L	3	1	VA	Sa	Jul-04	P*
+		Senecio vulgaris	Asw	R	Th	L	3	1	+	Ep	Apr-12	P*
		Sherardia arvensis	Aw	SR	Th	L	3	1	?	Sh	May-06	D*
		Silene alba	P/A	R/CR	H	L	4	2	-	Ea	May-05	D*
+		Silene dioica	P/M	CSR	H/Ch	L	3	2	-	Ea	May-02	D
		Silene vulgaris	P	CSR	H	L	3	2	-	Sa	Jun-03	D*
		Sinapis arvensis	Asw	R	Th	S	3	1	?	Sa	May-03	D*
	*	Sisymbrium altissimum	As	R/CR	Th	S	4	1	?	Sa	Jun-03	D
		Sisymbrium officinale	A/B	R/CR	Th	L	3	1	+	Ep	Jun-02	D
+		Solanum dulcamara	P	C/CSR	Ch/Ph	L	4	5	VA	Sa	Jun-04	D*
+		Solidago virgaurea	P	S/CSR	H	S	3	2	VA	Sa	Jul-03	D*
		Sonchus arvensis	P	CR	H	L	5	4	+	Sa	Jul-04	P*
+		Sonchus asper	Aws	R/CR	Th	S	5	1	+	Ep	May-06	D*
+		Sonchus oleraceus	Aws	R/CR	Th	S	5	1	VA	Ep	May-06	P*
		Sparganium emersum	P	CR	Wet	LA,F	3	4	?	Sa	Jun-02	R
+		Sparganium erectum	P	C/CR	Wet/G	LA	4	4	VA	Sa	Jul-03	R
+		Spergula arvensis	As	R/SR	Th	L	2	1	-	Sa	Jun-03	D*
		Spergularia rubra	Asw	SR	Th	L	1	1	?	Sa	May-05	P*
		Stachys arvensis	Asw	R	Th	L	2	1	?	Ep	Apr-08	D*
+		Stachys officinalis	P	S	H	S	1	2	VA	Ea	Jun-04	D*
		Stachys palustris	P	CR	G	L	4	4	VA	Sa	Jul-03	D*
+		Stachys sylvatica	P	C/CR	H	L	3	4	VA	Sa	Jul-02	P*
+		Stellaria alsine	P	CR/CSR	Hel	L	1	3	-	Ea	May-02	D*
		Stellaria graminea	P	CSR	H	L	2	4	-	Ea	May-04	D*F
+		Stellaria holostea	P	CSR	Ch	L	3	5	-	Ea	Apr-03	D*
+		Stellaria media	Aws	R	Th	L	2	1	-	Ep	Mar-12	P*
+		Succisa pratensis	P	S	H	S	1	2	VA	Ea	Jul-04	D*
		Symphytum officinale	P	C/CR	H	S	5	3	VA	Sa	May-02	D*F
	*	Symphytum x uplandicum	P	C/CR	H	S	5	3	?	Sa	Jun-03	?
+		Tamus communis	P	C/CR	G	L	5	2	VA	Sa	May-03	R
	*	Tanacetum parthenium	P	CR/CSR	H	L	3	2	+	Ea	Jul-02	D*
		Tanacetum vulgare	P	C/CR	H	L	4	3	-	Ea	Jul-03	D*
+		Taraxacum agg.	P	R/CSR	H	B	3	2	VA	Ea	Mar-08	D*F
		Teucrium scorodonia	P	S/CSR	H	L	3	3	VA	Ea	Jul-03	R*
		Thlaspi arvense	As	R	Th	L	2	1	-	Sa	May-03	D
+		Thymus praecox	P	S	Chw	L	1	4	?	Ea	May-03	rP*
+		Torilis japonica	A/B	SR/CSR	Th	S	4	1	?	Ep	Jul-02	D
		Tragopogon pratensis	P/M	CR/CSR	H/G	S	3	2	VA	Sa	Jun-02	D*
		Trifolium arvense	Aw	SR	Th	L	2	1	?	Sh	Jun-04	D

	Life history	Established strategy	Life form	Canopy structure	Canopy height	Lateral spread	Mycorrhizas	Leaf phenology	Flowering time and duration	Polyploidy
Trifolium campestre	Aw	SR	Th	L	2	1	?	Sh	Jun-04	D
Trifolium dubium	Aws	R/SR	Th	L	2	1	VA	Sh	May-06	P
Trifolium hybridum	P	CSR	H	S	3	2	VA	Ea	Jun-04	D
Trifolium medium	P	SC/CSR	H	L	2	4	VA	?S	Jun-04	P*
Trifolium pratense	P	CSR	H	L	2	3	VA	Ea	May-05	D*
Trifolium repens	P	CR/CSR	H/Ch	B	1	4	VA	Ea	Jun-04	P*
Triglochin palustris	P	SR/CSR	Hel/H	B	3	4	-	Sa	Jun-03	P
Tripleurospermum inodorum	Aws	R	Th	S	3	1	VA	Ep	Jul-04	D*P*
Trisetum flavescens	P	CSR	H	L	2	3	VA	Ea	May-02	DP
Tussilago farfara	P	CR	G	LA	2	5	VA	Sa	Feb-03	R*
Typha latifolia	P	C	Hyd	LA	5	5	?	Sa	Jun-02	R
Umbilicus rupestris	P	S	H	S	1	2	?	Ea	Jun-03	R
Urtica dioica	P	C	Ch/H	L	5	4	-	Ep	Jun-02	P*
Urtica urens	As	R/CR	Th	L	3	1	-	Sa	Jun-04	D*P
Vaccinium myrtillus	P	SC	Ch	L	3	5	ER	Sa	Apr-03	D*
Vaccinium vitis-idaea	P	S/SC	Ch/Ph	L	2	4	ER	Ea	Jun-03	D*
Valeriana officinalis	P	CSR	H	S	3	2	+	Sa	Jun-03	DP*
Valerianella locusta	Aw	SR	Th	L	3	1	?	Sh	Apr-03	D
Verbascum thapsus	M	R/CSR	H	S	5	1	VA	Ea	Jun-03	R
Veronica agrestis	Asw	R	Th	L	1	1	+	Ep	Apr-12	P*
Veronica anagallis-aquatica	P/M	R/CSR	Hel/H	L	2	2	VA	Ea	Jun-03	P
Veronica arvensis	Aws	SR	Th	S	1	1	VA	Sh	Apr-03	D
Veronica beccabunga	P	CR	Hel/H	L	2	4	-	Ea	May-03	D*P
Veronica chamaedrys	P	CSR	Ch	L	1	3	+	Ea	Apr-04	DP
Veronica filiformis	P	R/CR	Ch	L	1	4	VA	Ea	Apr-03	D
Veronica hederifolia	As	R/SR	Th	L	2	1	?	Sv	Mar-03	DP
Veronica montana	P	S/CSR	Ch	L	1	4	?	Ea	Apr-04	D
Veronica officinalis	P	S/CSR	Ch	L	1	3	+	Ea	May-04	Dp
Veronica persica	Aws	R	Th	L	1	1	VA	Ep	Apr-12	P
Veronica polita	A	R	Th	L	1	1	?	Ep	Apr-12	D
Veronica scutellata	P	CR/CSR	Hel/H	L	3	2	?	Ea	Jun-03	D
Veronica serpyllifolia	P	R/CSR	Ch	L	1	3	VA	Ea	Mar-08	D
Vicia cracca	P	C/CSR	H	L	5	4	VA	Sa	Jun-03	DP*
Vicia hirsuta	Aw	R/CSR	Th	L	2	1	VA	Sh	Jun-03	D
Vicia sativa ssp nigra	Aw	R/CSR	Th	L	5	1	VA	Sh	May-05	D*
Vicia sepium	P	C/CSR	H	L	4	4	VA	Sa	May-04	D
Viola arvensis	As	R	Th	S	3	1	+	Ep	Apr-07	P*
Viola hirta	P	S	H	B	3	2	VA	Ea	Apr-02	D*
Viola odorata	P	CSR	H	B	2	4	VA	Ea	Feb-03	D*
Viola palustris	P	S/CSR	H	B	2	4	VA	Sa	May-04	D
Viola riviniana	P	S	H	S	2	2	VA	Ea	Apr-03	P*
Viola tricolor	A/P	R/SR	Th/H	S	2	1	VA	Ep	Apr-06	D*
Vulpia bromoides	Aw	SR	Th	S	2	1	?	Sh	May-03	D*

(b) Woody species more than 1.5 m tall

	Life history	Established strategy	Life form	Canopy structure	Canopy height	Lateral spread	Mycorrhizas	Leaf phenology	Flowering time and duration	Polyploidy
Acer campestre (canopy)	P	SC	Ph	L	7	5	VA	Sa	May-02	D
Acer pseudoplatanus (canopy)	P	C/SC	Ph	L	8	5	VA	Sa	Apr-03	P
ditto juvenile										
Alnus glutinosa (canopy)	P	SC	Ph	L	8	5	EC	Sa	Feb-03	R
ditto juvenile										
Betula spp. (canopy)	P	C/SC	Ph/Ch	L	8	5	EC	Sa	Apr-02	R*rP*
ditto juvenile										

Autecological account	Introduced		Life history	Established strategy	Life form	Canopy structure	Canopy height	Lateral spread	Mycorrhizas	Leaf phenology	Flowering time and duration
		Corylus avellana (canopy)	P	SC	Ph	L	6	5	EC	Sa	Jan-04
	+	Crataegus monogyna (canopy)	P	SC	Ph	L	7	5	EC	Sa	May-02
		ditto juvenile									
		Cytisus scoparius (canopy / juv)	P	SC	Ph	L	5	5	VA	Ep	May-02
		Euonymus europaeus (canopy)	P	SC	Ph	L	6	5	VA	Sa	May-02
	+	Fagus sylvatica (canopy)	P	SC	Ph	L	8	5	EC	Sa	Apr-02
		ditto juvenile									
	+	Fraxinus excelsior (canopy)	P	C	Ph	L	8	5	VA	Sa	Apr-02
		ditto juvenile									
		Ilex aquifolium (canopy)	P	SC	Ph	L	7	5	VA	Ea	May-04
		Ligustrum vulgare (canopy)	P	SC	Ph	L	6	5	VA	Ep	Jun-02
		Malus sylvestris (canopy)	P	SC	Ph	L	6	5	VA	Sa	May-01
		Populus tremula (canopy)	P	SC	Ph	L	8	5	EC	Sa	Feb-02
		Prunus avium (canopy)	P	SC	Ph	L	8	5	VA	Sa	Apr-02
		Prunus spinosa (canopy)	P	SC	Ph	L	6	5	+		Mar-03
	+	Quercus agg. (canopy)	P	SC	Ph	L	8	5	EC	Sa	Apr-02
		ditto juvenile									
*	*	Rhododendron ponticum (cnpy)	P	SC	Ph	L	5	5	ER	Ea	May-02
		Rosa spp. (canopy)	P	SC	Ph	L	5	5	VA	Sa	May-03
		ditto juvenile									
	+	Salix cinerea agg. (canopy)	P	C	Ph	L	7	5	EC	Sa	Mar-02
		ditto juvenile									
	*	Salix fragilis agg. (canopy)	P	C	Ph	L	8	5	EC	Sa	Apr-02
		Salix purpurea (canopy)	P	C/SC	Ph	L	5	5	EC	Sa	Mar-02
	*	Salix viminalis (canopy)	P	C/SC	Ph	L	6	5	EC	Sa	Apr-02
	+	Sambucus nigra (canopy)	P	C	Ph	L	7	5	+	Sa	Jun-02
		ditto juvenile									
	+	Sorbus aucuparia (canopy)	P	SC	Ph	L	7	5	VA	Sa	May-02
		ditto juvenile									
	+	Ulex europaeus (canopy / juv)	P	SC	Ph	-	5	5	VA	Ea	Mar-03
	+	Ulmus glabra (canopy)	P	C	Ph	L	8	5	VA	Sa	Feb-03
		ditto juvenile									
		Ulmus procera (canopy)	P	C	Ph	L	8	5	VA	Sa	Feb-02
		Viburnum opulus (canopy)	P	SC	Ph	L	6	5	VA	Sa	Jun-02
	(c)	Pteridophytes									
		Asplenium adiantum-nigrum	P	S	H	B	3	2	-	Ea	Jun-05
	+	Asplenium ruta-muraria	P	S/SR	H	B	1	2	-	Ea	Jun-03
	+	Asplenium trichomanes	P	S	H	B	2	2	-	Ea	Aug-03
	+	Athyrium filix-femina	P	C/SC	H	LA	5	3	+	Sa	Jul-02
		Blechnum spicant	P	S	H	B	4	2	VA	Ea	Jul-04
	+	Cystopteris fragilis	P	S/SR	H	B	2	2	+	Sa	Jun-03
		Dryopteris affinis	P	SC/CSR	H	LA	5	2	?	Ep	Jul-03
	+	Dryopteris dilatata	P	SC/CSR	H	LA	5	2	VA	Ep	Jun-04
	+	Dryopteris filix-mas	P	SC/CSR	H	LA	5	2	VA	Ep	Aug-04
	+	Equisetum arvense	P	CR	G	-	4	5	-	Sa	Apr-01
	+	Equisetum fluviatile	P	SC	Wet	-	5	5	-	Sa	Jun-02
	+	Equisetum palustre	P	CR/CSR	G	-	3	5	-	Sa	Jun-02
		Phyllitis scolopendrium	P	S	H	LA	3	2	+	Ea	Aug-03
		Polypodium vulgare grp	P	S	G/Ch	B	3	3	-	Ea	Aug-09
	+	Pteridium aquilinum	P	C	G	LA	5	5	VA	Sa	Aug-03

	Regenerative strategies	Seed bank	Agency of dispersal	Dispersule/germinule form	Dispersule weight	Dispersule shape	Germination requirements	Family
Herbs and woody species up to 1.5 m tall								
Achillea millefolium	V,?S	1	WINDw	Fr	1	2	-	Com
Achillea ptarmica	V,	1	WINDw	Fr	2	2	-	Com
Aegopodium podagraria	V,S	1	*UNSP	Fr	4	3	Chill	Umb
Aethusa cynapium	S,Bs	3	UNSPag	Fr	4	1	Chill	Umb
Agrimonia eupatoria	S	2	ANIMb	Fr	5	2	Chill	Ros
Agrostis canina	V,Bs	3	*ANIMa	Fr	1	3	Dry	Gra
Agrostis capillaris	V,Bs	3	UNSP	Fr	1	2	-	Gra
Agrostis gigantea	V,Bs	3	*UNSP	Fr	1	2	-	Gra
Agrostis stolonifera	V,Bs	3	*UNSP	Fr	1	2	-	Gra
Agrostis vinealis	V,Bs	3	UNSP	Fr	1	3	-	Gra
Aira caryophyllea	S	?	ANIMa	Fr	1	2	Dry	Gra
Aira praecox	S	2	ANIMa	Fr	1	3	Dry	Gra
Ajuga reptans	V	2	*ANIMe	Fr	4	2	Unclassified	Lab
Alchemilla vulgaris agg.	?	?	ANIMa	Fr	2	1	Chill	Ros
Alisma plantago-aquatica	V,Bs	3	AN/AQ	Fr	2	1	Scar	Ali
Alliaria petiolata	S	2	UNSP	Sd	5	3	Chill	Cru
Allium ursinum	V,S	1	UNSP	Sd	5	1	Warm+Chill	Lil
Allium vineale	Sv	1	* -	Bul	X	-	X;Dry(bulbil)	Lil
Alopecurus geniculatus	V,?S	2	*UNSP	Fr	2	2	Dry	Gra
Alopecurus pratensis	?V,S	1	UNSPag	Fr	3	2	-	Gra
Anacamptis pyramidalis	W	?	WINDcm	Sd	3	3	Orchid	Orc
Anagallis arvensis	Bs	3	UNSPag	Sd	2	1	Chill	Pri
Anchusa arvensis	Bs	3	ANIMe	Fr	5	2	Unclassified	Bor
Anemone nemorosa	V,S	1	ANIMa	Fr	3	3	?Chill	Ran
Angelica sylvestris	S	1	AQUAT	Fr	4	1	Chill	Umb
Anthoxanthum odoratum	S	2	ANIMa	Fr	2	2	-	Gra
Anthriscus sylvestris	Sv,S	1	UNSP	Fr	5	2	Chill	Umb
Anthyllis vulneraria	?	1	?WIND	F/S	5	1	Scar	Leg
Aphanes arvensis	S,Bs	3	UNSPag	Fr	1	2	Dry	Ros
Apium nodiflorum	(V),?S	?	*AQUAT	Fr	2	2	-	Umb
Arabidopsis thaliana	S,Bs	3	WINDw	Sd	1	1	Dry	Cru
Arabis hirsuta	S,Bs	3	UNSP	Sd	1	2	-	Cru
Arctium lappa	S,Bs	3	ANIMb	Fr	6	2	-	Com
Arctium minus agg.	S,Bs	3	ANIMb	Fr	6	2	-/Scar	Com
Arenaria serpyllifolia	S,Bs	3	WINDc	Sd	1	1	-	Car
Armoracia rusticana	V	-	* -	X	X	-	X	Cru
Arrhenatherum elatius	(V),S	1	ANIMa	Fr	5	2	-	Gra
Artemisia absinthium	S,Bs	3	UNSP	Fr	1	2	Dry	Com
Artemisia vulgaris	V,S,Bs	3	UNSP	Fr	1	3	Dry	Com
Arum maculatum	V,?S	1	ANIMi	F/S	6	1	Chill	Ara
Atriplex patula	S,Bs	3	UNSP	f/S	4	1	Chill	Che
Atriplex prostrata	Bs	3	UNSP	f/S	3	1	Chill	Che
Avenula pratensis	S	1	ANIMa	Fr	5	3	-	Gra
Avenula pubescens	(V),S	1	ANIMa	Fr	4	3	-	Gra
Barbarea vulgaris	Bs	3	WINDw	Sd	1	1	Scar	Cru
Bellis perennis	V	2	UNSP	Fr	1	2	-	Com
Berula erecta	(V),?S	1	*AQUAT	Fr	2	1	-	Umb
Brachypodium pinnatum	V,?S	1	UNSP	Fr	5	3	-	Gra
Brachypodium sylvaticum	V,S	1	ANIMa	Fr	3	3	Dry	Gra
Brassica rapa	Bs	3	UNSP	Sd	4	1	-	Cru
Briza media	V,S	1	UNSP	Fr	2	2	-	Gra

Autecological account	Introduced		Regenerative strategies	Seed bank	Agency of dispersal	Dispersule/germinule form	Dispersule weight	Dispersule shape	Germination requirements
+		Bromus erectus	V,S	1	ANIMa	Fr	5	2	-
+		Bromus hordeaceus	S	1	ANIMa	Fr	5	3	-
+		Bromus ramosus	S	1	ANIMa	Fr	5	3	Chill
+		Bromus sterilis	S	1	ANIMa	Fr	5	3	-
+		Callitriche stagnalis	V	2	*UNSP	f/S	1	3	-
+		Calluna vulgaris	Bs	3	WINDcm	Sd	1	1	-
+		Caltha palustris	V,?S	1	*AQUAT	Sd	3	2	Chill
+		Calystegia sepium	(V),Bs	3	*UNSP	Sd	6	1	Scar
+		Campanula rotundifolia	V,Bs	3	WINDc	Sd	1	2	-
+		Capsella bursa-pastoris	Bs	3	WINDw	Sd	1	2	Chill,Scar
+		Cardamine amara	V,Bs	3	*UNSP	Sd	2	1	Dry
+		Cardamine flexuosa	S,Bs	3	WINDw	Sd	1	2	-
+		Cardamine hirsuta	S,Bs	3	WINDw	Sd	1	1	Dry
+		Cardamine pratensis	(V),Bs	3	*WINDw	Sd	3	2	-
	*	Cardaria draba	(V)	1	*UNSP	Sd	4	1	Dry
		Carduus acanthoides	W	?	WINDp	Fr	5	2	-
+		Carex acutiformis	V,?	?	AQUAT	Fr	4	2	?Dry
		Carex binervis	V	?2	UNSP	Fr	4	2	?Chill
+		Carex caryophyllea	V,?	1	UNSP	Fr	4	2	Chill
		Carex demissa	?	?	AQUAT	Fr	3	1	?Chill
		Carex echinata	?	2	AQUAT	Fr	3	1	Dry
+		Carex flacca	V,Bs	3	UNSP	Fr	2	1	Chill
		Carex hirta	V,?	?	UNSP	Fr	5	2	?Chill
		Carex hostiana	?	?	UNSP	Fr	4	1	?Chill
+		Carex nigra	V	2	UNSP	Fr	3	2	Chill
		Carex otrubae	?	2	AQUAT	Fr	4	1	Dry
		Carex ovalis	Bs	3	UNSP	Fr	2	1	Dry
		Carex pallescens	Bs	3	UNSP	Fr	4	1	?Chill
+		Carex panicea	V,?S	?	AQUAT	Fr	4	1	Chill
		Carex pendula	?	2	UNSP	Fr	3	2	?Chill
+		Carex pilulifera	V, Bs	3	ANIMe	Fr	4	1	?Chill
		Carex remota	Bs	3	AQUAT	Fr	2	1	Chill
		Carex sylvatica	Bs	3	UNSP	Fr	4	2	?Chill
+		Carlina vulgaris	S	2	WINDp	Fr	4	3	-
+		Centaurea nigra	V,S	2	UNSP	Fr	5	2	Dry
+		Centaurea scabiosa	S	2	UNSP	Fr	5	2	-
+		Centaurium erythraea	S, Bs	3	WINDc	Sd	1	1	-
	*	Centranthus ruber	W	?	WINDp	Fr	4	3	-
+		Cerastium fontanum	(V), Bs	3	UNSPc	Sd	1	1	-
		Cerastium glomeratum	S, Bs	3	WINDc	Sd	1	1	Dry
		Cerastium semidecandrum	S, Bs	3	WINDc	Sd	1	1	Dry
	*	Cerastium tomentosum	V,?	?	*WINDc	Sd	2	1	-
+		Chaenorhinum minus	S, Bs	3	WINDc	Sd	1	2	Chill
		Chaerophyllum temulentum	S	1	UNSP	Fr	5	3	Chill
+		Chamerion angustifolium	V,W	1	WINDp	Sd	1	3	Dry
	*	Cheiranthus cheiri	?	?	WINDw	Sd	4	1	-
		Chelidonium majus	?S,Bs	3	ANIMe	Sd	3	1	Chill
+		Chenopodium album	Bs	3	UNSP	f/S	4	1	-/Chill,Dry
	*	Chenopodium bonus-henricus	Bs	3	UNSP	f/S	4	1	-
+		Chenopodium rubrum	Bs	3	UNSP	f/S	1	1	Fluct
	*	Chrysanthemum segetum	Bs	3	UNSPag	Fr	4	1	Scar
+		Chrysosplenium oppositifolium	(V)	1	*UNSP	Sd	1	1	-

EXPLANATION OF TABLES   389

	Regenerative strategies	Seed bank	Agency of dispersal	Dispersule/germinule form	Dispersule weight	Dispersule shape	Germination requirements	Family
Circaea lutetiana	(V),S	1	*ANIMb	F/S	3	3	?Chill	Ona
Cirsium arvense	V,W, Bs	3	*WINDp	Fr	4	2	-/Unclassified	Com
Cirsium palustre	W	2	WINDp	Fr	4	2	-	Com
Cirsium vulgare	W	2	WINDp	Fr	5	2	Dry	Com
Clinopodium vulgare	V,?Bs	?3	UNSP	Fr	2	1	Dry	Lab
Conium maculatum	S	2	?AQUAT	Fr	5	1	?Dry	Umb
Conopodium majus	S	2	UNSP	Fr	5	3	Chill	Umb
Convolvulus arvensis	(V), Bs	3	*UNSP	Sd	6	2	Scar	Con
Coronopus squamatus	Bs	3	UNSP	F/S	3	2	Unclassified	Cru
Corydalis claviculata	S	?	ANIMe	Sd	4	1	-	Fum
Corydalis lutea	?S	?	ANIMe	Sd	4	1	Unclassified	Fum
Crepis capillaris	W, Bs	3	WINDp	Fr	2	3	-	Com
Cymbalaria muralis	?	?	UNSPc	Sd	1	2	?Chill	Scr
Cynosurus cristatus	S	1	UNSPag	Fr	3	3	-	Gra
Dactylis glomerata	S	1	UNSPag	Fr	3	2	Dry	Gra
Dactylorhiza fuchsii	W	?	WINDcm	Sd	S	3	Orchid	Orc
Dactylorhiza incarnata	W	?	WINDcm	Sd	S	3	Orchid	Orc
Dactylorhiza maculata	W	?	WINDcm	Sd	S	3	Orchid	Orc
Danthonia decumbens	Bs	3	ANIMe	Fr	3	2	Dry	Gra
Daucus carota	?S, Bs	3	ANIMb	Fr	3	1	Chill	Umb
Deschampsia cespitosa	V,S	2	ANIMa	Fr	2	3	-	Gra
Deschampsia flexuosa	V,S	1	ANIMa	Fr	2	2	-	Gra
Desmazeria rigida	S	1	UNSP	Fr	1	3	Dry	Gra
Digitalis purpurea	V,S, Bs	3	WINDc	Sd	1	2	-	Scr
Eleocharis palustris	V	2	*ANIMb	Fr	3	2	Chill	Cyp
Elodea canadensis	V	-	*-	X	X	-	X	Hyd
Elymus caninus	S	?	ANIMa	Fr	5	3	-	Gra
Elymus repens	(V)	2	*UNSP	Fr	5	3	Fluct	Gra
Empetrum nigrum ssp. nigrum	V,S	1	ANIMi	F/S	3	1	Chill	Emp
Epilobium ciliatum	(V),W	2	WINDp	Sd	1	3	-	Ona
Epilobium hirsutum	V,W, Bs	3	*WINDp	Sd	1	2	-	Ona
Epilobium montanum	(V),W, Bs	3	WINDp	Sd	1	2	-	Ona
Epilobium obscurum	(V),W, Bs	3	WINDp	Sd	1	2	-	Ona
Epilobium palustre	(V),W,?Bs	?3	WINDp	Sd	1	3	-	Ona
Epilobium parviflorum	(V),W, Bs	3	WINDp	Sd	1	2	-	Ona
Epipactis helleborine	W	?	WINDcm	Sd	S	3	Orchid	Orc
Erica cinerea	Bs	3	WINDcm	Sd	1	2	Chill,Heat,Scar	Eri
Erica tetralix	Bs	3	WINDcm	Sd	1	2	?Dry	Eri
Erigeron acer	W	?1	WINDp	Fr	1	3	-	Com
Eriophorum angustifolium	V,W	1	WINDp	Fr	2	3	-	Cyp
Eriophorum vaginatum	W, Bs	3	WINDp	Fr	4	2	-	Cyp
Erodium cicutarium	S,?Bs	?3	ANIMa	Sd	4	3	Scar	Ger
Erophila verna	S	2	WINDw	Sd	1	1	Dry	Cru
Eupatorium cannabinum	W, Bs	?3	WINDp	Fr	2	3	Unclassified	Com
Euphorbia helioscopia	Bs	3	ANIMe	Sd	5	1	Unclassified	Eup
Euphorbia peplus	Bs	3	ANIMe	Sd	2	2	Unclassified	Eup
Euphrasia officinalis s.l.	S	2	WINDc	Sd	1	2	Chill	Scr
Fallopia convolvulus	Bs	3	UNSPag	Fr	4	2	Chill	Pgo
Festuca arundinacea	V,S	1	ANIMa	Fr	4	3	-	Gra
Festuca gigantea	S	1	ANIMa	Fr	5	3	Chill	Gra
Festuca ovina agg.	V,S	1	ANIMa	Fr	2	3	-	Gra
Festuca pratensis	V,S	1	ANIMa	Fr	4	3	-	Gra

Autecological account	Introduced		Regenerative strategies	Seed bank	Agency of dispersal	Dispersule/germinule form	Dispersule weight	Dispersule shape	Germination requirements	Family
+		Festuca rubra	V,S	1	ANIMa	Fr	3	3	-	Gr
+		Filipendula ulmaria	V	1	AQUAT	Fr	3	2	-	Ro
+		Fragaria vesca	V, Bs	3	ANIMi	Fr	2	1	-	Ro
		Fumaria muralis	?	?	UNSPag	Sd	-	1	?Chill	Fu
		Fumaria officinalis	Bs	3	ANIMe	Sd	5	2	?Chill	Fu
+		Galeopsis tetrahit s.l.	Bs	3	UNSPag	Fr	5	1	Chill	La
+		Galium aparine	S	1	ANIMb	Fr	5	1	Chill	Ru
+		Galium cruciata	(V),S	1	UNSP	Fr	5	1	Chill	Ru
		Galium odoratum	V,S	1	ANIMb	Fr	5	1	?Chill	Ru
+		Galium palustre gp	V, Bs	3	AQUAT	Fr	3	1	-	Ru
+		Galium saxatile	V, Bs	3	UNSP	Fr	3	1	-	Ru
+		Galium sterneri	V	?	UNSP	Fr	2	1	-	Ru
		Galium uliginosum	V,?	?1	UNSP	Fr	2	1	?	Ru
+		Galium verum	V	1	UNSP	Fr	2	1	-	Ru
		Gentianella amarella	?	?1	WINDc	Sd	1	1	?	Ge
		Geranium columbinum	S	?	ANIMa	f/S	5	1	Scar	Ge
		Geranium dissectum	S	2	ANIMa	f/S	5	1	Scar	Ge
		Geranium lucidum	S	?	ANIMa	f/S	1	2	Scar	Ge
+		Geranium molle	S	2	ANIMa	f/S	4	1	Scar	Ge
		Geranium pyrenaicum	?	2	ANIMa	f/S	4	2	Scar	Ge
+		Geranium robertianum	?	2	ANIMa	f/S	4	1	Scar	Ge
		Geum rivale	S	1	ANIMa	Fr	4	3	-	Ro
+		Geum urbanum	S	2	ANIMa	Fr	3	3	-	Ro
+		Glechoma hederacea	V	2	*UNSP	Fr	3	2	Dry	La
		Glyceria declinata	V	?	*UNSP	Fr	4	2	Dry	Gr
+		Glyceria fluitans	V	2	*UNSP	Fr	4	2	Dry	Gr
+		Glyceria maxima	V	2	*AQUAT	Fr	3	2	Dry	Gr
		Glyceria plicata Fries	V	2	*UNSP	Fr	3	2	-	Gr
		Gnaphalium uliginosum	Bs	3	WINDp	Fr	1	2	-	Co
		Gymnadenia conopsea	W	?	WINDcm	Sd	S	3	Orchid	Or
+		Hedera helix	V	1	ANIMi	F/S	6	2	?	Ar
+		Helianthemum nummularium	?	2	ANIMm	Sd	4	2	Scar	Cis
+		Heracleum sphondylium	S	1	WINDw	Fr	5	1	Chill	Un
	*	Hesperis matronalis	?	?	UNSP	Sd	4	3	?	Cr
+		Hieracium pilosella	V,W	2	WINDp	Fr	1	3	Dry	Co
+		Hieracium subgen. Hieracium	v,W	?1	WINDp	Fr	2	3	-	Co
		Hippuris vulgaris	(V),?	?	*AQUAT	Fr	3	2	-/Scar	Hip
+		Holcus lanatus	V,S Bs	3	UNSP	Fr	2	2	-	Gr
+		Holcus mollis	V	1	*ANIMa	Fr	2	3	?	Gr
		Humulus lupulus	?	?	WINDw	Fr	5	1	Unclassified	Ca
+		Hyacinthoides non-scripta	V,S	1	UNSP	Sd	5	1	Warm+Chill	Lil
+		Hydrocotyle vulgaris	V	2	*AQUAT	Fr	2	2	Unclassified	Un
		Hypericum androsaemum	?	?	ANIMi	F/S	1	-	?	Hy
+		Hypericum hirsutum	V, Bs	3	WINDc	Sd	1	3	Dry,Wash	Hy
		Hypericum humifusum	Bs	3	UNSPc	Sd	1	2	-	Hy
		Hypericum maculatum	Bs	3	WINDc	Sd	1	2	Wash	Hy
+		Hypericum perforatum	V, Bs	3	WINDc	Sd	1	2	Wash	Hy
		Hypericum pulchrum	Bs	3	WINDc	Sd	1	2	Dry	Hy
		Hypericum tetrapterum	Bs	3	WINDc	Sd	1	2	Dry	Hy
+		Hypochaeris radicata	V,W	2	WINDp	Fr	3	3	-	Co
+	*	Impatiens glandulifera	S	1	?AQUAT	Sd	5	1	Chill	Ba
+		Inula conyza	W, Bs	3	WINDp	Fr	2	3	-	Co

	Regenerative strategies	Seed bank	Agency of dispersal	Dispersule/germinule form	Dispersule weight	Dispersule shape	Germination requirements	Family
Iris pseudacorus	V	1	*AQUAT	Sd	5	2	Scar	Iri
Isolepis setacea	Bs	3	UNSP	Fr	1	1	-	Cyp
Juncus acutiflorus	V, Bs	3	ANIMm	Sd	1	2	-	Jce
Juncus articulatus	V, Bs	3	ANIMm	Sd	1	2	-	Jce
Juncus bufonius	Bs	3	ANIMm	Sd	1	1	Dry	Jce
Juncus bulbosus	V, Bs	3	ANIMm	Sd	1	3	-	Jce
Juncus conglomeratus	V, Bs	3	ANIMm	Sd	1	2	Dry	Jce
Juncus effusus	V, Bs	3	ANIMm	Sd	1	2	-	Jce
Juncus inflexus	V, Bs	3	ANIMm	Sd	1	2	-	Jce
Juncus squarrosus	V, Bs	3	ANIMm	Sd	1	2	-	Jce
Knautia arvensis	S	1	ANIMa	Fr	5	2	Unclassified	Dip
Koeleria macrantha	S	1	UNSP	Fr	2	3	-	Gra
Lamiastrum galeobdolon	V,S	1	ANIMe	Fr	4	2	Chill	Lab
Lamium album	V, Bs	3	*ANIMe	Fr	5	2	?	Lab
Lamium purpureum	Bs	3	ANIMe	Fr	3	2	Dry	Lab
Lapsana communis	S, Bs	3	UNSPag	Fr	4	3	Dry	Com
Lathyrus montanus	V	1	UNSP	Sd	6	1	Scar	Leg
Lathyrus pratensis	V	1	UNSP	Sd	6	1	Scar	Leg
Lemna minor	V	-	* -	X	X	-	X	Lem
Lemna trisulca	V	-	* -	X	X	-	X	Lem
Leontodon autumnalis	W	2	WINDp	Fr	3	3	Dry	Com
Leontodon hispidus	V,W	2	WINDp	Fr	3	3	-	Com
Leontodon taraxacoides	W	2	WINDp	Fr	2	3	-	Com
Lepidium campestre	Bs	3	ANIMm	Sd	5	2	Dry	Cru
Lepidium heterophyllum	?	?	ANIMm	Sd	4	2	?	Cru
Leucanthemum vulgare	V,S, Bs	3	UNSPag	Fr	2	3	-	Com
Linaria vulgaris	(V), Bs	3	WINDcw	Fr	1	1	Dry	Scr
Linum catharticum	S, Bs	3	UNSP	Sd	1	2	Chill	Lin
Listera ovata	V,W	?	WINDcm	Sd	S	3	Orchid	Orc
Lolium perenne ssp perenne	S	1	UNSPag	Fr	4	3	-	Gra
Lonicera periclymenum	V,S	1	ANIMi	F/S	5	1	Chill	Cap
Lotus corniculatus	Bs	3	UNSP	Sd	4	1	Scar	Leg
Lotus uliginosus	V,?Bs	?3	UNSP	Sd	2	1	Scar	Leg
Luzula campestris	V, Bs	3	ANIMe	Sd	3	2	-	Jce
Luzula multiflora	V, Bs	3	ANIMe	Sd	3	2	Dry	Jce
Luzula pilosa	V,S, Bs	3	ANIMe	Sd	3	1	-	Jce
Luzula sylvatica	V	?1	ANIMe	Sd	3	2	Dry	Jce
Lychnis flos-cuculi	V, Bs	3	WINDc	Sd	2	1	-	Car
Lycopus europaeus	V	1	AQ/AN	Fr	2	1	Fluct	Lab
Lysimachia nemorum	V	2	*UNSPc	Sd	2	1	?Chill	Pri
Lysimachia nummularia	V	1	*UNSPc	Sd	-	-	?	Pri
Lysimachia vulgaris	V	2	AQUATc	Sd	2	2	Chill	Pri
Lythrum portula	Bs	3	UNSPc	Sd	1	2	-	Lyt
Lythrum salicaria	Bs	3	AQUAT	Sd	1	2	-	Lyt
Malva moschata	Bs	3	UNSP	Fr	5	1	Scar	Mal
Malva neglecta	Bs	3	UNSP	Fr	4	1	Scar	Mal
Malva sylvestris	Bs	3	UNSP	Fr	5	1	Scar	Mal
Matricaria matricarioides	Bs	3	UNSP	Fr	1	2	Dry	Com
Medicago lupulina	Bs	3	UNSP	f/S	5	1	Scar	Leg
Medicago sativa	Bs	3	UNSPag	f/S	5	1	Scar	Leg
Melampyrum pratense	S	1	ANIMe	Sd	5	3	Chill	Scr
Melica uniflora	V,S	1	ANIMe	Fr	5	-	Chill	Gra

Autecological account	Introduced		Regenerative strategies	Seed bank	Agency of dispersal	Dispersule/germinule form	Dispersule weight	Dispersule shape	Germination requirements	Family
+		Mentha aquatica	V, Bs	3	*AQUAT	Fr	1	2	Chill,Dry	La
		Mentha arvensis	V, Bs	3	*AQUAT	Fr	2	1	-	La
	*	Mentha spicata	V,?	?	*UNSP	Fr	-	-	?	La
		Mentha x verticillata	V	-	*UNSP	Fr	2	-	X(part sterile)	La
		Menyanthes trifoliata	V	1	*AQUAT	Sd	5	1	Scar	M
+		Mercurialis perennis	V,S	1	ANIMe	Sd	5	1	Chill	Eu
+		Milium effusum	V, Bs	3	UNSP	Fr	4	2	-	Gr
	*	Mimulus guttatus	V	?	*AQUAT	Sd	1	1	-	Sc
+		Minuartia verna	V, Bs	3	WINDc	Sd	1	1	-	Ca
+		Moehringia trinervia	?	2	UNSPc	Sd	2	1	Dry	Ca
+		Molinia caerulea	V	2	UNSP	Fr	3	2	Chill	Gr
		Montia fontana	V	2	UNSP	Sd	2	1	Scar	Po
+		Mycelis muralis	W	2	ANIMp	Fr	2	3	-	Co
+		Myosotis arvensis	S, Bs	3	ANIMa	Fr	2	2	Dry	Bo
		Myosotis discolor	S	?	ANIMa	Fr	1	1	Dry	Bo
		Myosotis laxa ssp caespitosa	S,?Bs	?3	AQUAT	Fr	2	1	Dry	Bo
+		Myosotis ramosissima	S, Bs	3	ANIMa	Fr	1	1	Dry	Bo
+		Myosotis scorpioides	V	1	*UNSP	Fr	2	2	Dry	Bo
		Myosotis secunda	V, Bs	3	*UNSP	Fr	2	1	Dry	Bo
		Myriophyllum spicatum	(V)	1	*UNSP	Fr	4	1	-/Scar	Ha
+	*	Myrrhis odorata	S	1	UNSP	Fr	6	3	Chill	Um
+		Nardus stricta	V,S	2	ANIMa	Fr	2	3	Chill	Gr
+		Nasturtium officinale agg.	V,?Bs	?3	*UNSP	Sd	1	1	-/Dry	Cr
		Nuphar lutea	V,?S	1	*AQUAT	Sd	6	1	?Chill	Ny
		Odontites verna	S, Bs	3	WINDc	Sd	2	2	Chill	Sc
		Oenanthe crocata	?	?	AQUAT	Fr	5	3	Dry	Um
		Oenanthe fistulosa	V	1	*AQ/AN	Fr	5	1	?	Um
		Ononis repens	V	?	UNSP	Sd	5	1	Scar	Le
		Orchis mascula	W	?	WINDcm	Sd	S	3	Orchid	Or
		Orchis morio	W	?	WINDm	Sd	S	3	Orchid	Or
+		Origanum vulgare	V, Bs	3	UNSP	Fr	1	1	-	La
+		Oxalis acetosella	V,S	1	ANIMm	Sd	4	2	Chill	Ox
		Papaver argemone	Bs	3	WINDc	Sd	1	2	Scar	Pa
		Papaver dubium	Bs	3	WINDc	Sd	1	1	Chill	Pa
+		Papaver rhoeas	Bs	3	WINDc	Sd	1	1	Chill	Pa
		Parietaria judaica	?	?	ANIMa	Fr	1	1	Dry	Ur
		Pedicularis sylvatica	S	1	ANIMe	Sd	4	2	Chill	Sc
	*	Pentaglottis sempervirens	?	?	UNSP	Fr	5	2	Dry	Bo
	*	Petasites fragrans	V	-	* -	X	X	-	X	Co
+		Petasites hybridus	V,W	1	*WINDp	Fr	2	3	-	Co
+		Phalaris arundinacea	V, Bs	3	*AQUAT	Fr	3	2	Dry,Scar	Gr
+		Phleum pratense	S, Bs	3	ANIMa	Fr	2	2	-	Gr
+		Phragmites australis	V,W	1	*WINDp	Fr	1	2	?	Gr
+		Pimpinella saxifraga	S	1	UNSP	Fr	4	2	Chill	Um
		Plantago coronopus	Bs	3	ANIMm	Sd	1	2	-	Pl
+		Plantago lanceolata	V, Bs	3	ANIMm	Sd	4	2	Dry	Pl
+		Plantago major	Bs	3	ANIMm	Sd	2	2	Chill	Pl
		Platanthera chlorantha	W	?	WINDcm	Sd	S	3	Orchid	Or
+		Poa annua	V, S, Bs	3	UNSPag	Fr	2	2	-	Gr
		Poa nemoralis	?	?1	UNSP	Fr	2	3	Dry	Gr
+		Poa pratensis agg.	V, Bs	3	UNSPag	Fr	2	3	-	Gr
+		Poa trivialis	V, Bs	3	UNSPag	Fr	1	3	-	Gr

	Regenerative strategies	Seed bank	Agency of dispersal	Dispersule/germinule form	Dispersule weight	Dispersule shape	Germination requirements	Family
Polygala serpyllifolia	?S	1	ANIMe	Sd	4	2	Chill	Pga
Polygala vulgaris	?S	1	ANIMe	Sd	4	2	Chill	Pga
Polygonum amphibium	(V),?Bs	?1	*AQUAT	Fr	5	1	Chill	Pgo
Polygonum aviculare grp	Bs	3	UNSPag	Fr	4	2	Chill	Pgo
Polygonum hydropiper	Bs	3	AQUAT	Fr	4	2	Chill	Pgo
Polygonum lapathifolium	Bs	3	UNSPag	Fr	5	1	Chill,Dry	Pgo
Polygonum persicaria	Bs	3	UNSPag	Fr	5	1	Chill,Dry	Pgo
Potamogeton crispus	S	?	*AQ/AN	Fr	-	2	Scar	Pot
Potamogeton natans	(V),?	?	*AQ/AN	Fr	5	1	Scar	Pot
Potentilla anglica	V,?	?2	*UNSP	Fr	2	1	Unclassified	Ros
Potentilla anserina	V	2	*UNSP	Fr	3	1	Unclassified	Ros
Potentilla erecta	V, Bs	3	UNSP	Fr	3	1	Warm	Ros
Potentilla palustris	V	2	UNSP	Fr	2	1	Unclassified	Ros
Potentilla reptans	V,Bs	3	*UNSP	Fr	2	1	Unclassified	Ros
Potentilla sterilis	V, Bs	3	UNSP	Fr	3	1	-	Ros
Primula veris	V	1	WINDc	Sd	3	1	Chill,Wash	Pri
Primula vulgaris	Bs	3	ANIMe	Sd	2	1	Wash	Pri
Prunella vulgaris	(V), Bs	3	ANIMm	Fr	3	2	-	Lab
Pulicaria dysenterica	V,W	?1	WINDp	Fr	1	3	?Dry	Com
Ranunculus acris	V	2	ANIMa	Fr	4	1	(-)	Ran
Ranunculus auricomus	?	?	UNSP	Fr	5	1	?Chill	Ran
Ranunculus bulbosus	Bs	3	ANIMa	Fr	5	1	Dry	Ran
Ranunculus ficaria	Sv,S	1	*ANIMe	Fr	4	2	Chill	Ran
Ranunculus flammula	V, Bs	3	*AQUAT	Fr	2	1	Dry,Freeze	Ran
Ranunculus peltatus	V	2	*AQUAT	Fr	2	1	?	Ran
Ranunculus penicillatus	V,?	?	*AQUAT	Fr	2	2	?	Ran
Ranunculus repens	(V), Bs	3	*AQ/AN	Fr	5	2	(Dry)	Ran
Ranunculus sceleratus	Bs	3	*AQUAT	Fr	1	2	-	Ran
Raphanus raphanistrum	Bs	3	UNSPag	Sd	5	1	Unclassified	Cru
Reseda luteola	Bs	3	WINDc	Sd	2	1	Unclassified	Res
Reynoutria japonica	V	-	*UNSP	Fr	3	2	(Dry)	Pgo
Rhinanthus minor s.l.	S	2	WINDcw	Sd	5	1	Chill	Scr
Ribes uva-crispa	S	?1	ANIMi	F/S	5	1	Chill	Gro
Rorippa palustris	S, Bs	3	AQUAT	Sd	1	1	Dry,Fluct,Scar	Cru
Rubus caesius	?	?2	ANIMi	Fr	5	2	Scar/Chill	Ros
Rubus fruticosus s.l.	V, Bs	3	ANIMi	Fr	5	1	Scar/Chill	Ros
Rubus idaeus	V, Bs	3	ANIMi	Fr	4	2	Scar/Chill	Ros
Rumex acetosa	V,S	2	WINDw	Fr	3	1	-	Pgo
Rumex acetosella agg.	V, Bs	3	UNSP	Fr	2	2	Dry	Pgo
Rumex conglomeratus	?	2	AQUAT	Fr	4	1	-	Pgo
Rumex crispus	(V), Bs	3	UNSP	Fr	4	2	Dry	Pgo
Rumex hydrolapathum	?	?	AQUAT	Fr	5	1	-	Pgo
Rumex obtusifolius	Bs	3	ANIMa	Fr	4	1	-	Pgo
Rumex sanguineus	Bs	3	UNSP	Fr	4	1	Dry	Pgo
Sagina apetala	Bs	3	WINDc	Sd	1	1	Dry	Car
Sagina nodosa	Sv, Bs	3	*UNSPc	Sd	1	1	Unclassified	Car
Sagina procumbens	(V), Bs	3	UNSPc	Sd	1	2	-	Car
Salix repens	V,W	1	WINDp	Sd	1	2	-	Sal
Sanguisorba minor	?Bs	?3	UNSP	Fr	4	2	Dry	Ros
Sanicula europaea	V,S	1	ANIMb	Fr	5	1	?Chill	Umb
Saponaria officinalis	V,?	?	WINDc	Sd	4	1	Chill,Scar	Car
Saxifraga tridactylites	S, Bs	3	WINDc	Sd	1	2	Dry	Sax

Autecological account	Introduced		Regenerative strategies	Seed bank	Agency of dispersal	Dispersule/germinule form	Dispersule weight	Dispersule shape	Germination requirements
+		Scabiosa columbaria	S	2	ANIMa	Fr	4	1	-/Chill
		Schoenoplectus lacustris	V,?S	?1	*ANIMa	Fr	4	2	Unclassified
		Scleranthus annuus	Bs	3	UNSP	Fr	4	3	Unclassified
		Scrophularia auriculata	S, Bs	3	AQ/WIc	Sd	1	2	-
		Scrophularia nodosa	Bs	3	WINDc	Sd	1	2	-
		Scutellaria galericulata	V,?	?	AQUAT	Fr	3	1	Unclassified
+		Sedum acre	V	?	WINDcm	Sd	1	2	Dry
		Sedum anglicum	V	?	WINDcm	Sd	1	2	?
		Sedum telephium	V	?	WINDcm	Sd	1	3	Dry
		Senecio aquaticus	W	?1	WINDp	Fr	2	3	-
+		Senecio jacobaea	(V),W, Bs	3	WINDp	Fr	1	3	-
+	*	Senecio squalidus	W, Bs	3	WINDp	Fr	2	2	Dry
		Senecio sylvaticus	W, Bs	3	WINDp	Fr	2	3	Dry
+		Senecio viscosus	W, Bs	3	WINDp	Fr	3	3	Dry
+		Senecio vulgaris	W, Bs	3	WINDp	Fr	2	3	Dry
		Sherardia arvensis	S	2	ANIMa	Fr	4	2	Dry
		Silene alba	Bs	3	WINDc	Sd	3	1	Dry
+		Silene dioica	Bs	3	WINDc	Sd	3	1	-
		Silene vulgaris	Bs	3	WINDc	Sd	4	1	Dry
+		Sinapis arvensis	Bs	3	UNSPag	Sd	4	1	Dry
	*	Sisymbrium altissimum	Bs	3	WINDw	Sd	2	1	Scar
		Sisymbrium officinale	Bs	3	WINDw	Sd	2	1	Scar
+		Solanum dulcamara	(V),S	2	ANIMi	F/S	4	1	Chill,Fluct
+		Solidago virgaurea	W	1	WINDp	Fr	3	3	-
		Sonchus arvensis	V	2	WINDp	Fr	3	2	Chill
+		Sonchus asper	W, Bs	3	WINDp	Fr	2	2	Dry
+		Sonchus oleraceus	W, Bs	3	WINDp	Fr	2	3	-
		Sparganium emersum	(V)	?	*AQ/AN	Fr	5	2	Chill,Scar
+		Sparganium erectum	(V)	?1	*AQ/AN	Fr	6	2	Chill,Scar
+		Spergula arvensis	Bs	3	WINDc	Sd	2	1	Dry
		Spergularia rubra	?Bs	?3	UNSPc	Sd	1	1	-
		Stachys arvensis	Bs	3	UNSPag	Fr	3	1	Dry
+		Stachys officinalis	V,S	2	UNSP	Fr	4	1	-
		Stachys palustris	V, Bs	3	*AQUAT	Fr	4	1	Unclassified
+		Stachys sylvatica	(V), Bs	3	*ANIMa	Fr	4	1	Chill
+		Stellaria alsine	(V), Bs	3	*UNSPc	Sd	1	2	-
		Stellaria graminea	V,?Bs	?3	UNSPc	Sd	2	1	Dry
+		Stellaria holostea	V	1	UNSPc	Sd	5	1	?Chill
+		Stellaria media	Bs	3	UNSPag	Sd	2	1	Dry
+		Succisa pratensis	S	1	ANIMa	Fr	4	3	-
		Symphytum officinale	V,?	?	AQ/ANe	Fr	-	-	?
	*	Symphytum x uplandicum	V,?	?	?AQUAT	Fr	5	2	Unclassified
+		Tamus communis	S	1	ANIMi	F/S	6	1	Chill
	*	Tanacetum parthenium	?	?	UNSP	Fr	1	3	-
		Tanacetum vulgare	V	1	UNSP	Fr	1	3	Dry
+		Taraxacum agg.	W	2	*WINDp	Fr	3	3	-
+		Teucrium scorodonia	V, Bs	3	UNSP	Fr	3	1	-
		Thlaspi arvense	Bs	3	WINDw	Sd	3	1	Fluct,Scar
+		Thymus praecox	V, Bs	3	UNSP	Fr	1	1	-
+		Torilis japonica	S, Bs	3	ANIMb	Fr	4	2	Chill
		Tragopogon pratensis	W	1	WINDp	Fr	6	3	Dry
		Trifolium arvense	S, Bs	3	ANIMa	f/S	2	1	Scar

Table 4.3

Introduced			Regenerative strategies	Seed bank	Agency of dispersal	Dispersule/germinule form	Dispersule weight	Dispersule shape	Germination requirements	Family
		Trifolium campestre	S, Bs	3	ANIMa	f/S	2	2	Scar	Leg
+		Trifolium dubium	S, Bs	3	ANIMa	f/S	2	2	Scar	Leg
	*	Trifolium hybridum	Bs	3	ANIMa	f/S	3	1	Scar	Leg
+		Trifolium medium	V	1	ANIMa	f/S	5	1	Scar	Leg
+		Trifolium pratense	S, Bs	3	ANIMa	f/S	4	1	Scar	Leg
+		Trifolium repens	(V), Bs	3	ANIMa	f/S	3	1	Scar	Leg
		Triglochin palustris	V,?	?	ANIMa	f/S	2	3	?	Jcg
+		Tripleurospermum inodorum	S, Bs	3	UNSPag	Fr	2	2	Dry	Com
+		Trisetum flavescens	V,S	1	ANIMa	Fr	1	3	-	Gra
+		Tussilago farfara	(V),W	1	*WINDp	Fr	2	3	-	Com
+		Typha latifolia	V,W, Bs	3	*WINDp	Fr	1	3	Fluct,Scar	Typ
		Umbilicus rupestris	?	?	WINDcm	Sd	1	3	Dry	Cra
+		Urtica dioica	V, Bs	3	*ANIMa	Fr	1	1	-	Urt
+		Urtica urens	Bs	3	ANIMa	Fr	2	1	-	Urt
+		Vaccinium myrtillus	V, Bs	3	ANIMi	F/S	2	2	-	Eri
+		Vaccinium vitis-idaea	V	1	ANIMi	F/S	2	2	-	Eri
+		Valeriana officinalis	V,S	1	WINDpw	Fr	3	2	Dry	Val
		Valerianella locusta	S	?	UNSP	Fr	3	2	Dry	Val
		Verbascum thapsus	Bs	3	WINDc	Sd	1	2	Dry	Scr
		Veronica agrestis	Bs	3	ANIMe	Sd	2	2	Dry	Scr
		Veronica anagallis-aquatica	?	?2	WINDc	Sd	1	2	-	Scr
+		Veronica arvensis	Bs	3	UNSP	Sd	1	1	Dry	Scr
+		Veronica beccabunga	V, Bs	3	*ANIMm	Sd	2	1	-	Scr
+		Veronica chamaedrys	V, Bs	3	UNSPcw	Sd	1	1	-	Scr
	*	Veronica filiformis	V	-	* -	X	X	-	X	Scr
		Veronica hederifolia	Bs	3	ANIMe	Sd	5	2	Chill	Scr
+		Veronica montana	V, Bs	3	*UNSPcw	Sd	2	1	Chill	Scr
		Veronica officinalis	V, Bs	3	*UNSPcw	Sd	1	1	Dry	Scr
+	*	Veronica persica	Bs	3	UNSPag	Sd	3	2	Dry	Scr
		Veronica polita	Bs	3	UNSPag	Sd	2	2	Unclassified	Scr
		Veronica scutellata	V,?Bs	?3	UNSPcw	Sd	1	1	Dry	Scr
		Veronica serpyllifolia	V, Bs	3	UNSPcw	Sd	1	2	Dry	Scr
+		Vicia cracca	V,?S	?1	UNSP	Sd	6	1	Scar	Leg
		Vicia hirsuta	Bs	3	UNSP	Sd	5	1	Scar	Leg
		Vicia sativa ssp nigra	Bs	3	UNSP	Sd	6	1	Scar	Leg
+		Vicia sepium	V	1	UNSP	Sd	6	1	Scar	Leg
		Viola arvensis	Bs	3	UNSPag	Sd	2	2	Unclassified	Vio
+		Viola hirta	V,?S	2	ANIMe	Sd	5	3	Chill	Vio
		Viola odorata	V,?S	?	*ANIMe	Sd	5	1	?Chill	Vio
		Viola palustris	V,S	1	*ANIMe	Sd	3	2	Chill	Vio
+		Viola riviniana	V,S	1	ANIMe	Sd	4	1	Chill	Vio
		Viola tricolor	Bs	3	ANIMm	Sd	3	2	Dry	Vio
		Vulpia bromoides	S	1	ANIMa	Fr	2	3	Dry	Gra
		(b) Woody species more than 1.5 m tall								
		Acer campestre (canopy)	W	1	WINDw	Fr	6	1	Chill	Ace
+	*	Acer pseudoplatanus (canopy)	W	1	WINDw	Fr	6	1	Chill	Ace
		ditto juvenile								
+		Alnus glutinosa (canopy)	W	2	AQ/WIw	Fr	4	1	Dry	Bet
		ditto juvenile								
+		Betula spp. (canopy)	W	2	WINDw	Fr	1	1	-	Bet
		ditto juvenile								

Autecological account	Introduced		Regenerative strategies	Seed bank	Agency of dispersal	Dispersule/germinule form	Dispersule weight	Dispersule shape	Germination requirements	Family
		Corylus avellana (canopy)	V,S	1	ANIMn	Fr	6	1	Chill	Cor
+		Crataegus monogyna (canopy)	S	1	ANIMi	Fr	6	2	Warm+Chill	Ros
		ditto juvenile								
		Cytisus scoparius (canopy / juv)	Bs	3	ANIMe	Sd	5	1	Scar	Leg
		Euonymus europaeus (canopy)	S	1	ANIMie	F/S	6	2	Chill	Sap
+		Fagus sylvatica (canopy)	S	1	ANIMn	Fr	6	1	Chill	Fag
		ditto juvenile								
+		Fraxinus excelsior (canopy)	S	1	WINDw	Fr	6	3	Chill	Ole
		ditto juvenile								
		Ilex aquifolium (canopy)	S	1	ANIMi	Fr	6	2	Scar+Chill	Aqu
		Ligustrum vulgare (canopy)	V,S	1	ANIMi	F/S	6	2	Chill	Ole
		Malus sylvestris (canopy)	S	1	ANIMi	F/S	6	1	Chill	Ros
		Populus tremula (canopy)	V,W	1	WINDp	Sd	1	2	-	Sal
		Prunus avium (canopy)	V,S	1	ANIMi	Fr	6	1	Chill	Ros
		Prunus spinosa (canopy)	V,S	1	ANIMi	Fr	6	1	Chill	Ros
+		Quercus agg. (canopy)	S	1	ANIMn	Fr	6	2	Chill	Fag Eri
		ditto juvenile								
*	*	Rhododendron ponticum (cnpy)	V,W	?1	WINDcm	Sd	1	3	-	
		Rosa spp. (canopy)	V,S	1	ANIMi	Fr	6	2	?Warm+Chill/Chill	Ros
		ditto juvenile								
+		Salix cinerea agg. (canopy)	(V),W	1	WINDp	Sd	1	2	-	Sal
		ditto juvenile								
	*	Salix fragilis agg. (canopy)	(V),W	1	WINDp	Sd	1	2	-	Sal
		Salix purpurea (canopy)	(V),W	1	WINDp	Sd	1	2	-	Sal
	*	Salix viminalis (canopy)	(V),W	1	WINDp	Sd	1	2	-	Sal
+		Sambucus nigra (canopy)	S, Bs	3	ANIMi	F/S	5	2	?Chill	Cap
		ditto juvenile								
+		Sorbus aucuparia (canopy)	S	1	ANIMi	F/S	5	2	Chill	Ros
		ditto juvenile								
+		Ulex europaeus (canopy / juv)	Bs	3	ANIMe	Sd	5	1	Scar	Leg
+		Ulmus glabra (canopy)	W	1	WINDw	Fr	5	1	-	Ulm
		ditto juvenile								
		Ulmus procera (canopy)	V,W	1	WINDw	Fr	-	1	-	Ulm
		Viburnum opulus (canopy)	S	1	ANIMi	F/S	5	1	Chill	Cap
		(c) Pteridophytes								
		Asplenium adiantum-nigrum	W	?	WINDm	Sp	S	1	-	Apl
+		Asplenium ruta-muraria	W	?	WINDm	Sp	S	1	-	Apl
+		Asplenium trichomanes	W	?	WINDm	Sp	S	1	-	Apl
+		Athyrium filix-femina	V,W	?	WINDm	Sp	S	2	-	Ath
		Blechnum spicant	W	?	WINDm	Sp	S	1	?-	Ble
+		Cystopteris fragilis	W	?	WINDm	Sp	S	2	-	Ath
		Dryopteris affinis	V,W,?Bs	?	WINDm	Sp	S	1	-	Api
+		Dryopteris dilatata	V,W,?Bs	?	WINDm	Sp	S	1	?-	Api
+		Dryopteris filix-mas	V,W,?Bs	?3	WINDm	Sp	S	1	-	Api
+		Equisetum arvense	V,W	1	WINDm	Sp	S	1	-	Equ
+		Equisetum fluviatile	V,W	1	WINDm	Sp	S	1	-	Equ
+		Equisetum palustre	V,W	1	WINDm	Sp	S	1	-	Equ
		Phyllitis scolopendrium	W	?	WINDm	Sp	S	1	-	Apl
		Polypodium vulgare grp	V,W	?	WINDm	Sp	S	1	-	Ppo
+		Pteridium aquilinum	V,W,?Bs	?3	WINDm	Sp	S	1	-	Hyo

schampsia flexuosa (L.) Trin.

rm Tufted or rhizomatous and mat-forming, polycarpic nnial, semi-rosette hemicryptophyte. Shoots erect; leaves -like, sometimes ›100 mm²; roots mainly shallow but some nding to ›500 mm (*Biol Fl*), with VA mycorrhizas.

ogy Winter green. Flowers June and July and sheds seed tly after ripening in August or September. Leaves long-d.

Foliage up to 200 mm; flowering shoots to 400 mm.

g RGR 0.5-0.9 week⁻¹.

r DNA amount 11.0 pg; $2n = 26, 28^*, 56$ (*BC*, *FES*); 8x.

shed strategy Intermediate between stress-tolerator and s-tolerant competitor.

rative strategies V, S. Regenerates effectively both vegetatively and by seed (see 'Synopsis'); SB I.

Flowers Silvery green, hermaphrodite, homogamous, wind-pollinated, self-incompatible (Weimarck 1968); ›100 in an open panicle (2 per spikelet).

Germinule 0.43 mg, 1.9 × 1.0 mm, caryopsis dispersed with an attached lemma, an hygroscopic bristle and a basal tuft of hairs (3.5 × 1.3 mm).

Germination Epigeal; t_{50} 4 days; 6-28°C; L = D.

Biological flora Scurfield (1954).

Geographical distribution The British Isles, especially in the N and W (98% of British vice-counties), *ABF* page 395; most of Europe, but rarer in the S and absent from much of the SE (85% of European territories); N Asia, N America and on mountains in warmer climates, e.g. America and E Africa.

s Of widespread occurrence in a range of unproductive, base-bitats. Abundant and widely distributed in upland pastures, cidic heaths, scrub, woodlands and plantations. Occurs on the ils of rock ledges in sandstone quarries, rock outcrops, cliffs and ps. Successful also in neglected hill pastures, the sides of railway and other waste places. Relatively frequent in bogs, in which it ows on the top of tussocks of other species, e.g. *Eriophorum m*. Uncommon in unleached calcareous habitats, and absent ble land. [An important component of many montane plant ities on acidic soils.]

Gregariousness A patch-forming species.

Altitude Widely distributed, but suitable habitats are more frequent in upland areas. [Up to 1150 m.]

Slope Found at all angles of slope.

Aspect More frequent on N-facing slopes, with a significant concentration in N-facing shaded habitats.

Hydrology Occurs in wetland habitats but infrequent on waterlogged soils.

Soil pH Mostly restricted to sites with pH ‹5.0.

Similarity in habitats

Vaccinium myrtillus	89%
Calluna vulgaris	84%
Vaccinium vitis-idaea	81%
Galium saxatile	80%
Erica cinerea	78%

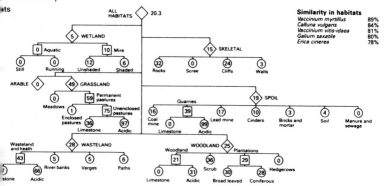

Frequency and abundance decline progressively with increasing e of bare soil.

ation Distribution strongly concentrated on the right-hand side of am, indicating restriction to unproductive, relatively undisturbed on. Excursions to other types of vegetation are extremely rare ably correspond to seedlings.

ted floristic diversity Low and declining slightly with increased ce of *D.f.*.

D.f. is a slow-growing, evergreen, clump- or carpet-forming erant of low pH, low mineral nutrient supply and high external ations of aluminium and manganese (Hackett 1965, Grime & 1969, Mahmoud & Grime 1974, Rorison 1985). The leaves ow amounts of N, P, Mg and Ca (*SIEF*) and *D.f.* responds to low nitrogen by more-efficient utilization and redistribution of dry the form of finer roots and root hairs (Robinson & Rorison The species is typical of cold climates and podzolic soils and

produces humus which is persistent and inhibitory to plant growth (Grime 1963b, Jarvis 1964). *D.f.* is the most successful calcifuge grass in Britain, and combines a very wide geographical distribution with the capacity to exploit a diversity of acidic dryland habitats. The species is particularly shade-tolerant, and seedlings are capable of persistence in total darkness for 3 months (Hutchinson 1967). However, vegetative vigour and flowering are both inhibited by dense shade, and *D.f.* is also rather vulnerable to submergence beneath deciduous tree litter. In unshaded habitats in Britain differences in performance may be associated with aspect. On N-facing slopes vegetative expansion may result in continuous monocultures, whilst on adjacent S-facing slopes tussocks are often stunted but produce abundant inflorescences. *D.f.* is eaten by sheep and rabbits, but in moorland habitats new shoots of *Calluna vulgaris* are preferred and grazing here can lead to an increase in *D.f.* (*Biol Fl*). However, under conditions of intense grazing, mowing or trampling, or if the soil is dry, *Festuca ovina* tends to be more prevalent. The species is

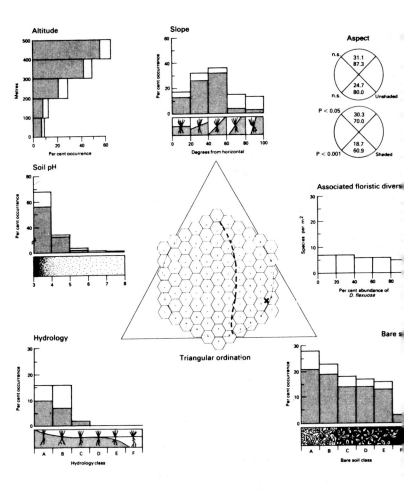

Altitude

Slope

Aspect

Soil pH

Associated floristic diversity

Hydrology

Triangular ordination

Bare soil

capable of forming large clonal patches as a result of rhizome growth. Freshly shed seed germinates in autumn. The seedlings so formed may be long-lived, often persisting in a stunted form until conditions become more favourable for growth, but no persistent seed bank is accumulated. In consequence, *D.f.* is not an effective colonist after severe fires. However, the species often persists vegetatively in habitats subject to

regular fires of low intensity. The species is occasionally planted shade-tolerant amenity grass (Shildrick 1980).

Current trends Although *D.f.* colonizes a number of artificial habi e.g. railway banks and coal-mine spoil, the species is decreasing in low areas, due to the destruction of heathland and acidic grassland. *D* likely to remain abundant in upland areas.

Index of common names

A brief key
to the contents of an
abridged autecological account

Nomenclature Principal name of the subject species; any alternative name(s); common name(s).

Established strategy Classification with respect to the seven primary or secondary C–S–R strategies of the established phase: C, competitive; S, stress-tolerant; R, ruderal; C–R, competitive ruderal; S–R, stress-tolerant ruderal; C–S, stress-tolerant competitor; C–S–R, C–S–R strategist.

Gregariousness Whether a sparse, intermediate, or patch-forming species.

Flowers Floral structure; pollination.

Regenerative strategies The regenerative strategies exhibited by the subject; any of: V, vegetative expansion; S, seasonal regeneration in gaps; B_s, regeneration from a persistent seed bank; W, regeneration involving numerous widely-dispersed propagules.

Seed Air-dry weight; physical dimensions and form; fruit characters.

British distribution Notes on the British distribution of the subject.

Commonest habitats Data from a standard 32-habitat classification.

Species most similar in habitats The five species most similar to the subject in distribution and abundance in the 32-habitat classification, 100% indicates an identical distribution and abundance.

Full Autecological Account Page reference to an Account in the parent volume, *Comparative Plant Ecology*.

Synopsis An ecological overview of the more important field, screening and literary information.

Current trends Estimates of the response of the subject, and of its future status, in relation to current changes in land use in Britain.